CAMBRIDGE MONOGRAPHS ON MATHEMATICAL PHYSICS

EXACT SOLUTIONS OF EINSTEIN'S FIELD EQUATIONS

EXACT SOLUTIONS OF EINSTEIN'S FIELD EQUATIONS

D. KRAMER, H. STEPHANI, E. HERLT

Department of Relativistic Physics, Friedrich-Schiller University, Jena, German Democratic Republic

M. MacCALLUM

Department of Mathematics, Queen Mary College, London, England

Edited by

E. SCHMUTZER

Department of Relativistic Physics, Friedrich-Schiller University, Jena, German Democratic Republic

CAMBRIDGE UNIVERSITY PRESS

CAMBRIDGE

LONDON · NEW YORK · NEW ROCHELLE · MELBOURNE · SYDNEY

Published by arrangement with VEB Deutscher Verlag der Wissenschaften, Berlin, German Democratic Republic

Published by the Press Syndicate of the University of Cambridge
The Pitt Building, Trumpington Street, Cambridge CB 2 1 RP
32 East 57th Street, New York, NY 10022, USA
296 Beaconsfield Parade, Middle Park, Melbourne 3206, Australia

First published 1980

Printed in the German Democratic Republic by VEB Druckhaus 'Maxim Gorki',
7400 Altenburg

British Library Cataloguing in Publication Data

Exact solutions of Einstein's field equations
(Cambridge monographs on mathematical physics; no. 6)
1. General relativity (Physics)
I. Kramer, D II. Schmutzer, E III. Series
530.1-43 QC173.585 80-40704

ISBN 0 521 23041 1

Editor's Preface

For a period of more than a decade some members of the Jena Department of Relativistic Physics have been studying the literature on the exact solutions of Einstein's gravitational field equations. Following this line of work, they also did a considerable amount of research on the classification of these solutions and on new methods of finding such solutions (H. Stephani on the embedding method, D. Kramer and G. Neugebauer on the invariance transformation method, E. Herlt on stationary axisymmetric solutions). After some years, conditions seemed favourable for a systematic treatment of these subjects. As is well known among specialists, many interesting exact solutions have been independently discovered two or more times.

This situation inspired H. Stephani and D. Kramer to write a monograph on exact solutions of Einstein's field equations, presenting the most important methods and solutions in a systematic way, with the aim of accomplishing a kind of catalogue which would give a good survey of the solutions already discovered and help to avoid duplications of discovery. It was originally hoped to offer this monograph to the international relativity community on the occasion of Einstein's 100th birthday, but soon an enormous amount of material accumulated. Therefore the target date was amended to that of the 9th International Conference on General Relativity and Gravitation in July 1980 at Jena, and to achieve even this aim required very substantial support from the Jena Department of Relativistic Physics and the Sektion Physik of the Friedrich Schiller University. We were all very happy that, in addition to E. Herlt, M. MacCallum (Queen Mary College, London) could be enlisted as an author. Furthermore, M. MacCallum agreed to revise the English text. As the head of the department and as the editor of this monograph I would like to thank the authorities of the Friedrich Schiller University, Jena, for their constant help.

Jena, February 1979

Ernst Schmutzer

Friedrich Schiller
University Jena

Authors' Preface

When, in 1975, two of the authors (D. K. and H. S.) proposed to change their field of research back to the subject of exact solutions of Einstein's field equations, they of course felt it necessary to make a careful study of the papers published in the meantime, so as to avoid duplication of known results. A fairly comprehensive review or book on the exact solutions would have been a great help, but no such book was available. This prompted them to ask "Why not use the preparatory work we have to do in any case to write such a book?" After some discussion, they agreed to go ahead with this idea, and then they looked for co-authors. They succeeded in finding two.

The first was E. H., a member of the Jena relativity group, who had been engaged before on the exact solutions and was also inclined to return to them.

The second, M. M., became involved by responding to the existing authors' appeal for information and then (during a visit by H. S. to London) agreeing to look over the English text. Eventually he also agreed to write some parts of the book. He wishes to record that any infelicities remaining in the English arose because the generally good standard of his colleagues' English lulled him into a false sense of security.

Our original optimism somewhat diminished when references to over 2000 papers had been collected and the magnitude of the task became all too clear. How could we extract even the most important information from this mound of literature? How could we avoid constant re-writing to incorporate new information, which would have made the job akin to the proverbial painting of the Forth bridge? How could we decide which topics to include and which to omit? How could we check the calculations, put the results together in a readable form, and still finish in a reasonable time?

Looking back now at the result of three years' work, we cannot really feel that we solved any of these questions in a completely convincing manner. In particular, we feel sure we must have accidentally overlooked many useful results and solutions. However, we did manage to produce an outcome in a finite time, largely because the labour of reading those papers conceivably relevant to each chapter, and then drafting the related manuscript, was divided. (Roughly, D. K. was responsible for most of the introductory Part I., M. M., D. K. and H. S. dealt with groups (Part II.), H. S., D. K. and E. H. with algebraically special solutions, and D. K. and H. S. with Part IV. (special methods) and Part V. (tables).) Each draft was then criticized by the other authors, so that its writer could not be held wholly responsible for any errors or omissions. (Since we hope to maintain up-to-date information, we will be glad to hear from any reader who detects such errors or omissions; we will also be pleased to answer as best we can any requests for further information.)

This book could not have been written, of course, without the efforts of the many scientists whose work is recorded here, and especially the many contemporaries who sent preprints, reprints, references and advice. More immediately, it would not have appeared without the help of Frau Kaschlik and Frau Reichardt in Jena, and Mrs. Smith in London, who did all the secretarial work including typing the illegible and apparently interminable manuscript, of the students in Jena who maintained our reference files, of Prof. Schmutzer, who supported the project from the beginning, and of the Sektion Physik in Jena and the Department of Applied Mathematics at Queen Mary College, London. Last but not least, we thank wives, families and colleagues for tolerating our incessant brooding and discussions.

January 1979

Dietrich Kramer
Hans Stephani
Eduard Herlt
Jena

Malcolm MacCallum
London

Contents

Notation

All symbols are explained in the text. Here we list only some important conventions which are frequently used throughout he book.

Complex conjugation is denoted by a bar over the symbol.

Indices

Small Latin indices run, in an n-dimensional space, from 1 to n, in space-time V_4 from 1 to 4. Indices from the first part of the alphabet ($a, b, \ldots, h$) are tetrad indices, i.e. they refer to a general basis $\{\boldsymbol{e}_a\}$ or its dual $\{\boldsymbol{\omega}^a\}$; $i, j, \ldots$ are reserved for a coordinate basis $\{\partial/\partial x^i\}$ or its dual $\{\mathrm{d}x^i\}$. For a vector $\boldsymbol{v}$ and a 1-form σ we write $\boldsymbol{v} = v^a \boldsymbol{e}_a = v^i\, \partial/\partial x^i$, $\sigma = \sigma_a \omega^a = \sigma_i\, \mathrm{d}x^i$. Small Greek indices run from 1 to 3, if not otherwise stated. Capital Latin indices are either spinor indices ($A, B = 1, 2$) or indices in group space ($A, B = 1 \ldots r$), or they label the coordinates in a Riemannian 2-space V_2 ($M, N = 1, 2$).

Symmetrization and antisymmetrization of index pairs:

$$v_{(ab)} \equiv \frac{1}{2}(v_{ab} + v_{ba}), \qquad v_{[ab]} \equiv \frac{1}{2}(v_{ab} - v_{ba}).$$

Metric and tetrads

Line element in terms of dual basis $\{\boldsymbol{\omega}^a\}$: $\mathrm{d}s^2 = g_{ab}\boldsymbol{\omega}^a\boldsymbol{\omega}^b$.
Signature of space-time metric: $(+ + + -)$.
Commutation coefficients: $D^c{}_{ab}$; $[\boldsymbol{e}_a, \boldsymbol{e}_b] = D^c{}_{ab}\boldsymbol{e}_c$.
(Complex) null tetrad: $\{\boldsymbol{e}_a\} = (\boldsymbol{m}, \overline{\boldsymbol{m}}, \boldsymbol{l}, \boldsymbol{k})$, $g_{ab} = 2m_{(a}\overline{m}_{b)} - 2k_{(a}l_{b)}$,

$$\mathrm{d}s^2 = 2\omega^1\omega^2 - 2\omega^3\omega^4.$$

Orthonormal basis: $\{\boldsymbol{E}_a\}$.
Projection tensor: $h_{ab} \equiv g_{ab} + u_a u_b$, $u_a u^a = -1$.

Bivectors

Levi-Civita tensor: ε_{abcd}; $\varepsilon_{abcd} m^a \overline{m}^b l^c k^d = i$.

Dual bivector: $\tilde{X}_{ab} \equiv \frac{1}{2}\varepsilon_{abcd}X^{cd}$.

(Complex) self-dual bivector: $X^*_{ab} \equiv X_{ab} + \mathrm{i}\tilde{X}_{ab}$.
Basis of self-dual bivectors: $U_{ab} \equiv 2\overline{m}_{[a}l_{b]}$, $V_{ab} \equiv 2k_{[a}m_{b]}$, $W_{ab} \equiv 2m_{[a}\overline{m}_{b]} - 2k_{[a}l_{b]}$.

Derivatives

Partial derivative: comma in front of index or coordinate, e.g.

$$f_{,i} \equiv \partial f/\partial x^i \equiv \partial_i f, \qquad f_{,\zeta} \equiv \partial f/\partial \zeta .$$

Directional derivative: denoted by stroke or comma, $f_{|a} \equiv f_{,a} \equiv \boldsymbol{e}_a(f)$, if followed by a numerical (tetrad) index, we prefer the stroke, e.g. $f_{|4} = f_{,i}k^i$. Directional derivatives with respect to the null tetrad $(\boldsymbol{m}, \overline{\boldsymbol{m}}, \boldsymbol{l}, \boldsymbol{k})$ are symbolized by $\delta f = f_{|1}$, $\overline{\delta} f = f_{|2}$, $\Delta f = f_{|3}$, $Df = f_{|4}$.

Covariant derivative: ∇; in component calculus, semicolon. (Sometimes other symbols are used to indicate that in V_4 a metric different from g_{ab} is used, e.g. $h_{ab\|c} = 0$, $\gamma_{ab=c} = 0$.)

Lie derivative of a tensor $\boldsymbol{T}$ with respect to a vector $\boldsymbol{v}$: $\pounds_{\boldsymbol{v}}\boldsymbol{T}$.

Exterior derivative: d.

Connection and curvature

Connection coefficients: $\Gamma^a{}_{bc}$, $v^a{}_{;c} = v^a{}_{,c} + \Gamma^a{}_{bc} v^b$.

Connection 1-forms: $\boldsymbol{\Gamma}^a{}_b \equiv \Gamma^a{}_{bc}\boldsymbol{\omega}^c$, $\mathrm{d}\boldsymbol{\omega}^a = -\boldsymbol{\Gamma}^a{}_b \wedge \boldsymbol{\omega}^b$.

Riemann tensor: $R^d{}_{abc}$, $2v_{a;[bc]} = v_d R^d{}_{abc}$.

Curvature 2-forms: $\boldsymbol{\Theta}^a{}_b \equiv \frac{1}{2} R^a{}_{bcd}\boldsymbol{\omega}^c \wedge \boldsymbol{\omega}^d = \mathrm{d}\boldsymbol{\Gamma}^a{}_b + \boldsymbol{\Gamma}^a{}_c \wedge \boldsymbol{\Gamma}^c{}_b$.

Ricci tensor, Einstein tensor, and scalar curvature:

$$R_{ab} \equiv R^c{}_{acb}, \quad G_{ab} \equiv R_{ab} - \frac{1}{2} R g_{ab}, \quad R \equiv R_a{}^a.$$

Weyl tensor in V_4:

$$C_{abcd} \equiv R_{abcd} + \frac{R}{3} g_{a[c}g_{d]b} - g_{a[c}R_{d]b} + g_{b[c}R_{d]a}.$$

Null tetrad components of the Weyl tensor:

$$\Psi_0 \equiv C_{abcd}k^a m^b k^c m^d, \qquad \Psi_1 \equiv C_{abcd}k^a l^b k^c m^d,$$

$$\Psi_2 \equiv \frac{1}{2} C_{abcd}k^a l^b (k^c l^d - m^c \overline{m}^d),$$

$$\Psi_3 \equiv C_{abcd}l^a k^b l^c \overline{m}^d, \qquad \Psi_4 \equiv C_{abcd}l^a \overline{m}^b l^c \overline{m}^d.$$

Metric of a 2-space of constant curvature:

$$\mathrm{d}\sigma^2 = \mathrm{d}x^2 \pm \Sigma^2(x, \varepsilon)\,\mathrm{d}y^2,$$

$$\Sigma(x, \varepsilon) = \sin x,\ x,\ \sinh x \text{ resp. when } \varepsilon = 1,\ 0 \text{ or } -1.$$

Gaussian curvature: K.

Physical fields

Energy-momentum tensor: T_{ab}, $T_{ab}u^a u^b \geqq 0$ if $u_a u^a = -1$.

Electromagnetic field: Maxwell tensor F_{ab}, $T_{ab} = \frac{1}{2} F^{*c}_{a}\overline{F^{*}_{bc}}$.

Null tetrad components of F_{ab}:

$$\Phi_0 \equiv F_{ab}k^a m^b, \qquad \Phi_1 \equiv \frac{1}{2} F_{ab}(k^a l^b + \overline{m}^a m^b), \qquad \Phi_2 \equiv F_{ab}\overline{m}^a l^b.$$

Perfect fluid: pressure p, energy density μ, 4-velocity $\boldsymbol{u}$,

$$T_{ab} = (\mu + p)\, u_a u_b + p g_{ab}.$$

Cosmological constant: Λ, gravitational constant: $\varkappa_0$.

Einstein's field equations: $R_{ab} - \dfrac{1}{2} R g_{ab} + \Lambda g_{ab} = \varkappa_0 T_{ab}$.

Symmetries

Group of motions (r-dim.), G_r; isotropy group (s-dim.), H_s.
Killing vectors: $\boldsymbol{\xi}_A$, $A = 1 \dots r$; $\boldsymbol{\xi}, \boldsymbol{\eta}, \boldsymbol{\zeta}$.
Killing equation: $(\pounds_\xi \boldsymbol{g})_{ab} = \xi_{a;b} + \xi_{b;a} = 0$.
Structure constants: $C^C{}_{AB}$; $[\boldsymbol{\xi}_A, \boldsymbol{\xi}_B] = C^C{}_{AB}\boldsymbol{\xi}_C$.
Orbits (m-dim.) of G_r: S_m (spacelike), T_m (timelike), N_m (null).

Chapter 1. Introduction

1.1. What are exact solutions, and why study them?

The theories of modern physics generally involve a mathematical model, defined by a certain set of differential equations, and supplemented by a set of rules for translating the mathematical results into meaningful statements about the physical world. In the case of theories of gravitation, it is generally accepted that the most successful is Einstein's theory of General Relativity. Here the differential equations consist of purely geometric requirements imposed by the idea that space and time can be represented by a Riemannian (Lorentzian) manifold, together with the description of the interaction of matter and gravitation contained in Einstein's famous field equations

$$R_{ab} - \frac{1}{2} R g_{ab} + \Lambda g_{ab} = \varkappa_0 T_{ab}. \tag{1.1}$$

(The full definition of the quantities used here appears later in the book.) This book will be concerned only with Einstein's theory.

We do not, of course, set out to discuss all aspects of General Relativity. For the basic problem of understanding the fundamental concepts (measurements using rods and clocks, etc.) used in the translation between the mathematics and the physics, we refer the reader to other texts. For any physical theory, there is also the purely mathematical problem of analysing, as far as possible, the set of differential equations and of finding as many exact solutions, or as complete a general solution, as possible. Next comes the mathematical and physical interpretation of the solutions thus obtained; in the case of General Relativity this requires global analysis and topological methods rather than just the purely local solution of the differential equations. In the case of gravity theories, because they deal with the most universal of physical interactions, one has an additional class of problems concerning the influence of the gravitational field on other fields and matter; these are often studied by working within a fixed gravitational field, usually an exact solution.

This book deals primarily with the solutions of the Einstein equations, (1.1), and only tangentially with the other subjects. The strongest reason for excluding the omitted topics is that each would fill (and some do fill) another book of equal length; we do, of course, give some references to the relevant literature. Unfortunately, one cannot say that the study of exact solutions has always maintained good contact with work on more directly physical problems. Kinnersley (1975) wrote 'Most of the known exact solutions describe situations which are frankly unphysical, and these do have a tendency to distract attention from the more useful ones. But the situation is also partially the fault of those of us who work in this field. We toss in null currents, macroscopic neutrino fields and tachyons for the sake of greater "generality"; we seem to take delight at the invention of confusing anti-intuitive notation; and when all is done we leave our newborn metric wobbling on its vierbein without any visible means of interpretation.'

In defence of work on exact solutions, it may be pointed out that certain solutions have played a very important role in the discussion of physical problems. Obvious examples are the Schwarzschild and Kerr solutions for black holes, the Friedmann solutions for cosmology, and the plane wave solutions which resolved some of the controversies about the existence of gravitational radiation. It should also be noted that because General Relativity is a highly non-linear theory, it is not always easy to understand what qualitative features solutions might possess, and here the exact solutions, including many such as the Taub-NUT solutions which may be thought unphysical, have proved an invaluable guide. Though the fact is not always appreciated, the non-linearities also mean that perturbation schemes in General Relativity can run into hidden dangers (see e.g. Ehlers et al. (1976)). Exact solutions which could be compared with approximate results would be very useful in checking the validity of approximation techniques.

In addition to the above reasons for devoting this book to the classification and construction of exact solutions, one may note that although much is known, it is often not generally known, because of the plethora of journals, languages, and mathematical notations in which it has appeared. We hope that one beneficial effect of our efforts will be to save colleagues from wasting their time rediscovering known results; in particular we hope our attempt to invariantly characterize the known solutions will help readers to identify any new examples that arise.

One surprise for the reader may lie in the enormous number of known exact solutions. Those who do not work in the field often suppose that the intractability of the full Einstein equations means that very few solutions are known. In a certain sense this is true: we know relatively few exact solutions for real physical problems. In most solutions, for example, there is no complete description of the relation of field to sources. Problems which are without an exact solution include the two-body problem, the realistic description of our inhomogeneous universe, the gravitational field of a stationary rotating star, and the generation and propagation of gravitational radiation from a realistic bounded source. There are, on the other hand, some problems where the known exact solutions may be the unique answer, for instance, the Kerr and Schwarzschild solutions for the final collapsed state of massive bodies.

Any metric whatsoever is a "solution" of (1.1) if no restriction is imposed on the energy-momentum tensor, since (1.1) then becomes just a definition of T_{ab}; so we must first make some assumptions about T_{ab}. Beyond this we may proceed, for example, by imposing symmetry conditions on the metric, by restricting the algebraic structure of the Riemann tensor, by adding field equations for the matter variables, or by imposing initial and boundary conditions. The exact solutions that are known have all been obtained by making some such restrictions.

We have used the term "exact solution" without a definition, and we do not intend to provide one. Clearly a metric would be called an exact solution if its components could be given, in suitable coordinates, in terms of the well-known analytic functions (polynomials, trigonometric and hyperbolic functions, and so on). It is then hard to find grounds for excluding any analytic function, even one defined only by some system of differential equations. Thus "exact solution" has a less clear meaning than one might like, although it conveys the impression that in some sense the properties of the metric are fully determined; no generally-agreed precise definition exists. We have proceeded rather on the basis that what we chose to include was, by definition, an exact solution.

1.2. The development of the subject

In the first few years (or decades) of research in General Relativity, only rather a small number of exact solutions were discussed. These mostly arose from highly idealized physical problems, and had very high symmetry. As examples, one may cite the well-known spherically-symmetric solutions of Schwarzschild, Reissner and Nordström, Tolman, and Friedmann (this last using the spatially-homogeneous metric form now associated with the names of Robertson and Walker), the axisymmetric static electromagnetic and vacuum solutions of Weyl, and the plane wave metrics. Although such a limited range of solutions was studied, we must, in fairness, point out that it includes nearly all the exact solutions which are of importance in physical applications; perhaps the only one of comparable importance which was a post-war discovery is the Kerr solution.

In the early period there were comparatively few people actively working on General Relativity, and it seems to us that the general belief at that time was that exact solutions would be of little importance, except perhaps as cosmological and stellar models, because of the extreme weakness of the relativistic corrections to Newtonian gravity. Of course, a wide variety of physical problems were attacked, but in a large number of cases they were treated only by some approximation scheme, especially the weak-field, slow-motion approximation.

Moreover, many of the techniques now in common use were either unknown or at least unknown to most relativists. The first to become popular was the use of groups of motions, especially in the construction of cosmologies more general than Friedmann's. The next, which was in part motivated by the study of gravitational radiation, was the algebraic classification of the Weyl tensor (into Petrov types) and the understanding of the properties of algebraically special metrics. Both these developments led in a natural way to the use of invariantly-defined tetrad bases, rather than coordinate components. The null tetrad methods, and some ideas from the theory of group representations and algebraic geometry, gave rise to the spinor techniques, and equivalent methods, now usually employed in the form given by Newman and Penrose.

Using these methods, it was possible to obtain many new solutions, and this growth is still continuing; there are now over 100 new papers on exact solutions every year. The same influences, of course, led us to the writing of this book.

1.3. The contents and arrangement of this book

Naturally, we begin by introducing differential geometry (Chapter 2.) and Riemannian geometry (Chapter 3.). We cannot provide a formal textbook of these subjects; our aim is to give just the notation, computational methods and (usually without proof) standard results we need for later chapters. After this point, the way ahead becomes more debatable.

There are (at least) four schemes for classification of the known exact solutions which could be regarded as having more or less equal importance; these four are the algebraic classification of conformal curvature (Petrov types), the algebraic classification of the Ricci tensor (Plebański types) and the physical characterization of the energy-momentum tensor, the existence and structure of preferred vector fields, and the groups of symmetry "admitted by" (i.e. which exist for) the metric. We have devoted a chapter (respectively, Chapters 4., 5., 6. and 8.) to each of these, introducing the terminology and methods used later and some general theorems. Among these chapters we have interpolated one (Chapter 7.) which gives the Newman-Penrose formalism; its position is due to the fact that this formalism can be applied immediately to elucidating some of the relationships between the considerations in the preceding three chapters.

The four-dimensional presentation of the solutions that would arise from the classification schemes outlined above may be acceptable to relativists but is impractical for authors. We could have worked through each classification in turn, but this would have been lengthy and repetitive (as it is, the reader will find certain solutions recurring in various disguises). We have therefore chosen to give pride of place to the two schemes which seem to have had the widest use in discovery and construction of new solutions, namely symmetry groups (Part II. of the book) and Petrov types (Part III.). The other main classifications have been used in subdividing the various classes of solutions discussed in Parts II. and III., and they are covered by the tables in Part V. We also found it necessary to add a number of chapters (Part IV.) which deal with less general ways of classifying and constructing exact solutions.

The specification of the energy-momentum tensor played a very important part because we decided at an early stage that it would be impossible to provide a comprehensive survey of all energy-momentum tensors that have ever been considered. We therefore restricted ourselves to the following energy-momentum tensors: vacuum, electromagnetic fields, pure radiation, dust and perfect fluids. (The term "pure radiation" is used here for an energy-momentum-tensor representing a situation in which all the energy is transported in one direction with the speed of light: such tensors are also referred to in the literature as null fields, null fluids and null dust.) Combinations of these, and matching of solutions with equal or different energy-momentum tensors (e.g. the Schwarzschild vacuoli in a Friedmann universe) are in general not considered, and the cosmological constant Λ, although sometimes introduced, is not treated systematically throughout. These limitations on the scope of our work may be disappointing to some, especially those working on solutions containing charged perfect fluids, scalar, Dirac and neutrino fields, or solid elastic bodies. They were made not only because some limits on the task we set ourselves were necessary, but also because most of the known solutions are for the energy-momentum tensors listed and because it is possible to give a fairly full systematic treatment for these cases. Ultimately, of course, it is a matter of taste.

The arrangement within Part II. is outlined more fully in § 9.1. Here we remark only that we treated first non-null and then null group orbits (as defined in Chapter 8.), arranging each in decreasing order of dimension of the orbit and thereafter (usually) in decreasing order of dimension of the group. Certain special cases of physical or mathematical interest were separated out of this orderly progression and given chapters of their own, for example, spatially homogeneous cosmologies and spherically symmetric solutions. Within each chapter we tried to give first the differential geometric results (i.e. general forms of the metric and curvature) and then the actual solutions for each type of energy-momentum in turn; this arrangement is followed in Parts III. and IV. also.

In Part III. we aimed to give a rather detailed account of the well-developed theory that is available for algebraically special solutions for vacuum, electromagnetic and pure radiation fields. Only a few classes, mostly very special cases, of algebraically special perfect-fluid solutions have been thoroughly discussed in the literature; a short review of these classes is given in Chapter 29. Quite a few of the algebraically special solutions also admit groups of motions. Where this is known (and, as far as we are aware, it has not been systematically studied for all cases), it is of course indicated in the text and in the tables.

Finally, Part IV. gives some discussion of the classification of space-times with special vector or tensor fields. It also covers solutions found by embedding, and the generation techniques developed by various authors in the last ten years (although most of these rely on the existence of a group of motions, and in some sense therefore belong in Part II.).

The weight of material, even with all the limitations described above, made it necessary to omit some proofs and details and give only the necessary references.

1.4. Using this book as a catalogue

This book has not been written simply as a catalogue. Nevertheless, we intended that it should be possible for the book to be used for this purpose. In arranging the information here, we have assumed that a reader who wishes to find (or, at least, search for) a solution will either know the original author (if the reader is aware the solution is not new) or know some of its invariant properties.

If the author is known, the reader should turn to the alphabetically-organized bibliography. He should then be able to identify the relevant paper(s) of that author, since the titles, and, of course, journals and dates, are given in full. Following each reference is a list of all places in the book where it is cited.

If the reader knows the (maximal) group of motions, he can find the relevant chapter in Part II. by consulting the contents list or the tables. If he knows the Petrov type, he can again consult the contents list or the tables by Petrov type; if he knows only the energy-momentum tensor, he can still consult the relevant tables. If none of this information is known, he can turn to Part IV., if he has used one of the special methods described there. If still in doubt, he will have to read the whole book.

If the solution is known (and not accidentally omitted) it will in general be given in full, possibly only in the sense of appearing contained in a more general form for a whole class of solutions; some solutions of great complexity or (to us) lesser importance have been given only in the sense of a reference to the literature. Each solution may, of course, be found in a great variety of coordinate forms and be characterized invariantly in several ways. We have tried to eliminate duplications, i.e. to identify those solutions that are really the same but appear in the literature as separate, and we give cross-references between sections where possible. The solutions are usually given in coordinates adapted to some invariant properties, and it should therefore be feasible (if non-trivial) for the reader to transform to any other coordinate system he has discovered. Solutions which are neither given nor quoted are either new or accidentally omitted, and in either case the authors would be interested to hear about them. (We should perhaps note here that not all papers containing frequently-rediscovered solutions have been cited; in such a case only the earliest papers, and those rediscoveries with some special importance, have been given. Moreover, if a general class of solutions is known, rediscoveries of special cases belonging to this class have been mentioned only occasionally.)

We have checked most of the solutions given in the book. This was done by hand and we may, therefore, have simply repeated the authors' errors; an additional check using an algebraic manipulation programme would be helpful. It is not always explicitly stated where we did not check solutions.

In addition to references within the text, cited by author and year, we have sometimes put at the ends of sections some references to parallel methods, or to generalizations, or to applications. We would also draw the reader's attention to previous review articles, such as those by Ehlers and Kundt (1962), Kinnersley (1975) and Bashkov (1976), which may offer him a useful alternative view.

References added in proof are indicated by the symbol □ in front of the author's name; they are listed at the end of the bibliography.

PART I. GENERAL METHODS

Chapter 2. Differential geometry without a metric

2.1. Introduction

The concept of a tensor is usually based on the law of transformation of the components under coordinate transformations, so that coordinates are explicitly used from the very beginning. This calculus provides adequate methods for many situations, but other techniques are sometimes more effective. In the modern literature on exact solutions coordinate-free geometric concepts, such as forms and exterior differentiation, are frequently used. Moreover, the underlying mathematical structure often becomes more evident when expressed in coordinate-free terms.

Hence this chapter will present a brief survey of some of the basic ideas of differential geometry. Most of these are independent of the introduction of a metric, although, of course, this is of fundamental importance in the space-times of General Relativity: the discussion of manifolds with metrics will therefore be deferred until the next chapter. Here we shall introduce vectors, p-forms, tensors of arbitrary rank, exterior differentiation and Lie differentiation, all of which follow naturally from the definition of a differentiable manifold. We then consider an additional structure, a covariant derivative, and its associated curvature; even this does not necessarily involve a metric. The absence of any metric will, however, mean that it will not be possible to convert 1-forms to vectors, or vice versa.

Since we are primarily concerned with specific applications, we shall emphasize the rules of manipulation and calculation in differential geometry. We do not attempt to provide a substitute for standard texts on the subject, e.g. Flanders (1963), Stoker (1969), Brickell and Clark (1970), Sternberg (1964), Helgason (1962), Kobayashi and Nomizu (1969), □ Choquet-Bruhat et al. (1978), to which the reader is referred for fuller information and for the proofs of many of the theorems. Useful introductions for our purpose can be found in texts on relativity, e.g. Hawking and Ellis (1973), Misner et al. (1973), and Israel (1970).

For the benefit of those familiar with the traditional approach to tensor calculus, certain formulae are displayed both in coordinate-free form and in the usual component formalism.

2.2. Differentiable manifolds

Differentiable manifolds are the most basic structures in differential geometry. Intuitively, an (n-dimensional) manifold is a space $\mathscr{M}$ such that any point $p \in \mathscr{M}$ has a neighbourhood $\mathscr{U} \subset \mathscr{M}$ which is homeomorphic to the interior of the

($(n-1)$-dimensional) unit sphere. To give a mathematically precise definition of a differentiable manifold we need to introduce some additional terminology.

A *chart* $(\mathcal{U}, \Phi)$ in $\mathcal{M}$ consists of a subset $\mathcal{U}$ of $\mathcal{M}$ together with a one-to-one map Φ from $\mathcal{U}$ onto the n-dimensional Euclidean space, E^n (or an open subset of E^n); Φ assigns to every point $p \in \mathcal{U}$ an n-tuple of real variables, the *local coordinates* $(x^1, \ldots, x^n)$.

Two charts $(\mathcal{U}, \Phi)$, $(\mathcal{U}', \Phi')$ are said to be *compatible* if the combined map $\Phi' \circ \Phi^{-1}$ on the image $\Phi(\mathcal{U} \cap \mathcal{U}')$ of the overlap of $\mathcal{U}$ and $\mathcal{U}'$ is a homeomorphism (i.e. continuous, one-to-one, and having a continuous inverse): see Fig. 2.1.

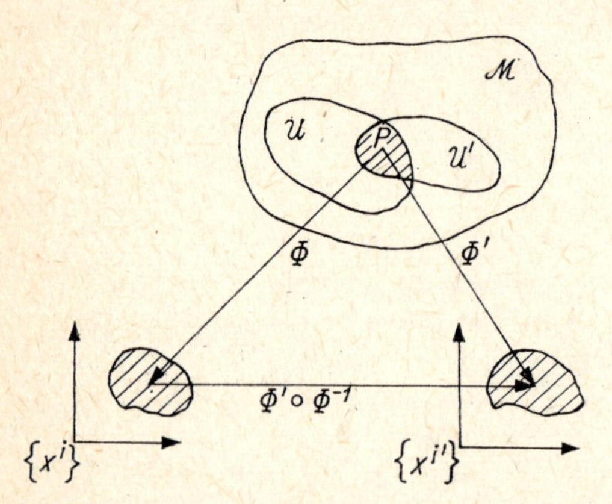

Fig. 2.1 Two compatible charts of a differentiable manifold

An *atlas* on $\mathcal{M}$ is a collection of compatible charts $(\mathcal{U}_\alpha, \Phi_\alpha)$ such that every point of $\mathcal{M}$ lies in at least one chart neighbourhood $\mathcal{U}_\alpha$. In most cases, it is impossible to cover the manifold with a single chart (an example of this is the n-dimensional sphere).

An n-dimensional (topological) *manifold* consists of a space $\mathcal{M}$ together with an atlas on $\mathcal{M}$. It is a (C^k or analytic) *differentiable manifold* $\mathcal{M}$ if the maps $\Phi' \circ \Phi^{-1}$ relating different charts are not just continuous but differentiable (respectively, C^k or analytic). Then the coordinates are related by n differentiable (C^k, analytic) functions, with non-vanishing Jacobian at each point of the overlap:

$$x^{i'} = x^{i'}(x^j), \qquad \det(\partial x^{i'}/\partial x^j) \neq 0. \tag{2.1}$$

Definitions of manifolds often include additional topological restrictions, such as paracompactness and Hausdorffness, and these are indeed essential for the rigorous proof of a few of the results we state. For brevity, we shall omit any consideration of these questions, which are of course fully discussed in the literature cited earlier.

A differentiable manifold $\mathcal{M}$ is called *orientable* if there exists an atlas such that the Jacobian is positive throughout the overlap of any pair of charts.

If $\mathcal{M}$ and $\mathcal{N}$ are manifolds, of dimensions m and n, respectively, the $(m+n)$-dimensional product $\mathcal{M} \times \mathcal{N}$ can be defined in a natural way.

A *map* $\Phi : \mathscr{M} \to \mathscr{N}$ is said to be *differentiable* if the coordinates $(y^1, \ldots, y^n)$ on $\mathscr{V} \in \mathscr{N}$ are differentiable functions of the coordinates $(x^1, \ldots, x^m)$ of the corresponding points in $\mathscr{U} \subset \mathscr{M}$ where Φ maps (a part of) the neighbourhood $\mathscr{U}$ into the neighbourhood $\mathscr{V}$.

A smooth *curve* $\gamma(t)$ in $\mathscr{M}$ is defined by a differentiable map of an interval of the real line into $\mathscr{M}$, $\gamma(t) : -\varepsilon < t < \varepsilon \to \mathscr{M}$.

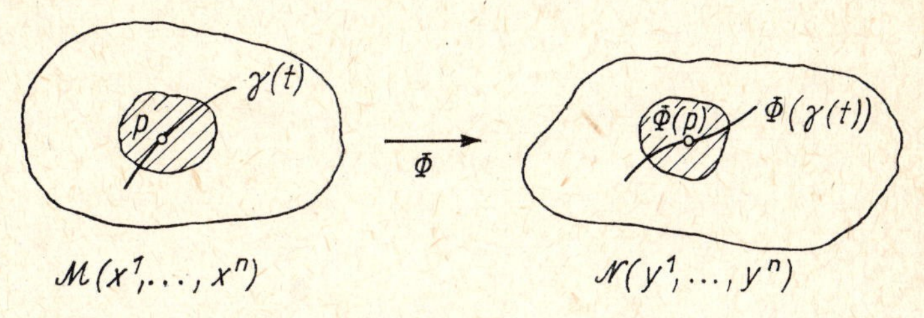

Fig. 2.2 The map of a smooth curve $\gamma(t)$ to $\Phi(\gamma(t))$

2.3. Tangent vectors

In general a vector cannot be considered as an arrow connecting two points of the manifold. To get a consistent generalization of the concept of vectors in E^n, one identifies vectors on $\mathscr{M}$ with tangent vectors. A *tangent vector* $\boldsymbol{v}$ at p is an *operator* (*linear functional*) which assigns to each differentiable function f on $\mathscr{M}$ a real number $\boldsymbol{v}(f)$. This operator satisfies the axioms

$$\begin{aligned} &\text{(i)} && \boldsymbol{v}(f + h) = \boldsymbol{v}(f) + \boldsymbol{v}(h), \\ &\text{(ii)} && \boldsymbol{v}(fh) = h\boldsymbol{v}(f) + f\boldsymbol{v}(h), \\ &\text{(iii)} && \boldsymbol{v}(cf) = c\boldsymbol{v}(f), \quad c = \text{constant}. \end{aligned} \tag{2.2}$$

It follows from these axioms that $\boldsymbol{v}(c) = 0$ for any constant function c. The definition (2.2) is independent of the choice of coordinates. A tangent vector is just a *directional derivative* along a curve $\gamma(t)$ through p: expanding any function f in a Taylor series at p, and using the axioms (2.2), one can easily show that any tangent vector $\boldsymbol{v}$ at p can be written as

$$\boldsymbol{v} = v^i \, \partial/\partial x^i. \tag{2.3}$$

The real coefficients v^i are the *components* of $\boldsymbol{v}$ at p with respect to the local coordinate system $(x^1, \ldots, x^n)$ in a neighbourhood of p. According to (2.3), the directional derivatives along the coordinate lines at p form a basis of an n-dimensional vector space whose elements are the tangent vectors at p. This space is called the tangent space T_p. The basis $\{\partial/\partial x^i\}$ is called a *coordinate basis* or *holonomic frame*.

A *general basis* $\{\boldsymbol{e}_a\}$ is formed by n linearly independent vectors $\boldsymbol{e}_a$; any vector $\boldsymbol{v} \in T_p$ is a linear combination of these basis vectors, i.e.

$$\boldsymbol{v} = v^a \boldsymbol{e}_a. \tag{2.4}$$

The action of a basis vector $\boldsymbol{e}_a$ on a function f is denoted by the symbol $f_{|a} \equiv \boldsymbol{e}_a(f)$. In a coordinate basis we use a comma in place of a stroke, $f_{,i} \equiv \partial f/\partial x^i$. A non-singular linear transformation of the basis $\{\boldsymbol{e}_a\}$ induces a change of the components v^a of the vector $\boldsymbol{v}$,

$$\boldsymbol{e}_{a'} = L_{a'}{}^b \boldsymbol{e}_b, \qquad v^{a'} = L^{a'}{}_b v^b, \quad L^{a'}{}_b L_{a'}{}^c = \delta_b{}^c. \tag{2.5}$$

A coordinate basis $\{\partial/\partial x^i\}$ represents a special choice of $\{\boldsymbol{e}_a\}$. In the older literature on General Relativity the components with respect to coordinate bases were preferred for actual computations. However, for many purposes it is more convenient to use a general basis (a tetrad). Well-known examples are the Petrov classification (Chapter 4.) and the Newman-Penrose formalism (Chapter 7.).

We can construct a *vector field* $\boldsymbol{v}(p)$ on $\mathcal{M}$ by assigning to each point $p \in \mathcal{M}$ a tangent vector $\boldsymbol{v} \in T_p$ so that the components v^i are differentiable functions of the local coordinates.

From the identification of vectors with directional derivatives one concludes that in general the result of the successive application of two basis vectors $\boldsymbol{e}_a$ to a function depends on the order in which the operators are applied. The *commutator* $[\boldsymbol{u}, \boldsymbol{v}]$ of two vector fields $\boldsymbol{u}$ and $\boldsymbol{v}$ is defined by $[\boldsymbol{u}, \boldsymbol{v}]\,(f) = \boldsymbol{u}(\boldsymbol{v}(f)) - \boldsymbol{v}(\boldsymbol{u}(f))$. For a given basis $\{\boldsymbol{e}_a\}$, the commutators

$$[\boldsymbol{e}_a, \boldsymbol{e}_b] = D^c{}_{ab}\boldsymbol{e}_c, \qquad D^c{}_{ab} = -D^c{}_{ba}, \tag{2.6}$$

define the commutation coefficients $D^c{}_{ab}$, which obviously vanish for a coordinate basis: $[\partial/\partial x^i, \partial/\partial x^j] = 0$. Commutators satisfy the *Jacobi identity*

$$[\boldsymbol{u}, [\boldsymbol{v}, \boldsymbol{w}]] + [\boldsymbol{v}, [\boldsymbol{w}, \boldsymbol{u}]] + [\boldsymbol{w}, [\boldsymbol{u}, \boldsymbol{v}]] = 0 \tag{2.7}$$

for arbitrary $\boldsymbol{u}, \boldsymbol{v}, \boldsymbol{w}$, from which one infers, for constant $D^c{}_{ab}$, the identity

$$D^f{}_{d[a}D^d{}_{bc]} = 0. \tag{2.8}$$

2.4. One-forms

By definition, a 1-*form* (Pfaffian form) σ maps a vector $\boldsymbol{v}$ into a real number, the *contraction*, denoted by the symbol $\langle\sigma, \boldsymbol{v}\rangle$, and this mapping is linear:

$$\langle\sigma, a\boldsymbol{u} + b\boldsymbol{v}\rangle = a\langle\sigma, \boldsymbol{u}\rangle + b\langle\sigma, \boldsymbol{v}\rangle \tag{2.9}$$

for real a, b, and $\boldsymbol{u}, \boldsymbol{v} \in T_p$. Linear combinations of 1-forms σ, $\boldsymbol{\tau}$ are defined by the rule

$$\langle a\sigma + b\boldsymbol{\tau}, \boldsymbol{v}\rangle = a\langle\sigma, \boldsymbol{v}\rangle + b\langle\boldsymbol{\tau}, \boldsymbol{v}\rangle, \tag{2.10}$$

for real a, b. The n linearly independent 1-forms ω^a which are uniquely determined by

$$\langle\omega^a, \boldsymbol{e}_b\rangle = \delta^a_b \tag{2.11}$$

form a basis $\{\omega^a\}$ of the *dual space* T^*_p of the tangent space T_p. This basis $\{\omega^a\}$ is said to be dual to the basis $\{\boldsymbol{e}_b\}$ of T_p. Any 1-form $\sigma \in T^*_p$ is a linear combination of the basis 1-forms ω^a;

$$\sigma = \sigma_a \omega^a. \tag{2.12}$$

For any $\sigma \in T_p^*$, $\boldsymbol{v} \in T_p$ the contraction $\langle \sigma, \boldsymbol{v} \rangle$ can be expressed in terms of the components σ_a, v^a of σ, $\boldsymbol{v}$ with respect to the bases $\{\omega^a\}$, $\{\boldsymbol{e}_a\}$ by

$$\langle \sigma, \boldsymbol{v} \rangle = \sigma_a v^a. \tag{2.13}$$

The *differential* $\mathrm{d}f$ of an arbitrary function f is a 1-form defined by the property

$$\langle \mathrm{d}f, \boldsymbol{v} \rangle = \boldsymbol{v}(f) \equiv v^a f_{|a}. \tag{2.14}$$

Specializing this definition to the functions $f = x^1, \ldots, x^n$ one obtains the relation

$$\langle \mathrm{d}x^i, \partial/\partial x^j \rangle = \delta^i_j, \tag{2.15}$$

indicating that the basis $\{\mathrm{d}x^i\}$ of T_p^* is dual to the coordinate basis $\{\partial/\partial x^j\}$ of T_p. Any 1-form $\sigma \in T_p^*$ can be written with respect to the basis $\{\mathrm{d}x^i\}$ as

$$\sigma = \sigma_i \, \mathrm{d}x^i. \tag{2.16}$$

In local coordinates, the differential $\mathrm{d}f$ has the usual form

$$\mathrm{d}f = f_{|a}\omega^a = f_{,i} \, \mathrm{d}x^i. \tag{2.17}$$

A *field of 1-forms* on $\mathcal{M}$ is defined analogously to a vector field, the components σ_i being differentiable functions of the local coordinates. In tensor calculus the components σ_i are often called "components of a covariant vector".

2.5. The exterior product

Let $\alpha^1, \ldots, \alpha^p$ denote p 1-forms. We define an algebraic operation $\wedge$, the *exterior product* (or wedge product) by the axioms: the exterior product $\alpha^1 \wedge \alpha^2 \wedge \cdots \wedge \alpha^p$

(i) is linear in each variable,

(ii) vanishes if any two factors coincide,

(iii) changes sign if any two factors are interchanged.

Thus the exterior product is completely antisymmetric. From the basis 1-forms $\omega^1, \ldots, \omega^n$ we obtain $\binom{n}{p}$ independent *p-forms*

$$\omega^{a_1} \wedge \cdots \wedge \omega^{a_p}, \qquad 1 \leqq a_1 < a_2 < \cdots < a_p \leqq n, \qquad p \leqq n. \tag{2.18}$$

Axiom (ii) implies that these exterior products vanish for $p > n$.

Any p-form $\underset{(p)}{\alpha}$ is a linear combination of the p-forms (2.18),

$$\underset{(p)}{\alpha} = \alpha_{a_1 \cdots a_p} \omega^{a_1} \wedge \cdots \wedge \omega^{a_p}. \tag{2.19}$$

where all indices run from 1 to n, the restriction (2.18) for the indices having been dropped. The expansion of $\underset{(p)}{\alpha}$ in terms of the dual coordinate basis $\{dx^i\}$ has the form

$$\underset{(p)}{\alpha} = \alpha_{i_1 \cdots i_p} \, dx^{i_1} \wedge \cdots \wedge dx^{i_p}. \tag{2.20}$$

A p-form $\underset{(p)}{\alpha}$ is said to be *simple* if it admits a representation as an exterior product of p linearly independent 1-forms,

$$\underset{(p)}{\alpha} = \alpha^1 \wedge \alpha^2 \wedge \cdots \wedge \alpha^p. \tag{2.21}$$

The exterior product can be extended to forms of arbitrary degree by the rule that

$$(\alpha^1 \wedge \cdots \wedge \alpha^p) \wedge (\beta^1 \wedge \cdots \wedge \beta^q) = \alpha^1 \wedge \cdots \wedge \alpha^p \wedge \beta^1 \wedge \cdots \wedge \beta^q. \tag{2.22}$$

Exterior multiplication is associative and distributive. However, the commutative law is slightly changed:

$$\underset{(p)}{\alpha} \wedge \underset{(q)}{\beta} = (-1)^{pq} \underset{(q)}{\beta} \wedge \underset{(p)}{\alpha}. \tag{2.23}$$

This property can easily be derived from the axioms defining exterior products of 1-forms and from expansions like (2.19). Let $\underset{(p)}{\alpha}$ and $\underset{(q)}{\beta}$ have the components $\alpha_{a_1 \ldots a_p}$ and $\beta_{b_1 \ldots b_q}$, respectively. Then the exterior product (2.23) has the components

$$\left(\underset{(p)}{\alpha} \wedge \underset{(q)}{\beta}\right)_{a_1 \ldots a_p b_1 \ldots b_q} = \alpha_{[a_1 \ldots a_p} \beta_{b_1 \ldots b_q]}. \tag{2.24}$$

2.6. Tensors

A tensor $\boldsymbol{T}$ of type (r, s), or of order $(r + s)$, at p is an element of the product space

$$T_p(r, s) = \underbrace{T_p \otimes \cdots \otimes T_p}_{r \text{ factors}} \otimes \underbrace{T_p^* \otimes \cdots \otimes T_p^*}_{s \text{ factors}}$$

and maps any ordered set of r 1-forms and s vectors,

$$(\sigma^1, \ldots, \sigma^r; \boldsymbol{v}_1, \ldots, \boldsymbol{v}_s), \tag{2.25}$$

at p into a real number. In particular, the tensor $\boldsymbol{u}_1 \otimes \cdots \otimes \boldsymbol{u}_r \otimes \boldsymbol{\tau}^1 \otimes \cdots \otimes \boldsymbol{\tau}^s$ maps the ordered set (2.25) into the product of contractions, $\langle \sigma^1, \boldsymbol{u}_1 \rangle \cdots \langle \sigma^r, \boldsymbol{u}_r \rangle \langle \boldsymbol{\tau}^1, \boldsymbol{v}_1 \rangle \cdots \langle \boldsymbol{\tau}^s, \boldsymbol{v}_s \rangle$. The map is linear in each argument. In terms of the bases $\{\boldsymbol{e}_a\}$, $\{\boldsymbol{\omega}^b\}$ an arbitrary tensor $\boldsymbol{T}$ of type (r, s) can be expressed as a sum of tensor products,

$$\boldsymbol{T} = T^{a_1 \cdots a_r}{}_{b_1 \ldots b_s} \, \boldsymbol{e}_{a_1} \otimes \cdots \otimes \boldsymbol{e}_{a_r} \otimes \boldsymbol{\omega}^{b_1} \otimes \cdots \otimes \boldsymbol{\omega}^{b_s}, \tag{2.26}$$

where all indices run from 1 to n. The coefficients $T^{a_1 \cdots a_r}{}_{b_1 \ldots b_s}$ with covariant indices $b_1 \ldots b_s$ and contravariant indices $a_1 \ldots a_r$ are the *components* of $\boldsymbol{T}$ with respect to the bases $\{\boldsymbol{e}_a\}$, $\{\boldsymbol{\omega}^b\}$. For a general tensor, the factors in the individual tensor product terms in (2.26) may not be interchanged.

The exterior product $\wedge$ is just the antisymmetrization of the tensor product $\otimes$. Consequently, the antisymmetric tensors of type $(0, p)$ (antisymmetric covariant tensors) are precisely the p-forms introduced in the previous section.

Non-singular linear transformations of the bases,

$$\boldsymbol{e}_{a'} = L_{a'}{}^{a}\boldsymbol{e}_a, \qquad \boldsymbol{\omega}^{a'} = L^{a'}{}_{a}\boldsymbol{\omega}^a, \qquad L^{a'}{}_{b}L_{a'}{}^{c} = \delta^c_b, \tag{2.27}$$

change the components of the tensor $\boldsymbol{T}$ according to the transformation law

$$T^{a'_1 \dots a'_r}{}_{b'_1 \dots b'_s} = L^{a'_1}{}_{a_1} \dots L^{a'_r}{}_{a_r} L_{b'_1}{}^{b_1} \dots L_{b'_s}{}^{b_s} T^{a_1 \dots a_r}{}_{b_1 \dots b_s}. \tag{2.28}$$

For transformations connecting two coordinate bases $\{\partial/\partial x^i\}$, $\{\partial/\partial x^{i'}\}$, the $(n \times n)$ matrices $L^{a'}{}_a$, $L_{a'}{}^a$ take the special forms $L^{a'}{}_a \to \partial x^{i'}/\partial x^i$, $L_{a'}{}^a \to \partial x^i/\partial x^{i'}$.

The following algebraic operations are independent of the basis used in (2.26): addition of tensors of the same type, multiplication by a real number, tensor product of two tensors, contraction on any pair of one contravariant and one covariant ndex, and formation of the (anti)symmetric part of a tensor.

Maps of tensors. The map Φ (Fig. 2.2) sending $p \in \mathcal{M}$ to $\Phi(p) \in \mathcal{N}$ induces in a natural way a map Φ^* of the real-valued functions f defined on $\mathcal{N}$ to functions on $\mathcal{M}$,

$$\Phi^* f(p) = f(\Phi(p)). \tag{2.29}$$

Moreover, induced maps of vectors and 1-forms

$$\begin{aligned} \Phi_*: &\quad \boldsymbol{v} \in T_p \to \Phi_* \boldsymbol{v} \in T_{\Phi(p)}, \\ \Phi^*: &\quad \sigma \in T^*_{\Phi(p)} \to \Phi^* \sigma \in T^*_p \end{aligned} \tag{2.30}$$

are defined by the postulates:

(i) the image of a vector satisfies

$$\Phi_* \boldsymbol{v}(f)|_{\Phi(p)} = \boldsymbol{v}(\Phi^* f)|_p, \tag{2.31a}$$

$\Phi_* \boldsymbol{v}$ being the tangent vector to the image curve $\Phi(\gamma(t))$ at $\Phi(p)$, if $\boldsymbol{v}$ is the tangent vector to $\gamma(t)$ at p;

(ii) the maps (2.30) preserve the contractions,

$$\langle \Phi^* \sigma, \boldsymbol{v} \rangle|_p = \langle \sigma, \Phi_* \boldsymbol{v} \rangle|_{\Phi(p)}. \tag{2.31b}$$

It follows immediately from (2.31a) that, for any $\boldsymbol{u}$ and $\boldsymbol{v}$,

$$[\Phi_* \boldsymbol{u}, \Phi_* \boldsymbol{v}] = \Phi_* [\boldsymbol{u}, \boldsymbol{v}]. \tag{2.32}$$

Let us denote local coordinates in corresponding neighbourhoods of p and $\Phi(p)$ by $(x^1, \dots, x^m)$ and $(y^1, \dots, y^n)$ respectively. The map of a 1-form σ is given simply by coordinate substitution,

$$\Phi^*: \; \sigma = \sigma_i(y)\, \mathrm{d}y^i \to \Phi^* \sigma = \sigma_i(y(x)) \frac{\partial y^i}{\partial x^k} \mathrm{d}x^k = \tilde{\sigma}_k(x)\, \mathrm{d}x^k, \qquad i = 1 \dots n, \quad k = 1 \dots m. \tag{2.33}$$

These maps can immediately be extended to tensors of arbitrary type (r, s) provided that the inverse map Φ^{-1} exists, i.e. that Φ is a one-to-one map. In this case, equation (2.31 b) can be rewritten in the form

$$\langle \Phi^* \sigma, \Phi_*^{-1} w \rangle|_p = \langle \sigma, w \rangle|_{\Phi(p)}. \tag{2.34}$$

Note that Φ^* maps tensors on $\mathcal{N}$ to tensors on $\mathcal{M}$, starting from a map Φ of $\mathcal{M}$ to $\mathcal{N}$. Equation (2.34) thus defines *new* tensors, $\Phi^*\sigma$ etc. In contrast, under the transformations (2.27) of the basis of a given manifold $\mathcal{M}$ any tensor remains the same object; only its components are changed. Tensors are invariantly defined.

Up to now we have considered tensors at a given point p. The generalization to *tensor fields* is straightforward. As special cases we have defined fields of 1-forms and vector fields on $\mathcal{M}$ at the ends of §§ 2.3., and 2.4. In the next few sections we shall introduce various derivatives of tensor fields. For brevity, we shall call tensor fields simply tensors.

2.7. The exterior derivative

In § 2.4. we defined the differential $\mathrm{d}f$ of a function f by the equation (2.14). The operator d generates a 1-form $\mathrm{d}f$ from a 0-form f by

$$\mathrm{d}: \; f \to \mathrm{d}f = f_{,i}\,\mathrm{d}x^i. \tag{2.35}$$

We generalize this differentiation to apply to any p-form. The *exterior derivative* d maps a p-form into a $(p+1)$-form and is completely determined by the axioms:

(i) $$\mathrm{d}(\alpha + \beta) = \mathrm{d}\alpha + \mathrm{d}\beta, \tag{2.36a}$$

(ii) $$\mathrm{d}\Big(\underset{(p)}{\alpha} \wedge \underset{(q)}{\beta}\Big) = \mathrm{d}\underset{(p)}{\alpha} \wedge \underset{(q)}{\beta} + (-1)^p \underset{(p)}{\alpha} \wedge \mathrm{d}\underset{(q)}{\beta}, \tag{2.36b}$$

(iii) $$\mathrm{d}f = f_{,i}\,\mathrm{d}x^i, \tag{2.36c}$$

(iv) $$\mathrm{d}(\mathrm{d}f) = 0. \tag{2.36d}$$

Because of the axiom (2.36a) it is sufficient to verify the existence and uniqueness of the exterior derivative for the p-form $\underset{(p)}{\alpha} = f\,\mathrm{d}x^{i_1} \wedge \cdots \wedge \mathrm{d}x^{i_p}$.

For the *proof* (see e.g. Flanders (1963)) we use a coordinate basis. Recalling the fact that coordinates are functions on $\mathcal{M}$, one first shows by induction that

$$\mathrm{d}(\mathrm{d}x^{i_1} \wedge \cdots \wedge \mathrm{d}x^{i_p}) = \mathrm{d}[\mathrm{d}(x^{i_1}\,\mathrm{d}x^{i_2} \ldots \mathrm{d}x^{i_p})] = 0. \tag{2.37}$$

Then the axiom (2.36b) requires that we have

$$\mathrm{d}(f\,\mathrm{d}x^{i_1} \wedge \cdots \wedge \mathrm{d}x^{i_p}) = \mathrm{d}f \wedge \mathrm{d}x^{i_1} \wedge \cdots \wedge \mathrm{d}x^{i_p}. \tag{2.38}$$

It is easy to show that all the axioms (2.36) are now satisfied. For instance, we demonstrate the validity of the axiom (2.36b) for the forms $\underset{(p)}{\alpha} = f\,\mathrm{d}x^{i_1} \wedge \cdots \wedge \mathrm{d}x^{i_p}$ and $\underset{(q)}{\beta} = h\,\mathrm{d}x^{j_1} \wedge \cdots \wedge \mathrm{d}x^{j_q}$:

$$\begin{aligned}\mathrm{d}\Big(\underset{(p)}{\alpha} \wedge \underset{(q)}{\beta}\Big) &= \mathrm{d}(fh) \wedge \mathrm{d}x^{i_1} \wedge \cdots \wedge \mathrm{d}x^{i_p} \wedge \mathrm{d}x^{j_1} \wedge \cdots \wedge \mathrm{d}x^{j_q} \\ &= (\mathrm{d}f \wedge \mathrm{d}x^{i_1} \wedge \cdots \wedge \mathrm{d}x^{i_p}) \wedge (h\,\mathrm{d}x^{j_1} \wedge \cdots \wedge \mathrm{d}x^{j_q}) \\ &\quad + (-1)^p\,(f\,\mathrm{d}x^{i_1} \wedge \cdots \wedge \mathrm{d}x^{i_p}) \wedge (\mathrm{d}h \wedge \mathrm{d}x^{j_1} \wedge \cdots \wedge \mathrm{d}x^{j_q}) \\ &= \mathrm{d}\underset{(p)}{\alpha} \wedge \underset{(q)}{\beta} + (-1)^p\,\underset{(p)}{\alpha} \wedge \mathrm{d}\underset{(q)}{\beta},\end{aligned}$$

where we have used equation (2.23). From a general p-form (2.20) we obtain the $(p+1)$-form

$$\mathrm{d}\underset{(p)}{\alpha} = \alpha_{i_1\ldots i_p,j}\,\mathrm{d}x^j \wedge \mathrm{d}x^{i_1} \wedge \cdots \wedge \mathrm{d}x^{i_p}. \tag{2.39}$$

by exterior differentiation. The (fully antisymmetric) components of $\mathrm{d}\underset{(p)}{\alpha}$ involve only partial derivatives of the components $\alpha_{i_1\ldots i_p}$. We remark that the axiom (2.36d) is just the equality of the mixed second partial derivatives,

$$\mathrm{d}(\mathrm{d}f) = \mathrm{d}(f_{,i}\,\mathrm{d}x^i) = f_{,i,j}\,\mathrm{d}x^j \wedge \mathrm{d}x^i = 0. \tag{2.40}$$

The expression (2.39) implies

$$\mathrm{d}(\mathrm{d}\alpha) = 0. \tag{2.41}$$

for any p-form α.

The following theorems, for whose proofs we refer the reader to the literature, hold locally, i.e. in the neighbourhood of a point p.

Theorem 2.1 (Poincaré's Theorem). *If α is a p-form ($p \geqq 1$) and $\mathrm{d}\alpha = 0$, then there is a $(p-1)$-form β such that $\alpha = \mathrm{d}\beta$.* In components,

$$\alpha_{[i_1\ldots i_p,j]} = 0 \Leftrightarrow \alpha_{i_1\ldots i_p} = \beta_{[i_1\ldots i_{p-1},i_p]}. \tag{2.42}$$

Theorem 2.2 (Frobenius' Theorem). *Let $\sigma^1, \ldots, \sigma^r$ be r 1-forms linearly independent at a point $p \in \mathcal{M}$. Suppose there are 1-forms $\tau^A{}_B$ ($A, B = 1\ldots r$) satisfying $\mathrm{d}\sigma^A = \tau^A{}_B \wedge \sigma^B$. Then in a neighbourhood of p there are functions $f^A{}_B$, h^A such that $\sigma^A = f^A{}_B\,\mathrm{d}h^B$.* (For a proof, see Flanders (1963).)

Other formulations of this theorem: introducing the r-form $\Sigma \equiv \sigma^1 \wedge \cdots \wedge \sigma^r$, we can replace the condition $\mathrm{d}\sigma^A = \tau^A{}_B \wedge \sigma^B$ by either of the two equivalent conditions:

(i) $\mathrm{d}\sigma^A \wedge \Sigma = 0$,

(ii) there exists a 1-form λ such that $\mathrm{d}\Sigma = \lambda \wedge \Sigma$.

In the case of a single 1-form σ we have the result

$$\sigma \wedge \mathrm{d}\sigma = 0 \Leftrightarrow \sigma = f\,\mathrm{d}h, \tag{2.43}$$

or in components,

$$\sigma_{[a,b}\sigma_{c]} = 0 \Leftrightarrow \sigma_a = fh_{,a}. \tag{2.44}$$

The surfaces $h =$ constant are called the integral surfaces of the equation $\sigma = 0$, f^{-1} being the integrating factor.

The Frobenius Theorem is important in the construction of exact solutions because it allows us to introduce local coordinates f, h adapted to given normal 1-forms (see e.g. § 23.1.1.).

The *rank* q of a 2-form α is defined by

$$\underbrace{\alpha \wedge \cdots \wedge \alpha}_{q \text{ factors}} \neq 0, \qquad \underbrace{\alpha \wedge \cdots \wedge \alpha}_{(q+1) \text{ factors}} = 0, \qquad 2q \leqq n. \tag{2.45}$$

Using this definition we can generalize the statement (2.43) to

Theorem 2.3 (Darboux's Theorem). *Let σ be a 1-form and let the 2-form $\mathrm{d}\sigma$ have rank q. Then we can find local coordinates $x^1, \ldots, x^q, \xi^1, \ldots, \xi^{n-q}$ such that*

$$\text{if} \quad \sigma \wedge \underbrace{\mathrm{d}\sigma \wedge \cdots \wedge \mathrm{d}\sigma}_{q \text{ factors}} \begin{cases} = 0: & \sigma = x^1\, \mathrm{d}\xi^1 + \cdots + x^q\, \mathrm{d}\xi^q, \\ \neq 0: & \sigma = x^1\, \mathrm{d}\xi^1 + \cdots + x^q\, \mathrm{d}\xi^q + \mathrm{d}\xi^{q+1}. \end{cases} \tag{2.46}$$

(For a proof, see Sternberg (1964).)

This theorem gives the possible normal forms of a 1-form σ. Specializing Darboux's Theorem to a 4-dimensional manifold one obtains the following classification of a 1-form σ in terms of its components:

$$\begin{array}{llllll} q = 0: & \sigma_{[a,b]} = 0: & & & \sigma_a = \xi_{,a} \\ q = 1: & \sigma_{[a,b]} \neq 0, & \sigma_{[a,b}\sigma_{c,d]} = 0, & \sigma_{[a,b}\sigma_{c]} = 0: & \sigma_a = x\xi_{,a} \\ & \sigma_{[a,b]} \neq 0, & \sigma_{[a,b}\sigma_{c,d]} = 0, & \sigma_{[a,b}\sigma_{c]} \neq 0: & \sigma_a = x\xi_{,a} + \eta_{,a} \\ q = 2: & \sigma_{[a,b}\sigma_{c,d]} \neq 0: & & & \sigma_a = x\xi_{,a} + y\eta_{,a}. \end{array} \tag{2.47}$$

The real functions x, y, ξ, η denote independent functions. The second subcase is just the Frobenius Theorem applied to a single 1-form σ.

Now we give a theorem concerning 2-forms.

Theorem 2.4. *For any 2-form α of rank q there exists a basis $\{\omega^a\}$ such that*

$$\alpha = (\omega^1 \wedge \omega^2) + (\omega^3 \wedge \omega^4) + \cdots + (\omega^{2q-1} \wedge \omega^{2q}). \tag{2.48}$$

If $\mathrm{d}\alpha = 0$, then we can introduce local coordinates $x^1, \ldots, x^q, \xi^1, \ldots, \xi^{n-q}$ such that

$$\alpha = \mathrm{d}x^1 \wedge \mathrm{d}\xi^1 + \cdots + \mathrm{d}x^q \wedge \mathrm{d}\xi^q. \tag{2.49}$$

(For a proof see Sternberg (1964).)

To conclude this series of theorems, we consider a map $\Phi : \mathscr{M} \to \mathscr{N}$ between two manifolds, as in (2.30), and show by induction the

Theorem 2.5. *For the exterior derivative $\mathrm{d}\alpha$ of a p-form α we have*

$$\mathrm{d}(\Phi^*\alpha) = \Phi^*(\mathrm{d}\alpha). \tag{2.50}$$

Proof: Let us denote local coordinates in corresponding neighbourhoods of $p \in \mathcal{M}$ and $\Phi(p) \in \mathcal{N}$ by $(x^1, \ldots, x^m)$ and $(y^1, \ldots, y^n)$ respectively. Obviously, the relation (2.50) is true for a 0-form f:

$$\mathrm{d}(\Phi^* f) = \frac{\partial(\Phi^* f)}{\partial x^k}\, \mathrm{d}x^k = \frac{\partial f(y(x))}{\partial y^i} \frac{\partial y^i}{\partial x^k}\, \mathrm{d}x^k = \Phi^* (\mathrm{d}f). \tag{2.51}$$

Suppose the relation is valid for the $(p-1)$-form β and let $\alpha = f\, \mathrm{d}\beta$. (This is sufficiently general.) Then,

$$\mathrm{d}(\Phi^* \alpha) = \mathrm{d}[(\Phi^* f)\, \mathrm{d}(\Phi^* \beta)] = \mathrm{d}(\Phi^* f) \wedge \mathrm{d}(\Phi^* \beta) = \Phi^* (\mathrm{d}\alpha). \tag{2.52}$$

We do not consider integration on manifolds, except to note that the operator d of exterior derivation has been defined so that *Stokes' Theorem* can be written in the simple form

$$\int_{\partial \mathcal{M}} \alpha = \int_{\mathcal{M}} \mathrm{d}\alpha, \tag{2.53}$$

where α is any $(n-1)$-form and $\partial \mathcal{M}$ denotes the oriented boundary of $\mathcal{M}$. (A manifold with boundary is defined by charts which map their neighbourhoods $\mathcal{U}$ into the half space H^n defined by $x^n \geqq 0$ rather than into E^n, the *boundary* then being the set of points mapped to $x^n = 0$.)

2.8. The Lie derivative

For each point $p \in \mathcal{M}$, a vector field $\boldsymbol{v}$ on $\mathcal{M}$ determines a unique curve $\gamma_p(t)$ such that $\gamma_p(0) = p$ and $\boldsymbol{v}$ is the tangent vector to the curve. The family of these curves is called the congruence associated with the vector field. Along a curve $\gamma_p(t)$ the local coordinates $(y^1, \ldots, y^n)$ are solutions of the system of ordinary differential equations

$$\frac{\mathrm{d}y^i}{\mathrm{d}t} = v^i\big(y^1(t) \ldots y^n(t)\big) \tag{2.54}$$

with the initial values $y^i(0) = x^i(p)$.

To introduce a new type of differentiation we consider the map Φ_t dragging each point p, with coordinates x^i, along the curve $\gamma_p(t)$ through p into the image point $q = \Phi_t(p)$, with coordinates $y^i(t)$. For sufficiently small values of the parameter t the map Φ_t is a one-to-one map which induces a map $\Phi_t^* \boldsymbol{T}$ of any tensor $\boldsymbol{T}$. The *Lie derivative* of $\boldsymbol{T}$ with respect to $\boldsymbol{v}$ is defined by

$$\pounds_v \boldsymbol{T} \equiv \lim_{t \to 0} \frac{1}{t} (\Phi_t^* \boldsymbol{T} - \boldsymbol{T}). \tag{2.55}$$

The tensors $\boldsymbol{T}$ and $\Phi_t^* \boldsymbol{T}$ have the same type (r, s) and are both evaluated at the same point p. Therefore the Lie derivative (2.55) is also a tensor of type (r, s) at p. The Lie derivative vanishes if the tensors $\boldsymbol{T}$ and $\Phi_t^* \boldsymbol{T}$ coincide. In this case the tensor field $\boldsymbol{T}$ remains in a sense the "same" under Lie transport along the integral curves of the vector field $\boldsymbol{v}$. However, the components of $\boldsymbol{T}$ with respect to the coordinate basis $\{\partial/\partial x^i\}$ may vary along the curves. Using coordinate bases $\{\partial/\partial x^i\}$ and $\{\partial/\partial y^i\}$, we compute the *components* of the Lie derivative. The relations

$$\left.\frac{\partial y^i}{\partial x^k}\right|_{t=0} = \delta^i_k, \qquad \left.\frac{\mathrm{d}y^i}{\mathrm{d}t}\right|_{t=0} = v^i, \qquad \left.\frac{\mathrm{d}x^i}{\mathrm{d}t}\right|_{t=0} = -v^i \tag{2.56}$$

will be used. We start with the Lie derivatives of functions, 1-forms, and vectors:

function f: $$\pounds_v f = v^i f_{,i}. \tag{2.57}$$

Proof: $\Phi^*_t f|_p = f(y(x,t)), \qquad \pounds_v f|_p = \left.\dfrac{\partial f}{\partial y^i}\dfrac{\mathrm{d}y^i}{\mathrm{d}t}\right|_p.$

1-form σ: $$\pounds_v \sigma = (v^m \sigma_{i,m} + \sigma_m v^m{}_{,i})\,\mathrm{d}x^i. \tag{2.58}$$

Proof: $$\Phi^*_t \sigma|_p = \sigma_j(y(x,t))\,\frac{\partial y^j}{\partial x^i}\,\mathrm{d}x^i,$$

$$\pounds_v \sigma|_p = \left[\frac{\partial \sigma_j}{\partial y^m}\frac{\mathrm{d}y^m}{\mathrm{d}t}\frac{\partial y^j}{\partial x^i} + \sigma_j \frac{\partial}{\partial x^i}\left(\frac{\mathrm{d}y^j}{\mathrm{d}t}\right)\right]_{t=0}\mathrm{d}x^i.$$

vector u: $$\pounds_v \boldsymbol{u} = (v^m u^i{}_{,m} - u^m v^i{}_{,m})\,\frac{\partial}{\partial x^i}. \tag{2.59}$$

Proof: $$\Phi^*_t \boldsymbol{u}|_p = u^j((y(x,t))\,\frac{\partial x^i}{\partial y^j}\frac{\partial}{\partial x^i},$$

$$\pounds_v \boldsymbol{u}|_p = \left[\frac{\partial u^j}{\partial y^m}\frac{\mathrm{d}y^m}{\mathrm{d}t}\frac{\partial x^i}{\partial y^j} + u^j \frac{\partial}{\partial y^j}\left(\frac{\mathrm{d}x^i}{\mathrm{d}t}\right)\right]_{t=0}\frac{\partial}{\partial x^i}.$$

The Lie derivative of $\boldsymbol{u}$ with respect to $\boldsymbol{v}$ is equal to the commutator $[\boldsymbol{v}, \boldsymbol{u}]$,

$$\pounds_v \boldsymbol{u} = [\boldsymbol{v}, \boldsymbol{u}] = v^m \frac{\partial}{\partial x^m}\left(u^i \frac{\partial}{\partial x^i}\right) - u^m \frac{\partial}{\partial x^m}\left(v^i \frac{\partial}{\partial x^i}\right). \tag{2.60}$$

Two *commuting* vector fields generate a family of 2-dimensional submanifolds of $\mathscr{M}$ on which the parameters of the integral curves of both vector fields can be taken as coordinates.

From the Leibniz product rule and the formulae (2.58), (2.59) one obtains the components of the Lie derivative of an arbitrary tensor,

$$(\pounds_v \boldsymbol{T})^{ij\ldots}{}_{kl\ldots} = v^m T^{ij\ldots}{}_{kl\ldots,m} - T^{mj\ldots}{}_{kl\ldots} v^i{}_{,m} - T^{im\ldots}{}_{kl\ldots} v^j{}_{,m} - \cdots + T^{ij\ldots}{}_{ml\ldots} v^m{}_{,k} + T^{ij\ldots}{}_{km\ldots} v^m{}_{,l} + \cdots \tag{2.61}$$

From the equations (2.50), (2.55), it follows that the Lie derivative applied to forms commutes with the exterior derivative;

$$\mathrm{d}(\pounds_v \alpha) = \pounds_v(\mathrm{d}\alpha) \tag{2.62}$$

for any p-form α. Of course, this rule can also be verified by using the expressions (2.39), (2.61) in terms of components.

As will be seen later, the Lie derivative plays an important role in describing symmetries of gravitational fields and other physical fields.

The exterior derivative and the Lie derivative are operations defined on a differentiable manifold without imposing additional structures. Both operations are generalizations of the partial derivative. The exterior derivative is a limited generalization acting only on forms. The Lie derivative depends on the vector $\boldsymbol{v}$ not only at p, but also at neighbouring points. To introduce covariant derivatives which have neither of these defects we have to impose a new structure on $\mathcal{M}$, and we proceed to do so in the following section.

Ref.: Schouten (1954), Yano (1955).

2.9. The covariant derivative

The covariant derivative $\nabla_{\boldsymbol{v}}$ in the direction of the vector $\boldsymbol{v}$ at p maps an arbitrary tensor into a tensor of the same type. If $\boldsymbol{v}$ is unspecified, the covariant derivative ∇ generates a tensor of type $(r, s+1)$ from a tensor of type (r, s). In particular, for a vector $\boldsymbol{u}$ we have the expansion

$$\nabla \boldsymbol{u} = u^a{}_{;b} \boldsymbol{e}_a \otimes \boldsymbol{\omega}^b \tag{2.63}$$

with components $u^a{}_{;b}$ as yet unspecified. The directional covariant derivative is given by the vector

$$\nabla_{\boldsymbol{v}} \boldsymbol{u} = (u^a{}_{;b}\, v^b)\, \boldsymbol{e}_a . \tag{2.64}$$

The covariant derivative of the basis vector $\boldsymbol{e}_a$ in the direction of the basis vector $\boldsymbol{e}_b$ can be expanded in terms of basis vectors:

$$\nabla_b \boldsymbol{e}_a = \Gamma^c{}_{ab} \boldsymbol{e}_c , \qquad \Gamma^c{}_{ab} = \langle \boldsymbol{\omega}^c, \nabla_b \boldsymbol{e}_a \rangle . \tag{2.65}$$

We assume for the covariant derivative of a dual basis $\{\boldsymbol{\omega}^a\}$:

$$\nabla_b \boldsymbol{\omega}^a = -\Gamma^a{}_{cb} \boldsymbol{\omega}^c , \tag{2.66}$$

which is consistent with (2.11) and (2.65). The coefficients $\Gamma^c{}_{ab}$, called the *connection coefficients*, relate the bases at different points of $\mathcal{M}$, and they have to be imposed as an extra structure on $\mathcal{M}$. We restrict ourselves to covariant derivatives satisfying

$$\nabla_{\boldsymbol{u}} \boldsymbol{v} - \nabla_{\boldsymbol{v}} \boldsymbol{u} = [\boldsymbol{u}, \boldsymbol{v}] \tag{2.67}$$

for two arbitrary vectors $\boldsymbol{u}$ and $\boldsymbol{v}$. This relation is equivalent to the equation

$$2\Gamma^c{}_{[ab]} = -D^c{}_{ab} , \tag{2.68}$$

where the commutation coefficients are defined by (2.6). In a coordinate basis, the connection coefficients $\Gamma^c{}_{ab}$ have a symmetric index pair (ab). Therefore a covariant derivative satisfying (2.67) is called *symmetric* (or *torsion-free*).

Once the connection coefficients are prescribed, the components $u^a{}_{;c}$ of the covariant derivative of $\boldsymbol{u}$ in the direction of the basis vector $\boldsymbol{e}_c$ are completely determined,

$$\nabla_c \boldsymbol{u} = \nabla_c(u^a \boldsymbol{e}_a) = (u^a{}_{|c} + \Gamma^a{}_{dc} u^d)\, \boldsymbol{e}_a = u^a{}_{;c} \boldsymbol{e}_a, \tag{2.69a}$$

and the components of the covariant derivative $\nabla \boldsymbol{T}$ of a tensor (2.26) are

$$\begin{aligned} T^{a_1 \ldots a_r}{}_{b_1 .. b_s;c} = (T^{a_1 .. a_r}{}_{b_1 .. b_s})_{|c} + \Gamma^{a_1}{}_{dc} T^{d .. a_r}{}_{b_1 .. b_s} + \cdots + \Gamma^{a_r}{}_{dc} T^{a_1 .. d}{}_{b_1 .. b_s} \\ - \Gamma^d{}_{b_1 c} T^{a_1 .. a_r}{}_{d .. b_s} - \cdots - \Gamma^d{}_{b_s c} T^{a_1 .. a_r}{}_{b_1 .. d}, \end{aligned} \tag{2.69b}$$

where the symbol $f_{|a} \equiv \boldsymbol{e}_a(f) = f_{,i} e_a{}^i$ has been used. Note that the formula (2.69b) is valid for a general basis $\{\boldsymbol{e}_a\}$.

Using the symmetry axiom (2.68), we may replace the partial derivatives in the expressions (2.39) and (2.61) for the components of, respectively, the exterior derivative and the Lie derivative by covariant derivatives, so that the commas can be replaced by semicolons.

The bases $\{\boldsymbol{e}_a\}$, $\{\boldsymbol{\omega}^a\}$ are linear combinations of coordinate bases,

$$\boldsymbol{e}_a = e_a{}^i\, \partial/\partial x^i, \qquad \boldsymbol{\omega}^a = \omega^a{}_i\, \mathrm{d}x^i. \tag{2.70}$$

The connection coefficients (2.65) can be written as

$$\Gamma^c{}_{ab} = \omega^c{}_k e_b{}^i e_a{}^k{}_{;i} = -\omega^c{}_{k;i} e_b{}^i e_a{}^k. \tag{2.71}$$

For the exterior derivative of the basis 1-forms we get

$$\mathrm{d}\boldsymbol{\omega}^a = \omega^a{}_{i,j}\, \mathrm{d}x^j \wedge \mathrm{d}x^i = \omega^a{}_{i;j}\, \mathrm{d}x^j \wedge \mathrm{d}x^i = \Gamma^a{}_{bc} \boldsymbol{\omega}^b \wedge \boldsymbol{\omega}^c. \tag{2.72}$$

Introducing the *connection 1-forms*

$$\boldsymbol{\Gamma}^a{}_b \equiv \Gamma^a{}_{bc} \boldsymbol{\omega}^c \tag{2.73}$$

we can write equation (2.72) in the form

$$\mathrm{d}\boldsymbol{\omega}^a = -\boldsymbol{\Gamma}^a{}_b \wedge \boldsymbol{\omega}^b \tag{2.74}$$

due to Cartan (the first Cartan equation). For a given basis, the *antisymmetric part* $\Gamma^c{}_{[ab]}$ of the connection coefficients can be computed from this.

The tangent vector $\boldsymbol{v}$ of a *geodesic* curve satisfies

$$\nabla_{\boldsymbol{v}} \boldsymbol{v} = f\boldsymbol{v}; \qquad \text{in components, } v^b v^a{}_{;b} = f v^a. \tag{2.75}$$

By a suitable scaling of the parameter of the geodesic curve we can make the function f in (2.75) vanish. Such a parameter is called an *affine parameter* μ and the geodesic equation reads

$$\frac{\mathrm{D}v^i}{\mathrm{d}\mu} \equiv v^k v^i{}_{;k} = 0. \tag{2.76}$$

2.10. The curvature tensor

The *curvature tensor* (*Riemann tensor*), $\boldsymbol{R} = R^a{}_{bcd}\boldsymbol{e}_a \otimes \boldsymbol{\omega}^b \otimes \boldsymbol{\omega}^c \otimes \boldsymbol{\omega}^d$, is a tensor of type (1.3) mapping the ordered set $(\sigma; \boldsymbol{w}, \boldsymbol{u}, \boldsymbol{v})$ of a 1-form σ and three vectors $\boldsymbol{w}, \boldsymbol{u}, \boldsymbol{v}$ into the real number

$$\begin{aligned} \sigma_a w^b u^c v^d R^a{}_{bcd} &= \langle \sigma, (\nabla_{\boldsymbol{u}}\nabla_{\boldsymbol{v}} - \nabla_{\boldsymbol{v}}\nabla_{\boldsymbol{u}} - \nabla_{[\boldsymbol{u},\boldsymbol{v}]})\,\boldsymbol{w}\rangle \\ &= \sigma_a[(w^a{}_{;c}v^c)_{;d}\,u^d - (w^a{}_{;c}\,u^c)_{;d}\,v^d - w^a{}_{;c}(u^d v^c{}_{;d} - v^d u^c{}_{;d})] \\ &= \sigma_a(w^a{}_{;cd} - w^a{}_{;dc})\,v^c u^d. \end{aligned} \tag{2.77}$$

As the components σ_a, v^c, u^d can be chosen arbitrarily we arrive at the *Ricci identity*

$$w^a{}_{;cd} - w^a{}_{;dc} = w^b R^a{}_{bdc}. \tag{2.78}$$

The general rule (2.69) for the components of the covariant derivative of a tensor implies the formula

$$R^a{}_{bcd} = \Gamma^a{}_{bdc} - \Gamma^a{}_{bcd} + \Gamma^e{}_{bd}\Gamma^a{}_{ec} - \Gamma^e{}_{bc}\Gamma^a{}_{ed} - D^e{}_{cd}\Gamma^a{}_{be}. \tag{2.79}$$

In a coordinate basis, the last term vanishes. The components (2.79) of the curvature tensor satisfy the symmetry relations

$$R^a{}_{bcd} = -R^a{}_{bdc}, \qquad R^a{}_{[bcd]} = 0. \tag{2.80}$$

The covariant derivatives of the curvature tensor obey the *Bianchi identities*

$$R^a{}_{b[cd;e]} = 0. \tag{2.81}$$

By contraction we obtain the identities

$$R^a{}_{bcd;a} + 2R_{b[c;d]} = 0, \tag{2.82}$$

where the components R_{bd} of the *Ricci tensor* are defined by

$$R_{bd} \equiv R^a{}_{bad}. \tag{2.83}$$

A compact and efficient method for calculating the components (2.79) with respect to a general basis is provided by Cartan's procedure. Defining the *curvature 2-forms* $\Theta^a{}_b$ by

$$\Theta^a{}_b \equiv \frac{1}{2} R^a{}_{bcd}\,\omega^c \wedge \omega^d, \tag{2.84}$$

the equation (2.79) is completely equivalent to the second Cartan equation

$$\mathrm{d}\boldsymbol{\Gamma}^a{}_b + \boldsymbol{\Gamma}^a{}_c \wedge \boldsymbol{\Gamma}^c{}_b = \boldsymbol{\Theta}^a{}_b, \tag{2.85}$$

which gives an algorithm for the calculation of the curvature from the connection. We collect the relations between the various quantities in Fig. 2.3.

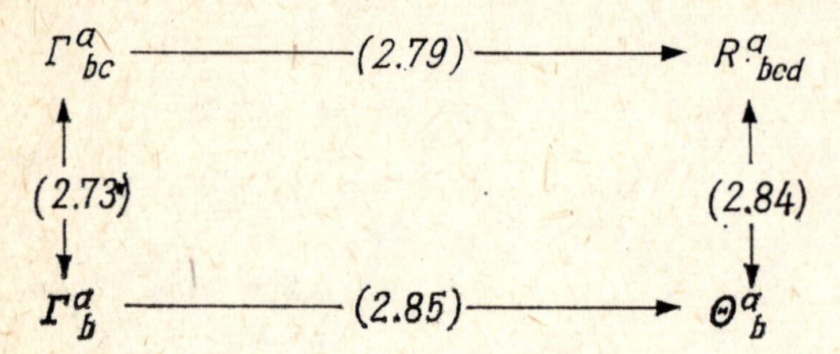

Fig. 2.3 How to get the curvature from the connection

Chapter 3. Some topics in Riemannian geometry

3.1. Introduction

In Chapter 2. we have treated differential geometry without a metric. In order to define covariant derivatives we imposed an extra structure on the differentiable manifold $\mathscr{M}$, the connection. Adding a further structure, the metric g_{ab}, and postulating $g_{ab;c} = 0$, we arrive at Riemannian geometry. The metric will be introduced in the next section.

General Relativity is based on the concept of *space-time*, which is a four-dimensional differentiable (C^∞, Hausdorff) manifold $\mathscr{M}$ endowed with a *Lorentzian metric* g_{ab} which can be transformed to

$$g_{ab} = \eta_{ab} \equiv \mathrm{diag}\,(1, 1, 1, -1) \tag{3.1}$$

at any point of $\mathscr{M}$, i.e. space-time is a normal-hyperbolic Riemannian space V_4. In what follows, a knowledge of fundamental facts about Riemannian geometry as given in most textbooks on General Relativity is presumed; we give here only some notation and results used in the remainder of this book. For further details the reader is referred to standard texts on Riemannian geometry, e.g. Eisenhart (1949), Schouten (1954).

3.2. The metric tensor and null tetrads

We introduce as a new structure a symmetric tensor of type (0, 2), called the *metric tensor* $\boldsymbol{g}$, which endows each vector space T_p with a scalar product (inner product)

$$\boldsymbol{e}_a \cdot \boldsymbol{e}_b = g_{ab}. \tag{3.2}$$

The tensor $\boldsymbol{g}$, sometimes called the line element $\mathrm{d}s^2$, is written

$$\boldsymbol{g} = \mathrm{d}s^2 = g_{ab}\omega^a\omega^b. \tag{3.3}$$

The scalar product of two vectors $\boldsymbol{v}$, $\boldsymbol{w}$ is given by

$$\boldsymbol{v} \cdot \boldsymbol{w} = g_{ab}v^a w^b. \tag{3.4}$$

Two vectors $\boldsymbol{v}$, $\boldsymbol{w}$, are *orthogonal* if their scalar product vanishes. A non-zero vector $\boldsymbol{v}$ is said to be *spacelike, timelike*, or *null*, respectively, when the product $\boldsymbol{v} \cdot \boldsymbol{v} = g_{ab}v^a v^b$ is positive, negative, or zero. In a coordinate basis, we write the line element $\mathrm{d}s^2$ as

$$\mathrm{d}s^2 = g_{ij}\,\mathrm{d}x^i\,\mathrm{d}x^j. \tag{3.5}$$

The contravariant components, g^{ab}, form the matrix inverse to g_{ab}. Raising and lowering the indices of tensor components has to be performed in the usual manner,

$$v_a = g_{ab}v^b, \qquad v^a = g^{ab}v_b. \tag{3.6}$$

In this sense, the vector $v^a\boldsymbol{e}_a$ and the 1-form $v_a\omega^a$ represent the same geometric object.

In space-time, an *orthonormal basis* $\{\boldsymbol{E}_a\}$ consists of three spacelike vectors $\boldsymbol{E}_\alpha$ and one timelike vector $\boldsymbol{E}_4 \equiv \boldsymbol{t}$, such that

$$\begin{aligned} &\{\boldsymbol{E}_a\} = \{\boldsymbol{E}_\alpha, \boldsymbol{t}\} = \{\boldsymbol{x}, \boldsymbol{y}, \boldsymbol{z}, \boldsymbol{t}\}, \quad g_{ab} = x_a x_b + y_a y_b + z_a z_b - t_a t_b \\ &\Leftrightarrow \boldsymbol{E}_\alpha \cdot \boldsymbol{E}_\beta = \delta_{\alpha\beta}, \quad \boldsymbol{t}\cdot\boldsymbol{t} = -1, \quad \boldsymbol{E}_\alpha \cdot \boldsymbol{t} = 0. \end{aligned} \tag{3.7}$$

In the literature, and in many places in this book, the symbol $\boldsymbol{t}$ is replaced by $\boldsymbol{u}$. However there are times when one is treating a fluid with four-velocity $\boldsymbol{u}$ in terms of a tetrad with $\boldsymbol{t} \neq \boldsymbol{u}$, and it is for this reason that we have introduced $\boldsymbol{t}$ as the general symbol for $\boldsymbol{E}_4$.

Complex null tetrads play an important role. A complex null tetrad consists of two real null vectors $\boldsymbol{k}$, $\boldsymbol{l}$ and two complex conjugate null vectors $\boldsymbol{m}$, $\overline{\boldsymbol{m}}$:

$$\begin{aligned} &\{\boldsymbol{e}_a\} = (\boldsymbol{m}, \overline{\boldsymbol{m}}, \boldsymbol{l}, \boldsymbol{k}), \\ &g_{ab} = 2m_{(a}\overline{m}_{b)} - 2k_{(a}l_{b)} = \begin{pmatrix} 0 & 1 & 0 & 0 \\ 1 & 0 & 0 & 0 \\ 0 & 0 & 0 & -1 \\ 0 & 0 & -1 & 0 \end{pmatrix}, \end{aligned} \tag{3.8}$$

(g_{ab} are the components of $\boldsymbol{g}$ with respect to the complex null tetrad), i.e. the scalar products of the tetrad vectors vanish apart from

$$k^a l_a = -1, \qquad m^a \overline{m}_a = 1. \tag{3.9}$$

In terms of a coordinate basis, a complex null tetrad $\{\boldsymbol{e}_a\}$ and its dual basis $\{\omega^a\}$ will take the form

$$\begin{aligned} &\boldsymbol{e}_1 = m^i \frac{\partial}{\partial x^i}, \quad \boldsymbol{e}_2 = \overline{m}^i \frac{\partial}{\partial x^i}, \quad \boldsymbol{e}_3 = l^i \frac{\partial}{\partial x^i}, \quad \boldsymbol{e}_4 = k^i \frac{\partial}{\partial x^i}; \\ &\omega^1 = \overline{m}_i\,\mathrm{d}x^i, \quad \omega^2 = m_i\,\mathrm{d}x^i, \quad \omega^3 = -k_i\,\mathrm{d}x^i, \quad \omega^4 = -l_i\,\mathrm{d}x^i. \end{aligned} \tag{3.10}$$

The explicit expressions for the directional derivatives $f_{|a}$ of a function f with respect to the complex null tetrad (3.8) are

$$f_{|1} = f_{,i}m^i, \quad f_{|2} = f_{,i}\overline{m}^i, \quad f_{|3} = f_{,i}l^i, \quad f_{|4} = f_{,i}k^i. \tag{3.11}$$

An orthonormal basis (3.7) and a complex null tetrad (3.8) may be related by

$$\sqrt{2}\,\boldsymbol{m} = \boldsymbol{E}_1 - \mathrm{i}\boldsymbol{E}_2, \quad \sqrt{2}\,\overline{\boldsymbol{m}} = \boldsymbol{E}_1 + \mathrm{i}\boldsymbol{E}_2, \quad \sqrt{2}\,\boldsymbol{l} = \boldsymbol{E}_4 - \boldsymbol{E}_3, \quad \sqrt{2}\,\boldsymbol{k} = \boldsymbol{E}_4 + \boldsymbol{E}_3. \tag{3.12}$$

In flat space-time, (3.12) implies the relation

$$\zeta = \frac{1}{\sqrt{2}}(x + \mathrm{i}y), \quad \overline{\zeta} = \frac{1}{\sqrt{2}}(x - \mathrm{i}y), \quad u = \frac{1}{\sqrt{2}}(t - z), \quad v = \frac{1}{\sqrt{2}}(t + z) \tag{3.13}$$

between the null coordinates $\zeta, \bar{\zeta}, u, v$ (adapted to the basis vectors $\boldsymbol{m} = \partial_\zeta$, $\overline{\boldsymbol{m}} = \partial_{\bar{\zeta}}$, $\boldsymbol{l} = \partial_u$, $\boldsymbol{k} = \partial_v$) and the Minkowski coordinates x, y, z, t (adapted to the basis vectors $\boldsymbol{E}_1 = \partial_x$, $\boldsymbol{E}_2 = \partial_y$, $\boldsymbol{E}_3 = \partial_z$, $\boldsymbol{E}_4 = \partial_t$). Here we have adopted the convention $\partial_\zeta \equiv \partial/\partial_\zeta$, etc.

Lorentz transformations give rise to the following changes of the basis (3.8):

null rotations (*$\boldsymbol{l}$ fixed*),

$$\boldsymbol{l}' = \boldsymbol{l}, \boldsymbol{m}' = \boldsymbol{m} + E\boldsymbol{l}, \qquad \boldsymbol{k}' = \boldsymbol{k} + E\overline{\boldsymbol{m}} + \bar{E}\boldsymbol{m} + E\bar{E}\boldsymbol{l}, \quad E \text{ complex}; \tag{3.14}$$

null rotations (*$\boldsymbol{k}$ fixed*),

$$\boldsymbol{k}' = \boldsymbol{k}, \ \boldsymbol{m}' = \boldsymbol{m} + B\boldsymbol{k}, \quad \boldsymbol{l}' = \boldsymbol{l} + B\overline{\boldsymbol{m}} + \bar{B}\boldsymbol{m} + B\bar{B}\boldsymbol{k}, \quad B \text{ complex}; \tag{3.15}$$

spatial rotations in the $\boldsymbol{m} - \overline{\boldsymbol{m}}$-plane,

$$\boldsymbol{m}' = \mathrm{e}^{i\Theta}\boldsymbol{m}, \quad \Theta \text{ real}; \tag{3.16}$$

special Lorentz transformations (*boosts*) *in the $\boldsymbol{k} - \boldsymbol{l}$-plane,*

$$\boldsymbol{k}' = A\boldsymbol{k}, \qquad \boldsymbol{l}' = A^{-1}\boldsymbol{l}, \quad A > 0. \tag{3.17}$$

The transformations (3.14)—(3.17) contain six real parameters. The transformations preserving the $\boldsymbol{k}$-direction are

$$\boldsymbol{k}' = A\boldsymbol{k}, \quad \boldsymbol{m}' = \mathrm{e}^{i\Theta}(\boldsymbol{m} + B\boldsymbol{k}), \quad \boldsymbol{l}' = A^{-1}(\boldsymbol{l} + B\overline{\boldsymbol{m}} + \bar{B}\boldsymbol{m} + B\bar{B}\boldsymbol{k}). \tag{3.18}$$

Symmetric connection coefficients (2.65) are uniquely determined by adding the *metric condition*

$$\begin{aligned} &\nabla \boldsymbol{g} = 0 \Leftrightarrow g_{ab;c} = 0 = g_{ab|c} - 2\Gamma_{(ab)c}, \\ &\Gamma_{abc} \equiv g_{ad}\Gamma^d{}_{bc}. \end{aligned} \tag{3.19}$$

Combining the metric condition (3.19) and the symmetry condition (2.68) one obtains the general formula

$$\Gamma_{cab} = \frac{1}{2}(g_{ca|b} + g_{cb|a} - g_{ab|c} + D_{bca} + D_{acb} - D_{cab}), \ D_{abc} \equiv g_{ad}D^d{}_{bc}, \tag{3.20}$$

expressing the connection coefficients in terms of the metric tensor and the commutation coefficients. We mention two cases of special interest:

coordinate basis (*holonomic frame*), $D_{ijk} = 0$: $\Gamma_{i[jk]} = 0$,
$\Gamma^i_{jk} \equiv \left\{ {i \atop jk} \right\}$ (= Christoffel symbols);

constant metric (*rigid frame*), $g_{ab|c} = 0$: $\Gamma_{(ab)c} = 0$.

In a holonomic frame, the connection coefficients Γ_{abc} are symmetric in the index pair (bc), while in a rigid frame they are antisymmetric in the index pair (ab).

3.3. Calculation of curvature from the metric

The components of the curvature tensor are given by equation (2.79). In a coordinate basis, one may simply substitute the Christoffel symbols into this expression. However, Cartan's method for the calculation of curvature is more compact and efficient in many applications. It immediately yields the tetrad components. The algorithm is divided into two steps.

(i) *Calculation of the connection* 1-*forms* (2.73), $\Gamma^a_b = \Gamma^a{}_{bc}\omega^c$, from the first Cartan equation (2.74) and the metric condition (3.19),

$$\mathrm{d}\omega^a = -\Gamma^a{}_b \wedge \omega^b, \quad \mathrm{d}g_{ab} = \Gamma_{ab} + \Gamma_{ba}, \tag{3.21}$$

which determine Γ_{ab} uniquely. In a rigid frame ($\mathrm{d}g_{ab} = 0$), at most six independent connection 1-forms survive.

(ii) *Calculation of the curvature* 2-*forms* (2.84) from the second Cartan equation (2.85),

$$\Theta^a{}_b = \mathrm{d}\Gamma^a{}_b + \Gamma^a{}_c \wedge \Gamma^c{}_b = \frac{1}{2} R^a{}_{bcd}\omega^c \wedge \omega^d. \tag{3.22}$$

This calculus gives the *components* $R^a{}_{bcd}$ *with respect to a general basis* $\{\boldsymbol{e}_a\}$.

For the complex null tetrad $\{\boldsymbol{e}_a\} = (\boldsymbol{m}, \overline{\boldsymbol{m}}, \boldsymbol{l}, \boldsymbol{k})$, the second Cartan equation (3.22) takes the form of three complex equations,

$$\mathrm{d}\Gamma_{41} + \Gamma_{41} \wedge (\Gamma_{21} + \Gamma_{43}) = \frac{1}{2} R_{41cd}\, \omega^c \wedge \omega^d, \tag{3.23a}$$

$$\mathrm{d}\Gamma_{32} - \Gamma_{32} \wedge (\Gamma_{21} + \Gamma_{43}) = \frac{1}{2} R_{32cd}\, \omega^c \wedge \omega^d, \tag{3.23b}$$

$$\mathrm{d}(\Gamma_{21} + \Gamma_{43}) + 2\Gamma_{32} \wedge \Gamma_{41} = \frac{1}{2} (R_{21cd} + R_{43cd})\, \omega^c \wedge \omega^d, \tag{3.23c}$$

where the indices refer to the basis vectors $\boldsymbol{e}_1 = \boldsymbol{m}$, $\boldsymbol{e}_2 = \overline{\boldsymbol{m}}$, $\boldsymbol{e}_3 = \boldsymbol{l}$, and $\boldsymbol{e}_4 = \boldsymbol{k}$; $\Gamma_{41} = \overline{\Gamma}_{42} = \Gamma_{ab}k^a m^b$, $\Gamma_{21} = \overline{\Gamma}_{12} = \Gamma_{ab}\overline{m}^a m^b$, etc. (Exchanging the indices 1 and 2 implies complex conjugation.)

As an *example* of the method based on equations (3.21), (3.22) let us take the metric $g_{ab} = \eta_{ab}$ and the dual basis

$$\omega^1 = \frac{1}{ax^1}\,\mathrm{d}x^1, \quad \omega^2 = \frac{1}{ax^1}\,\mathrm{d}x^2, \quad \omega^3 = \left(\frac{3x^1}{2}\right)^{1/2} \mathrm{d}x^3 + \left(\frac{2}{3x^1}\right)^{1/2} \mathrm{d}x^4,$$
$$\omega^4 = \left(\frac{2}{3x^1}\right)^{1/2} \mathrm{d}x^4, \quad a \equiv (x^1)^{1/2}/x^4, \quad x^1 > 0. \tag{3.24}$$

The corresponding line element reads (cf. (33.1))

$$\mathrm{d}s^2 = \frac{(x^4)^2}{(x^1)^3}\,[(\mathrm{d}x^1)^2 + (\mathrm{d}x^2)^2] + 2\mathrm{d}x^3\,\mathrm{d}x^4 + \frac{3}{2}\,x^1(\mathrm{d}x^3)^2. \tag{3.25}$$

We obtain the following explicit expressions for the exterior derivatives of the basis 1-forms, $d\omega^a$, for the connection 1-forms $\Gamma_{ab} = -\Gamma_{ba}$, and for the curvature 2-forms $\Theta_{ab} = -\Theta_{ba}$:

$$d\omega^1 = \sqrt{\frac{3}{2}}\,a\omega^4 \wedge \omega^1, \quad d\omega^2 = \sqrt{\frac{3}{2}}\,a\omega^4 \wedge \omega^2 - \frac{3}{2}\,a\omega^1 \wedge \omega^2,$$
$$d\omega^3 = \frac{a}{2}\,\omega^1 \wedge \omega^3 - a\omega^1 \wedge \omega^4, \quad d\omega^4 = -\frac{a}{2}\,\omega^1 \wedge \omega^4. \tag{3.26a}$$

$$\Gamma_{14} = \sqrt{\frac{3}{2}}\,a\omega^1 + \frac{a}{2}\,\omega^3 - \frac{a}{2}\,\omega^4, \quad \Gamma_{31} = \frac{a}{2}\,(\omega^3 - \omega^4),$$
$$\Gamma_{24} = \sqrt{\frac{3}{2}}\,a\omega^2, \quad \Gamma_{23} = 0, \quad \Gamma_{12} = \frac{3}{2}\,a\omega^2, \quad \Gamma_{34} = \frac{a}{2}\,\omega^1 \tag{3.26b}$$

$$\Theta_{14} = \Theta_{31} = a^2\left(\frac{3}{4}\,\omega^1 \wedge \omega^3 - \frac{3}{4}\,\omega^1 \wedge \omega^4 + \frac{1}{2}\sqrt{\frac{3}{2}}\,\omega^3 \wedge \omega^4\right),$$
$$\Theta_{32} = \Theta_{24} = a^2\left(\frac{3}{4}\,\omega^2 \wedge \omega^4 - \frac{3}{4}\,\omega^2 \wedge \omega^3 + \frac{1}{2}\sqrt{\frac{3}{2}}\,\omega^1 \wedge \omega^2\right), \tag{3.26c}$$
$$\Theta_{21} = \frac{1}{2}\sqrt{\frac{3}{2}}\,a^2(\omega^2 \wedge \omega^3 - \omega^2 \wedge \omega^4), \quad \Theta_{34} = \frac{1}{2}\sqrt{\frac{3}{2}}\,a^2(\omega^1 \wedge \omega^4 - \omega^1 \wedge \omega^3).$$

These forms give immediately the components of the Riemann curvature tensor with respect to the basis (3.24). One can verify that the Ricci tensor (2.83) vanishes: $R_{ab} = 0$; i.e. the metric (3.25) describes a vacuum solution.

3.4. Bivectors

Bivectors are antisymmetric tensors of second order, or 2-forms,

$$\boldsymbol{X} = X_{ab}\,\omega^a \wedge \omega^b. \tag{3.27}$$

A *simple bivector*, $X_{ab} = u_{[a}v_{b]}$, represents a 2-surface element spanned by the two tangent vectors $\boldsymbol{u} = u^a\boldsymbol{e}_a$ and $\boldsymbol{v} = v^a\boldsymbol{e}_a$. This surface element is spacelike, timelike or null according as $X_{ab}X^{ab}$ is positive, negative or zero, respectively.

Taking a particular orientation of (a neighbourhood in) $\mathcal{M}$, we define the Levi-Civita 4-form $\boldsymbol{\varepsilon}$ to be $\sqrt{-g}\,\omega^1 \wedge \omega^2 \wedge \omega^3 \wedge \omega^4$, where g is the determinant of the matrix g_{ab} of metric tensor components with respect to a positively-oriented basis $\{\boldsymbol{e}_a\}$. Its components are written ε_{abcd}.

With the aid of the Levi-Civita tensor we define the dual bivector $\tilde{\boldsymbol{X}}$, in index notation, by

$$\tilde{X}_{ab} \equiv \frac{1}{2}\,\varepsilon_{abcd}X^{cd}. \tag{3.28}$$

To avoid confusion we emphasize that the concepts of dual basis and dual bivector have entirely distinct meanings. Repeated application of the duality operation (3.28) gives

$$(\tilde{X}_{ab})^{\sim} = -X_{ab}. \tag{3.29}$$

Two bivectors $\boldsymbol{X}$ and $\boldsymbol{Y}$ satisfy the identities

$$X_{ac}Y_b{}^c - \tilde{X}_{bc}\tilde{Y}_a{}^c = \frac{1}{2}g_{ab}X_{cd}Y^{cd}, \qquad \tilde{X}_{ab}Y^{ab} = X_{ab}\tilde{Y}^{ab}, \tag{3.30}$$

which can be verified from the well-known formula

$$\varepsilon_{abcd}\varepsilon^{fghd} = -6\delta^f_{[a}\delta^g_b\delta^h_{c]}. \tag{3.31}$$

The complex bivector defined by

$$X^*_{ab} \equiv X_{ab} + \mathrm{i}\tilde{X}_{ab} \tag{3.32}$$

is self-dual, i.e. it fulfils the condition

$$(X^*_{ab})^{\sim} = -\mathrm{i}X^*_{ab}. \tag{3.33}$$

In a positively-oriented basis (3.7) the Levi-Civita tensor would have components defined by

$$\varepsilon_{1234} = -1. \tag{3.34}$$

This amounts to a choice of orientation of the four-dimensional manifold in which the basis $\{\boldsymbol{E}_a\}$ represents a Lorentz frame with $\boldsymbol{E}_4$ pointing toward the future and with $\{\boldsymbol{E}_\alpha\}$ as a righthanded spatial triad. If that basis is related to the complex null tetrad (3.8) by the formula (3.12), then (3.34) can be written as

$$\varepsilon_{abcd}m^a\overline{m}^bl^ck^d = \mathrm{i}. \tag{3.35}$$

A self-dual bivector is completely determined by a timelike unit vector $\boldsymbol{u}$ and the projection

$$X_a \equiv X^*_{ab}u^b, \qquad u_cu^c = -1, \tag{3.36}$$

according to the equation

$$X^*_{ab} = 2u_{[a}X_{b]} + \mathrm{i}\varepsilon_{abcd}u^cX^d = 2(u_{[a}X_{b]})^*. \tag{3.37}$$

In virtue of the orthogonality $X_au^a = 0$, the vector X_a is in fact a complex 3-vector. As consequences of this important relation we get

$$X^*_{ab}X^{*ab} = -4X_aX^a, \qquad X^*_{\alpha\beta} = \mathrm{i}\varepsilon_{\alpha\beta\gamma}X^\gamma. \tag{3.38}$$

A general *self-dual* bivector can be expanded in terms of the basis $\boldsymbol{Z}^{\mu} = (\boldsymbol{U}, \boldsymbol{V}. \boldsymbol{W})$ constructed from the complex null tetrad (3.8) by

$$\begin{aligned} \boldsymbol{Z}^1 &\equiv \boldsymbol{U}\colon\ U_{ab} = -l_a\overline{m}_b + l_b\overline{m}_a, \\ \boldsymbol{Z}^2 &\equiv \boldsymbol{V}\colon\ V_{ab} = k_a m_b - k_b m_a, \\ \boldsymbol{Z}^3 &\equiv \boldsymbol{W}\colon\ W_{ab} = m_a\overline{m}_b - m_b\overline{m}_a - k_a l_b + k_b l_a. \end{aligned} \tag{3.39}$$

All contractions vanish except

$$U_{ab}V^{ab} = 2, \qquad W_{ab}W^{ab} = -4. \tag{3.40}$$

With the aid of equation (3.35) we can verify that the bivectors (3.39) are self-dual: $\tilde{Z}^{\alpha}_{ab} = -\mathrm{i}Z^{\alpha}_{ab}$. The complex conjugate bivectors $\overline{\boldsymbol{U}}, \overline{\boldsymbol{V}}, \overline{\boldsymbol{W}}$ form a basis $\{\overline{\boldsymbol{Z}}^{\alpha}\}$ of the space of antiself-dual bivectors, i.e. those obeying $\tilde{\overline{Z}}^{\alpha}_{ab} = \mathrm{i}\overline{Z}^{\alpha}_{ab}$.

The null rotations (3.14), (3.15) induce the following transformations of the bivectors (3.39).

$$\boldsymbol{l}\ \text{fixed:}\ U'_{ab} = U_{ab},\ V'_{ab} = V_{ab} - EW_{ab} + E^2U_{ab},\ W'_{ab} = W_{ab} - 2EU_{ab}, \tag{3.41 a}$$

$$\boldsymbol{k}\ \text{fixed:}\ V'_{ab} = V_{ab},\ U'_{ab} = U_{ab} - \overline{B}W_{ab} + \overline{B}^2V_{ab},\ W'_{ab} = W_{ab} - 2\overline{B}V_{ab}. \tag{3.41 b}$$

A general bivector can be expanded in terms of the basis $\{\boldsymbol{Z}^{\alpha}, \overline{\boldsymbol{Z}}^{\alpha}\}$;

$$X_{ab} = c_{\alpha}Z^{\alpha}_{ab} + d_{\alpha}\overline{Z}^{\alpha}_{ab}. \tag{3.42}$$

Finally, we mention the relation

$$\overline{Z}^{\alpha}_{a[b}Z^{\beta}_{c]d} = 0. \tag{3.43}$$

Ref.: Jordan et al. (1960), Debever (1966), Israel (1970), Zund and Brown (1971), Greenberg (1972a, b).

3.5. Decomposition of the curvature tensor

The curvature tensor, with components (2.79) with respect to a basis $\{\boldsymbol{e}_a\}$, can be uniquely decomposed into parts which are irreducible representations of the full Lorentz group,

$$R_{abcd} = C_{abcd} + E_{abcd} + G_{abcd}, \tag{3.44}$$

where the following abbreviations have been used:

$$E_{abcd} \equiv \frac{1}{2}\left(g_{ac}S_{bd} + g_{bd}S_{ac} - g_{ad}S_{bc} - g_{bc}S_{ad}\right), \tag{3.45}$$

$$G_{abcd} \equiv \frac{R}{12}\left(g_{ac}g_{bd} - g_{ad}g_{bc}\right) \equiv \frac{R}{12}\,g_{abcd}, \tag{3.46}$$

$$S_{ab} \equiv R_{ab} - \frac{1}{4}Rg_{ab}, \qquad R \equiv R^a_a. \tag{3.47}$$

R and S_{ab} respectively denote the trace and the traceless part of the Ricci tensor R_{ab} defined by (2.83).

The decomposition (3.44) defines *Weyl's conformal tensor* C_{abcd}, which has the symmetries

$$C_{abcd} = -C_{bacd} = -C_{abdc} = C_{cdab}, \qquad C_{a[bcd]} = 0. \tag{3.48}$$

The other parts in the decomposition (3.44) have the same symmetries. Moreover, we have the relations

$$C^a{}_{bad} = 0, \qquad E^a{}_{bad} = S_{bd}, \qquad G^a{}_{bad} = \frac{1}{4} g_{bd} R. \tag{3.49}$$

The Weyl tensor is completely traceless, i.e. the contraction with respect to each pair of indices vanishes, and it has 10 independent components. A space-time with zero Weyl tensor is said to be *conformally flat*.

We give an equivalent form of the decomposition (3.44)–(3.47):

$$R^{ab}{}_{cd} = C^{ab}{}_{cd} - \frac{R}{3} \delta^a_{[c}\delta^b_{d]} + 2\delta^{[a}_{[c} R^{b]}_{d]}. \tag{3.50}$$

Because the tensors C_{abcd}, E_{abcd}, G_{abcd} have two pairs of bivector indices we can introduce the notions of the left dual and the right dual, e.g.

$$\tilde{C}_{abcd} \equiv \frac{1}{2} \varepsilon_{abef} C^{ef}{}_{cd}, \qquad C\tilde{}_{abcd} \equiv \frac{1}{2} \varepsilon_{cdef} C_{ab}{}^{ef}. \tag{3.51}$$

It turns out that these dual tensors obey the relations

$$\tilde{C}_{abcd} = C\tilde{}_{abcd}, \qquad \tilde{E}_{abcd} = -E\tilde{}_{abcd}, \qquad \tilde{G}_{abcd} = G\tilde{}_{abcd}. \tag{3.52}$$

For algebraic classification (Chapter 4.) it is convenient to introduce the complex tensors

$$C^*_{abcd} \equiv C_{abcd} + \mathrm{i} C\tilde{}_{abcd}, \qquad \tilde{C}^*_{abcd} = -\mathrm{i} C^*_{abcd}, \tag{3.53}$$

$$E^*_{abcd} \equiv E_{abcd} + \mathrm{i} E\tilde{}_{abcd}, \qquad \tilde{E}^*_{abcd} = +\mathrm{i} E^*_{abcd}, \tag{3.54}$$

$$G^*_{abcd} \equiv G_{abcd} + \mathrm{i} G\tilde{}_{abcd}, \qquad \tilde{G}^*_{abcd} = -\mathrm{i} G^*_{abcd}. \tag{3.55}$$

The "unit tensor" defined by

$$I_{abcd} \equiv \frac{1}{4} (g_{abcd} + \mathrm{i}\varepsilon_{abcd}) = \frac{1}{2} (V_{ab} U_{cd} + U_{ab} V_{cd}) - \frac{1}{4} W_{ab} W_{cd} \tag{3.56}$$

(so that $I_{abcd} Z^{\alpha cd} = Z^\alpha_{ab}$, cp. (3.39), and $G^*_{abcd} = (R/3)\, I_{abcd}$) is self-dual with respect to both pairs of bivector indices. Therefore I_{abcd} admits the double expansion, given in (3.56), in terms of the basis $\{\mathbf{Z}^\alpha\}$. The decompositions

$$C^*_{abcd} = c_{\alpha\beta} Z^\alpha_{ab} Z^\beta_{cd}, \qquad E^*_{abcd} = e_{\alpha\beta} \bar{Z}^\alpha_{ab} Z^\beta_{cd}, \tag{3.57}$$

are valid for the tensors defined by (3.53), (3.54).

Because of the tracelessness of C^*_{abcd} we have the explicit expansion

$$\frac{1}{2}C^*_{abcd} = \Psi_0 U_{ab}U_{cd} + \Psi_1(U_{ab}W_{cd} + W_{ab}U_{cd}) + \Psi_2(V_{ab}U_{cd} + U_{ab}V_{cd} + W_{ab}W_{cd}) + \Psi_3(V_{ab}W_{cd} + W_{ab}V_{cd}) + \Psi_4 V_{ab}V_{cd}, \tag{3.58}$$

the five complex coefficients $\Psi_0, \ldots, \Psi_4$ being defined by

$$\begin{aligned} \Psi_0 &\equiv C_{abcd}k^a m^b k^c m^d, & \Psi_3 &\equiv C_{abcd}l^a k^b l^c \overline{m}^d,\\ \Psi_1 &\equiv C_{abcd}k^a l^b k^c m^d, & \Psi_4 &\equiv C_{abcd}l^a \overline{m}^b l^c \overline{m}^d,\\ \Psi_2 &\equiv \frac{1}{2}C_{abcd}k^a l^b(k^c l^d - m^c\overline{m}^d). \end{aligned} \tag{3.59}$$

In these definitions, $\frac{1}{2}C^*_{abcd}$ may be substituted for C_{abcd}. The various terms in (3.58) admit the following physical interpretation (Szekeres (1966b)): the Ψ_4-term represents a transverse wave in the $\boldsymbol{k}$-direction, the Ψ_3-term a longitudinal wave component, and the Ψ_2-term a "Coulomb" component; the Ψ_0- and Ψ_1-terms represent transverse and longitudinal wave components in the $\boldsymbol{l}$-direction.

With the aid of the equations (3.41) we find the transformation laws of $\Psi_0, \ldots, \Psi_4$ under the null rotations (3.14), (3.15):

$$\boldsymbol{l}\text{ fixed:}\quad \begin{aligned} \Psi_4' &= \Psi_4,\\ \Psi_3' &= \Psi_3 + E\Psi_4,\\ \Psi_2' &= \Psi_2 + 2E\Psi_3 + E^2\Psi_4,\\ \Psi_1' &= \Psi_1 + 3E\Psi_2 + 3E^2\Psi_3 + E^3\Psi_4,\\ \Psi_0' &= \Psi_0 + 4E\Psi_1 + 6E^2\Psi_2 + 4E^3\Psi_3 + E^4\Psi_4. \end{aligned} \tag{3.60}$$

$$\boldsymbol{k}\text{ fixed:}\quad \begin{aligned} \Psi_0' &= \Psi_0,\\ \Psi_1' &= \Psi_1 + \overline{B}\Psi_0,\\ \Psi_2' &= \Psi_2 + 2\overline{B}\Psi_1 + \overline{B}^2\Psi_0,\\ \Psi_3' &= \Psi_3 + 3\overline{B}\Psi_2 + 3\overline{B}^2\Psi_1 + \overline{B}^3\Psi_0,\\ \Psi_4' &= \Psi_4 + 4\overline{B}\Psi_3 + 6\overline{B}^2\Psi_2 + 4\overline{B}^3\Psi_1 + \overline{B}^4\Psi_0. \end{aligned} \tag{3.61}$$

Generalizing equations (3.36), (3.37), we can express C^*_{abcd} in terms of the complex tensor

$$-Q_{ac} \equiv C^*_{abcd}u^b u^d \equiv E_{ac} + \mathrm{i}B_{ac}, \qquad u_c u^c = -1, \tag{3.62}$$

according to the formula

$$-\frac{1}{2}C^*_{abcd} = 4u_{[a}Q_{b][d}u_{c]} + g_{a[c}Q_{d]b} - g_{b[c}Q_{d]a} + \mathrm{i}\varepsilon_{abef}u^e u_{[c}Q_{d]}^{\ f} + \mathrm{i}\varepsilon_{cdef}u^e u_{[a}Q_{b]}^{\ f}. \tag{3.63}$$

E_{ac} and B_{ac} respectively denote the "electric" and "magnetic" parts of the Weyl tensor. The components Q_{ab} satisfy the relations

$$Q^a{}_a = 0, \qquad Q_{ab} = Q_{ba}, \qquad Q_{ab}u^b = 0, \tag{3.64}$$

and can be considered as a symmetric complex (3×3) matrix $\mathbf{Q}$ with zero trace. The matrix $\mathbf{Q}$ determines 10 real numbers corresponding to the 10 independent components of the Weyl tensor.

Ref.: Lanczos (1962), Misra and Singh (1967), Cahen et al. (1967).

3.6. Spinors

Spinor formalism provides a very compact and elegant framework for numerous calculations in General Relativity, e.g. algebraic classification of the Weyl tensor (Chapter 4.) and the Newman-Penrose technique (Chapter 7.).

It can be shown that the (connected) group $SL(2, C)$ of linear transformations in two complex dimensions, with determinant of modulus 1, has a two-to-one homomorphism onto the group $L_+^\uparrow$. The space on which $SL(2, C)$ acts is called *spinor space,* and its elements are (one-index) *spinors* with components φ^A. Spinor indices like A obviously range over 1 and 2. Every proper Lorentz transformation defines an element of $SL(2, C)$ up to overall sign. Since the defining property of $L_+^\uparrow$ (within all linear transformations in four dimensions) is that it is the (connected) group preserving the Minkowski metric, and since $SL(2, C)$ is defined (within all linear transformations of two complex dimensions) as the (connected) group that preserves determinants, we expect that the determinant-forming two-form in spin space, with components

$$\varepsilon_{AB} = \begin{pmatrix} 0 & 1 \\ -1 & 0 \end{pmatrix} = \varepsilon^{AB} \tag{3.65}$$

will play the role of the metric. Spinor indices are raised and lowered according to the rule

$$\varphi^A = \varepsilon^{AB}\varphi_B \Leftrightarrow \varphi_A = \varphi^B\varepsilon_{BA}. \tag{3.66}$$

Note that $\varphi_A\varepsilon^{AB} \neq \varepsilon^{BA}\varphi_A$. The scalar product of two spinors (with components φ^A and ψ^A) is then defined by

$$\varepsilon_{AB}\varphi^A\psi^B = \varphi_A\psi^A = -\varphi^A\psi_A. \tag{3.67}$$

If φ^B transforms under $S^A{}_B \in SL(2, C)$, the complex conjugate spinor $\bar{\varphi}^{\dot{B}}$ must, for consistency, transform under the complex conjugate $\bar{S}^{\dot{A}}{}_{\dot{B}}$, and similarly φ_A transforms under the inverse of $S^A{}_B$. Dotted indices are used to indicate that the complex conjugate transformations are to be applied. The order of dotted and undotted indices is clearly irrelevant. One can obviously build multi-index spinors, in just the same way that tensors are developed from vectors.

It is now natural to seek a correspondence between the vectors $\boldsymbol{v}$ of Minkowski space and spinors. To do so we shall need not one-index spinors, but two-index spinors $v^{A\dot{B}}$, because the sign ambiguity arising from the map of $SL(2, C)$ to $L_+^\uparrow$ must be removed and it must be possible to relate the length of a vector in the Minkowski metric (quadratic in the components v^a) to a determinant (also quadratic in the entries of a (2×2) matrix). Such a map will be given by any set of *spin tensors* $\sigma_{aA\dot{B}}$ satisfying

$$\sigma_{aA\dot{B}}\sigma^{aC\dot{D}} = -\delta^C_A\delta^{\dot{D}}_{\dot{B}} \Leftrightarrow \sigma_{aA\dot{B}}\sigma^{bA\dot{B}} = -\delta_a{}^b. \tag{3.68}$$

Then v^a corresponds to $v^{A\dot{B}}$ by

$$v^a = -\sigma^a{}_{A\dot{B}}v^{A\dot{B}} \Leftrightarrow v^{A\dot{B}} = \sigma_a{}^{A\dot{B}}v^a. \tag{3.69}$$

(Note that the formulae here exhibit some sign changes as compared with Penrose (1960), due to a change of convention about the signature of the space-time metric.) The spin tensors will be Hermitian,

$$\sigma_a{}^{A\dot{B}} = \overline{\sigma_a{}^{B\dot{A}}} \equiv \bar{\sigma}_a{}^{A\dot{B}}. \tag{3.70}$$

Given a null vector, its spinor counterpart must be an outer product $\zeta_A\bar{\eta}_{\dot{B}}$ since the matrix $v_{A\dot{B}}$ must have determinant zero and thus be of rank 1. Given a null tetrad $(\boldsymbol{m}, \overline{\boldsymbol{m}}, \boldsymbol{l}, \boldsymbol{k})$ and a pair of basis spinors o^A, ι^A such that $o_A\iota^A = 1$ (a *dyad*), one can choose the map $\sigma_a{}^{A\dot{B}}$ so that in the orthonormal basis associated with $(\boldsymbol{m}, \overline{\boldsymbol{m}}, \boldsymbol{l}, \boldsymbol{k})$ by (3.12), and in the spin basis consisting of o^A, $\bar{\iota}^{\dot{A}}$ themselves (so $o^A = (1, 0)$ and $\iota^A = (0, 1)$), one has

$$\begin{aligned} \sigma_1{}^{A\dot{B}} &= \frac{1}{\sqrt{2}}\begin{pmatrix}0 & 1\\ 1 & 0\end{pmatrix}, \quad & \sigma_2{}^{A\dot{B}} &= \frac{1}{\sqrt{2}}\begin{pmatrix}0 & -\mathrm{i}\\ \mathrm{i} & 0\end{pmatrix}, \\ \sigma_3{}^{A\dot{B}} &= \frac{1}{\sqrt{2}}\begin{pmatrix}1 & 0\\ 0 & -1\end{pmatrix}, \quad & \sigma_4{}^{A\dot{B}} &= \frac{1}{\sqrt{2}}\begin{pmatrix}1 & 0\\ 0 & 1\end{pmatrix}. \end{aligned} \tag{3.71}$$

Then

$$m^a \leftrightarrow o^A\bar{\iota}^{\dot{B}}, \qquad \overline{m}^a \leftrightarrow \iota^A\bar{o}^{\dot{B}}, \qquad l^a \leftrightarrow \iota^A\bar{\iota}^{\dot{B}}, \qquad k^a \leftrightarrow o^A\bar{o}^{\dot{B}}. \tag{3.72}$$

One can check from (3.68), (3.69) that this is consistent with the normalization of $(\boldsymbol{m}, \overline{\boldsymbol{m}}, \boldsymbol{l}, \boldsymbol{k})$. Conversely (3.72) could be used to define a dyad so that (3.71) arises.

The null rotations (3.14), (3.15) correspond to the transformations

$$o'^A = o^A + E\iota^A, \qquad \iota'^A = \iota^A + Bo^A. \tag{3.73}$$

In Table 3.1 we give some examples of spinor equivalents of tensors, constructed according to the relation (3.69). The spinor form of the decomposition (3.44) of the curvature tensor is obtained from the spinor equivalent of R_{abcd} by using the relation

$$\chi_{ABCD} = \Psi_{ABCD} + \frac{1}{12}R(\varepsilon_{AC}\varepsilon_{BD} + \varepsilon_{AD}\varepsilon_{BC}). \tag{3.74}$$

The reason why spinors are frequently used in General Relativity is that the spinor formalism simplifies some relations involving null vectors and bivectors. For example, Table 3.1 shows that the Weyl tensor has a completely symmetric spinor equivalent Ψ_{ABCD} while the corresponding tensorial symmetry relations (3.48), (3.49) are much more complicated. In addition, the definitions (3.59) of

Table 3.1 Examples of spinor equivalents, defined as in (3.69). Indices a, b, c, d correspond to index pairs $A\dot W$, $B\dot X$, $C\dot Y$, $D\dot Z$, respectively

Tensor	Spinor equivalent
Metric g_{ab}:	$-\varepsilon_{AB}\varepsilon_{\dot W\dot X}$; ε_{AB} from (3.65)
Levi-Civita tensor ε_{abcd}:	$\varepsilon_{A\dot W B\dot X}{}^{C\dot Y D\dot Z} \equiv \mathrm{i}(\delta_A{}^C\delta_B{}^D\delta_{\dot W}{}^{\dot Z}\delta_{\dot X}{}^{\dot Y} - \delta_A{}^D\delta_B{}^C\delta_{\dot W}{}^{\dot Y}\delta_{\dot X}{}^{\dot Z})$
Null vector n_a:	$\zeta_A\bar\eta_{\dot W}$
Real null vector, $\bar n_a = n_a$	$\pm\zeta_A\bar\zeta_{\dot W}$
Bivector X_{ab}:	$\varepsilon_{AB}\bar\xi_{\dot W\dot X} + \varepsilon_{\dot W\dot X}\eta_{AB}$, $\eta_{[AB]} = 0 = \bar\xi_{[\dot A\dot B]}$
Real bivector F_{ab}:	$\varepsilon_{AB}\Phi_{\dot W\dot X} + \varepsilon_{\dot W\dot X}\Phi_{AB}$, $\Phi_{[AB]} = 0$
Dual bivector $\tilde F_{ab}$:	$\mathrm{i}(\varepsilon_{AB}\bar\Phi_{\dot W\dot X} - \varepsilon_{\dot W\dot X}\Phi_{AB})$
Complex self-dual bivector F^*_{ab}:	$2\Phi_{AB}\varepsilon_{\dot W\dot X}$
V_{ab}, U_{ab}, W_{ab}:	$o_A o_B\varepsilon_{\dot W\dot X}$, $\iota_A\iota_B\varepsilon_{\dot W\dot X}$, $-2o_{(A}\iota_{B)}\varepsilon_{\dot W\dot X}$
Curvature tensor R_{abcd}:	$\chi_{ABCD}\varepsilon_{\dot W\dot X}\varepsilon_{\dot Y\dot Z} + \varepsilon_{AB}\varepsilon_{CD}\bar\chi_{\dot W\dot X\dot Y\dot Z}$ $+ \Phi_{AB\dot Y\dot Z}\varepsilon_{CD}\varepsilon_{\dot W\dot X} + \varepsilon_{AB}\varepsilon_{\dot Y\dot Z}\bar\Phi_{\dot W\dot X CD}$,
	$\chi_{ABCD} = \chi_{(AB)(CD)} = \chi_{CDAB}$,
	$\Phi_{AB\dot C\dot D} = \Phi_{(AB)(\dot C\dot D)} = \overline{\Phi_{CD\dot A\dot B}} = \bar\Phi_{AB\dot C\dot D}$
Weyl tensor C_{abcd}:	$\Psi_{ABCD}\varepsilon_{\dot W\dot X}\varepsilon_{\dot Y\dot Z} + \varepsilon_{AB}\varepsilon_{CD}\bar\Psi_{\dot W\dot X\dot Y\dot Z}$, $\Psi_{ABCD} = \chi_{(ABCD)}$
C^*_{abcd}:	$2\Psi_{ABCD}\varepsilon_{\dot W\dot X}\varepsilon_{\dot Y\dot Z}$
Traceless Ricci tensor S_{ab}:	$2\Phi_{AB\dot W\dot X}$

the complex tetrad components $\Psi_0, \ldots, \Psi_4$ are very symmetric in the spinor calculus,

$$\begin{aligned} \Psi_0 &= \Psi_{ABCD}o^A o^B o^C o^D, \quad \Psi_1 = \Psi_{ABCD}o^A o^B o^C \iota^D, \quad \Psi_2 = \Psi_{ABCD}o^A o^B \iota^C \iota^D, \\ \Psi_3 &= \Psi_{ABCD}o^A \iota^B \iota^C \iota^D, \quad \Psi_4 = \Psi_{ABCD}\iota^A \iota^B \iota^C \iota^D. \end{aligned} \tag{3.75}$$

Up to now we have been concerned only with algebraic relations. The covariant derivatives also have their spinor equivalents,

$$\nabla_{A\dot B} = \sigma^a{}_{A\dot B}\nabla_a \Leftrightarrow \nabla_a = -\sigma_a{}^{A\dot B}\nabla_{A\dot B}. \tag{3.76}$$

The Bianchi identities (2.81)

$$R_{ab[cd;e]} = 0 \Leftrightarrow \tilde{R}_{abcd}{}^{;d} = 0 \tag{3.77}$$

(Lanczos (1962)), transcribed into spinor language, become

$$\nabla^D{}_{\dot E}\chi_{ABCD} = \nabla_C{}^{\dot F}\Phi_{AB\dot E\dot F}. \tag{3.78}$$

For vacuum fields ($R_{ab} = 0$), in virtue of the relation (3.74), these equations take the simpler form

$$\nabla^D{}_{\dot E}\Psi_{ABCD} = 0. \tag{3.79}$$

The directional derivatives along the null tetrad $(\boldsymbol{m}, \overline{\boldsymbol{m}}, \boldsymbol{l}, \boldsymbol{k})$ are denoted by the symbols

$$D \equiv k^a\nabla_a = -o^A\bar{o}^{\dot{B}}\nabla_{A\dot{B}}, \qquad \Delta \equiv l^a\nabla_a = -\iota^A\bar{\iota}^{\dot{B}}\nabla_{A\dot{B}}, \tag{3.80}$$

$$\delta \equiv m^a\nabla_a = -o^A\bar{\iota}^{\dot{B}}\nabla_{A\dot{B}}, \qquad \bar{\delta} \equiv \overline{m}^a\nabla_a = -\iota^A\bar{o}^{\dot{B}}\nabla_{A\dot{B}}.$$

Ref.: Witten (1959), Jordan et al. (1960), Pirani (1965), Penrose (1968), Schmutzer (1968).

3.7. Conformal transformations

A special type of map of metric spaces is given by dilatation (or contraction) of all lengths by a common factor which varies from point to point,

$$\hat{g}_{ab} = \mathrm{e}^{2U}g_{ab}, \qquad \hat{g}^{ab} = \mathrm{e}^{-2U}g^{ab}, \qquad U = U(x). \tag{3.81}$$

The connection coefficients and the covariant derivative of a 1-form σ are transformed to

$$\hat{\Gamma}^c{}_{ab} = \Gamma^c{}_{ab} + 2\delta^{\,c}_{(a}U_{,b)} - g_{ab}U^{\,c},$$
$$\hat{\nabla}_a\sigma_b = \nabla_a\sigma_b - U_{,b}\sigma_a - U_{,a}\sigma_b + g_{ab}U^{,c}\sigma_{,c}, \qquad \hat{\sigma}_a = \sigma_a. \tag{3.82}$$

The curvature tensors of the two spaces with metrics $\hat{g}_{ab}$ and g_{ab} are connected by the relation

$$\mathrm{e}^{2U}\hat{R}^{da}{}_{bc} = R^{da}{}_{bc} + 4Y^{[a}_{[b}\delta^{d]}_{c]},$$
$$Y^a_b \equiv U^{,a}{}_{;b} - U^{,a}U_{,b} + \frac{1}{2}\,\delta^a_bU_{,e}U^{,e} \tag{3.83}$$

which holds for n-dimensional Riemannian spaces $\hat{V}_n$ and V_n. From (3.83) one obtains the equation

$$\hat{R}_{ab} = R_{ab} + (2 - n)\,Y_{ab} - g_{ab}Y_c{}^c \tag{3.84}$$

for the Ricci tensors in $\hat{V}_n$ and V_n. In three dimensions, this equation takes the form

$$\hat{R}_{\alpha\beta} = R_{\alpha\beta} - U_{,\alpha;\beta} + U_{,\alpha}U_{,\beta} - g_{\alpha\beta}(U_{,\gamma}{}^{;\gamma} + U_{,\gamma}U^{,\gamma}). \tag{3.85}$$

The application of formula (3.84) to a space V_2 yields

$$\mathrm{d}\hat{s}^2 = \mathrm{e}^{2U}(\mathrm{d}x^2 + \mathrm{d}y^2): \qquad \hat{R}_{AB} = K\delta_{AB}, \qquad K = -\mathrm{e}^{-2U}U_{,A,A}. \tag{3.86}$$

A space is called conformally flat if it can be related, by a conformal transformation, to flat space. A space V_2 is always conformally flat. Any space V_3 is conformally flat if and only if the Cotton-York tensor

$$C^{\alpha\beta} \equiv 2\varepsilon^{\alpha\gamma\delta}\left(R^\beta{}_\gamma - \frac{1}{4}\,\delta^\beta{}_\gamma R\right)_{;\delta} \tag{3.87}$$

vanishes. A space V_n, $n > 3$, is conformally flat if and only if the conformal tensor

$$C_{abcd} \equiv R_{abcd} + \frac{1}{(n-1)(n-2)} R(g_{ac}g_{bd} - g_{ad}g_{bc})$$
$$- \frac{1}{(n-2)} (g_{ac}R_{bd} - g_{bc}R_{ad} + g_{bd}R_{ac} - g_{ad}R_{bc}) \qquad (3.88)$$

(for V_4, cf. (3.50)) vanishes. The conformal tensor components $C^a{}_{bcd}$ are unchanged by (3.81).

The special case $U =$ constant of (3.81) gives a (trivial) new solution of the Einstein field equations (1.1) if applied to an already known solution (see (3.84)). All exact solutions given hereafter could therefore be multiplied by an unspecified constant $k^2 = e^{2U}$. In many cases this is already allowed for, in that the metrics given can absorb such a change by redefinition of coordinates and constant parameters. In some cases the one-parameter family of metrics generated by varying k are in fact all isometric to each other; this happens when the space-time admits a (non-trivial) homothety, cp. Chapter 8. and § 31.4. When neither of the last two sentences applies, the unspecified constant should appear explicitly in the metrics given below.

Ref.: Eisenhart (1949), Jordan et al. (1960), Szekeres (1963), Debever (1966).

Chapter 4. The Petrov classification

4.1. The eigenvalue problem

We are interested in invariant characterizations of a gravitational field, independent of any special coordinate system. For this purpose we investigate the algebraic structure of the tensors C_{abcd} and E_{abcd} introduced in (3.44)—(3.47). The classification of S_{ab} (which is equivalent to E_{abcd}) will be treated in Chapter 5. Here we consider the classification of the Weyl tensor C_{abcd} (Petrov classification).

The starting point is the *eigenvalue equation*

$$\frac{1}{2} C_{abcd}X^{cd} = \lambda X_{ab} \tag{4.1}$$

with eigenbivectors X_{ab} and eigenvalues λ. With each solution (X_{ab}, λ) of this eigenvalue equation is associated its complex conjugate solution $(\overline{X}_{ab}, \overline{\lambda})$. Without loss of generality we can rewrite the equation (4.1) in the form

$$\frac{1}{4} C^*_{abcd}X^{*cd} = \lambda X^*_{ab}. \tag{4.2}$$

We note that an analogous equation with E_{abcd} in place of C_{abcd} would be inconsistent because of the property (3.54).

Multiplying the eigenvalue equation (4.2) by a timelike unit vector u^a, and taking into account the definitions (3.36), (3.62), and the expressions (3.37), (3.63), we reduce the eigenvalue problem to the simple form

$$Q_{ab}X^b = \lambda X_a, \tag{4.3}$$

which is completely equivalent to the original formulation. In 3-dimensional vector notation suggested by (3.64) we write

$$\mathbf{Q}\boldsymbol{r} = \lambda \boldsymbol{r}. \tag{4.4}$$

We have to determine the eigenvectors $\boldsymbol{r}$ and the eigenvalues λ of the *complex* (3×3) matrix $\mathbf{Q}$, which is symmetric and traceless. From the 4-dimensional Lorentz frame we have passed to a 3-dimensional complex space with Euclidean metric.

The group $SO(3, C)$ of proper orthogonal transformations in this complex 3-space is isomorphic to the group $L^\uparrow_+$ of proper orthochronous Lorentz transformations. The transformation matrices of these two groups,

$$\begin{aligned} SO(3, C)&: \quad X_{\beta'} = A^{\alpha}{}_{\beta'}X_{\alpha}, \qquad A^{\alpha}{}_{\gamma'}A_{\beta}{}^{\gamma'} = \delta^{\alpha}_{\beta}, \\ L^\uparrow_+&: \quad X^*_{a'b'} = \Lambda^{c}{}_{a'}\Lambda^{d}{}_{b'}X^*_{cd}, \qquad \Lambda^{a}{}_{c'}\Lambda^{c'}{}_{b} = \delta^{a}_{b}, \end{aligned} \tag{4.5}$$

are related by the formula

$$A^{\alpha}{}_{\beta'} = \Lambda^{\alpha}{}_{\beta'}\Lambda^{4}{}_{4'} - \Lambda^{4}{}_{\beta'}\Lambda^{\alpha}{}_{4'} + \mathrm{i}\varepsilon^{\alpha}{}_{\gamma\delta}\Lambda^{\gamma}{}_{\beta'}\Lambda^{\delta}{}_{4'} \tag{4.6}$$

which follows from (3.38): each Lorentz transformation induces a unique orthogonal transformation in the complex 3-space. The isomorphism is explicitly verified in Synge (1964).

The eigenvalue problem (4.4) leads to the characteristic equation $\det(\mathbf{Q} - \lambda\boldsymbol{I}) = 0$, and determines the orders $[m_1 \dots m_k]$ of the elementary divisors $(\lambda - \lambda_1)^{m_1}, \dots, (\lambda - \lambda_k)^{m_k}$, $m_1 + \dots + m_k = 3$, belonging to the eigenvalues $\lambda_1, \dots, \lambda_k$.

Ref.: Debever (1964).

4.2. The Petrov types

The distinct algebraic structures studied by Petrov (1966) are characterized by the elementary divisors and the multiplicities of the eigenvalues (Table 4.1). The algebraic type of the matrix $\mathbf{Q}$ provides an invariant characterization of the gravitational field at a given point p; these characteristics are independent of the coordinate system and of the choice of the tetrad at p.

Table 4.1 also gives matrix criteria for the distinct Petrov types. At a given point p, the field is of the Petrov type corresponding, in Table 4.1, to the most restrictive of the criteria which the matrix $\mathbf{Q}$ satisfies; for instance

$$\text{Type } III \;\Leftrightarrow\; \mathbf{Q}^3 = 0,\ \mathbf{Q}^2 \neq 0. \tag{4.7}$$

A real matrix $\mathbf{Q}$ has simple elementary divisors (Petrov types I, D or O).

In order to determine the Petrov type of a given metric we can calculate the complex matrix $\mathbf{Q}$ with respect to an *arbitrary* orthonormal basis $\{\boldsymbol{E}_a\}$ and use the invariant criteria listed in Table 4.1.

Table 4.1 The Petrov types. Round brackets indicate that the corresponding eigenvalues coincide, e.g., [(11) 1] means: simple elementary divisors, and $\lambda_1 = \lambda_2 \neq \lambda_3$

Petrov types	Orders of the elementary divisors $[m_1 \dots m_k]$	Matrix criterion
I	[1 1 1]	$(\mathbf{Q} - \lambda_1\boldsymbol{I})(\mathbf{Q} - \lambda_2\boldsymbol{I})(\mathbf{Q} - \lambda_3\boldsymbol{I}) = 0$
D	[(1 1) 1]	$\left(\mathbf{Q} + \dfrac{\lambda\boldsymbol{I}}{2}\right)(\mathbf{Q} - \lambda\boldsymbol{I}) = 0$
II	[2 1]	$\left(\mathbf{Q} + \dfrac{\lambda\boldsymbol{I}}{2}\right)^2(\mathbf{Q} - \lambda\boldsymbol{I}) = 0$
N	[(2 1)]	$\mathbf{Q}^2 = 0$
III	[3]	$\mathbf{Q}^3 = 0$
O	—	$\mathbf{Q} = 0$

Gravitational fields of Petrov types D, II, N, III, and O are said to be *algebraically special.*

The purpose of applying Lorentz rotations, or, equivalently, elements of the group $SO(3, C)$, is to find simple *normal forms* for the various types (Table 4.2). The normal forms of $\mathbf{Q}$ and C_{abcd} are uniquely associated.

Apart from reflections, which are not considered here, the basis $\{\boldsymbol{E}_a\}$ (cp. (3.7)) of the normal form of a *non-degenerate* Petrov type (I, distinct λ_α; II, $\lambda \neq 0$; III) is uniquely determined; for the non-degenerate types there is no subgroup of $L_+^\uparrow$ preserving the corresponding normal forms given in Table 4.2. We call the uniquely determined basis $\{\boldsymbol{E}_a\}$ the *Weyl principal tetrad.*

In the case of Petrov type I we have spacelike and timelike 2-planes ("blades") associated with the complex self-dual eigenbivectors

$$\begin{aligned}
\boldsymbol{V} - \boldsymbol{U} &= 2(\boldsymbol{E}_{[4}\boldsymbol{E}_{1]} + \mathrm{i}\boldsymbol{E}_{[2}\boldsymbol{E}_{3]}), \\
\mathrm{i}(\boldsymbol{V} + \boldsymbol{U}) &= 2(\boldsymbol{E}_{[4}\boldsymbol{E}_{2]} + \mathrm{i}\boldsymbol{E}_{[3}\boldsymbol{E}_{1]}), \\
\boldsymbol{W} &= 2(\boldsymbol{E}_{[4}\boldsymbol{E}_{3]} + \mathrm{i}\boldsymbol{E}_{[1}\boldsymbol{E}_{2]}),
\end{aligned} \tag{4.8}$$

and the intersections of these 2-planes determine the principal tetrad.

The principal tetrad is only partially determined by the metric for the two *degenerate* Petrov types D (I, $\lambda_1 = \lambda_2$), and N (II, $\lambda = 0$). It is not difficult to find the remaining subgroups of $L_+^\uparrow$ which preserve the normal forms

$$C^*_{abcd} = -\frac{\lambda}{2}\,(g_{abcd} + \mathrm{i}\varepsilon_{abcd}) - \frac{3}{2}\,\lambda W_{ab} W_{cd}, \qquad \text{for type } D, \tag{4.9}$$

$$C^*_{abcd} = -4 V_{ab} V_{cd}, \qquad \text{for type } N. \tag{4.10}$$

In type D metrics this invariance group consists of special Lorentz transformations in the $\boldsymbol{E}_3 - \boldsymbol{E}_4$-plane and spatial rotations in the $\boldsymbol{E}_1 - \boldsymbol{E}_2$-plane. In terms of the complex null tetrad, these transformations are given by (3.16), (3.17). In type N metrics the invariance group with $V'_{ab} = V_{ab}$ is just the 2-parameter subgroup (3.15).

The spinor form (§ 3.6) of the eigenvalue equation (4.2) with the eigenbivector $X^*_{ab} \leftrightarrow \eta_{AB}\varepsilon_{\dot{C}\dot{D}}$ reads

$$\Psi_{ABCD}\eta^{CD} = \lambda\eta_{AB}. \tag{4.11}$$

The invariants

$$\begin{aligned}
I &\equiv \frac{1}{2}\,\Psi_{ABCD}\Psi^{ABCD} = \frac{1}{2}\,(\lambda_1^2 + \lambda_2^2 + \lambda_3^2), \\
J &\equiv \frac{1}{6}\,\Psi_{ABCD}\Psi^{CDEF}\Psi_{EF}{}^{AB} = \frac{1}{6}\,(\lambda_1^3 + \lambda_2^3 + \lambda_3^3)
\end{aligned} \tag{4.12}$$

of the Weyl tensor are useful in Petrov classification; algebraically special fields (all Petrov types except type I) satisfy the relation

$$I^3 = 27 J^2 \tag{4.13}$$

Table 4.2 Normal forms of the Weyl tensor, and Petrov types

Normal forms of the matrix $\mathbf{Q}$	Eigenvalues λ_α and corresponding eigenvectors $\boldsymbol{r}_\alpha$ of $\mathbf{Q}$	Eigenbivectors X^*_{ab} of C^*_{abcd}	Coefficients $\Psi_0, \ldots, \Psi_4$ in the expansions (3.58) and normal forms of C^*_{abcd}	Petrov types
$\mathbf{Q} = \begin{pmatrix} \lambda_1 & & \\ & \lambda_2 & \\ & & \lambda_3 \end{pmatrix}$ $\lambda_1 + \lambda_2 + \lambda_3 = 0$	$\lambda_1 : \boldsymbol{r}_1 = (1, 0, 0)$ $\lambda_2 : \boldsymbol{r}_2 = (0, 1, 0)$ $\lambda_3 : \boldsymbol{r}_3 = (0, 0, 1)$	$X^*_{(1)ab} = V_{ab} - U_{ab}$ $X^*_{(2)ab} = \mathrm{i}(V_{ab} + U_{ab})$ $X^*_{(3)ab} = W_{ab}$	$\Psi_0 = \Psi_4 = (\lambda_2 - \lambda_1)/2,$ $\Psi_1 = \Psi_3 = 0,\ \Psi_2 = -\lambda_3/2;$ $C^*_{abcd} = -\sum_{\alpha=1}^{3} \lambda_\alpha X^*_{(\alpha)ab} X^*_{(\alpha)cd}$	I; D (for $\lambda_1 = \lambda_2$)
$\mathbf{Q} = \begin{pmatrix} -\frac{\lambda}{2} + 1 & -\mathrm{i} & 0 \\ -\mathrm{i} & -\frac{\lambda}{2} - 1 & 0 \\ 0 & 0 & \lambda \end{pmatrix}$	$\lambda_1 = \lambda_2 = -\frac{\lambda}{2}:$ $\boldsymbol{r}_1 = \left(\frac{1}{2}, -\frac{\mathrm{i}}{2}, 0\right)$ $\lambda_3 = \lambda : \boldsymbol{r}_2 = (0, 0, 1)$	$X^*_{(1)ab} = V_{ab}$ $X^*_{(2)ab} = W_{ab}$	$\Psi_0 = \Psi_1 = \Psi_3 = 0,$ $\Psi_2 = -\frac{\lambda}{2},\ \Psi_4 = -2;$ $C^*_{abcd} = -4V_{ab}V_{cd} - \lambda(V_{ab}U_{cd} + U_{ab}V_{cd} + W_{ab}W_{cd})$	II; N (for $\lambda = 0$)
$\mathbf{Q} = \begin{pmatrix} 0 & 0 & \mathrm{i} \\ 0 & 0 & 1 \\ \mathrm{i} & 1 & 0 \end{pmatrix}$	$\lambda_1 = \lambda_2 = \lambda_3 = 0:$ $\boldsymbol{r} = \left(\frac{1}{2}, -\frac{\mathrm{i}}{2}, 0\right)$	$X^*_{cb} = V_{ab}$	$\Psi_0 = \Psi_1 = \Psi_2 = \Psi_4 = 0,$ $\Psi_3 = -\mathrm{i};$ $C^*_{abcd} = -2\mathrm{i}(V_{ab}W_{cd} + W_{ab}V_{cd})$	III

and, in particular, for the types *III*, *N*, and *O* both invariants (4.12) vanish, $I = J = 0$.

Next we give two theorems valid for vacuum fields ($R_{ab} = 0$).

Theorem 4.1 (Brans (1975)). *A type I vacuum solution for which one of the eigenvalues λ_α of the Weyl tensor vanishes over an open region must be flat space-time.*

Theorem 4.2 (Zakharov (1965, 1970)). *Vacuum fields satisfying the equation $R_{abcd;e}{}^{e} = \omega R_{abcd}$ are either type N ($\omega = 0$) or type D ($\omega \neq 0$).*

The *proof* of Theorem 4.2 follows immediately from the identity (Zakharov (1972))

$$R_{abcd;m}{}^{m} = R^{m}{}_{nab}R^{n}{}_{mcd} + 2(R^{m}{}_{adn}R^{n}{}_{cbm} - R^{m}{}_{bdn}R^{n}{}_{cam}) \tag{4.14}$$

($R_{ab} = 0$), written down with respect to a principal tetrad (cf. the normal forms in Table 4.2).

Ref.: Synge (1964), Case (1968), Roche and Dowker (1968), Thorpe (1969), Ludwig (1969), Polishchuk (1970), Narain (1970), Brans (1974).

4.3. Principal null directions

For Petrov types *II* and *III*, the real and imaginary parts of the eigenbivector $V_{ab} = 2k_{[a}m_{b]}$ represent 2-spaces containing the real null vector $\boldsymbol{k}$. The normal forms (Table 4.2) for these types are adapted to this preferred null vector, which is significant in what follows.

The classification based on the distinct possible solutions of the eigenvalue problem (4.4) is equivalent to the characterization of the Weyl (conformal) tensor in terms of *principal null directions* $\boldsymbol{k}$ with the property (Penrose (1960))

$$k_{[e}C_{a]bc[d}k_{f]}k^{b}k^{c} = 0 \Leftrightarrow \Psi_0 \equiv C_{abcd}k^{a}m^{b}k^{c}m^{d} = 0. \tag{4.15}$$

There are at most four such null vectors; to determine them we apply the inverse of the null rotation (3.14) to an arbitrary complex null tetrad ($\boldsymbol{m}', \overline{\boldsymbol{m}}', \boldsymbol{l}', \boldsymbol{k}'$) defined by (3.8). By this means the null vector $\boldsymbol{k}'$ can be transformed into any other real null vector except $\boldsymbol{l}'$. The coefficients $\Psi_0, \ldots, \Psi_4$ defined by (3.59) then undergo the transformations (3.60), in particular,

$$\Psi_0 = \Psi_0' - 4E\Psi_1' + 6E^2\Psi_2' - 4E^3\Psi_3' + E^4\Psi_4', \tag{4.16}$$

and so, from the condition (4.15), we obtain an algebraic equation of at most fourth order for the complex number E:

$$\Psi_0' - 4E\Psi_1' + 6E^2\Psi_2' - 4E^3\Psi_3' + E^4\Psi_4' = 0. \tag{4.17}$$

Starting with the normal forms for the various Petrov types given in Table 4.2 and in equations (4.9), (4.10), we calculate the roots E given in Table 4.3.

In types *D* and *III* there is an additional principal null direction $\boldsymbol{l}$ not obtainable with the aid of the null rotations (3.14); $\boldsymbol{l}$ fulfils the condition

$$l_{[e}C_{a]bc[d}l_{f]}l^{b}l^{c} = 0 \Leftrightarrow \Psi_4 \equiv C_{abcd}l^{a}\overline{m}^{b}l^{c}\overline{m}^{d} = 0. \tag{4.18}$$

Table 4.3 The roots of the algebraic equation (4.17) and their multiplicities. The corresponding multiplicities of the principal null directions are symbolically depicted on the right side of this table.

Type	Roots E	Multiplicities	
I	$\dfrac{\sqrt{\lambda_2 + 2\lambda_1} \pm \sqrt{\lambda_1 + 2\lambda_2}}{\sqrt{\lambda_1 - \lambda_2}}$	(1, 1, 1, 1)	
D	$0, \infty$	(2, 2)	
II	$0, \pm \mathrm{i}\sqrt{\frac{3}{2}}\,\lambda$	(2, 1, 1)	
III	$0, \infty$	(3, 1)	
N	0	(4)	

A Weyl tensor is said to be *algebraically special* if it admits at least one *multiple* principal null direction. (The multiplicity of a null direction is equal to the multiplicity of the corresponding root of the algebraic equation (4.17).)

One can show the validity of the following equations (Jordan et al. (1961))

$$k_{[e}C_{a]bc[d}k_{f]}k^b k^c = 0 \Leftrightarrow \Psi_0 = 0, \qquad \Psi_1 \neq 0, \tag{4.19}$$

$$C_{abc[d}k_{f]}k^b k^c = 0 \Leftrightarrow \Psi_0 = \Psi_1 = 0, \qquad \Psi_2 \neq 0, \tag{4.20}$$

$$C_{abc[d}k_{f]}k^c = 0 \Leftrightarrow \Psi_0 = \Psi_1 = \Psi_2 = 0, \qquad \Psi_3 \neq 0, \tag{4.21}$$

$$C_{abcd}k^c = 0 \Leftrightarrow \Psi_0 = \Psi_1 = \Psi_2 = \Psi_3 = 0, \qquad \Psi_4 \neq 0 \tag{4.22}$$

for principal null directions $\boldsymbol{k}$ of multiplicity 1, 2, 3 and 4 respectively. An equivalent formulation of the criterion (4.20) is

$$k^a k^c C^*_{abcd} = \lambda k_b k_d, \qquad \lambda \neq 0. \tag{4.23}$$

Any two of the following conditions for a null direction $\boldsymbol{k}$ imply the third (Hall (1973))

$$C_{abc[d}k_{f]}k^b k^c = 0, \qquad R_{abc[d}k_{f]}k^b k^c = 0, \qquad R_{a[b}k_{c]}k^a = 0. \tag{4.24}$$

The last condition in (4.24) means that $\boldsymbol{k}$ is a Ricci eigendirection (§ 5.1).

Type D is characterized by the existence of *two double* principal null directions, $\boldsymbol{k}$ and $\boldsymbol{l}$,

$$\begin{aligned} C_{abc[d}k_{f]}k^b k^c = 0 \ &\Leftrightarrow\ \Psi_0 = \Psi_1 = 0, \qquad \Psi_2 \neq 0, \\ C_{abc[d}l_{f]}l^b l^c = 0 \ &\Leftrightarrow\ \Psi_4 = \Psi_3 = 0, \qquad \Psi_2 \neq 0. \end{aligned} \tag{4.25}$$

Type O (zero Weyl tensor) does not single out any null directions.

In the *Penrose Diagram* (Fig. 4.1) the arrows point in the direction of increasing multiplicity of the principal null directions; every arrow indicates one additional degeneration.

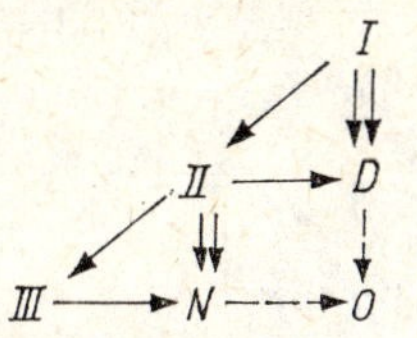

Fig. 4.1 Penrose diagram

The classification in terms of principal null directions can be formulated in terms of spinors (Penrose (1960)). The completely symmetric spinor Ψ_{ABCD} can be written as a symmetrized product of one-index spinors, which are uniquely determined apart from factors. The proof is an application of the fundamental theorem of algebra: any polynomial may be factorized into linear forms,

$$\Psi_{ABCD}\zeta^A\zeta^B\zeta^C\zeta^D = (\alpha_A\zeta^A)(\beta_B\zeta^B)(\gamma_C\zeta^C)(\delta_D\zeta^D). \tag{4.26}$$

The Petrov types are then characterized by the criteria:

$$\begin{aligned}
\text{Type} \quad I:&\quad \Psi_{ABCD} \sim o_{(A}\beta_B\gamma_C\iota_{D)},\\
D:&\quad \Psi_{ABCD} \sim o_{(A}o_B\iota_C\iota_{D)},\\
II:&\quad \Psi_{ABCD} \sim o_{(A}o_B\gamma_C\iota_{D)},\\
III:&\quad \Psi_{ABCD} \sim o_{(A}o_Bo_C\iota_{D)},\\
N:&\quad \Psi_{ABCD} \sim o_Ao_Bo_Co_D,
\end{aligned} \tag{4.27}$$

$(k^a \leftrightarrow o^A\bar{o}^{\dot{B}},\ l^a \leftrightarrow \iota^A\bar{\iota}^{\dot{B}})$.

The two approaches to the classification of the Weyl tensor — the eigenvalue problem for the matrix $\mathbf{Q}$ and the principal null directions — are completely equivalent. This classification enables one to divide the gravitational fields in an invariant way into distinct types; the Petrov types. (We use this term although other authors obtained analogous results, see Debever (1959), Géhéniau (1957), Pirani (1957), Penrose (1960).)

Ref.: Bel (1958, 1959), Penrose (1968), Adler and Sheffield (1973), Perjes (1976).

4.4. Determination of the Petrov type

The simplest method of finding the Petrov type of (the Weyl tensor of) a gravitational field at a given point p is to determine the *roots of the quartic algebraic equation* (4.17),

$$\Psi_0 - 4E\Psi_1 + 6E^2\Psi_2 - 4E^3\Psi_3 + E^4\Psi_4 = 0, \tag{4.28}$$

where the coefficients $\Psi_0, \ldots, \Psi_4$ defined by (3.59) can be calculated with respect to an *arbitrary* complex null tetrad at p. If the order of the algebraic equation (4.28) is $(4 - m)$, then there are $(4 - m)$ principal null directions $\boldsymbol{k}$ and $\boldsymbol{l}$ represents an m-fold principal null direction. The Petrov type can be obtained immediately, by inspection of Table 4.3, once the multiplicities of the roots of (4.28) are known.

An algorithm for determining the Petrov type from $\Psi_0, \ldots, \Psi_4$ is displayed in Fig. 4.2 (d'Inverno and Russell-Clark (1971)). Provided that $\Psi_4 \neq 0$, the definitions used in the flow diagram are

$$I \equiv \Psi_0\Psi_4 - 4\Psi_1\Psi_3 + 3\Psi_2^2, \qquad J \equiv \begin{vmatrix} \Psi_4 & \Psi_3 & \Psi_2 \\ \Psi_3 & \Psi_2 & \Psi_1 \\ \Psi_2 & \Psi_1 & \Psi_0 \end{vmatrix} \tag{4.29}$$

$$K \equiv \Psi_1\Psi_4^2 - 3\Psi_4\Psi_3\Psi_2 + 2\Psi_3^3, \qquad L \equiv \Psi_2\Psi_4 - \Psi_3^2,$$

$$N \equiv 12L^2 - \Psi_4^2 I.$$

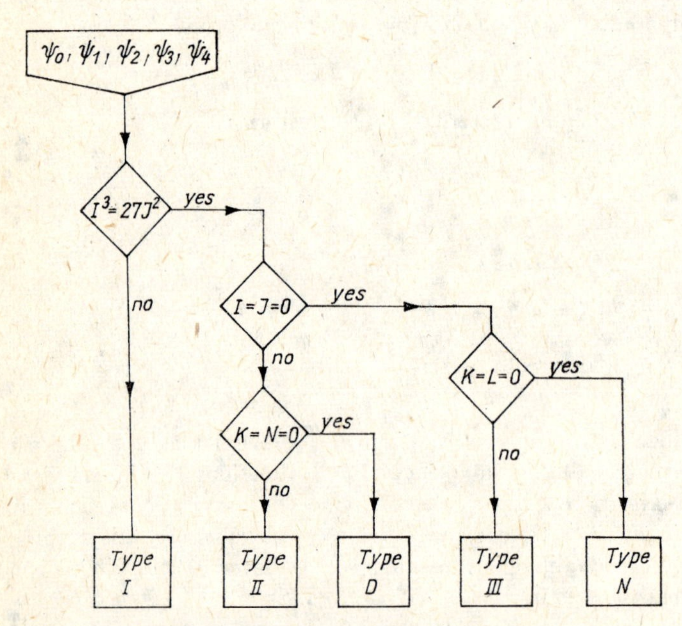

Fig. 4.2 Flow diagram for determining the Petrov type

If $\Psi_4 = 0$, but $\Psi_0 \neq 0$, one has to interchange Ψ_0 with Ψ_4 and Ψ_1 with Ψ_3 in these definitions, and the algorithm proceeds as before. For $\Psi_0 = \Psi_4 = 0$, the multiplicities of the roots of the equation (4.28) are very simple to determine.

If both the invariants I and J vanish, then at least three of the principal null directions coincide. From the diagram and the definitions (4.29) it follows that gravitational fields with a repeated principal null direction $\boldsymbol{k}$ ($\Psi_0 = \Psi_1 = 0$) are type D if and only if the remaining tetrad components of the Weyl tensor satisfy the condition

$$3\Psi_2\Psi_4 = 2\Psi_3^2. \tag{4.30}$$

As an *example* we take the field considered at the end of § 3.3. Using the null tetrad related by (3.12) to the basis (3.24) we get

$$\Psi_0 = \Psi_1 = \Psi_2 = 0, \qquad \Psi_3 = \frac{1}{2}\sqrt{\frac{3}{2}}\,a^2, \qquad \Psi_4 = -\frac{3}{2}\,a^2, \tag{4.31}$$
$$I = J = 0, \qquad K = 2\Psi_3{}^2 \neq 0.$$

The triple root $E = 0$ confirms that the field is Petrov type *III*. The original basis was accidentally adapted to the triple principal null direction $\boldsymbol{k}$.

An *equivalent method* for determining the Petrov type is based on the eigenvalue equation (4.4). One can use the invariant criteria for the matrix $\mathbf{Q}$ which are listed in Table 4.1, $\mathbf{Q}$ being calculated with respect to an arbitrary orthonormal basis $\{\boldsymbol{E}_a\}$.

In § 3.3. we evaluated the curvature for the example considered above. The resulting matrix

$$\mathbf{Q} = \sqrt{\frac{3}{2}}\,\frac{a^2}{2}\begin{pmatrix} \sqrt{\frac{3}{2}} & -\mathrm{i}\sqrt{\frac{3}{2}} & -1 \\ -\mathrm{i}\sqrt{\frac{3}{2}} & -\sqrt{\frac{3}{2}} & \mathrm{i} \\ -1 & \mathrm{i} & 0 \end{pmatrix} \tag{4.32}$$

satisfies precisely the criterion (4.7) for type *III*, and $\mathbf{Q}$ can be transformed into the normal form from Table 4.2 with the aid of a complex orthogonal transformation.

Chapter 5. Classification of the Ricci tensor and the energy-momentum tensor

5.1. The algebraic types of the Ricci tensor

In § 3.5. we decomposed the curvature tensor into irreducible parts. The invariant classification of the Weyl tensor has been treated in Chapter 4. Now we consider the algebraic classification of the remaining part, the traceless Ricci tensor S_{ab}.

Every second-order symmetric tensor defines a linear mapping which takes a vector $\boldsymbol{v}$ into another vector $\boldsymbol{w}$. It is natural to examine the eigenvalue equation

$$S^a{}_b v^b = \hat{\lambda} v^a. \tag{5.1}$$

Because a term proportional to g^a_b merely shifts all eigenvalues by the same amount, we may as well consider the eigenvalue equation for the Ricci tensor $R^a{}_b$

$$R^a{}_b v^b = \lambda v^a; \qquad \lambda = \hat{\lambda} + \frac{1}{4} R. \tag{5.2}$$

In a positive definite metric, a real symmetric matrix can always be diagonalized by a real orthogonal transformation. However, the Lorentz metric g_{ab} leads to a more complicated algebraic structure; the elementary divisors can be non-simple, and the eigenvalues can be complex. The eigenvalue equation (5.2) determines the orders $m_1 \dots m_k$ of the elementary divisors belonging to the various eigenvalues. The Segré notation gives just these orders, and round brackets indicate that the corresponding eigenvalues coincide. If two eigenvalues are complex conjugate they are symbolized by Z and $\bar{Z}$.

The Plebański (1964) notation indicates whether the space spanned by the eigenvectors belonging to a certain real eigenvalue is timelike (T), null (N) or spacelike (S). The multiplicity of the eigenvalue is written in front of this symbol. Finally, the numbers $m_1, \dots, m_k$, are added as indices enclosed in brackets.

Table 5.1 gives a complete list of all possible types for the Ricci tensor $R^a{}_b$ of a space-time in both Segré and Plebański notation. In accordance with the conventions established in §§ 3.1., 3.2. we have arranged the eigenvalues in such an order that those whose corresponding eigenvectors are null or timelike (or complex) appear last, and we have used a comma to separate these from the eigenvalues with spacelike eigenvectors. The table also gives the most important of the possible physical interpretations of the various types; these are discussed further in § 5.2. Further refinements of the classification given here are possible (see e.g. Ludwig and Scanlan (1971)).

One can establish this list of possibilities by the following sequence of results (Churchill (1932), Hall (1976b)). We first define an *invariant 2-plane* (at any point) as a 2-dimensional subspace of the tangent space which is mapped to itself by $R^a{}_b$; any vector $\boldsymbol{v}$ lying in an invariant 2-plane is mapped by $R^a{}_b$ to a vector $\boldsymbol{w}$ in the same plane.

Table 5.1 The algebraic types of the Ricci tensor (for explanation, see the text)

Segré notation	Plebański notation	Important physical interpretations
$A1$ [1 1 1, 1]	$[S_1 - S_2 - S_3 - T]_{(1111)}$	
[1 1(1, 1)]	$[S_1 - S_2 - 2T]_{(111)}$	
[(1 1) 1, 1]	$[2S_1 - S_2 - T]_{(111)}$	
[(1 1) (1, 1)]	$[2S - 2T]_{(11)}$	electromagnetic non-null field
[1(1 1, 1)]	$[S - 3T]_{(11)}$	(tachyon fluid)
[(1 1 1), 1]	$[3S - T]_{(11)}$	perfect fluid
[(1 1 1, 1)]	$[4T]_{(1)}$	Λ-term
$A2$ [1 1, $Z\bar{Z}$]	$[S_1 - S_2 - Z - \bar{Z}]_{(1111)}$	
[(1 1), $Z\bar{Z}$]	$[2S - Z - \bar{Z}]_{(111)}$	
$A3$ [11, 2]	$[S_1 - S - 2N]_{(112)}$	
[1 (1, 2)]	$[S - 3N]_{(12)}$	
[(1 1), 2]	$[2S - 2N]_{(12)}$	pure radiation and
[(1 1, 2)]	$[4N]_{(2)}$	electromagnetic null field
B [1, 3]	$[S - 3N]_{(13)}$	
[(1, 3)]	$[4N]_{(3)}$	

1) There is always an invariant 2-plane.

Proof: Either there are two real eigenvectors, or (at least) one complex eigenvector. In the former case the two eigenvectors give the required plane; in the latter case, the real and imaginary parts of the eigenvector do the same.

2) The orthogonal 2-plane to that given in 1) is also invariant.

Proof: If the 2-plane of 1) is timelike or spacelike, taking an orthonormal basis with $\boldsymbol{E}_1$, $\boldsymbol{E}_2$, lying in this 2-plane breaks $R^a{}_b$ into block diagonal form, and so shows the orthogonal 2-plane is invariant. If the 2-plane of 1) is null, let $\boldsymbol{k}, \boldsymbol{x}$ span it. Then in an expansion of R_{ab} using a null tetrad $(\boldsymbol{m}, \bar{\boldsymbol{m}}, \boldsymbol{l}, \boldsymbol{k})$ with $\sqrt{2}\,\boldsymbol{m} = \boldsymbol{x} + \mathrm{i}\boldsymbol{y}$ only terms in $k_a k_b$, $k_{(a}l_{b)}$, $k_{(a}x_{b)}$, $k_{(a}y_{b)}$, $x_a x_b$ and $y_a y_b$ survive. Thence $\boldsymbol{k}, \boldsymbol{y}$ span another invariant 2-plane.

3) If $R^a{}_b$ has an invariant timelike (or spacelike) 2-plane, it has 2 distinct spacelike eigenvectors.

Proof: By 2) above the spacelike and timelike cases are the same and $R^a{}_b$ takes block diagonal form in an orthonormal tetrad. The (2×2) matrix acting on the spacelike 2-plane can be diagonalized by spatial rotation in the usual way for symmetric matrices.

4) If $R^a{}_b$ has an invariant null plane, it has a null eigenvector.

Proof: In the proof of 2), we see that $\boldsymbol{k}$ is a null eigenvector.

Following Hall (1976b), one can now enumerate the cases listed in Table 5.1 by taking a null tetrad basis and systematically considering first the cases where $R^a{}_b$ has a null eigenvector, and then the cases where there is no null eigenvector. In the table the different cases are divided into classes according to whether or not there is a timelike (and hence also a spacelike) invariant 2-plane. The distinct types are

A: *Timelike invariant 2-plane*
- $A1$: Two real orthogonal eigenvectors exist in this plane
- $A2$: No real eigenvectors exist in this plane
- $A3$: One double null real eigenvector exists in this plane

B: *Null invariant 2-plane:* one triple null real eigenvector exists.

In each case one can transform R_{ab} to a canonical form. Since the classes $A2$ and B have no physical interpretation, because they necessarily violate the energy condition (§ 5.3), we give the forms only for $A1$ and $A3$.

$$A1: \quad R_{ab} = \lambda_1 x_a x_b + \lambda_2 y_a y_b + \lambda_3 z_a z_b - \lambda_4 u_a u_b, \tag{5.3a}$$

$$A3: \quad R_{ab} = \lambda_1 x_a x_b + \lambda_2 y_a y_b - 2\lambda_3 k_{(a} l_{b)} + k_a k_b. \tag{5.3b}$$

Here $(\boldsymbol{x}, \boldsymbol{y}, \boldsymbol{z}, \boldsymbol{u})$ is an orthonormal tetrad, and $\left(\frac{\boldsymbol{x} + i\boldsymbol{y}}{\sqrt{2}}, \frac{\boldsymbol{x} - i\boldsymbol{y}}{\sqrt{2}}, \boldsymbol{k}, \boldsymbol{l}\right)$ a complex null tetrad. In each case the associated orthonormal tetrad is called the *Ricci principal tetrad.*

The Ricci tensor types are called *degenerate* when there is more than one elementary divisor with the same eigenvalue; in the Segré notation these degeneracies are indicated by round brackets (in Table 5.1).

If the Ricci tensor is non-degenerate and the elementary divisors are simple, the type is said to be *algebraically general.* Otherwise it is called *algebraically special.* These ideas are analogous to those used for Petrov types (Chapter 4.). As we see in the next section, the physically most important types are algebraically special.

Table 5.2 Invariance groups of the Ricci tensor types

Invariance group	Ricci tensor type
none	[111, 1], [11, $Z\bar{Z}$], [11, 2], [1, 3]
Spatial rotations (3.16)	[(11)1, 1] [(11), $Z\bar{Z}$] [(11), 2]
Boosts (3.17)	[11(1, 1)]
Boosts (3.17) and rotations (3.16)	[(11) (1, 1)]
$SO(3)$ rotations	[(111), 1]
$SO(2, 1)$: 3-dimensional Lorentz group	[1(11, 1)]
One-parameter group of null rotations	[(1(1, 2)], [(1, 3)]
Null rotations (3.15) and rotations (3.16)	[(11, 2)]
Full Lorentz group	[(111, 1), and of course, $R^a{}_b = 0$.

The Ricci principal tetrads of the non-degenerate types (where the eigenvalues of different elementary divisors are distinct) are uniquely determined (cf. § 4.2), but in other cases some freedom is allowed. We can list the possibilities as in Table 5.2.

The Riemann tensor has in general 14 independent algebraic invariants (Géhéniau (1956), Géhéniau and Debever (1956b); Peres (1960a), Witten (1959)). These can be considered to be the four independent real eigenvalues of each of the Weyl and Ricci tensors, together with the six parameters specifying the Lorentz transformation between the Weyl and Ricci principal tetrads (Ehlers and Kundt (1962)).

Ref.: Petrov (1963a), Plebański and Stachel (1968), Thompson and Schrank (1969), Misra (1969), Collinson and Shaw (1972), Penrose (1972), Dozmorov (1973), Barnes (1974), Hall (1976a), Siklos (1976b), □ Crade and Hall (1979).

5.2. The energy-momentum tensor

The *Einstein field equations*

$$R_{ab} - \frac{1}{2} R g_{ab} + \Lambda g_{ab} = \varkappa_0 T_{ab} \tag{5.4}$$

($\varkappa_0$ being Einstein's gravitational constant and Λ the cosmological constant) connect the Ricci tensor (with components R_{ab}) with the energy-momentum tensor (with components T_{ab}). The Bianchi identities (2.81) imply the important relation

$$\varkappa_0 T^{ab}{}_{;b} = \left(R^{ab} - \frac{1}{2} R g^{ab}\right)_{;b} = 0. \tag{5.5}$$

In this book we try to give all known exact solutions of the *vacuum field* equations

$$R_{ab} = 0 \tag{5.6}$$

(empty spaces). Besides this case of zero T_{ab} (and Λ) we shall consider solutions of the field equations (5.4) for the following physically relevant energy-momentum tensors:

(i) *electromagnetic field* (*Maxwell field*):

$$\begin{aligned} T_{ab} &= F_{ac} F_b{}^c - \frac{1}{4} g_{ab} F_{cd} F^{cd} \\ &= \frac{1}{2} (F_{ac} F_b{}^c + \tilde{F}_{ac} \tilde{F}_b{}^c) = \frac{1}{2} F^{*c}_a \overline{F^*_{bc}}, \end{aligned} \tag{5.7}$$

$$F^*_{ab} \equiv F_{ab} + \mathrm{i}\tilde{F}_{ab}, \qquad \tilde{F}_{ab} \equiv \frac{1}{2} \varepsilon_{abcd} F^{cd}, \qquad \overset{*}{F}{}^{ab}{}_{;b} = 0.$$

(ii) *pure radiation field:*

$$T_{ab} = \Phi^2 k_a k_b, \qquad k_a k^a = 0, \tag{5.8}$$

(iii) *perfect fluid type:*

$$T_{ab} = (\mu + p)\, u_a u_b + p g_{ab}, \qquad u_a u^a = -1 \tag{5.9}$$

$(\mu + p \neq 0, \mu > 0)$.

In most cases the *cosmological constant* Λ is set equal to zero; occasionally solutions including Λ are listed. The Λ-term can be incorporated into an energy-momentum tensor (5.9) of the perfect fluid type, by substituting $(p - \Lambda/\varkappa_0)$ for p, and $(\mu + \Lambda/\varkappa_0)$ for μ; of course this may violate the condition $\mu > 0$.

In the particular case where $T_{ab} = 0$ and $\Lambda \neq 0$, or where T_{ab} is of perfect fluid type (5.9) but with $\mu + p = 0$, we shall say the Ricci tensor is of *Λ-term type*, but in view of the preceding remark such solutions are often considered as special cases of perfect fluid solutions. *Einstein spaces* are characterized by $R_{ab} = \Lambda g_{ab}$.

In general we do not consider superpositions of these energy-momentum tensors.

In virtue of the field equations (5.4), T_{ab} has the same algebraic type as R_{ab}. We shall now determine these types for the energy-momentum tensors (5.7)—(5.9).

(i) The complex self-dual *electromagnetic field* tensor F^*_{ab} can be expanded in terms of the basis $(\boldsymbol{U}, \boldsymbol{V}, \boldsymbol{W})$ (see § 3.4) as

$$\frac{1}{2} F^*_{ab} = \Phi_0 U_{ab} + \Phi_1 W_{ab} + \Phi_2 V_{ab}, \tag{5.10}$$

Φ_0, Φ_1, Φ_2 being complex functions. There is an invariant, $F^*_{ab}F^{*ab} = 16(\Phi_0\Phi_2 - \Phi_1{}^2)$. If it is non-zero, the electromagnetic field is said to be *non-null* (or non-singular), while if the invariant is zero, the electromagnetic field is said to be *null* (or singular). In either case one can, with the aid of a tetrad rotation, set $\Phi_0 = 0$.

A *non-null electromagnetic field* and the corresponding energy-momentum tensor (5.7) can be transformed into

$$F^*_{ab} = 2\Phi_1 W_{ab} = 4\Phi_1(m_{[a}\overline{m}_{b]} - k_{[a}l_{b]}) \tag{5.11}$$

$$T_{ab} = 4\Phi_1\overline{\Phi}_1(m_{(a}\overline{m}_{b)} + k_{(a}l_{b)}). \tag{5.12}$$

(The null tetrad is adapted to the two null eigendirections of the Maxwell field). In terms of an orthonormal tetrad related to the null tetrad by (3.12), the expression (5.12) can be rewritten in the canonical form (5.3a),

$$T_{ab} = \Phi^2(x_a x_b + y_a y_b - z_a z_b + u_a u_b), \qquad \Phi^2 = 2\Phi_1\overline{\Phi}_1. \tag{5.13}$$

In the principal tetrad $(\boldsymbol{x}, \boldsymbol{y}, \boldsymbol{z}, \boldsymbol{u})$ so defined, the electric and magnetic fields (E_a and B_a) are parallel to each other,

$$E_a + \mathrm{i}B_a \equiv F^*_{ab}u^b = (E + \mathrm{i}B)\, z_a = 2\Phi_1 z_a. \tag{5.14}$$

From the canonical form (5.13) one infers that for gravitational fields produced by a non-null electromagnetic field the Ricci tensor has the type [(11) (1, 1)] with $\lambda_1 = \lambda_2 = -\lambda_3 = -\lambda_4 = 2\varkappa_0\Phi_1\overline{\Phi}_1 = \frac{\varkappa_0}{2}(E^2 + B^2)$; the double roots have equal magnitude and opposite sign. The Ricci tensor obeys the relation

$$(R^b_a - \lambda g^b_a)(R^c_b + \lambda g^c_b) = 0 \tag{5.15}$$

(compare the similar equations for the **Q**-matrix in Table 4.1).

A *null electromagnetic field* and the corresponding energy-momentum tensor (5.7) can be transformed into

$$F^*_{ab} = 2\Phi_2 V_{ab} = 4\Phi_2 k_{[a} m_{b]}, \tag{5.16}$$

$$T_{ab} = \Phi^2 k_a k_b, \qquad \Phi^2 = 2\Phi_2\bar{\Phi}_2. \tag{5.17}$$

Obviously, the Ricci tensor is type [(11, 2)] with eigenvalue zero.

ii) The energy-momentum tensor (5.8) of a *pure radiation field* has the same algebraic type as (5.17). However, a pure radiation field need not obey the Maxwell equations. For any particular solution with $T_{ab} = \Phi^2 k_a k_b$ one has to check whether or not there is a null electromagnetic field satisfying the (source-free) Maxwell equations.

This energy-momentum tensor also arises from other types of directed massless radiation, for example, massless scalar fields, neutrino fields, or (high-frequency) gravitational waves. It may be considered as representing the incoherent superposition of waves with random phases and polarizations but the same propagation direction; otherwise, as with the electromagnetic null field interpretation, one would need to check that one could actually solve the equations for the underlying physical field.

(iii) The energy-momentum tensor (5.9) of *perfect fluid type* has the algebraic type [(1 11), 1]; three eigenvalues coincide. For *dust* solutions ($p = 0$) the triple eigenvalue of T_{ab} is equal to zero and the eigenvalues of R_{ab} are $\lambda_1 = \lambda_2 = \lambda_3 = -\lambda_4 = \varkappa_0\mu/2$. We intend to give all known dust solutions and to at least give references to all known perfect fluid solutions.

(iv) Energy-momentum tensors of Λ-term type are clearly of algebraic type [(1 1 1, 1)].

The energy-momentum tensors (5.7)—(5.9) (and the associated Ricci tensors) have very simple algebraic types, namely [(11) (1, 1)], [(11, 2)] and [(111), 1].

Ref.: Woolley (1973a), Goodinson (1969), Singh and Roy (1972).

5.3. The energy conditions

A physically reasonable energy-momentum tensor has to obey the *dominant energy condition:* the local energy density as measured by any observer with 4-velocity $\boldsymbol{u}$ is non-negative and the local energy flow vector $\boldsymbol{q}$ is non-spacelike,

$$T_{ab}u^a u^b \geqq 0, \tag{5.18a}$$

$$q^a q_a \leqq 0, \qquad q^a \equiv T^a_b u^b. \tag{5.18b}$$

For a discussion of energy conditions, see Hawking and Ellis (1973). The dominant energy condition (5.18) should hold for *all* timelike (unit) vectors $\boldsymbol{u}$ and, by continuity, these inequalities must still be true if we replace $\boldsymbol{u}$ by a null vector $\boldsymbol{k}$.

For type [111, 1] (and its degeneracies), T_{ab} can be diagonalized, $T_{ab} = \text{diag}\,(p_1, p_2, p_3, \mu)$, and (5.18) is then satisfied if

$$\mu \geqq 0, \qquad -\mu \leqq p_\alpha \leqq \mu \qquad (\alpha = 1, 2, 3). \tag{5.19}$$

These inequalities hold for a non-null electromagnetic field (see (5.13)) and impose reasonable restrictions on the energy density μ and pressure p ($p = p_1 = p_2 = p_3$) of a perfect fluid. The dominant energy condition (5.18) is also satisfied by the energy-momentum tensors of pure radiation fields and null electromagnetic fields.

The types $[11, Z\bar{Z}]$ and [1, 3] (and their degeneracies) in Table 5.1 violate even the weak energy condition (5.18a). Therefore these types are not considered physically significant.

5.4. The Rainich conditions

Locally, a gravitational field originates in a non-null electromagnetic field (outside matter and charges) if and only if the space-time metric and its derivatives satisfy the *Rainich conditions* (Rainich (1925), Misner and Wheeler (1957)):
algebraic part:

$$\begin{aligned} R^a_b R^b_c &= \frac{1}{4}\, \delta^a_c R_{bd} R^{bd} \neq 0, \\ R \equiv R^a_a &= 0, \qquad u^a u_a < 0 \Rightarrow R_{ab} u^a u^b > 0, \end{aligned} \tag{5.20}$$

analytic part:

$$\begin{aligned} &\alpha_{a,b} - \alpha_{b,a} = 0, \\ &\alpha_a \equiv (R_{mn} R^{mn})^{-1}\, \varepsilon_{abcd} R^b_e R^{ed;c}. \end{aligned} \tag{5.21}$$

Such a space-time V_4 is said to be a Rainich geometry. Let us give a brief summary of Rainich's "already unified theory".

The idea of the proof is to first find the so-called "*extremal*" *field* f_{ab} which satisfies

$$f_{ab}\tilde{f}^{ab} = 0, \qquad f_{ab}f^{ab} < 0, \qquad \tilde{f}_{ab} \equiv \frac{1}{2}\, \varepsilon_{abcd} f^{cd}, \tag{5.22}$$

and then obtain from f_{ab} a solution F_{ab} of the Maxwell equations with the aid of a duality rotation

$$\begin{aligned} F_{ab} &= f_{ab} \cos\alpha - \tilde{f}_{ab} \sin\alpha, \\ \tilde{F}_{ab} &= f_{ab} \sin\alpha + \tilde{f}_{ab} \cos\alpha \end{aligned} \tag{5.23}$$

at each point p of V_4.

The Einstein-Maxwell system of simultaneous equations (outside the charge and current distribution) reads

$$R_{ab} = \frac{\varkappa_0}{2}\,(F_{ac}F_b{}^c + \tilde{F}_{ac}\tilde{F}_b{}^c) \;\Leftrightarrow\; R_{ab} = \frac{\varkappa_0}{2}\,(f_{ac}f_b{}^c + \tilde{f}_{ac}\tilde{f}_b{}^c), \tag{5.24}$$

and

$$F^{ab}{}_{;b} = 0 = \tilde{F}^{ab}{}_{;b} \Leftrightarrow f^{ab}{}_{;b} - \alpha_{,b}\tilde{f}^{ab} = 0 = \tilde{f}^{ab}{}_{;b} + \alpha_{,b}f^{ab}. \tag{5.25}$$

The Einstein equations (5.24) can be rewritten in the form

$$\begin{aligned} E_{abcd} &= \frac{\varkappa_0}{2}\,(f_{ab}f_{cd} + \tilde{f}_{ab}\tilde{f}_{cd}),\\ E_{abcd} &\equiv \frac{1}{2}\,(g_{ac}R_{bd} - g_{bc}R_{ad} + g_{bd}R_{ac} - g_{ad}R_{bc}) \end{aligned} \tag{5.26}$$

and, together with (5.22), imply the formula

$$\frac{\varkappa_0}{2}\,f_{ab}f_{cd} = \frac{1}{2}\,E_{abcd} - \frac{1}{2}\,(R_{mn}R^{mn})^{-1/2}\,E_{abef}E_{cd}{}^{ef} \tag{5.27}$$

which enables one to find f_{ab} (up to an overall sign) provided that (5.20) is satisfied. The explicit determination of f_{ab} can most easily be carried out in a tetrad system in which $R_a{}^b$ has the diagonal form

$$R_a{}^b = \operatorname{diag}(\lambda, \lambda, -\lambda, -\lambda), \qquad \lambda > 0, \tag{5.28}$$

in accordance with the algebraic Rainich conditions (5.20).

As the next step one has to determine the scalar field α in the duality rotation (5.23) so that F_{ab} is a solution of the Maxwell equations (5.25), from which one obtains

$$\alpha_{,b} = 2(f_{mn}f^{mn})^{-1}\,(\tilde{f}_{ba}f^{ac}{}_{;c} + f_{ba}\tilde{f}^{ac}{}_{;c}). \tag{5.29}$$

(The identity (3.30) for arbitrary bivectors has been applied to f_{ab} and $\tilde{f}_{ab}$.) In order to express the gradient $\alpha_{,a}$ in terms of geometrical quantities, one uses the relations

$$\begin{aligned} E^{\sim}_{abcd} &= \frac{1}{2}\,\varepsilon_{cdef}(\delta^e_a R^f_b - \delta^e_b R^f_a) = \frac{\varkappa_0}{2}\,(f_{ab}\tilde{f}_{cd} - \tilde{f}_{ab}f_{cd}),\\ E^{abcd}{}_{;d} &= \frac{1}{2}\,(R^{ac;b} - R^{bc;a}) = \frac{\varkappa_0}{2}\,(f^{ab}f^{cd} + \tilde{f}^{ab}\tilde{f}^{cd})_{;d} \end{aligned} \tag{5.30}$$

which follow from (5.26). The resulting formula

$$\alpha_{,a} = -2\left(\frac{\varkappa_0}{2}\,f_{mn}f^{mn}\right)^{-2} E^{\sim}_{abcd}E^{cdbf}{}_{;f} = (R_{mn}R^{mn})^{-1}\,\varepsilon_{abcd}R^b_e R^{ed;c} = \alpha_a \tag{5.31}$$

enables one to find the "*complexion*" α (up to an additive constant) provided that (5.21) is satisfied. Hence, a Rainich geometry determines the associated electromagnetic field F_{ab} uniquely up to a *constant* duality rotation.

The problem of determining a *null* electromagnetic field from the geometry has not yet been completely solved (Jordan and Kundt (1961), Ludwig (1970)).

Ref.: Goodinson and Newing (1968).

5.5. Perfect fluids

In order to describe a perfect fluid completely, the energy-momentum tensor (5.9) of perfect fluid type has to be supplemented by an equation of state expressing, say, the density ϱ of rest mass as a function of the energy density μ and the pressure p (taking a picture of a fluid composed of conserved microscopic particles). Conservation of the number of particles can be formulated as

$$(\varrho u^a)_{;a} = 0. \tag{5.32}$$

The conservation law $T^{ab}{}_{;b} = 0$ and the thermodynamic relation

$$\mathrm{d}h = \frac{1}{\varrho}\,\mathrm{d}p + T\,\mathrm{d}s, \qquad h \equiv (\mu + p)/\varrho, \tag{5.33}$$

where h and s denote respectively the specific enthalpy and entropy, lead to the equation

$$\Omega_{ab}u^b = -Ts_{,a}, \qquad \Omega_{ab} \equiv (hu_a)_{;b} - (hu_b)_{;a}. \tag{5.34}$$

From (5.34) we obtain $s_{,a}u^a = 0$ (isentropic motion).

For constant specific entropy s, the relation (5.33) reads

$$\mathrm{d}\mu = (\mu + p)\,\mathrm{d}\varrho/\varrho, \tag{5.35}$$

that is, both μ and p depend only on ϱ, or, p is a function of $\mu: p = p(\mu)$.

In surveying perfect fluid solutions with special properties (symmetries etc.) we found that two approaches had been used: either a relation $p = p(\mu)$ was prescribed, or, more frequently, μ and p were evaluated from the field equations, there being, in general, no relation $p = p(\mu)$ between them. We will not investigate the question of whether or not solutions of the latter type allow a thermodynamic interpretation in accordance with (5.32); such an interpretation must, of course, be established before one can really treat the resulting solutions as having physical significance. On the other hand one can construct physical situations, such as the presence of viscous fluids, in which it becomes imperative to consider more general Ricci tensor types than those treated in this book.

In those cases where an equation of state of a perfect fluid is prescribed before the field equations are solved, it has frequently taken the simple form

$$p = (\gamma - 1)\,\mu, \tag{5.36}$$

where γ is a constant. Cases regarded as of particular interest are the "*dust*" case,

$$p = 0, \qquad \gamma = 1, \tag{5.37}$$

the "(*incoherent*) *radiation*" case

$$p = \mu/3, \quad \gamma = 4/3, \tag{5.38}$$

so-called because it represents the superposition of waves of a massless field (e.g. the electromagnetic field) with random propagation direction, and the "*stiff matter*" case

$$p = \mu, \quad \gamma = 2. \tag{5.39}$$

Because (5.39) leads to a sound speed equal to the velocity of light, the characteristics of its governing equations are the same as those of the gravitational field, and consequently such solutions can often be derived from vacuum solutions (see e.g. Wainwright et al. (1979)).

A class of isentropic twisting perfect fluid solutions has been constructed by starting with equation (5.34) and using the functions τ, ξ, η in the expansion $\sigma_a \equiv hu_a = \tau_{,a} + \eta\xi_{,a}$ (cf. Theorem 2.3) of a 1-form σ as local coordinates (Krasiński (1974), § 20.2.).

Chapter 6. Vector fields

6.1. Vector fields and their invariant classification

Vector fields in four-dimensional Riemannian spaces are frequently characterized by the properties of their first covariant derivatives and the invariants which can be built from these derivatives. The methods being standard, we give only the definitions and some simple applications for further reference. The physical meaning and the interpretation of the invariants in question can be found in the literature (Ehlers (1961), Greenberg (1970), Jordan et al. (1961)).

A vector field $\boldsymbol{v}(x^i)$ is said to be *hypersurface-orthogonal* or non-rotating or normal, if it is proportional to a gradient,

$$v_a = \lambda f_{,a}, \tag{6.1}$$

i.e. if and only if the rotation ω^a,

$$\omega^a \equiv \varepsilon^{abcd} v_{b;c} v_d, \tag{6.2}$$

vanishes.

A vector field $\boldsymbol{v}(x^i)$ is said to be *geodesic* if it is proportional to the tangent vector $\boldsymbol{t}$ of a geodesic (see (2.76)),

$$v_a = \lambda t_a, \qquad \mathrm{D}t_a/\mathrm{d}\mu = 0, \tag{6.3}$$

μ being an affine parameter. Equation (6.3) is equivalent to

$$v_{[a} v_{b];c} v^c = 0. \tag{6.4}$$

A vector field $\boldsymbol{v}$ is said to be *recurrent* (or parallel), if its covariant derivative is proportional to itself,

$$v_{a;b} = v_a K_b, \tag{6.5}$$

$\boldsymbol{K} = K^i\, \partial/\partial x^i$ being the recurrence vector. As can easily be checked, a recurrent vector is geodesic and non-rotating and its components can therefore be written as $v_a = \lambda f_{,a}$. Together with (6.5) this gives

$$f_{,a;b} = f_{,a}(K_b - \lambda_{,b}/\lambda). \tag{6.6}$$

If $\boldsymbol{v}$ is non-null ($f^{,a} f_{,a} \neq 0$), then (6.6) implies

$$K_a = \frac{\lambda_{,a}}{\lambda} + \frac{f^{,b} f_{,b;a}}{f^{,c} f_{,c}} = \frac{1}{2}\,(\ln \lambda^2 f^{,b} f_{,b})_{,a}. \tag{6.7}$$

$\boldsymbol{K}$ is a gradient, and a function $\alpha = (\lambda^2 f^{,b} f_{,b})^{-1/2}$ can be found such that

$$\alpha v_a = w_a, \qquad w_{a;b} = 0. \tag{6.8}$$

Consequently, a non-null recurrent vector field is proportional to a (covariantly) constant vector field.

If $\boldsymbol{v}$ is null, we can infer from (6.6) that the null vector $\boldsymbol{k}$ ($f_{,a} \equiv k_a$) obeys

$$k_{a;b} = \alpha k_a k_b, \tag{6.9}$$

because $f_{,a;b}$ is symmetric in a and b, and the right-hand side must be symmetric too. This null vector is porportional to a constant vector only if $\alpha_{,[a}k_{b]} = 0$. All space-times admitting a recurrent null vector were discussed by Debever and Cahen (1961), and Öktem (1976).

A *Killing vector field* $\boldsymbol{\xi}$ satisfies Killing's equation

$$\xi_{a;b} + \xi_{b;a} = 0. \tag{6.10}$$

Because of its considerable importance, we devote a separate chapter (Chapter 8.) to the equation of Killing and groups of motions.

Two vector fields $\boldsymbol{v}, \boldsymbol{w}$ are said to be *surface-forming* if the Lie derivative of one vector with respect to the other lies in the plane defined by $\boldsymbol{v}$ and $\boldsymbol{w}$,

$$v^a w^b{}_{;a} - w^a v^b{}_{;a} = (\pounds_{\boldsymbol{v}} \boldsymbol{w})^b = -(\pounds_{\boldsymbol{w}} \boldsymbol{v})^b = \lambda v^b + \mu w^b \tag{6.11}$$

or, equivalently,

$$\varepsilon_{abcd} v^b w^c (\pounds_{\boldsymbol{v}} \boldsymbol{w})^d = 0. \tag{6.12}$$

Time-like unit vector fields

The covariant derivative of a timelike vector field $\boldsymbol{u}(x^i)$, $u^a u_a = -1$, can be decomposed as follows:

$$\begin{aligned}
u_{a;b} &= -\dot{u}_a u_b + \omega_{ab} + \sigma_{ab} + \Theta h_{ab}/3, \\
\dot{u}_a &\equiv u_{a;b} u^b = \mathrm{D}u_a/\mathrm{d}\tau, \qquad \dot{u}_a u^a = 0, \\
\omega_{ab} &\equiv u_{[a;b]} + \dot{u}_{[a} u_{b]}, \qquad \omega_{ab} u^b = 0, \\
\sigma_{ab} &\equiv u_{(a;b)} + \dot{u}_{(a} u_{b)} - \Theta h_{ab}/3, \qquad \sigma_{ab} u^b = 0, \\
h_{ab} &\equiv g_{ab} + u_a u_b, \qquad h_{ab} u^b = 0, \\
\Theta &\equiv u^a{}_{;a}.
\end{aligned} \tag{6.13}$$

Physically, the timelike vector field $\boldsymbol{u}$ is often taken to be the 4-velocity of a fluid, and the quantities $\dot{u}_a$, Θ, ω_{ab} and σ_{ab} are accordingly called acceleration, expansion, rotation, and shear, respectively. (The analogous decomposition for a spacelike vector field is discussed by Greenberg (1970).)

When we study the conformal properties of space-time, it is interesting to know how these invariants change under a conformal transformation $\hat{g}_{ab} = \mathrm{e}^{2U} g_{ab}$, cp.

(3.81). The world lines $x^i(\tau)$ being the same, the 4-velocities are related to each other by $\hat{u}^a = \mathrm{e}^{-U}u^a$, $\hat{u}_b = \mathrm{e}^U u_b$, and we obtain

$$\begin{aligned} \hat{\sigma}_{ab} &= \mathrm{e}^U \sigma_{ab}, \qquad \hat{\Theta} = \mathrm{e}^{-U}\Theta - 3u^a(\mathrm{e}^{-U})_{,a}, \\ \hat{\omega}_{ab} &= \mathrm{e}^U \omega_{ab}, \qquad \hat{\dot{u}}_a = e^U(\dot{u}_a + u_a u_b U^{,b} + U_{,a}) \end{aligned} \tag{6.14}$$

for the invariant parts of their derivatives.

Geodesic null vector fields

To get the decomposition of the covariant derivative of a geodesic null vector field $\boldsymbol{k}(x^i)$ fulfilling

$$k_{a;b}k^b = 0, \tag{6.15}$$

one usually introduces the complex null tetrad $(\boldsymbol{m}, \overline{\boldsymbol{m}}, \boldsymbol{l}, \boldsymbol{k})$ defined in (3.8). The result is

$$\begin{aligned} &k_{a;b} = 2\,\mathrm{Re}\,[(\Theta + \mathrm{i}\omega)\,\overline{m}_a m_b - \sigma \overline{m}_a \overline{m}_b] + v_a k_b + k_a w_b, \\ &v^a k_a = 0 = w^a k_a, \\ &\Theta = \frac{1}{2}\,k^a{}_{;a}, \qquad \omega^2 = \frac{1}{2}\,k_{[a;b]}k^{a;b}, \\ &\sigma\overline{\sigma} = \frac{1}{2}\,k_{(a;b)}k^{a;b} - \frac{1}{4}\,(k^a{}_{;a})^2. \end{aligned} \tag{6.16}$$

$\boldsymbol{k}$ is given, but the vectors $\boldsymbol{m}$, $\overline{\boldsymbol{m}}$, $\boldsymbol{l}$ are not uniquely defined. Nevertheless, the quantities Θ, ω^2 and $\sigma\overline{\sigma}$ are invariants and independent of the choice of the null tetrad.

As a consequence of (3.35) and of (6.16), ω may also be defined by

$$\frac{1}{2}\,\varepsilon^{abcd}k_b k_{c;d} = \omega k^a. \tag{6.17}$$

Because of (6.16), $k_{a;b}$ obeys $\varepsilon^{abcd}k_{a;b}k_{c;d} = 0$ (the rank of the matrix $k_{a;b}$ is maximally two), this and the identity $\varepsilon^{abcd}k_{c;da} = 0$ allow one to conclude from (6.16) that

$$(\omega k^a)_{;a} = \omega_{,a}k^a + 2\Theta\omega = 0. \tag{6.18}$$

Physically, geodesic null vector fields can be interpreted as the tangent vectors of optical rays. Accordingly, the quantities Θ, ω and σ are called *expansion* (or divergence), *twist* (or rotation), and *shear*, respectively.

The decomposition (6.16) implies

$$\begin{aligned} &\varrho \equiv -(\Theta + \mathrm{i}\omega) = -k_{a;b}m^a\overline{m}^b, \\ &\sigma \equiv -k_{a;b}m^a m^b, \qquad \varkappa \equiv -k_{a;b}m^a k^b = 0. \end{aligned} \tag{6.19}$$

The expressions ϱ, σ and $\varkappa$ are three of the twelve (complex) Newman-Penrose spin coefficients, see § 7.1. Note that we have changed the sign of σ in (6.16) and (6.19) from that of Ehlers and Kundt (1962) to achieve conformity with the usual definition (§ 7.1.) of the spin coefficients.

The null congruence $\boldsymbol{k}$ is hypersurface-orthogonal if and only if ω vanishes. If an arbitrary null congruences is normal, it is also geodesic.

If we carry out a conformal transformation $\hat{g}_{ab} = \mathrm{e}^{2U} g_{ab}$, see (3.81), but retain the condition $k_{a;b}k^b = 0$ for the transformed vector, then the null vector field has to be transformed by $\hat{k}_a = k_a$; this implies

$$\hat{k}_{a;b} = k_{a;b} - k_a U_b - k_b U_{,a} + g_{ab} k^c U_{,c} \tag{6.20}$$

and

$$\hat{\omega} = \mathrm{e}^{-2U} \omega, \qquad \hat{\Theta} = \mathrm{e}^{-2U}(\Theta + k^a U_{,a}). \tag{6.21}$$

The importance of the invariant classification of vector fields in the context of exact solutions is twofold. If a solution defines a preferred vector field (velocity field, eigenvector field of the Weyl tensor, ...), a classification of the vector field is also a classification of the solution in question. On the other hand, the existence of a vector field with some special properties (shearfree, rotationfree, ...) imposes conditions on the metric via

$$2a_{a;[b,c]} = a_d R^d{}_{abc}. \tag{6.22}$$

We will discuss this in detail in the following section.

6.2. Vector fields and the curvature tensor

6.2.1. Time-like unit vector fields

The main idea of this section is to evaluate the Ricci identity (6.22),

$$u_{a;bc} - u_{a;cb} = R^d{}_{abc} u_d, \tag{6.23}$$

using the decomposition (6.13),

$$u_{a;b} = -\dot{u}_a u_b + \omega_{ab} + \sigma_{ab} + \Theta h_{ab}/3, \tag{6.24}$$

and see what limitations on the metric (curvature tensor) follow from the existence of a vector field $\boldsymbol{u}$ with special derivative properties.

Writing down (6.23) in detail, one easily gets

$$\begin{aligned} \frac{1}{2} R^d{}_{abc} u_d = {} & -\dot{u}_a(\omega_{bc} - \dot{u}_{[b} u_{c]}) - \dot{u}_{a;[c} u_{b]} \\ & + \omega_{a[b;c]} + \sigma_{a[b;c]} + \frac{1}{3} \Theta_{,[c} h_{b]a} \\ & + \frac{\Theta}{3} \left[u_a \omega_{bc} - u_a \dot{u}_{[b} u_{c]} + \omega_{a[c} u_{b]} + \sigma_{a[c} u_{b]} + \frac{\Theta}{3} h_{a[c} u_{b]} \right], \end{aligned} \tag{6.25}$$

and from this, by contraction and/or multiplication with u^b,

$$R^a{}_b u_a = \dot{u}^a \omega_{ab} - \dot{u}^a{}_{;a} u_b - \dot{u}^a \sigma_{ab} + \omega^a{}_{b;a}$$

$$+ \sigma^a{}_{b;a} - \frac{2}{3}\,\Theta_{,b} + \frac{\dot{\Theta}}{3}\,u_b + \frac{\Theta^2}{3}\,u_b\,, \tag{6.26}$$

$$R^d{}_{abc} u_d u^b = \dot{u}_a \dot{u}_c - \omega_{ab}\omega^b{}_c - \left(\frac{\dot{\Theta}}{3} + \frac{\Theta^2}{9}\right) h_{ca}$$

$$- \sigma_{ab}\sigma^b{}_c - \frac{2}{3}\,\Theta\sigma_{ac} + h^d_a h^e_c (\dot{u}_{(d;e)} - \dot{\sigma}_{de})\,, \tag{6.27}$$

$$R^a{}_b u_a u^b = \dot{u}^a{}_{;a} + \omega_{ab}\omega^{ab} - \sigma_{ab}\sigma^{ab} - \dot{\Theta} - \Theta^2/3\,. \tag{6.28}$$

Equation (6.28) is often called the Raychaudhuri equation (Raychaudhuri (1955)).

Formulae (6.26)–(6.28) may be considered as equations governing the temporal variation of u^a, σ_{ab} and Θ. In the context of exact solutions, however, we prefer to interpret them as equations determining some components of the curvature tensor if the properties of the vector field are prescribed or known. The implications of (6.25)–(6.28) for the self-dual Weyl tensor C^*_{abcd} can be calculated using the definition (3.50) and the previous equations. The result is

$$C^*{}_{abcd} u^a u^c = \dot{u}_b \dot{u}_d - \omega_{bc}\omega^c{}_d - \sigma_{bc}\sigma^c{}_d - \frac{2}{3}\,\Theta\sigma_{bd} + h^e_b h^f_d [\dot{u}_{(e;f)} - \dot{\sigma}_{ef}]$$

$$- \frac{1}{3}\,h_{bd}(\dot{u}^a{}_{;a} + \omega_{ae}\omega^{ae} - \sigma_{ae}\sigma^{ae}) + \frac{1}{2}\,h^e_b h^f_d R_{ef}$$

$$- \frac{1}{6}\,h_{bd} h^{ef} R_{ef} - \mathrm{i} h^g_{(b}\varepsilon_{d)c}{}^{ef} u^c [-\dot{u}_g \omega_{ef} + \omega_{ge;f} + \sigma_{ge;f}]\,. \tag{6.29}$$

The most remarkable property of this formula is the simplicity of the imaginary part of the tensor $Q_{ab} \equiv -C^*_{cadb} u^c u^d$. As shown in Chapter 4., the Petrov classification is the classification of this tensor Q_{ab}, and if Q_{ab} is real, then space-time is of Petrov type I, D or O.

Thus we can formulate the following

Theorem 6.1. *There is a timelike unit vector field* $\boldsymbol{u}$ *satisfying*

$$h^g_{(b}\varepsilon_{d)c}{}^{ef} u^c [\omega_{ge;f} + \sigma_{ge;f} - \dot{u}_g \omega_{ef}] = 0\,, \tag{6.30}$$

if and only if the space-time is of Petrov type I, D *or* O.

Simple examples of vector fields satisfying (6.30) are those which are hypersurface-orthogonal and shearfree ($\omega_{ab} = 0 = \sigma_{ab}$). For *static* metrics, characterized by the existence of a timelike non-rotating Killing vector ξ, the vector field $\boldsymbol{u} = \xi/\sqrt{-\xi_a \xi^a}$ obeys $u_{a;b} = -\dot{u}_a u_b$ because of the Killing equation (6.10). Consequently, all static metrics are of type I, D or O.

Rigid motions of a (test) body correspond to vector fields $\boldsymbol{u}$ satisfying $\Theta = 0 = \sigma_{ab}$, i.e. $\pounds_{\boldsymbol{u}} h_{ab} = 0$. Einstein spaces admitting rigid motions are either flat, or of constant curvature, or they are degenerate static metrics of class B (Wahlquist and Estabrook (1966)).

The energy-momentum tensor being specified, the equations (6.25)—(6.30) together with the Bianchi identities can be used to derive theorems on exact solutions.

6.2.2. Null vector fields

One can perform a detailed evaluation of

$$k_{a;bc} - k_{a;cb} = R^d{}_{abc} k_d, \tag{6.31}$$

similar to the one which was carried out for timelike vector fields in the preceding section. This is best done by means of the Newman-Penrose formalism, which we will introduce below. We only mention here one conclusion which can be drawn *without* specifying the tetrad ($\boldsymbol{m}, \overline{\boldsymbol{m}}, \boldsymbol{k}, \boldsymbol{l}$): from (6.31) and (6.16) we get (for geodesic null vector fields)

$$\Theta_{,a} k^a - \omega^2 + \Theta^2 + \sigma\bar{\sigma} = -\frac{1}{2} R_{ab} k^a k^b, \tag{6.32}$$

and, using (6.18), we can write this in the form

$$(\Theta + \mathrm{i}\omega)_{,a} k^a + (\Theta + \mathrm{i}\omega)^2 + \sigma\bar{\sigma} = -\frac{1}{2} R_{ab} k^a k^b. \tag{6.33}$$

Ref.: Debever and Cahen (1961), Trümper (1965), Takeno and Kitamura (1968), Glass (1975), Neugebauer and Sust (1975).

Chapter 7. The Newman-Penrose formalism

7.1. The spin coefficients and the field equations

The null tetrad formalism due to Newman and Penrose (1962), has proved very useful in the construction of exact solutions, and for other investigations. In particular, this formalism is convenient for studying algebraically special gravitational fields (e.g. Kinnersley (1969b), Lind (1974), Talbot (1969)). Simple applications, useful for beginners, were given by Davis (1976). Campbell and Wainwright (1977) wrote computer programs using the Newman-Penrose formalism as an efficient way of calculating the curvature tensor.

We give here an outline of this important approach to General Relativity.

Using the complex null tetrad $\{\boldsymbol{e}_a\} = (\boldsymbol{m}, \overline{\boldsymbol{m}}, \boldsymbol{l}, \boldsymbol{k})$, and recalling the definition (2.65),

$$\nabla_b \boldsymbol{e}_a = \Gamma^c{}_{ab} \boldsymbol{e}_c, \tag{7.1}$$

of the connection coefficients $\Gamma^c{}_{ab}$, we can define the so-called *spin coefficients*, 12 independent complex linear combinations of the connection coefficients. Explicitly, the spin coefficients are defined in tensor and spinor notation as follows

$$-\varkappa \equiv \Gamma_{144} = k_{a;b} m^a k^b = o^A \bar{o}^{\dot{B}} o^C \nabla_{A\dot{B}} o_C, \tag{7.2}$$

$$-\varrho \equiv \Gamma_{142} = k_{a;b} m^a \overline{m}^b = \iota^A \bar{o}^{\dot{B}} o^C \nabla_{A\dot{B}} o_C, \tag{7.3}$$

$$-\sigma \equiv \Gamma_{141} = k_{a;b} m^a m^b = o^A \bar{\iota}^{\dot{B}} o^C \nabla_{A\dot{B}} o_C, \tag{7.4}$$

$$-\tau \equiv \Gamma_{143} = k_{a;b} m^a l^b = \iota^A \bar{\iota}^{\dot{B}} o^C \nabla_{A\dot{B}} o_C, \tag{7.5}$$

$$\nu \equiv \Gamma_{233} = l_{a;b} \overline{m}^a l^b = -\iota^A \bar{\iota}^{\dot{B}} \iota^C \nabla_{A\dot{B}} \iota_C, \tag{7.6}$$

$$\mu \equiv \Gamma_{231} = l_{a;b} \overline{m}^a m^b = -o^A \bar{\iota}^{\dot{B}} \iota^C \nabla_{A\dot{B}} \iota_C, \tag{7.7}$$

$$\lambda \equiv \Gamma_{232} = l_{a;b} \overline{m}^a \overline{m}^b = -\iota^A \bar{o}^{\dot{B}} \iota^C \nabla_{A\dot{B}} \iota_C, \tag{7.8}$$

$$\pi \equiv \Gamma_{234} = l_{a;b} \overline{m}^a k^b = -o^A \bar{o}^{\dot{B}} \iota^C \nabla_{A\dot{B}} \iota_C, \tag{7.9}$$

$$-\varepsilon \equiv \frac{1}{2}(\Gamma_{344} - \Gamma_{214}) = \frac{1}{2}(k_{a;b} l^a k^b - m_{a;b} \overline{m}^a k^b) = o^A \bar{o}^{\dot{B}} \iota^C \nabla_{A\dot{B}} o_C, \tag{7.10}$$

$$-\beta \equiv \frac{1}{2}(\Gamma_{341} - \Gamma_{211}) = \frac{1}{2}(k_{a;b} l^a m^b - m_{a;b} \overline{m}^a m^b) = o^A \bar{\iota}^{\dot{B}} \iota^C \nabla_{A\dot{B}} o_C, \tag{7.11}$$

$$\gamma \equiv \frac{1}{2}(\Gamma_{433} - \Gamma_{123}) = \frac{1}{2}(l_{a;b} k^a l^b - \overline{m}_{a;b} m^a l^b) = -\iota^A \bar{\iota}^{\dot{B}} o^C \nabla_{A\dot{B}} \iota_C, \tag{7.12}$$

$$\alpha \equiv \frac{1}{2}(\Gamma_{432} - \Gamma_{122}) = \frac{1}{2}(l_{a;b} k^a \overline{m}^b - \overline{m}_{a;b} m^a \overline{m}^b) = -\iota^A \bar{o}^{\dot{B}} o^C \nabla_{A\dot{B}} \iota_C. \tag{7.13}$$

Some of these spin coefficients have already been introduced in (6.19). From the spinor expressions for the spin coefficients and from the relation

$$o_A\iota^A = 1 \Rightarrow \iota^C \nabla_{A\dot{B}} o_C = o^C \nabla_{A\dot{B}} \iota_C \tag{7.14}$$

it follows that all connection coefficients can be expressed in terms of the 12 complex spin coefficients (7.2)—(7.13). It is very convenient to work with the connection forms

$$\begin{aligned}
\Gamma_{14} &\equiv m^a k^b \Gamma_{abc}\omega^c = -\sigma\omega^1 - \varrho\omega^2 - \tau\omega^3 - \varkappa\omega^4, \\
\Gamma_{23} &\equiv \overline{m}^a l^b \Gamma_{abc}\omega^c = \mu\omega^1 + \lambda\omega^2 + \nu\omega^3 + \pi\omega^4, \\
\frac{1}{2}(\Gamma_{12} + \Gamma_{34}) &\equiv \frac{1}{2}(m^a\overline{m}^b + l^a k^b)\,\Gamma_{abc}\omega^c = -\beta\omega^1 - \alpha\omega^2 - \gamma\omega^3 - \varepsilon\omega^4.
\end{aligned} \tag{7.15}$$

Written in these combinations, the second Cartan equation has the special form (3.23). On the right-hand side of the equations (3.23) we insert the decomposition (3.44) of the curvature tensor, using the following abbreviations for the tetrad components of the traceless Ricci tensor ($S_{ab} \equiv R_{ab} - g_{ab}R/4$) and the Weyl tensor:

$$\Phi_{00} \equiv \frac{1}{2} S_{ab}k^a k^b = \Phi_{AB\dot{C}\dot{D}} o^A o^B \bar{o}^{\dot{C}} \bar{o}^{\dot{D}} = \overline{\Phi}_{00} = \frac{1}{2} R_{44}, \tag{7.16}$$

$$\Phi_{01} \equiv \frac{1}{2} S_{ab}k^a m^b = \Phi_{AB\dot{C}\dot{D}} o^A o^B \bar{o}^{\dot{C}} \bar{\iota}^{\dot{D}} = \overline{\Phi}_{10} = \frac{1}{2} R_{41}, \tag{7.17}$$

$$\Phi_{02} \equiv \frac{1}{2} S_{ab}m^a m^b = \Phi_{AB\dot{C}\dot{D}} o^A o^B \bar{\iota}^{\dot{C}} \bar{\iota}^{\dot{D}} = \overline{\Phi}_{20} = \frac{1}{2} R_{11}, \tag{7.18}$$

$$\Phi_{11} \equiv \frac{1}{4} S_{ab}(k^a l^b + m^a \overline{m}^b) = \Phi_{AB\dot{C}\dot{D}} o^A \iota^B \bar{o}^{\dot{C}} \bar{\iota}^{\dot{D}} = \overline{\Phi}_{11} = (R_{43} + R_{12})/4, \tag{7.19}$$

$$\Phi_{12} \equiv \frac{1}{2} S_{ab}l^a m^b = \Phi_{AB\dot{C}\dot{D}} o^A \iota^B \bar{\iota}^{\dot{C}} \bar{\iota}^{\dot{D}} = \overline{\Phi}_{21} = \frac{1}{2} R_{31}, \tag{7.20}$$

$$\Phi_{22} \equiv \frac{1}{2} S_{ab}l^a l^b = \Phi_{AB\dot{C}\dot{D}} \iota^A \iota^B \bar{\iota}^{\dot{C}} \bar{\iota}^{\dot{D}} = \overline{\Phi}_{22} = \frac{1}{2} R_{33}; \tag{7.21}$$

$$\Psi_0 \equiv C_{abcd}k^a m^b k^c m^d = \Psi_{ABCD} o^A o^B o^C o^D, \tag{7.22}$$

$$\Psi_1 \equiv C_{abcd}k^a l^b k^c m^d = \Psi_{ABCD} o^A o^B o^C \iota^D, \tag{7.23}$$

$$\Psi_2 \equiv \frac{1}{2} C_{abcd}k^a l^b (k^c l^d - m^c \overline{m}^d) = \Psi_{ABCD} o^A o^B \iota^C \iota^D, \tag{7.24}$$

$$\Psi_3 \equiv C_{abcd}l^a k^b l^c \overline{m}^d = \Psi_{ABCD} o^A \iota^B \iota^C \iota^D, \tag{7.25}$$

$$\Psi_4 \equiv C_{abcd}l^a \overline{m}^b l^c \overline{m}^d = \Psi_{ABCD} \iota^A \iota^B \iota^C \iota^D. \tag{7.26}$$

The definitions (7.22)—(7.26) agree with (3.59), (3.75).

On the left-hand side of the equations (3.23) we calculate the exterior derivatives $\mathrm{d}\Gamma_{ab}$ and use the notation (3.80),

$$\begin{aligned} D &\equiv k^a\nabla_a = -o^A\bar{o}^{\dot{B}}\nabla_{A\dot{B}}, & \Delta &\equiv l^a\nabla_a = -\iota^A\bar{\iota}^{\dot{B}}\nabla_{A\dot{B}}, \\ \delta &\equiv m^a\nabla_a = -o^A\bar{\iota}^{\dot{B}}\nabla_{A\dot{B}}, & \bar{\delta} &\equiv \bar{m}^a\nabla_a = -\iota^A\bar{o}^{\dot{B}}\nabla_{A\dot{B}}, \end{aligned} \tag{7.27}$$

for the directional derivatives. We arrive at the *Newman-Penrose equations:*

$$D\varrho - \bar{\delta}\varkappa = (\varrho^2 + \sigma\bar{\sigma}) + (\varepsilon + \bar{\varepsilon})\,\varrho - \bar{\varkappa}\tau - \varkappa(3\alpha + \bar{\beta} - \pi) + \Phi_{00}, \tag{7.28}$$

$$D\sigma - \delta\varkappa = (\varrho + \bar{\varrho})\,\sigma + (3\varepsilon - \bar{\varepsilon})\,\sigma - (\tau - \bar{\pi} + \bar{\alpha} + 3\beta)\,\varkappa + \Psi_0, \tag{7.29}$$

$$D\tau - \Delta\varkappa = (\tau + \bar{\pi})\,\varrho + (\bar{\tau} + \pi)\,\sigma + (\varepsilon - \bar{\varepsilon})\,\tau - (3\gamma + \bar{\gamma})\,\varkappa + \Psi_1 + \Phi_{01}, \tag{7.30}$$

$$D\alpha - \bar{\delta}\varepsilon = (\varrho + \bar{\varepsilon} - 2\varepsilon)\,\alpha + \beta\bar{\sigma} - \bar{\beta}\varepsilon - \varkappa\lambda - \bar{\varkappa}\gamma + (\varepsilon + \varrho)\,\pi + \Phi_{10}, \tag{7.31}$$

$$D\beta - \delta\varepsilon = (\alpha + \pi)\,\sigma + (\bar{\varrho} - \bar{\varepsilon})\,\beta - (\mu + \gamma)\,\varkappa - (\bar{\alpha} - \bar{\pi})\,\varepsilon + \Psi_1, \tag{7.32}$$

$$\begin{aligned} D\gamma - \Delta\varepsilon &= (\tau + \bar{\pi})\,\alpha + (\bar{\tau} + \pi)\,\beta - (\varepsilon + \bar{\varepsilon})\,\gamma - (\gamma + \bar{\gamma})\,\varepsilon \\ &\quad + \tau\pi - \nu\varkappa + \Psi_2 + \Phi_{11} - R/24, \end{aligned} \tag{7.33}$$

$$D\lambda - \bar{\delta}\pi = (\varrho\lambda + \bar{\sigma}\mu) + \pi^2 + (\alpha - \bar{\beta})\,\pi - \nu\bar{\varkappa} - (3\varepsilon - \bar{\varepsilon})\,\lambda + \Phi_{20}, \tag{7.34}$$

$$D\mu - \delta\pi = (\bar{\varrho}\mu + \sigma\lambda) + \pi\bar{\pi} - (\varepsilon + \bar{\varepsilon})\,\mu - \pi(\bar{\alpha} - \beta) - \nu\varkappa + \Psi_2 + R/12, \tag{7.35}$$

$$D\nu - \Delta\pi = (\pi + \bar{\tau})\,\mu + (\bar{\pi} + \tau)\,\lambda + (\gamma - \bar{\gamma})\,\pi - (3\varepsilon + \bar{\varepsilon})\,\nu + \Psi_3 + \Phi_{21}, \tag{7.36}$$

$$\Delta\lambda - \bar{\delta}\nu = -(\mu + \bar{\mu})\,\lambda - (3\gamma - \bar{\gamma})\,\lambda + (3\alpha + \bar{\beta} + \pi - \bar{\tau})\,\nu - \Psi_4, \tag{7.37}$$

$$\delta\varrho - \bar{\delta}\sigma = \varrho(\bar{\alpha} + \beta) - \sigma(3\alpha - \bar{\beta}) + (\varrho - \bar{\varrho})\,\tau + (\mu - \bar{\mu})\,\varkappa - \Psi_1 + \Phi_{01}, \tag{7.38}$$

$$\begin{aligned} \delta\alpha - \bar{\delta}\beta &= (\mu\varrho - \lambda\sigma) + \alpha\bar{\alpha} + \beta\bar{\beta} - 2\alpha\beta + \gamma(\varrho - \bar{\varrho}) + \varepsilon(\mu - \bar{\mu}) \\ &\quad - \Psi_2 + \Phi_{11} + R/24, \end{aligned} \tag{7.39}$$

$$\delta\lambda - \bar{\delta}\mu = (\varrho - \bar{\varrho})\,\nu + (\mu - \bar{\mu})\,\pi + \mu(\alpha + \bar{\beta}) + \lambda(\bar{\alpha} - 3\beta) - \Psi_3 + \Phi_{21}, \tag{7.40}$$

$$\delta\nu - \Delta\mu = (\mu^2 + \lambda\bar{\lambda}) + (\gamma + \bar{\gamma})\,\mu - \bar{\nu}\pi + (\tau - 3\beta - \bar{\alpha})\,\nu + \Phi_{22}, \tag{7.41}$$

$$\delta\gamma - \Delta\beta = (\tau - \bar{\alpha} - \beta)\,\gamma + \mu\tau - \sigma\nu - \varepsilon\bar{\nu} - \beta(\gamma - \bar{\gamma} - \mu) + \alpha\bar{\lambda} + \Phi_{12}, \tag{7.42}$$

$$\delta\tau - \Delta\sigma = (\mu\sigma + \bar{\lambda}\varrho) + (\tau + \beta - \bar{\alpha})\,\tau - (3\gamma - \bar{\gamma})\,\sigma - \varkappa\bar{\nu} + \Phi_{02}, \tag{7.43}$$

$$\Delta\varrho - \bar{\delta}\tau = -(\varrho\bar{\mu} + \sigma\lambda) + (\bar{\beta} - \alpha - \bar{\tau})\,\tau + (\gamma + \bar{\gamma})\,\varrho + \nu\varkappa - \Psi_2 - R/12, \tag{7.44}$$

$$\Delta\alpha - \bar{\delta}\gamma = (\varrho + \varepsilon)\,\nu - (\tau + \beta)\,\lambda + (\bar{\gamma} - \bar{\mu})\,\alpha + (\bar{\beta} - \bar{\tau})\,\gamma - \Psi_3. \tag{7.45}$$

The set of equations (7.28)—(7.45) is exactly equivalent to the formula (2.79) for the components of the Riemann curvature tensor, if the basis is the complex null tetrad $(\boldsymbol{m}, \bar{\boldsymbol{m}}, \boldsymbol{l}, \boldsymbol{k})$; the Newman-Penrose equations are appropriate linear combinations of equations (2.79) and all summations are written out explicitly. This is convenient for the practical applications of the formalism.

The definitions of $\{\boldsymbol{e}_a\}$, ∇_a, $\Gamma^a{}_{bc}$, and $R^a{}_{bcd}$ coincide with those of the original paper (Newman and Penrose (1962)). Whenever the metric is employed to move indices we have to remember the change of signature (for sign conventions see Ernst (1978)).

In the Newman-Penrose formalism, the Maxwell equations read

$$D\Phi_1 - \bar{\delta}\Phi_0 = (\pi - 2\alpha)\,\Phi_0 + 2\varrho\Phi_1 - \varkappa\Phi_2, \tag{7.46}$$

$$D\Phi_2 - \bar{\delta}\Phi_1 = -\lambda\Phi_0 + 2\pi\Phi_1 + (\varrho - 2\varepsilon)\,\Phi_2, \tag{7.47}$$

$$\delta\Phi_1 - \Delta\Phi_0 = (\mu - 2\gamma)\,\Phi_0 + 2\tau\Phi_1 - \sigma\Phi_2, \tag{7.48}$$

$$\delta\Phi_2 - \Delta\Phi_1 = -\nu\Phi_0 + 2\mu\Phi_1 + (\tau - 2\beta)\,\Phi_2, \tag{7.49}$$

where the notation

$$\Phi_0 \equiv F_{ab}k^a m^b = \frac{1}{4}\,F^*_{ab}V^{ab} = \Phi_{AB}o^A o^B, \tag{7.50}$$

$$\Phi_1 \equiv \frac{1}{2}\,F_{ab}(k^a l^b + \overline{m}^a m^b) = -\frac{1}{8}\,F^*_{ab}W^{ab} = \Phi_{AB}o^A \iota^B, \tag{7.51}$$

$$\Phi_2 \equiv F_{ab}\overline{m}^a l^b = \frac{1}{4}\,F^*_{ab}U^{ab} = \Phi_{AB}\iota^A \iota^B \tag{7.52}$$

is used for the tetrad components of the electromagnetic field tensor (cf. (5.10)). The Ricci tensor components (7.16)—(7.21) of Einstein-Maxwell fields are given by

$$\Phi_{\alpha\beta} = \varkappa_0\Phi_\alpha\overline{\Phi}_\beta, \qquad \alpha, \beta = 0, 1, 2. \tag{7.53}$$

7.2. The commutators and Bianchi identities

The *commutators* introduced previously,

$$[\boldsymbol{e}_a, \boldsymbol{e}_b] = D^c{}_{ab}\boldsymbol{e}_c, \qquad D^c{}_{ab} = -2\Gamma^c{}_{[ab]} \tag{7.54}$$

are given explicitly in the present notation as follows:

$$(\Delta D - D\Delta) = (\gamma + \bar{\gamma})\,D + (\varepsilon + \bar{\varepsilon})\,\Delta - (\tau + \bar{\pi})\,\bar{\delta} - (\bar{\tau} + \pi)\,\delta, \tag{7.55}$$

$$(\delta D - D\delta) = (\bar{\alpha} + \beta - \bar{\pi})\,D + \varkappa\Delta - \sigma\bar{\delta} - (\bar{\varrho} + \varepsilon - \bar{\varepsilon})\,\delta, \tag{7.56}$$

$$(\delta\Delta - \Delta\delta) = -\bar{\nu}D + (\tau - \bar{\alpha} - \beta)\,\Delta + \bar{\lambda}\bar{\delta} + (\mu - \gamma + \bar{\gamma})\,\delta, \tag{7.57}$$

$$(\bar{\delta}\delta - \delta\bar{\delta}) = (\bar{\mu} - \mu)\,D + (\bar{\varrho} - \varrho)\,\Delta - (\bar{\alpha} - \beta)\,\bar{\delta} - (\bar{\beta} - \alpha)\,\delta. \tag{7.58}$$

The application of the commutator relations (7.55)—(7.58) to scalar functions (e.g. the space-time coordinates x^i) yields information important in solving the field equations.

The *Bianchi identities* (3.77),

$$R_{ab[cd;e]} = 0, \tag{7.59}$$

form the remaining set of equations to be satisfied. They are the integrability conditions for the determination of the tetrad components $e_a{}^i$ with respect to a coordinate basis $\{\partial/\partial x^i\}$. This can be seen as follows (Papapetrou (1971a, b)). To obtain the components $e_a{}^i$ from (2.71), the integrability conditions $e_{ai,[j,k]} = 0$, which are precisely equivalent to the equation (2.79), have to be fulfilled. The equations (2.71), (2.79) contain the variables $e_a{}^i$, Γ_{abc}, R_{abcd}. Considering (2.71) as differential equations for Γ_{abi} we have, by Theorem 2.1, the integrability conditions $\Gamma_{ab[i,j,k]} = 0$, which are just the Bianchi identities (7.59), i.e., written in terms of the tetrad components and the directional derivatives,

$$R_{ab[cd|f]} = -2R_{abe[c}\Gamma^e{}_{df]} + \Gamma^e{}_{a[c}R_{df]eb} - \Gamma^e{}_{b[c}R_{df]ea}. \tag{7.60}$$

The integrability conditions for R_{abcd} following from (7.60) are identically satisfied as a consequence of the equations (2.79), (7.54), (7.60).

It is rather cumbersome to write the Bianchi identities (7.60) out in full detail (Pirani (1965), p. 350), but for completeness and for later use they are listed here:

$$\begin{aligned}\bar{\delta}\Psi_0 - D\Psi_1 + D\Phi_{01} - \delta\Phi_{00} &= (4\alpha - \pi)\,\Psi_0 - 2(2\varrho + \varepsilon)\,\Psi_1 + 3\varkappa\Psi_2\\ &\quad + (\bar{\pi} - 2\bar{\alpha} - 2\beta)\,\Phi_{00} + 2(\varepsilon + \bar{\varrho})\,\Phi_{01} + 2\sigma\Phi_{10} - 2\varkappa\Phi_{11} - \bar{\varkappa}\Phi_{02},\end{aligned} \tag{7.61}$$

$$\begin{aligned}\Delta\Psi_0 - \delta\Psi_1 + D\Phi_{02} - \delta\Phi_{01} &= (4\gamma - \mu)\,\Psi_0 - 2(2\tau + \beta)\,\Psi_1 + 3\sigma\Psi_2\\ &\quad + (2\varepsilon - 2\bar{\varepsilon} + \bar{\varrho})\,\Phi_{02} + 2(\bar{\pi} - \beta)\,\Phi_{01} + 2\sigma\Phi_{11} - 2\varkappa\Phi_{12} - \bar{\lambda}\Phi_{00},\end{aligned} \tag{7.62}$$

$$\begin{aligned}\bar{\delta}\Psi_3 - \Delta\Psi_4 + \bar{\delta}\Phi_{21} - \Delta\Phi_{20} &= (4\varepsilon - \varrho)\,\Psi_4 - 2(2\pi + \alpha)\,\Psi_3 + 3\lambda\Psi_2\\ &\quad + (2\gamma - 2\bar{\gamma} + \bar{\mu})\,\Phi_{20} + 2(\bar{\tau} - \alpha)\,\Phi_{21} + 2\lambda\Phi_{11} - 2\nu\Phi_{10} - \bar{\sigma}\Phi_{22},\end{aligned} \tag{7.63}$$

$$\begin{aligned}\Delta\Psi_3 - \delta\Psi_4 + \bar{\delta}\Phi_{22} - \Delta\Phi_{21} &= (4\beta - \tau)\,\Psi_4 - 2(2\mu + \gamma)\,\Psi_3 + 3\nu\Psi_2\\ &\quad + (\bar{\tau} - 2\bar{\beta} - 2\alpha)\,\Phi_{22} + 2(\gamma + \bar{\mu})\,\Phi_{21} + 2\lambda\Phi_{12} - 2\nu\Phi_{11} - \bar{\nu}\Phi_{20},\end{aligned} \tag{7.64}$$

$$\begin{aligned}&D\Psi_2 - \bar{\delta}\Psi_1 + \Delta\Phi_{00} - \bar{\delta}\Phi_{01} + \frac{1}{12}DR\\ &\quad = -\lambda\Psi_0 + 2(\pi - \alpha)\,\Psi_1 + 3\varrho\Psi_2 - 2\varkappa\Psi_3\\ &\quad + (2\gamma + 2\bar{\gamma} - \bar{\mu})\,\Phi_{00} - 2(\bar{\tau} + \alpha)\,\Phi_{01} - 2\tau\Phi_{10} + 2\varrho\Phi_{11} + \bar{\sigma}\Phi_{02},\end{aligned} \tag{7.65}$$

$$\begin{aligned}&\Delta\Psi_2 - \delta\Psi_3 + D\Phi_{22} - \delta\Phi_{21} + \frac{1}{12}\Delta R\\ &\quad = \sigma\Psi_4 + 2(\beta - \tau)\,\Psi_3 - 3\mu\Psi_2 + 2\nu\Psi_1\\ &\quad + (\bar{\varrho} - 2\varepsilon - 2\bar{\varepsilon})\,\Phi_{22} + 2(\bar{\pi} + \beta)\,\Phi_{21} + 2\pi\Phi_{12} - 2\mu\Phi_{11} - \bar{\lambda}\Phi_{20},\end{aligned} \tag{7.66}$$

$$\begin{aligned}&D\Psi_3 - \bar{\delta}\Psi_2 - D\Phi_{21} + \delta\Phi_{20} - \frac{1}{12}\bar{\delta}R\\ &\quad = -\varkappa\Psi_4 + 2(\varrho - \varepsilon)\,\Psi_3 + 3\pi\Psi_2 - 2\lambda\Psi_1\\ &\quad + (2\bar{\alpha} - 2\beta - \bar{\pi})\,\Phi_{20} - 2(\bar{\varrho} - \varepsilon)\,\Phi_{21} - 2\pi\Phi_{11} + 2\mu\Phi_{10} + \bar{\varkappa}\Phi_{22},\end{aligned} \tag{7.67}$$

$$\Delta\Psi_1 - \delta\Psi_2 - \Delta\Phi_{01} + \bar{\delta}\Phi_{02} - \frac{1}{12}\,\delta R$$
$$= \nu\Psi_0 + 2(\gamma - \mu)\,\Psi_1 - 3\tau\Psi_2 + 2\sigma\Psi_3$$
$$+ (\bar{\tau} - 2\bar{\beta} + 2\alpha)\,\Phi_{02} + 2(\bar{\mu} - \gamma)\,\Phi_{01} + 2\tau\Phi_{11} - 2\varrho\Phi_{12} - \bar{\nu}\Phi_{00}, \tag{7.68}$$

$$D\Phi_{11} - \delta\Phi_{10} - \bar{\delta}\Phi_{01} + \Delta\Phi_{00} + \frac{1}{8}\,DR = (2\gamma - \mu + 2\bar{\gamma} - \bar{\mu})\,\Phi_{00}$$
$$+ (\pi - 2\alpha - 2\bar{\tau})\,\Phi_{01} + (\bar{\pi} - 2\bar{\alpha} - 2\tau)\,\Phi_{10} + 2(\varrho + \bar{\varrho})\,\Phi_{11}$$
$$+ \bar{\sigma}\Phi_{02} + \sigma\Phi_{20} - \bar{\varkappa}\Phi_{12} - \varkappa\Phi_{21}, \tag{7.69}$$

$$D\Phi_{12} - \delta\Phi_{11} - \bar{\delta}\Phi_{02} + \Delta\Phi_{01} + \frac{1}{8}\,\delta R = (-2\alpha + 2\bar{\beta} + \pi - \bar{\tau})\,\Phi_{02}$$
$$+ (\bar{\varrho} + 2\varrho - 2\bar{\varepsilon})\,\Phi_{12} + 2(\bar{\pi} - \tau)\,\Phi_{11} + (2\gamma - 2\bar{\mu} - \mu)\,\Phi_{01}$$
$$+ \bar{\nu}\Phi_{00} - \bar{\lambda}\Phi_{10} + \sigma\Phi_{21} - \varkappa\Phi_{22}, \tag{7.70}$$

$$D\Phi_{22} - \delta\Phi_{21} - \bar{\delta}\Phi_{12} + \Delta\Phi_{11} + \frac{1}{8}\,\Delta R = (\varrho + \bar{\varrho} - 2\varepsilon - 2\bar{\varepsilon})\,\Phi_{22}$$
$$+ (2\bar{\beta} + 2\pi - \bar{\tau})\,\Phi_{12} + (2\beta + 2\bar{\pi} - \tau)\,\Phi_{21} - 2(\mu + \bar{\mu})\,\Phi_{11}$$
$$+ \nu\Phi_{01} + \bar{\nu}\Phi_{10} - \bar{\lambda}\Phi_{20} - \lambda\Phi_{02}. \tag{7.71}$$

The Newman-Penrose equations (7.28)–(7.45), the commutator relations (7.55)–(7.58), and the Bianchi identities (7.61)–(7.71) form a complete set of equations from which exact solutions of Einstein's field equations may be found.

Despite the fact that we have to solve a considerable number of equations, the Newman-Penrose formalism has great advantages. All differential equations are of first order. Gauge transformations of the tetrad can be used to simplify the field equations. One can extract invariant properties of the gravitational field without using a coordinate basis. The algebraic structure of the Weyl tensor can be specified from the very beginning. (In the normal form of Petrov type D and N fields, only one of the coefficients $\Psi_0, \ldots, \Psi_4$ is non-zero, see (4.22), (4.25).) Sometimes it is possible to choose special subcases for which the field equations can be solved exactly.

Ref.: Alekseev and Khlebnikov (1978), Frolov (1977).

7.3. The modified calculus

Geroch et al. (1973) have developed a modified calculus adapted to physical situations in which a pair of real null directions is naturally picked out at each space-time point. This new version of the spin coefficient method leads to even simpler formulae than the standard Newman-Penrose technique.

In spinor notation, the most general transformation preserving the two preferred null directions and the dyad normalization $o_A\iota^A = 1$ is given by

$$o^A \to Co^A, \qquad \iota^A \to C^{-1}\iota^A, \qquad C \text{ complex.} \tag{7.72}$$

The corresponding 2-parameter subgroup of the Lorentz group (boosts and spatial rotations), affects the complex null tetrad $(\boldsymbol{m}, \overline{\boldsymbol{m}}, \boldsymbol{l}, \boldsymbol{k})$ as follows (cf. equations (3.16), (3.17)):

$$\boldsymbol{k} \to A\boldsymbol{k}, \quad \boldsymbol{l} \to A^{-1}\boldsymbol{l}, \quad \boldsymbol{m} \to e^{i\Theta}\boldsymbol{m}; \quad A = C\overline{C} \quad e^{i\Theta} = C\overline{C}^{-1}. \tag{7.73}$$

A scalar η which undergoes the transformation

$$\eta \to C^p\overline{C}^q\eta \tag{7.74}$$

is called a spin- and boost-weighted *scalar of type* (p, q). The components of the Weyl and Ricci tensors, and the spin coefficients $\varkappa, \varrho, \sigma, \tau; \nu, \mu, \lambda, \pi$ have the types

$$\begin{aligned}
&\Psi_0: (4, 0), \quad \Psi_1: (2, 0), \quad \Psi_2: (0, 0), \quad \Psi_3: (-2, 0), \quad \Psi_4: (-4, 0),\\
&\Phi_{00}: (2, 2), \quad \Phi_{01}: (2, 0), \quad \Phi_{02}: (2, -2), \quad \Phi_{10}: (0, 2), \quad \Phi_{11}: (0, 0),\\
&\Phi_{22}: (-2, -2), \quad \Phi_{21}: (-2, 0), \quad \Phi_{20}: (-2, 2), \quad \Phi_{12}: (0, -2),\\
&\varkappa: (3, 1), \quad \varrho: (1, 1), \quad \sigma: (3, -1), \quad \tau: (1, -1),\\
&\nu: (-3, -1), \quad \mu: (-1, -1), \quad \lambda: (-3, 1), \quad \pi: (-1, 1).
\end{aligned} \tag{7.75}$$

In the modified calculus the replacement

$$\boldsymbol{k} \to \boldsymbol{l}, \quad \boldsymbol{l} \to \boldsymbol{k}, \quad \boldsymbol{m} \to \overline{\boldsymbol{m}}, \quad \overline{\boldsymbol{m}} \to \boldsymbol{m} \tag{7.76}$$

is indicated by a prime on the symbols, for instance

$$\varkappa' \equiv -\nu, \quad \sigma' \equiv -\lambda, \quad \varrho' \equiv -\mu, \quad \tau' \equiv -\pi, \quad \beta' \equiv -\alpha, \quad \varepsilon' \equiv -\gamma. \tag{7.77}$$

The operations given by the substitution (7.76) and by complex conjugation respectively produce scalars η' of type $(-p, -q)$ and $\overline{\eta}$ of type (q, p). The prime convention considerably reduces the notational effort; e.g. the Newman-Penrose equation (7.41) is simply the primed version of the equation (7.28), etc.

The spin coefficients $\beta, \beta', \varepsilon, \varepsilon'$ transform, under the tetrad change (7.72), (7.73), according to inhomogeneous laws containing derivatives of C. Therefore, these spin coefficients do not appear directly in the modified equations. However, they enter the new derivative operators acting on spin- and boost-weighted scalars η of type (p, q):

$$\begin{aligned}
\text{Þ}\eta &\equiv (D - p\varepsilon - q\overline{\varepsilon})\,\eta, & \text{Þ}'\eta &\equiv (\Delta + p\varepsilon' + q\overline{\varepsilon}')\,\eta,\\
\eth\eta &\equiv (\delta - p\beta + q\overline{\beta}')\,\eta, & \eth'\eta &\equiv (\overline{\delta} + p\beta' - q\overline{\beta})\,\eta.
\end{aligned} \tag{7.78}$$

The operators Þ and $\eth$ respectively map a scalar of type (p, q) into scalars of types $(p + 1, q + 1)$ and $(p + 1, q - 1)$. In consequence of the definitions (7.77), (7.78), the Newman-Penrose equations (7.28)–(7.45), the commutators (7.55)–(7.58), and the Bianchi identities (7.61)–(7.71) get a new explicit form. They contain only scalars and derivative operators of good weight, and split into two sets of equations, one being the primed version of the other.

This formalism has been used for an elegant discussion of a class of type D vacuum metrics (Held (1974b)). The first part of the integration is a coordinate-free procedure. Then the coordinate system can be based on an already known solution. This is one of the advantages of the present technique, which works well when applied to algebraically special metrics. It is also useful in dealing with fields in algebraically special background metrics, especially the Kerr metric.

Ref.: Breuer (1975), Held (1975, 1976a).

7.4. Geodesic null congruences

In § 6.1. we dealt with geodesic null congruences, whose tangent vector fields $\boldsymbol{k}$ satisfy

$$\varkappa \equiv -k_{a;b}m^a k^b = 0, \tag{7.79}$$

and introduced the complex divergence ϱ and the complex shear σ,

$$\varrho \equiv -k_{a;b}m^a\overline{m}^b = -(\Theta + \mathrm{i}\omega), \qquad \sigma \equiv -k_{a;b}m^a m^b, \tag{7.80}$$

(see (6.16) and (6.19)), in agreement with the definitions (7.3), (7.4), of these spin coefficients. If $\sigma = 0$, the congruence will be called *shearfree*.

Here we consider the possible simplifications of the Newman-Penrose equations when the null tetrad vector $\boldsymbol{k}$ is geodesic. Null vector fields which are both geodesic and shearfree will be the subject of the next section.

Choosing $\boldsymbol{k}$ so that the geodesics are affinely parametrized, we have

$$k_{a;b}k^b = 0 \;\Leftrightarrow\; \varkappa = 0, \qquad \varepsilon + \bar{\varepsilon} = 0. \tag{7.81}$$

Then the equation (7.28) is equivalent to the propagation equation (6.33),

$$D\varrho = \varrho^2 + \sigma\bar{\sigma} + \Phi_{00}. \tag{7.82}$$

If $\boldsymbol{k}$ is a double principal null direction of the Weyl tensor ($\Psi_0 = \Psi_1 = 0$), the Bianchi identity (7.65) takes the simple form

$$D\Psi_2 = 3\varrho\Psi_2 \tag{7.83}$$

for vacuum fields.

If the null tetrad $\{\boldsymbol{e}_a\}$ is parallelly propagated along the geodesic null congruence $\boldsymbol{k}$, we obtain

$$\varkappa = \varepsilon = \pi = 0, \tag{7.84}$$

i.e. three (complex) spin coefficients are zero. This choice of tetrad is very convenient for certain calculations; the left-hand sides of the equations (7.28)–(7.36) become directional derivatives $D\varrho, \ldots, D\nu$ of the remaining spin coefficients and thus in a coordinate system with $\boldsymbol{k} = \partial_r$, these equations determine the r-dependence of the spin coefficients.

A simple geodesic principal null direction $\boldsymbol{k}$ of a vacuum field is non-twisting, $k_{[a;b}k_{c]} = 0$ (Kammerer (1966)).

7.5. The Goldberg-Sachs theorem and its generalizations

The Goldberg-Sachs theorem is very useful in constructing algebraically special solutions. It exhibits a close connection between certain geometrical properties of a null congruence and Petrov type. The original paper (Goldberg and Sachs (1962)) presents the proof of two theorems:

Theorem 7.1. *If a gravitational field contains a shearfree geodesic null congruence* $\boldsymbol{k}$ $(\varkappa = 0 = \sigma)$ *and if*

$$R_{ab}k^a k^b = R_{ab}k^a m^b = R_{ab}m^a m^b = 0, \tag{7.85}$$

then the field is algebraically special, and $\boldsymbol{k}$ *is a degenerate eigendirection;*

$$C_{abc[d}k_{e]}k^b k^c = 0 \;\Leftrightarrow\; \Psi_0 = 0 = \Psi_1. \tag{7.86}$$

Remark: The conditions (7.85) are invariant with respect to the null rotations (3.15).

Theorem 7.2. *If a vacuum metric* $(R_{ab} = 0)$ *is algebraically special, then the multiple principal null vector is tangent to a shearfree geodesic null congruence.*

Combining these two theorems one obtains the well-known form of the Goldberg-Sachs theorem:

Theorem 7.3 (Goldberg-Sachs theorem). *A vacuum metric is algebraically special if and only if it contains a shearfree geodesic null congruence,*

$$\varkappa = 0 = \sigma \;\Leftrightarrow\; \Psi_0 = 0 = \Psi_1. \tag{7.87}$$

All statements in (7.87) remain unchanged under conformal transformations (3.81). This remark leads to an obvious generalization of the Goldberg-Sachs theorem, namely to any gravitational field that is conformal to a vacuum field (Robinson and Schild (1963)). The Goldberg-Sachs theorem has been proved in Newman and Penrose (1962) using the formalism outlined in §§ 7.1. and 7.2. From the Bianchi identities (7.61)–(7.71) it is easily seen that with the assumption $\Psi_0 = 0 = \Psi_1$ we obtain $\varkappa = 0 = \sigma$. The converse is more difficult to prove. In the special case $\varrho = 0$ we obtain $\Psi_0 = 0 = \Psi_1$ from the equations (7.29), (7.38). If $\varrho \neq 0$ one can always set $\alpha + \bar{\beta} = 0$ by tetrad rotations and from the equations (7.29), (7.31), (7.32), (7.81) we arrive at $\Psi_0 = 0$, $\Psi_1 = \varrho\pi$. Various steps using the commutators and Bianchi identities lead to the final result $\pi = 0 = \Psi_1$.

For empty space-times which are algebraically special on a given *submanifold* $\mathscr{S}$, which is either a spacelike hypersurface or a timelike world line, the vector field tangent to a principal null direction of the curvature tensor and pointing, on $\mathscr{S}$, in the repeated principal null direction is geodesic and shearfree on $\mathscr{S}$ (Collinson (1967)).

We give (without proof) an interesting theorem which allows a reformulation of the Goldberg-Sachs theorem.

Theorem 7.4 (Mariot (1954), Robinson (1961)). *An arbitrary space-time* V_4 *admits a geodesic shearfree null congruence if and only if* V_4 *admits an electromagnetic null field ("test field") satisfying the Maxwell equations in* V_4*:*

$$\varkappa = 0 = \sigma \;\Leftrightarrow\; F^*_{ab}k^b = 0, \qquad F^{*ab}{}_{;b} = 0. \tag{7.88}$$

The conditions (7.85) show that only a part of the vacuum field equations is needed to prove Theorem 7.1. Thus we have the

Corollary: *the Weyl tensor of Einstein-Maxwell fields with an electromagnetic null field is algebraically special.*
Proof: The Ricci tensor obeys the conditions (7.85) and the Maxwell equations demand that $\boldsymbol{k}$ is geodesic and shearfree (Theorem 7.4).

Now we give a generalization of the Goldberg-Sachs theorem due to Kundt and Thompson (1962), and Robinson and Schild (1963). This generalized theorem is not restricted to vacuum solutions.

Theorem 7.5 (Kundt-Thompson theorem). *Any two of the following imply the third:*

(A) *the Weyl tensor is algebraically special,* $\boldsymbol{k}$ *being the repeated principal null vector.*

(B) $\boldsymbol{k}$ *is shearfree and geodesic* ($\sigma = \varkappa = 0$).

(C) $V^{ab}C_{abcd}{}^{;d}V^{ce} = 0$ *for Petrov type II or D,*

$V^{ab}C_{abcd}{}^{;d} = 0$ *for Petrov type III,*

$U^{ab}C_{abcd}{}^{;d}V^{ce} = 0$ *for Petrov type N.*

Proof: By elementary calculation we derive the following equations (Szekeres (1966b))

$$\text{Type } II, D \quad (\Psi_0 = \Psi_1 = 0): \quad V^{ab}C_{abcd}{}^{;d}V^{ce} = 6\Psi_2(\sigma k^e - \varkappa m^e), \tag{7.89a}$$

$$\text{Type } III \quad (\Psi_0 = \Psi_1 = \Psi_2 = 0): \quad V^{ab}C_{abcd}{}^{;d} = 4\Psi_3(\sigma k_c - \varkappa m_c), \tag{7.89b}$$

$$\text{Type } N \quad (\Psi_0 = \Psi_1 = \Psi_2 = \Psi_3 = 0): \quad U^{ab}C_{abcd}{}^{;d}V^{ce} = 2\Psi_4(\sigma k^e - \varkappa m^e) \tag{7.89c}$$

from the decomposition (3.58). Here we have used the definitions (3.39) of the complex self-dual bivectors U_{ab}, V_{ab}, W_{ab}, and the normalization (3.9) of the complex null tetrad. From the equations (7.89) it is clear that (A),(B) $\Rightarrow$ (C) and (A),(C) $\Rightarrow$ (B). The proof that (B), (C) implies (A) is less trivial and is omitted here. The condition (C) of the Kundt-Thompson theorem may be replaced (Bell and Szekeres (1972)) by the condition (C′): there exists a null type solution ($\Phi_{AB\ldots M} = \Phi o_A o_B \ldots o_M$) of the zero rest-mass free field equation $\nabla^{A\dot{X}}\Phi_{AB\ldots M} = 0$ for some spin value $s > 1$.

For vacuum fields, condition (C) of Theorem 7.5 is automatically true and (A) $\Leftrightarrow$ (B) is just the Goldberg-Sachs theorem (Theorem 7.3). This follows from the Bianchi identities (3.77) written in the form

$$C_{abcd}{}^{;d} = R_{c[a;b]} - \frac{1}{6}\, g_{c[a}R_{,b]}. \tag{7.90}$$

(This form can be obtained by using the relations (3.50) and (3.52).)

Suppose we have an algebraically special Einstein-Maxwell field with a non-null electromagnetic field such that one of the eigendirections of the Maxwell tensor is *aligned* with the multiple principal null direction of the Weyl tensor. Then it follows from the Bianchi identities (7.61), (7.62), and the expressions (7.53), (5.12) that

$$(2\varkappa_0\Phi_1\overline{\Phi}_1 + 3\Psi_2)\,\sigma = 0, \qquad (-2\varkappa_0\Phi_1\overline{\Phi}_1 + 3\Psi_2)\,\varkappa = 0 \tag{7.91}$$

(Kundt and Trümper (1962)). If $\Psi_2 = 0$ (Petrov type *III* and more special types), then $\varkappa = \sigma = 0$. If $\Psi_2 \neq 0$, one obtains $\varkappa\sigma = 0$, i.e. either $\varkappa$ or σ must vanish. The relation (7.89a) also leads to the equations (7.91).

Unfortunately, the Kundt-Thompson theorem does not directly specify the most general matter distribution which would allow one to conclude that (A) $\Rightarrow$ (B). For instance, the assumption that the Ricci tensor is of pure radiation type,

$$R_{ab} = \varkappa_0 \Phi^2 k_a k_b = \varkappa_0 T_{ab}\,, \tag{7.92}$$

does not guarantee condition (C) for fields of Petrov type N; in general the shear of $\boldsymbol{k}$ does not vanish, cp. § 22.1. However, if T_{ab} in equation (7.92) is the energy-momentum tensor of an electromagnetic (null) field, the congruence $\boldsymbol{k}$ is necessarily shearfree because of the above corollary to Theorem 7.4.

Chapter 8. Continuous groups of transformations. Groups of motions.

8.1. Introduction: Lie groups and Lie algebras

In this chapter we shall summarize those elements of the theory of continuous groups of transformations which we require for the following chapters. As far as we know, the most extensive treatment of this subject is to be found in Eisenhart (1933), while more recent applications to General Relativity can be found in the works of Petrov (1966) and Defrise (1969), for example. General treatments of Lie groups and of transformation groups in coordinate-free terms can be found in, for example, Cohn (1957), Warner (1971), and Brickell and Clark (1970), but none of these cover the whole of the material contained in Eisenhart's treatise.

We begin by introducing the concepts of Lie groups and Lie algebras, and the relation between them. A *Lie group* G is (i) a group (in the usual sense of algebra), with elements $q_0, q_1, q_2, \ldots$ say, q_0 being the identity element, and (ii) a differentiable manifold such that the map $\Phi: G \times G \to G$ given by the algebraic product $(q_1 q_2) \to q_1 q_2$ is analytic. (See § 2.2. for definitions of manifolds). We do not concern ourselves with the minimum differentiability requirements which will ensure that a continuous group (and, in later sections, the action of a continuous transformation group) is analytic; for these see e.g. Cohn (1957). In our applications such conditions always hold.

Moreover, as in the rest of the book, many of the results stated apply only locally: for example, in what follows, we may say that a manifold is invariant under a group, whereas all that we really require is that a neighbourhood (on which we solve the Einstein equations) is isometric to a neighbourhood in a space with the stated symmetry. In a few places where the distinction is crucial we have reminded the reader of it by inserting "locally".

The coordinates in (a neighbourhood of the identity q_0 of) G are called group *parameters*, and the analytic functions describing Φ in such coordinates are called *composition functions.*

Our aim is to study transformation groups. The abstract Lie group naturally has associated with it two transformation groups. One of them consists of the *left translations*, the left translation associated with $q \in G$ being the map L_q of G to G such that

$$q' \to qq'; \tag{8.1a}$$

the other consists of *right translations* R_q defined similarly by

$$q' \to q'q. \tag{8.1b}$$

Each of these has associated with it a set of vector fields related to one-dimensional sets of transformations in the same way as $\boldsymbol{v}$ is related to Φ_t in § 2.8.

Right translations commute with left translations,

$$R_q L_{q'} = L_{q'} R_q, \tag{8.2}$$

as is easily seen.

If we follow the convention that maps are written on the left (e.g. $L_q(q') = qq'$), then the left translation group is isomorphic to G and is called the *parameter group*, while the right translation group is algebraically dual to G. (If maps are written on the right, "right" and "left" must be interchanged in all subsequent statements.) The vector fields related to left translations turn out to be right-invariant vector fields, which we now study.

A *right-invariant vector field* $\boldsymbol{v}$ on G is defined to be one satisfying

$$(R_q)_* \boldsymbol{v} = \boldsymbol{v} \tag{8.3}$$

(for definitions, see (2.30), (2.31)). The value $\boldsymbol{v}(q)$ of such a vector field at a point q gives, and is given by, its value $\boldsymbol{v}(q_0)$ at the group identity q_0:

$$\boldsymbol{v}(q) = (R_q)_* \boldsymbol{v}(q_0), \qquad \boldsymbol{v}(q_0) = (R_{q^{-1}})_* \boldsymbol{v}(q). \tag{8.4}$$

(8.4) shows that the group G has the same dimension r at all points, and that the set of all right-invariant vector fields and the tangent space T_{q_0} to G at q_0 are isomorphic vector spaces. An r-dimensional group is denoted by G_r and said to be of r parameters.

The transformations Φ_t generated by a right-invariant vector field $\boldsymbol{v}$, in the way described in § 2.8., clearly commute with right translations. If $\Phi_t q_0 = q(t)$, we find

$$\Phi_t q' = \Phi_t R_{q'} q_0 = R_{q'} \Phi_t q_0 = R_{q'} q(t) = q(t)\, q', \tag{8.5}$$

so that $\Phi_t = L_{q(t)}$; the right-invariant vector fields represent infinitesimal left translations. The equation (2.32) shows that the commutator of two right-invariant vector fields is also right-invariant, so that if we take a basis $\{\xi_A, A = 1, \ldots, r\}$ of the space of right-invariant vector fields we must have

$$[\xi_A, \xi_B] = C^C{}_{AB} \xi_C, \qquad C^C{}_{AB} = -C^C{}_{BA}. \tag{8.6}$$

The coefficients $C^C{}_{AB}$ are known as the *structure constants* of the group. A *Lie algebra* is defined to be a (finite-dimensional) vector space in which a bilinear operation $[\boldsymbol{u}, \boldsymbol{v}]$, obeying $[\boldsymbol{u}, \boldsymbol{v}] = -[\boldsymbol{v}, \boldsymbol{u}]$, and the Jacobi identity (2.7), is defined. Thus we have proved

Theorem 8.1. *A Lie group defines a unique Lie algebra.*

It is possible to show that the converse also holds.

Theorem 8.2. *Every Lie algebra defines a unique* (*simply-connected*) *Lie group.*

The elements of the Lie algebra, or a basis of them, are said to *generate* the group. For proof, see (e.g.) Cohn (1957). Theorem 8.2 can be rewritten by noting that the Jacobi identity (2.7) holds for (8.6) if and only if

$$C^E{}_{[AB} C^F{}_{C]E} = 0. \tag{8.7}$$

Thus we have

Theorem 8.3 (Lie's third fundamental theorem). *Any set of constants $C^A{}_{BC}$ satisfying $C^A{}_{BC} = C^A{}_{[BC]}$ and* (8.7) *are the structure constants of a group.*

Theorem 8.2 does not imply that a given Lie algebra arises from only one Lie group. For example the Lorentz group $L_+^\uparrow$ and the group $SL(2, C)$ (see § 3.6.) have the same Lie algebra. It is true, however, that all connected Lie groups with a given Lie algebra are homomorphic images of the one specified in Theorem 8.2.

All the above work can be repeated interchanging left and right. We shall denote a basis of the Lie algebra of left-invariant fields by $\{\eta_A;\, A = 1, \ldots, r\}$. For all A and B, (8.2) implies

$$[\xi_A, \eta_B] = 0. \tag{8.8}$$

Clearly we must have some position-dependent matrix $M_A{}^B(q)$ such that $\eta_A = M_A{}^B\xi_B$, with inverse $(M^{-1})^A{}_B$. Equations (8.8) and (8.6) show that

$$[\eta_A, \eta_B] = -M_A{}^C M_B{}^D C^E{}_{CD}(M^{-1})^F{}_E\, \eta_F, \tag{8.9}$$

so that the structure constants $D^A{}_{BE}$ of the basis $\{\eta_A\}$ are related to the $C^A{}_{BE}$. Choosing $\eta_A = -\xi_A$ at q_0 shows that the Lie algebras, and hence the Lie groups, of left and right translations are isomorphic. However, it is more usual to take $\eta_A = \xi_A$, leading to

$$D^A{}_{BC} = -C^A{}_{BC}. \tag{8.10}$$

The commutators $[\boldsymbol{u}, \boldsymbol{v}]$ of right-invariant vector fields are the infinitesimal generators of the commutator subgroup of G (i.e. that formed from all products of the form $q_1q_2(q_1)^{-1}\,(q_2)^{-1}$). This is also known as the (first) *derived group*, and its Lie algebra, which is spanned by $C^A{}_{BC}\xi_A$, is the *derived algebra*. A group is said to be *Abelian* if every pair of elements commutes: for Lie groups, this is true if and only if all the structure constants are zero. A subgroup H of a group G is said to be *normal* or *invariant* if $qhq^{-1} \in H$ for any $h \in H$ and $q \in G$; for Lie groups this is true if and only if the generators ζ_i $(i = 1, \ldots, p)$ of H obey

$$[\xi_A, \zeta_i] = C^j{}_{Ai}\zeta_j \tag{8.11}$$

for all A and i. A group is said to be *simple* if it has no invariant subgroup other than the group itself and the identity, and *semi-simple* if it similarly has no invariant Abelian subgroup. The derived group is always invariant. A group is said to be *solvable* (or *integrable*) when there is a sequence of subgroups $G_r, G_{r_1}, \ldots, G_{r_k}$ with $r > r_1 > \cdots > r_k = 0$ such that G_{r_t} is the derived group of $G_{r_{t-1}}$.

Any subalgebra of the Lie algebra of a Lie group generates a Lie subgroup, and a subalgebra with basis (ζ_i) satisfying (8.11), known as an *ideal*, generates an invariant subgroup.

It is possible to define *canonical coordinates* on a Lie group G in such a way that a given basis $\{\xi_A\}$ has $\xi_A = \partial/\partial x^A$ at q_0; actually, this can be done in more than one way (Cohn (1957)).

8.2. Enumeration of distinct group structures

Linear transformations of the basis $\{\xi_A\}$ transform the $C^A{}_{BC}$ of (8.6) as a tensor. To find distinct groups we need sets of constants $C^A{}_{BC}$ which cannot be related by such a linear transformation. The enumerations therefore naturally use properties, such as dimension of the derived algebra, which are invariant under these transformations. Methods of enumerating all complex Lie groups are well-known, being useful in pure mathematics and quantum physics, but the enumeration of the *real* Lie algebras, although its foundations have also long been known, is not so widely studied. We give here the distinct structures of groups G_2 and G_3, and indicate how one could enumerate all G_4. We omit the full list of G_4 because many cases do not arise in exact solutions (the list given by Petrov (1966), p. 72, has been refined by Patera and Winternitz (1977) and MacCallum (1979c).

In a G_2 there is only one (non-trivial) commutator; hence all G_2 are solvable. If the G_2 is Abelian, it is called type G_2I. If it is non-Abelian, one can choose ξ_1 in the derived algebra, and scale ξ_2 so that

$$[\xi_1, \xi_2] = \xi_1; \tag{8.12}$$

this case is called type G_2II.

The G_3 were originally enumerated by Bianchi (1897). There are nine types, Bianchi I to Bianchi IX, two of which, VI and VII, are one-parameter families of distinct group structures. Complex transformations relate type VIII to type IX, and type VI to type VII. Bianchi's method began, like that above for the G_2, by considering the dimension of the derived algebra, but we shall obtain the result a different way (Estabrook et al. (1968), Ellis and MacCallum (1969)). Taking any completely skew tensor ε^{ABC} on the Lie algebra we write

$$\frac{1}{2}\, C^D{}_{BC}\varepsilon^{BCE} = N^{DE} + \varepsilon^{DEF}A_F, \qquad A_D \equiv \frac{1}{2}\, C^B{}_{DB}, \qquad N^{DE} = N^{(DE)}. \tag{8.13}$$

The Jacobi identity (8.7) reduces to

$$N^{DE}A_E = 0. \tag{8.14}$$

N^{DE} is defined up to an overall factor (since ε^{ABC} is). Its invariant properties are its rank and the modulus of its signature. In types VI and VII there is a further invariant h, defined by

$$(1 - h)\, C^A{}_{BA}C^D{}_{CD} = -2hC^A{}_{DB}C^D{}_{AC}, \tag{8.15}$$

which supplies the one parameter required to subdivide these Bianchi types: in type VI $h \leqq 0$ and in type VII $h \geqq 0$. It is related to Bianchi's parameter q for types VI and VII by $h = -(1+q)^2/(1-q)^2$ and $h = q^2/(4-q^2)$ respectively. Bianchi type III is the same as VI ($h = -1$).

There are two main classes of G_3, Class G_3A ($A_E = 0$) and Class G_3B ($A_E \neq 0$). In all cases, by rotation and rescaling of the basis ξ_A, one can set $N^{DE} = \mathrm{diag}\,(N_1, N_2, N_3)$, $A_E = (A, 0, 0)$, with N_1, N_2, N_3 equal to 0 or ± 1 as appropriate, and $A = \sqrt{hN_2N_3}$ (for Bianchi types VI, VII, III). Thus one obtains Table 8.1, which lists all types and canonical forms of the structure constants. All types are solvable, except VIII and IX which are semi-simple. The canonical form does not uniquely specify the basis. The dimension of the subgroup of the linear transformations which preserves the canonical form is shown in Table 8.1 (for proof see Siklos (1976b)).

Table 8.1 Enumeration of, and canonical structure constants for, the Bianchi types

Class	G_3A						G_3B				
Type	I	II	VI_0	VII_0	VIII	IX	V	IV	III	VI_h	VII_h
Rank (N^{DE})	0	1	2	2	3	3	0	1	2	2	2
|Signature (N^{DE})|	0	1	0	2	1	3	0	1	0	0	2
A	0	0	0	0	0	0	1	1	1	$\sqrt{-h}$	$\sqrt{h}$
N_1	0	1	0	0	-1	1	0	0	0	0	0
N_2	0	0	-1	1	1	1	0	0	-1	-1	1
N_3	0	0	1	1	1	1	0	1	1	1	1
Dimensions of canonical basis freedom	9	6	4	4	3	3	6	4	4	4 (each $h < 0$)	4 (each $h > 0$)

The G_4 can be divided into two according to whether $A_E \equiv \frac{1}{2} C^B{}_{EB} = 0$ or not. In the first case one has (Farnsworth and Kerr (1966)):

Theorem 8.4. *If* $A_E = 0$, *then either* (A) *the structure constants of a* G_4 *can be written in the form*

$$C^A{}_{BC} = \Theta^A{}_{[B}P_{C]} \tag{8.16}$$

or (B) *if no form* (8.16) *exists, there is a non-zero vector* L^A *such that*

$$C^A{}_{BC}L^B = 0. \tag{8.17}$$

Proof: Take any non-zero totally skew tensor with components ε^{ABCD}. Since skewing over five indices automatically gives zero,

$$\varepsilon^{[BCDE}C^{A]}{}_{BC}C^F{}_{DE} = 0. \tag{8.18}$$

Expanding (8.18), and using the Jacobi identities, which are equivalent to $\varepsilon^{ABCD}C^E{}_{BC} \times C^F{}_{DE} = 0$, we obtain

$$\varepsilon^{BCDE}C^A{}_{BC}C^F{}_{DE} = 0. \tag{8.19}$$

Regarding $C^A{}_{BC}$ as the components of a two-form $\boldsymbol{C}^A$ on the Lie algebra, (8.19) reads

$$\boldsymbol{C}^A \wedge \boldsymbol{C}^F = 0. \tag{8.20}$$

With $A = F$, Theorem 2.4. shows $\boldsymbol{C}^A$ is simple, and using $A \neq F$ then shows that the 2-planes spanned by these bivectors must intersect. If all four $\boldsymbol{C}^A$ intersect (i.e. have a common one-form factor) we have case (A). If there are three which do not have a common intersection, they span a three-space in which the fourth C^A must lie, and we thus have a vector L^A obeying (B).

As a corollary of this theorem, we have

Theorem 8.5 (Egorov; see Petrov (1966), p. 180). *Every G_4 contains a G_3 (locally).*

Proof: If $A_E \neq 0$, (8.19) yields $A_B C^B{}_{CD} = 0$, showing that the derived algebra is three-dimensional (at most). If $A_E = 0$ the forms (8.16), (8.17), show clearly that the derived algebra is again at most three-dimensional. In all cases the derived algebra (together, if necessary, with enough linearly independent vectors to make the dimension three) generates a G_3.

A slightly different proof was found by Kantowski (see Collins (1977)). Egorov's proof is not known to us. Patera and Winternitz (1977) have explicitly calculated all subgroups G_2 and G_3 of the real G_4.

Another result due to Egorov (see Petrov (1966), p. 180) is

Theorem 8.6. *Every G_5 contains a subgroup G_4.*

8.3. Transformation groups

Let $\mathscr{M}$ be a differentiable (analytic) manifold and G a Lie group of r parameters. An *action* of G on $\mathscr{M}$ is an (analytic) map $\mu: G \times \mathscr{M} \to \mathscr{M}$; $(q, p) \to \tau_q p$. Each element q of G is associated with a transformation $\tau_q: \mathscr{M} \to \mathscr{M}$. It is assumed that the identity q_0 of G is associated with the identity map $I: p \to p$ of $\mathscr{M}$, and that

$$\tau_q \tau_{q'} p = \tau_{qq'} p \tag{8.21}$$

so that the transformations τ_q form a group isomorphic with G. The group is said to be *effective* (and the parameters *essential*) if $\tau_q = I$ implies $q = q_0$; only such groups need be considered.

The *orbit* (or *trajectory*, or *minimum invariant variety*) of G through a fixed p in $\mathscr{M}$ is defined to be $\mathscr{O}_p = \{p' : p' \in \mathscr{M}$ and $p' = \tau_q p$ for some $q \in G\}$. It is a submanifold of $\mathscr{M}$. The group G is said to be *transitive* on its orbits, and to be either *transitive* (when $\mathscr{O}_p = \mathscr{M}$) or *intransitive* ($\mathscr{O}_p \neq \mathscr{M}$) on $\mathscr{M}$. It is *simply-transitive* on an orbit if $\tau_q p = \tau_{q'} p$ implies $q = q'$; otherwise it is *multiply-transitive*. The set of q in G such that $\tau_q p = p$ forms a subgroup of G called the *stability group* $H(p)$ of p. If $p' \in \mathscr{O}_p$, so there is a q in G such that $\tau_q p = p'$, and if $q' \in H(p)$, then $\tau_q \tau_{q'} \tau_{q^{-1}} p' = p'$ and hence $qq'q^{-1} \in H(p')$. Thus $H(p)$ and $H(p')$ are conjugate subgroups of G, and have the same dimension, s say; for brevity, one often refers to the stability subgroup H_s of an orbit.

For an orbit, the map $\mu_p: G \to \mathscr{M}$; $q \to \tau_q p$ can be defined. The map $(\mu_p)_*$ then maps the right-invariant vector fields on G to vector fields tangent to the orbit. It can be shown that the choice of base point p in $\mathscr{O}_p$ does not affect $(\mu_p)_*$. Hence, using a map $(\mu_p)_*$ in each $\mathscr{O}_p$, we can define a Lie algebra of vector fields on $\mathscr{M}$ by taking the image of the Lie algebra of G. At the risk of some confusion we use $\{\xi_A\}$ to denote a basis of either Lie algebra. The two algebras are isomorphic because G is assumed to be effective, and so $(\mu_p)_* \boldsymbol{v} = \boldsymbol{0}$ for all $\boldsymbol{v}$ only if $p = \boldsymbol{0}$.

The stability group of p is generated by those $\boldsymbol{v}$ such that $(\mu_p)_*\boldsymbol{v} = \boldsymbol{0}$ at p; this is clearly the kernel of the map $(\mu_p)_*$ at q_0. Denoting the dimension of $\mathcal{O}_p$ by d we thus have

$$r = d + s. \tag{8.22}$$

The classical theorems on continuous transformation groups can be expressed as

Theorem 8.7 (Lie's first fundamental theorem). *An action $\mu: G \times \mathcal{M} \to \mathcal{M}$ of a continuous (Lie) group of transformations defines and is defined by a linear map of the right-invariant vector fields on G_r onto an r-dimensional set of (smooth) vector fields on $\mathcal{M}$.*

Theorem 8.8 (Lie's second fundamental theorem). *A set of r (smooth) linearly independent vector fields $\{\xi_A\}$ on $\mathcal{M}$ obeying* (8.6) *defines and is defined by a continuous (Lie) group of transformations on $\mathcal{M}$.*

A single generator ξ of a transformation group G_r gives rise to a one-parameter subgroup Φ_x (see § 2.8.) of G_r, and by choosing one point p in each orbit of this group as $x = 0$ we can find a coordinate x in $\mathcal{M}$ such that $\xi = \partial_x$ (the term *trajectory* is sometimes reserved for such one-dimensional orbits). If there are m commuting generators $\{\xi_A\}$ (forming an Abelian subgroup), all non-zero at p, then one can thus find m coordinates $(x^1, \ldots, x^m)$ such that $\xi_A = \partial/\partial x^A$.

8.4. Groups of motions

Manifolds with structure, such as Riemannian manifolds V_n, may admit (continuous) groups of transformations preserving this structure. A *conformal motion* of a V_n preserves the metric up to a factor, i.e. the map Φ_t associated with a generator $\boldsymbol{v}$ has the property $(\Phi_t\boldsymbol{g})_{ab} = e^U g_{ab}$ for the metric $\boldsymbol{g}$ (see §§ 2.8. and 3.7.). A *motion* (or *isometry*) has $U = 0$, i.e. it preserves the metric. For more general symmetries, we refer to § 31.4.

A *group G_r of motions* (or *isometry group*) has generators $\{\xi_A, A = 1, \ldots, r\}$ such that

$$0 = (\pounds_{\xi_A}\boldsymbol{g})_{ab} = g_{ab,c}\xi^c_A + g_{ac}\xi^c_{A,b} + g_{cb}\xi^c_{A,a} = \xi_{Aa;b} + \xi_{Ab;a}. \tag{8.23}$$

(8.23) is *Killing's equation* (6.10). Its solutions ξ_A are called *Killing vectors*. The set of all solutions of (8.23) can easily be seen to form a Lie algebra (8.6) and hence, by Theorem 8.8, to generate a Lie group of transformations. If we use the coordinate x adapted to a Killing vector ξ, so that $\xi = \partial_x$, then g_{ab} is independent of x.

Some special terminology is used for groups of motions. The stability group of p in a group G_r of motions is called the *isotropy group* H_s of p; it gives rise to a subgroup, the *linear isotropy subgroup* $\hat{H}$ of p, acting in the tangent space to the orbit $\mathcal{O}_p$ at p by (2.59). If we use the term *generalized orthogonal group* for the set of linear transformations of the tangent space at p which preserve scalar products formed with the metric $\boldsymbol{g}$ by (3.4). We see that $\hat{H}$ is a subgroup of this group. For a space-time, the generalized orthogonal group is the Lorentz group. The orbits (with dimension d) of a group of motions may be spacelike, null, or timelike submanifolds, and these are denoted by S_d, N_d and T_d respectively. If we use V_d it denotes either an S_d or a T_d. A space V_d on which a group of motions acts transitively is called *homogeneous*.

We now consider the question of the dimension r of the (maximal) group of motions admitted by a given Riemannian manifold. A useful step is provided by the following result.

Theorem 8.9. *If a Killing vector field* $\boldsymbol{\xi}$ *has* $\xi^a = 0$ *and* $\xi_{a;b} = 0$ *at a point* p, *then* $\boldsymbol{\xi} \equiv \boldsymbol{0}$.

Proof: Locally, any point p' may be joined to p by a geodesic, with tangent vector $\boldsymbol{v}$ at p, say. Then $\boldsymbol{\xi}$ fixes p and $\boldsymbol{v}$ (by (2.59)), and preserves the affine parameter distance along the geodesic with tangent vector $\boldsymbol{v}$ at p. It thus fixes p'. Thus $\boldsymbol{\xi} = \boldsymbol{0}$ at any p'.

From this result we see (i) that the isotropy and linear isotropy groups of p are isomorphic, (ii) that a Killing vector field will be completely specified given the $n(n+1)/2$ values of ξ_a and $\xi_{a;b}$ at a point p. Thus we have only to check if further restrictions have to be imposed on these values to make (8.23) soluble. It is a general result for systems of partial differential equations that all such restrictions are obtained by repeated differentiation (see e.g. Eisenhart (1933)). In the case of the Killing equation, the first differentiation of (8.23) gives

$$(\pounds_{\xi}\boldsymbol{\Gamma})^a{}_{bc} = 0 \Leftrightarrow \xi_{a;bc} = R_{abcd}\xi^d, \tag{8.24}$$

which, together with (8.23), gives a system of first-order differential equations for the quantities ξ_a, $\xi_{b;c}$. The integrability conditions given by further differentiation are exactly the equations

$$\pounds_{\xi}(\nabla_{a_1} \cdots \nabla_{a_N}\boldsymbol{R}) = 0, \qquad N = 0, 1, 2, \ldots \tag{8.25}$$

for the successive covariant derivatives of the Riemann tensor $\boldsymbol{R}$. Each of these gives an equation linear in ξ_a and $\xi_{b;c}$ (as we see from § 2.8.). Since there are at most $n(n+1)/2$ independent conditions, we see there must exist an integer Q such that the conditions (8.25) for $N > Q$ depend algebraically on those for $N \leqq Q$ at any point. From this argument, and similar considerations for the isotropy group ($\xi_a = 0$) and the group of conformal motions, one obtains the following results.

Theorem 8.10. *If the rank of the linear algebraic equations* (8.25) *for* ξ_a *and* $\xi_{a;b}$ *is* q, *then the maximal group* G_r *of motions of the* V_n *has* $r = \frac{1}{2}n(n+1) - q$ *parameters.*

Theorem 8.11 (Defrise (1969)). *For a* V_n *admitting a* G_r *of motions, the rank of the linear algebraic equations* (8.25) *for* $\xi_{a;b}$ *with* $\xi_c = 0$ *is* p *if and only if there is an isotropy subgroup* H_s, $s = \frac{1}{2}n(n-1) - p$.

Theorem 8.12 (Eisenhart (1949), Appendix 27). *The maximal order of a group* G_r *of conformal transformations in a* V_n *is* $r = \frac{1}{2}(n+1)(n+2)$.

To find the Killing vectors of a metric, or, alternatively, to find the restrictions on the metric and curvature of a space admitting a group G_r of motions with given r, one can use (8.25). Petrov (1966) largely worked by this method. As far as finding the Killing vectors of a given metric is concerned, one can usually derive the same information as would be provided by (8.25) by remarking that any invariantly-defined vector field (e.g. principal null direction of the Weyl tensor or velocity vector of a perfect fluid) or bivector field (e.g. those defined by the eigenblades of a type D Weyl tensor) or similar structure must be invariant under the isometries; the coordinates are usually adapted to some such invariant structure, and this facilitates the calculation.

In particular, scalar invariants of the Riemann tensor and its derivatives (see §§ 5.1. and 31.4.) must be invariant under isometries. Kerr (1963b) proved that in a four-dimensional Einstein space, the number of functionally independent scalar invariants is $4 - d$, where d is the dimension of the orbits of the maximal group of motions.

8.5. Spaces of constant curvature

A two-dimensional Riemannian space has only one independent component, R_{1212} say, of its curvature tensor. The tensor g_{abcd}, defined as in (3.46), has the same index symmetries as R_{abcd}, and is non-zero (being, in two dimensions, essentially the determinant of g_{ab}). Thus in two dimensions

$$R_{abcd} = K(g_{ac}g_{bd} - g_{ad}g_{bc}). \tag{8.26}$$

K is called the *Gaussian curvature.*

In a Riemannian space of more than two dimensions one can, at any point p, form a two-dimensional submanifold by taking all geodesics through p whose initial tangent vector is of the form $\alpha \boldsymbol{v} + \beta \boldsymbol{w}$, where α, β are real and $\boldsymbol{v}$ and $\boldsymbol{w}$ are fixed vectors at p. The equation (8.26) then defines the *sectional curvature* K of this two-dimensional manifold, and it can be shown that

$$K = \frac{R_{abcd}v^a w^b v^c w^d}{(g_{ac}g_{bd} - g_{ad}g_{bc})\, v^a w^b v^c w^d}. \tag{8.27}$$

The space V_n is said to be of *constant curvature* if K in (8.27) is independent of p and of $\boldsymbol{v}$ and $\boldsymbol{w}$. Then (8.27) leads to

$$\begin{aligned} &Q_{abcd} + Q_{adcb} + Q_{cbad} + Q_{cdab} = 0, \\ &Q_{abcd} \equiv R_{abcd} - K(g_{ac}g_{bd} - g_{ad}g_{bc}), \end{aligned} \tag{8.28}$$

and the Riemann tensor symmetries then yield (8.26) for the Riemann tensor of the V_n, with constant K.

If we take a space of constant curvature, the conditions (8.25) are all identically satisfied, so by Theorem 8.10 there is a group of motions of $\frac{1}{2}n(n+1)$ parameters.

If a space V_n admits an isotropy group of $\frac{1}{2}n(n-1)$ parameters, it is the whole of the relevant generalized orthogonal group (see Theorem 8.11). In this case (8.27) is independent of the choice of $\boldsymbol{v}$ and $\boldsymbol{w}$ and, as above, we obtain (8.28) and (8.26). The Bianchi identities $R_{ab[cd;e]} = 0$, contracted on b and d, yield $(n-2) \times (K_{,a}g_{ce} - K_{,c}g_{ae}) = 0$, and contracting again on a and e gives $(n-2)(n-1)K_{,c} = 0$. Thus, if $n \geqq 3$, K is constant and the space admits a G_r $\left(r = \frac{1}{2}n(n+1)\right)$

of motions. Conversely if a space V_n admits a G_r $\bigl(r = \frac{1}{2}\, n(n+1)\bigr)$ of motions then by (8.22) it admits an H_s $\bigl(s = \frac{1}{2}\, n(n-1)\bigr)$ of isotropies and is thus of constant curvature if $n \geqq 3$. If $n = 2$, a G_2 (or G_3) of motions must be transitive and then $\pounds_\xi \boldsymbol{R} = 0$ for the Riemann tensor leads to $K =$ constant.

Collecting together these arguments we find we have proved

Theorem 8.13. *A Riemannian space is of constant curvature if and only if it (locally) admits a group G_r of motions with $r = \frac{1}{2}\, n(n+1)$.*

Theorem 8.14. *A Riemannian space V_n $(n \geqq 3)$ is of constant curvature if and only if it (locally) admits an isotropy group H_s of $s = \frac{1}{2}\, n(n-1)$ parameters at each point.*

Theorem 8.15. *A two-dimensional Riemannian space admitting a G_2 of motions admits a G_3 of motions.*

Substituting (8.26) into the definition (3.50) of the Weyl tensor we find $C^a{}_{bcd} = 0$, and so the equations (3.83) can be solved to find the factor e^{2U} in (3.81) relating the metric to that of a flat space of the same dimension and signature, $\mathring{g}_{ab} = \mathrm{diag}$ $(\varepsilon_1, \ldots, \varepsilon_n)$, where $\varepsilon_1, \ldots, \varepsilon_n = \pm 1$ as appropriate. (3.83) is satisfied if

$$2(\mathrm{e}^{-U})_{,ab} = K\mathring{g}_{ab}\,, \qquad (\mathrm{e}^{-U})_{,a}\, (\mathrm{e}^{-U})^{,a} = K(e^{-U} - 1)\,, \tag{8.29}$$

the solution of which can be transformed to

$$\mathrm{e}^{-U} = 1 + \frac{K}{4}\, \mathring{g}_{ab} x^a x^b\,. \tag{8.30}$$

Hence the metric of a space V_n of constant curvature can always be written as

$$\mathrm{d}s^2 = \frac{\mathrm{d}x_a\, \mathrm{d}x^a}{\left(1 + \dfrac{K}{4}\ \ x_b x^b\right)^2} \tag{8.31}$$

(indices raised and lowered with $\mathring{g}_{ab}$), for any value of K or signature of V_n, and any two metrics of the same constant curvature and signature must be locally equivalent.

A space V_n of non-zero constant curvature, $K \neq 0$, can be considered as a hypersurface

$$Z_a Z^a + k(Z^{n+1})^2 = kY^2\,, \qquad K = kY^{-2}\,, \qquad k = \pm 1\,, \tag{8.32}$$

in an $(n+1)$-dimensional pseudo-Euclidean space with the metric

$$\mathrm{d}s^2 = \mathrm{d}Z_a\, \mathrm{d}Z^a + k(\mathrm{d}Z^{n+1})^2\,. \tag{8.33}$$

For each parametrization of Z^a and Z^{n+1} in terms of coordinates (e.g. angular coordinates) in V_n, in accordance with the surface equation (8.32), the metric of V_n can be obtained from (8.33), and the relation

$$x^a = \frac{2Z^a}{1 + (1 - KZ_b Z^b)^{1/2}} \tag{8.34}$$

yields the transformation from the x^a in (8.31) to the new coordinates in V_n.

In the remainder of this section we consider some special cases which play an important role in General Relativity. The metric

$$\mathrm{d}s^2 = \frac{\mathrm{d}x^2 + \mathrm{d}y^2 + \mathrm{d}z^2 - \mathrm{d}t^2}{\left[1 + \frac{K}{4}(x^2 + y^2 + z^2 - t^2)\right]^2} \tag{8.35}$$

of a space-time V_4 of constant curvature (*de Sitter space*) can be given in the equivalent form

$$\mathrm{d}s^2 = \frac{\mathrm{d}r^2}{1 - Kr^2} + r^2(\mathrm{d}\vartheta^2 + \sin^2\vartheta\,\mathrm{d}\varphi^2) - (1 - Kr^2)\,\mathrm{d}t^2. \tag{8.36}$$

In this metric, K may be related to a Λ-term (see § 5.2.) by $\Lambda = 3K$.

Gravitational fields often admit subspaces of constant curvature. On a single subspace, K is of course constant, but it may have differing values on different subspaces.

The metric

$$\mathrm{d}s^2 = \frac{\mathrm{d}x^2 + \mathrm{d}y^2 + \mathrm{d}z^2}{\left[1 + \frac{K}{4}(x^2 + y^2 + z^2)\right]^2} \tag{8.37}$$

of a three-dimensional positive-definite space (e.g. a spacelike hypersurface in a space-time) can be transformed (cp. (32.14)—(32.15)) to the form

$$\mathrm{d}s^2 = a^2[\mathrm{d}r^2 + \Sigma^2(r, \varepsilon)(\mathrm{d}\vartheta^2 + \sin^2\vartheta\,\mathrm{d}\varphi^2)], \qquad K = \varepsilon a^{-2}, \tag{8.38}$$

where $\Sigma(r, \varepsilon) = \sin r$, r, or $\sinh r$, respectively when $\varepsilon = 1$, 0, or -1.

The metrics of two-spaces of constant curvature have six distinct types

$$\mathrm{d}\sigma^2 = Y^2[(\mathrm{d}x^1)^2 \pm \Sigma^2(x^1, k)(\mathrm{d}x^2)^2], \qquad K = kY^{-2}. \tag{8.39}$$

with $\Sigma(x^1, k)$ chosen as in (8.38) for $k = \pm 1$ or 0. In the case of signature zero, and $k = -1$, the parametrization

$$Z^1 = Y\sin x^1 \sinh x^2, \qquad Z^2 = Y \sin x^1 \cosh x^2, \qquad Z^3 = Y\cos x^1 \tag{8.40}$$

leads (Barnes (1973b)) to the form

$$\mathrm{d}\sigma^2 = Y^2\big(-(\mathrm{d}x^1)^2 + \sin^2 x^1 (\mathrm{d}x^2)^2\big). \tag{8.41}$$

The specific case of spacelike surfaces S_2 frequently occurs. For these

$$\mathrm{d}\sigma^2 = \frac{2\,\mathrm{d}\zeta\,\mathrm{d}\bar{\zeta}}{\left(1 + \frac{K}{2}\zeta\bar{\zeta}\right)^2}, \qquad \zeta = \frac{1}{\sqrt{2}}(x^1 + \mathrm{i}x^2), \tag{8.42}$$

where x^1, x^2 are as in (8.31). For $k = -1$, the transformation $z = (1 + z')/(1 - z')$, $z' = \sqrt{-\frac{1}{2}K}\,\zeta$, leads to

$$\mathrm{d}\sigma^2 = 4Y^2 \frac{\mathrm{d}z\,\mathrm{d}\bar{z}}{(z + \bar{z})^2}. \tag{8.43}$$

Besides (8.43), other coordinate systems are frequently used in the literature, e.g. for S_2, $k = -1$,

$$d\sigma^2 = Y^2(d\vartheta^2 + \cosh^2 \vartheta \, d\varphi^2), \tag{8.44}$$

$$d\sigma^2 = Y^2(dx^2 + e^{2x} \, dy^2). \tag{8.45}$$

All the results given above follow from well-known classical methods and theorems and are described in many texts, e.g. Eisenhart (1933), Petrov (1966), Plebański (1967), Weinberg (1972).

8.6. Orbits of isometry groups

From the previous section we know a great deal about orbits V_n of groups of motions G_r with $r = \frac{1}{2} n(n + 1)$. In the present section we intend to discuss orbits of smaller groups of motions. We first note the following well-known theorems (see e.g. Eisenhart (1933)).

Theorem 8.16. *If the orbits of a group of motions are hypersurfaces then their normals are geodesic, and if the hypersurfaces are non-null they are geodesically parallel, i.e. taking an affine parameter along the normal geodesics as the coordinate x^n, the metric has the form*

$$ds^2 = g_{\mu\nu} \, dx^\mu \, dx^\nu + \varepsilon(dx^n)^2, \qquad \mu, \nu = 1, \ldots, (n - 1), \qquad \varepsilon = \pm 1. \tag{8.46}$$

Theorem 8.17 (Fubini's Theorem). *A Riemannian manifold V_n of dimension $n > 2$ cannot have a maximal group of motions of $\frac{1}{2} n(n + 1) - 1$ parameters.*

8.6.1. Simply-transitive groups

If a group is simply-transitive, the map $\mu_p : G \to \mathcal{O}_p$ used in § 8.3. is an isomorphism and thus $(\mu_p)_*$ can be used to map the left-invariant vector fields. Taking a basis of these on each orbit, the metric in the orbit is

$$d\sigma^2 = (\boldsymbol{\eta}_A \cdot \boldsymbol{\eta}_B) \, \boldsymbol{\omega}^A \boldsymbol{\omega}^B = g_{AB} \boldsymbol{\omega}^A \boldsymbol{\omega}^B, \tag{8.47}$$

where the $\boldsymbol{\omega}^A$ are dual to the $\boldsymbol{\eta}_A$ (this is unambiguous for a non-null submanifold if the basis is completed by adding vectors orthogonal to the orbits; if the orbits are null, the dual to the null normal will lie out of the orbit in general, but even then similar ideas can be applied). For any Killing vector $\boldsymbol{\xi}$, (8.23) and (8.8) show $\pounds_\xi \, g_{AB} = 0$, so the g_{AB} are constant in the orbit.

If a simply-transitive group G is intransitive on the V_n, one can choose the basis $\boldsymbol{e}_A = \boldsymbol{\eta}_A$, $(A = 1, \ldots, r)$, separately in each orbit, and complete a basis $\{\boldsymbol{e}_a; a = 1, \ldots, n\}$ of the tangent space at one point p in each orbit by adding $(n - r)$ arbitrary vectors. If vector fields are then defined throughout $\mathcal{O}_p$ by using the $(\tau_q)_*$ on $\{\boldsymbol{e}_a\}$ we find, similarly to the above arguments, that

$$ds^2 = g_{ab} \boldsymbol{\omega}^a \boldsymbol{\omega}^b \tag{8.48}$$

with g_{ab} *constant in each orbit.*

The vector fields η_A generate a group of transformations on each orbit, called the *reciprocal group*; it will not necessarily share any symmetry of the transformation group G_r, i.e. in the present case it will not in general consist of isometries.

It is often convenient to choose an orthonormal basis of reciprocal group generators in each orbit. Their Lie algebra cannot then be completely reduced to canonical form because only the generalized orthogonal group of linear transformations, and not the general linear group, is available. For a simply-transitive G_3 on S_3 we can reduce the commutators of such orthonormal reciprocal group generators $\{\boldsymbol{E}_\alpha; \alpha = 1, 2, 3\}$ to the form

$$[\boldsymbol{E}_\alpha, \boldsymbol{E}_\beta] = \gamma^\delta{}_{\alpha\beta}\boldsymbol{E}_\delta, \qquad \frac{1}{2}\gamma^\delta{}_{\alpha\beta}\varepsilon^{\alpha\beta\varphi} = n^{(\delta\varphi)} + \varepsilon^{\delta\varphi\nu}a_\nu, \tag{8.49}$$

$$(n^{\delta\varphi}) = \operatorname{diag}(n_1, n_2, n_3) \qquad \text{and} \qquad a^\nu = (a, 0, 0), \tag{8.50}$$

where $\varepsilon^{\alpha\beta\gamma}$ is the natural skew tensor defined by $g_{\alpha\beta}$ (up to sign). In the case of a G_4 simply-transitive on space-time the orthonormal reciprocal group generators have commutators

$$[\boldsymbol{E}_a, \boldsymbol{E}_b] = D^c{}_{ab}\boldsymbol{E}_c, \tag{8.51}$$

where $D^c{}_{ca} \neq 0$ or, using (8.10), (8.16) or (8.17) holds. (The proof of this is just like that of Theorem 8.4.)

In the case of a (non-null) orbit of a simply-transitive group the curvature tensor of the orbit is easily calculated from (3.20) and (2.79), remembering that here, using (8.47), the g_{AB} and $D^C{}_{AB}$ are constants. One gets

$$\Gamma_{CAB} = \frac{1}{2}(D_{ACB} + D_{BCA} - D_{CAB}), \tag{8.52}$$

$$R^D{}_{ABC} = \Gamma^E{}_{AC}\Gamma^D{}_{EB} - \Gamma^E{}_{AB}\Gamma^D{}_{EC} - D^E{}_{BC}\Gamma^D{}_{AE}, \tag{8.53}$$

$$R_{AB} = -\frac{1}{2}D^E{}_{DA}D^D{}_{EB} - \frac{1}{2}D^E{}_{DA}D_E{}^D{}_B + \frac{1}{4}D_{ADE}D_B{}^{DE} - \frac{1}{2}(D^D{}_{DE})(D_{AB}{}^E + D_{BA}{}^E), \tag{8.54}$$

where indices are to be raised and lowered by g^{AB} and g_{AB}. One can, by the choice giving (8.10), use the group structure constants directly. For space-time metrics with groups G_3 simply-transitive on hypersurfaces (e.g. spatially-homogenous cosmologies), it will be convenient to have expressions for ξ_A, η_A and ω^A obeying (8.10), (8.47) and Table 8.1 in terms of the canonical coordinates mentioned at the end of § 8.1. These are given as Table 8.2. Note that such coordinates can still be chosen in many different ways, owing to the initial basis freedom listed in Table 8.1 and the freedom of choice of p for μ_p in the orbit (§ 8.3.).

Table 8.2 Expressions for Killing vectors and reciprocal group generators in canonical coordinates for the Bianchi types: for full explanation, see text

	I	*II*	*IV*	*V*	*VI* (including *III*)	*VII*
ξ_A	∂_x ∂_y ∂_z	∂_x ∂_y $\partial_z + z\,\partial_x$	$\partial_x - y\,\partial_y - (y+z)\,\partial_z$ ∂_y ∂_z	$\partial_x - y\,\partial_y - z\,\partial_z$ ∂_y ∂_z	$\partial_x + (z - Ay)\,\partial_y$ $+ (y - Az)\,\partial_z$ ∂_y ∂_z	$\partial_x + (z - Ay)\,\partial_y$ $- (y + Az)\,\partial_z$ ∂_y ∂_z
η_A	∂_x ∂_y ∂_z	∂_x $\partial_y + z\,\partial_x$ ∂_z	∂_x $e^{-x}(\partial_y - x\,\partial_z)$ $e^{-x}\,\partial_z$	∂_x $e^{-x}\,\partial_y$ $e^{-x}\,\partial_z$	∂_x $e^{-Ax}(\cosh x\,\partial_y + \sinh x\,\partial_z)$ $e^{-Ax}(\sinh x\,\partial_y + \cosh x\,\partial_z)$	∂_x $e^{-Ax}(\cos x\,\partial_y - \sin x\,\partial_z)$ $e^{-Ax}(\sin x\,\partial_y + \cos x\,\partial_z)$
ω_A	dx dy dz	$dx - z\,dy$ dy dz	dx $e^{x}\,dy$ $e^{x}(dz + x\,dy)$	dx $e^{x}\,dy$ $e^{x}\,dz$	dx $e^{Ax}(\cosh x\,dy - \sinh x\,dz)$ $e^{Ax}(-\sinh x\,dy + \cosh x\,dz)$	dx $e^{Ax}(\cos x\,dy - \sin x\,dz)$ $e^{Ax}(\sin x\,dy + \cos x\,dz)$

	VIII	*IX*
ξ_A	∂_x $-\sinh x \tanh y\,\partial_x + \cosh x\,\partial_y - \sinh x\,\mathrm{sech}\,y\,\partial_z$ $\cosh x \tanh y\,\partial_x - \sinh x\,\partial_y + \cosh x\,\mathrm{sech}\,y\,\partial_z$	∂_x $\sin x \tan y\,\partial_x + \cos x\,\partial_y + \sin x \sec y\,\partial_z$ $\cos x \tan y\,\partial_x - \sin x\,\partial_y + \cos x \sec y\,\partial_z$
η_A	$\mathrm{sech}\,y \cos z\,\partial_x - \sin z\,\partial_y - \tanh y \cos z\,\partial_z$ $\mathrm{sech}\,y \sin z\,\partial_x + \cos z\,\partial_y - \tanh y \sin z\,\partial_z$ ∂_z	$\sec y \cos z\,\partial_x - \sin z\,\partial_y + \tan y \cos z\,\partial_z$ $\sec y \sin z\,\partial_x + \cos z\,\partial_y + \tan y \sin z\,\partial_z$ ∂_z
ω_A	$\cosh y \cos z\,dx - \sin z\,dy$ $\cosh y \sin z\,dx + \cos z\,dy$ $\sinh y\,dx + dz$	$\cos y \cos z\,dx - \sin z\,dy$ $\cos y \sin z\,dx + \cos z\,dy$ $-\sin y\,dx + dz$

8.6.2. Multiply-transitive groups

Here we consider only non-null orbits. Schmidt (1968) has shown how to calculate all possible Lie algebras for a given isotropy group and dimension and signature of orbit, and how to find the curvature of the resulting orbits.

The method is as follows. The isotropy subgroup H_s of a chosen point p must be a (known) subgroup of the generalized orthogonal group and hence the commutators of its generators, $\{\boldsymbol{Y}_i;\ i = 1, \ldots, s\}$, are known. The basis of generators of the complete group of motions can be completed by adding d non-zero Killing vectors $\{\xi_\alpha;\ \alpha = 1, \ldots, d\}$ which may be chosen at p in a way adapted to the isotropy group (e.g. if H_s consists of null rotations fixing a vector $\boldsymbol{k}$, choose $\xi_1 = \boldsymbol{k}$). The action of the (linear) isotropy group on vectors at p is known (from that of the generalized orthogonal group) and hence the commutators $[\xi_\alpha, \boldsymbol{Y}_i]$ are known up to terms in $\boldsymbol{Y}_j$, i.e. the structure constants $C^\beta{}_{\alpha i}$ are known. The unknown structure constants $C^j{}_{i\alpha}$, $C^j{}_{\alpha\beta}$, $C^\gamma{}_{\alpha\beta}$ must satisfy the Jacobi identities (8.7) and all possibilities can then be enumerated. One can add to Schmidt's remarks that if the group has a simply-transitive subgroup, this latter must have a basis $\{\boldsymbol{Z}_\alpha;\ \alpha = 1, \ldots, d\}$ agreeing with $\{\xi_\alpha\}$ at p. Therefore one must have

$$\boldsymbol{Z}_\alpha = \xi_\alpha + A_\alpha{}^i \boldsymbol{Y}_i, \tag{8.55}$$

where the $A_\alpha{}^i$ are constants. One can easily evaluate the commutators $[\boldsymbol{Z}_\alpha, \boldsymbol{Z}_\beta]$, and the condition that these should be spanned by the $\boldsymbol{Z}_\alpha$ (so that a subalgebra is generated) gives restrictions on the $A_\alpha{}^i$. All possible simply-transitive subgroups can thus be determined.

Using the basis vector fields $\{\xi_\alpha\}$ in a neighbourhood of p, it is possible to determine the curvature at p as follows (Schmidt (1971)). The connection coefficients are given by

$$\nabla_{\xi_\alpha}\xi_\beta = \Gamma^\gamma{}_{\alpha\beta}\xi_\gamma. \tag{8.56}$$

The commutator gives

$$2\Gamma^\gamma{}_{[\alpha\beta]} = C^\gamma{}_{\alpha\beta} \tag{8.57}$$

at p. The symmetric part $\Gamma^\gamma{}_{(\alpha\beta)}$ of the connection can be found using the Killing equations which yield $\Gamma_{\beta\gamma\alpha} + \Gamma_{\gamma\beta\alpha} = 0$ and thus

$$2\Gamma_{\gamma(\alpha\beta)} = C_{\beta\alpha\gamma} + C_{\alpha\beta\gamma}, \tag{8.58}$$

whence

$$2\nabla_{\xi_\alpha}\xi_\beta = [\xi_\alpha, \xi_\beta] + \{[\xi_\alpha, \xi_\gamma]\cdot\xi_\beta + [\xi_\beta, \xi_\gamma]\cdot\xi_\alpha\}\, g^{\gamma\delta}\xi_\delta. \tag{8.59}$$

Now to compute the next derivative, and hence the Riemann tensor, by (2.77), we need $\nabla_{\xi_\alpha}\boldsymbol{g}$ and $\nabla_{\xi_\alpha}\boldsymbol{Y}_i$. The first of these involves only $\nabla_{\xi_\alpha}\xi_\beta$ and $\nabla_{\xi_\alpha}\xi_\gamma$ since $\pounds_{\xi_\alpha}\boldsymbol{g} = 0$; these are already known at p. Also $\nabla_{\xi_\alpha}\boldsymbol{Y}_i = \nabla_{\boldsymbol{Y}_i}\xi_\alpha + [\xi_\alpha, \boldsymbol{Y}_i]$ and since $\boldsymbol{Y}_i = \boldsymbol{0}$ at p, this simplifies, at p, to $\nabla_{\xi_\alpha}\boldsymbol{Y}_i = [\xi_\alpha, \boldsymbol{Y}_i]$ which is known. Thus the components of the Riemann tensor, in the basis ξ_α, at p, can be evaluated.

We now give some examples of this method with applications in the sequel.

First we determine the possible isometry groups of two-dimensional positive definite spaces of constant curvature. The isotropy $\boldsymbol{Y}$ is a spatial rotation, and ξ_1, ξ_2, can be chosen at p so that

$$[\xi_1, \boldsymbol{Y}] = \xi_2 + \alpha \boldsymbol{Y}, \tag{8.60a}$$

$$[\xi_2, \boldsymbol{Y}] = \xi_1 + \beta \boldsymbol{Y}.$$

A linear transformation $\xi_1 \to \xi_1 + \beta \boldsymbol{Y}$, $\xi_2 \to \xi_2 - \alpha \boldsymbol{Y}$, eliminates α and β. There is a finite transformation $\Phi: (\xi_1, \xi_2) \to (-\xi_1, -\xi_2)$ so that $\Phi[\xi_1, \xi_2] = [\xi_1, \xi_2]$ implies

$$[\xi_1, \xi_2] = K\boldsymbol{Y}. \tag{8.60b}$$

By the method outlined above one can compute the Riemann tensor and show that K is its constant curvature. Clearly the isometry group is a G_3 of Bianchi type IX, VII_0, or VIII respectively when K is positive, zero, or negative. On changing the basis to $\boldsymbol{Z}_1 = \xi_1 + \alpha \boldsymbol{Y}$, $\boldsymbol{Z}_2 = \xi_2 + \beta \boldsymbol{Y}$ we find

$$[\boldsymbol{Z}_1, \boldsymbol{Z}_2] = -\alpha \boldsymbol{Z}_1 - \beta \boldsymbol{Z}_2 + (K + \alpha^2 + \beta^2)\, \boldsymbol{Y}. \tag{8.61}$$

Thus there is a simply-transitive subgroup of type G_2I if $K = 0$ (given by $\alpha = \beta = 0$), and a one-parameter family of simply-transitive subgroups of type G_2II (conjugate to one another within the G_3) if $K < 0$, given by $\alpha = |K| \sin \varphi$, $\beta = |K| \cos \varphi$ for arbitrary angle φ. The second of these results does not appear to be widely known. If $K > 0$ there are no simply-transitive subgroups; this is equivalent to the statement that the rotation group of three-dimensional space (whose orbits are the spheres centred at the origin, two-dimensional spaces of constant curvature) has no two-dimensional subgroup. The three-dimensional "Lorentz group" (Bianchi VIII), however, has simply-transitive G_2 subgroups; they are generated by the combinations of a null rotation and a boost.

It is quite useful to calculate the Killing vectors for the S_2 of constant curvature. For the form (8.42) they have components given by

$$\xi^\zeta = \frac{1}{2} \gamma K \zeta^2 + \mathrm{i} a \zeta + \bar{\gamma}, \tag{8.62}$$

where a is real and γ complex (three parameters).

A second application of Schmidt's method is to prove that the maximal isotropy group of a space-time cannot consist of a (non-trivial) combination of a boost and a spatial rotation. If it did, the full isometry group must be a G_5 on V_4 (because a G_3 on V_2 or G_4 on V_3 would have isotropies acting only in 2- or 3-dimensional subspaces of the tangent space). Using a basis at p in which ξ_1, ξ_2 are spacelike unit vectors in the rotation plane, and ξ_3, ξ_4 null vectors in the boosted plane such that $\xi_3 \cdot \xi_4 = -1$, Schmidt's calculation shows that the four Killing vectors $(\xi_1, \xi_2, \xi_3, \xi_4)$ form an Abelian group and thus the space is flat. Hence its isotropy group is really an H_6.

The result of Theorem 8.16 is a special case of a phenomenon known as *orthogonal transitivity*. This occurs when the orbits of a group of motions are submanifolds of a V_n which have orthogonal surfaces. Physically important examples arise in static and in stationary axisymmetric space-times. Schmidt proved a number of further theorems on this matter, including

Theorem 8.18 (Schmidt (1967)). *If a group G_r of motions of $r = \frac{1}{2}d(d+1)$ parameters has orbits of dimension d ($d > 1$) the orbits admit orthogonal surfaces.*

PART II. SOLUTIONS WITH GROUPS OF MOTIONS

Chapter 9. Classification of solutions with isometries

9.1. The cases to be considered

In specifying the symmetry properties of a metric one has to state the dimension of the maximal group of motions, its algebraic structure, and the nature and dimension of its orbits. For this purpose we shall, as in § 8.4., use the following notations: the symbols S, T and N will denote, respectively, spacelike, timelike and null orbits, and will be followed by a subscript giving the dimension. If the group is transitive on the whole manifold V_4, the space-time will be said to be *homogeneous*. If the group is transitive on S_3, T_3, or N_3, the space-time will be called *hypersurface-homogeneous* (or, respectively, spatially-homogeneous, time-homogeneous, or null-homogeneous).

It turns out that if the orbits are null, the construction of the metric and the understanding of its properties have to be achieved by a rather different method from that used when the orbits are non-null. Accordingly we give first the discussion of non-null orbits (Chapters 10.—20.) and later the discussion of null orbits (Chapter 21.). Within this broad division we proceed in order of decreasing dimension of the orbits. A further subdivision occurs because the group may be multiply- or simply-transitive on its orbits. We shall in general treat the higher-dimensional (multiply-transitive) case first. However, it frequently happens that the multiply-transitive group contains a subgroup simply-transitive on the same orbits, and in this case it may be advantageous to make use of the simply-transitive subgroup.

All these subdivisions give a rather long list of inequivalent possible structures. As far as possible we shall try to group these together when looking for exact solutions, and for each case we will first discuss the form of the metric and only subsequently nsert this into the Einstein field equations. As usual we will mainly deal with vacuum, electromagnetic, pure radiation and perfect fluid energy-momentum tensors.

Even a short search of the literature reveals that certain types of space-time symmetry have attracted much more attention than others, either because of their physical importance, or because they are mathematically tractable and interesting. We have therefore devoted separate chapters to some of these special cases, these chapters following the appropriate more general chapters.

One particular complication is that the maximal group of motions of a space-time may contain a considerable number of inequivalent subgroups. The metric may therefore be re-discovered as a special case of a space-time invariant under one of these subgroups, often in a form which makes it difficult to recognize. We shall try to indicate such possibilities.

For the energy-momentum tensors of perfect fluid fields, the quantities appearing in (5.9) clearly have the same symmetries as the metric. The analogous result for Einstein-Maxwell fields is not true, however. Instead we have

Theorem 9.1. *If a source-free Einstein-Maxwell field admits a Killing vector field* ξ, *then*

$$\pounds_\xi F_{ab} = C\tilde{F}_{ab}, \tag{9.1}$$

where C *is a constant for a non-null electromagnetic field, and* $C_{,[a}k_{b]} = 0$ *for a null field with repeated principal null direction* $\boldsymbol{k}$.

The proof for non-null fields (Ray and Thompson (1975)) uses the information obtained from the Rainich treatment (§ 5.4): the extremal field f_{ab} and the gradient $\alpha_{,a}$ of the complexion are determined by the metric; hence $\pounds_\xi f_{ab} = 0 = \pounds_\xi \alpha_{,a}$, and (5.23) then yields (9.1). The proof for null fields is given in Coll (1975a).

The metric (10.21) and special plane waves (§ 21.5.) provide us with examples where the Maxwell field does not share the symmetries of the metric ($C \neq 0$ in (9.1)). See also Hoenselaers (1978b), □ Ftaclas and Cohen (1978).

Petrov (1966) and his colleagues were the first to give a systematic treatment of metrics with isometries, and we therefore inevitably recover many of Petrov's results in the following chapters.

9.2. Isotropy and the curvature tensor

Following the remarks in § 8.4. we see that if a space-time admits a group G_r of motions transitive on orbits of dimension $d(< r)$, each point has an isotropy group of dimension $r - d$ (see equation (8.22)), which must be (isomorphic to) a subgroup of the Lorentz group.

Space-times with such isotropy have been particularly intensively studied. The term *locally isotropic* has been introduced for the cases where every point p has a non-trivial isotropy group (Cahen and Defrise (1968)); when this group consists of spatial rotations the space-time is called *locally rotationally symmetric* (Ellis (1967)). Some general remarks about the possible cases are given in this section. It can be shown that local isotropy of the Riemann tensor and its first derivatives is sufficient to ensure the existence of a group of motions (Ellis (1967), Cahen and Defrise (1968), Siklos (1976b)). Such a group must clearly be continuous (since every point in some neighbourhood has an isotropy subgroup of at least one parameter), and is at least a G_3 (acting on 2-surfaces).

Schmidt (1968) gave a calculation of all possible Lie algebras of isometry groups acting transitively on a V_4 with local isotropy, and Defrise (1969) determined all locally isotropic metric forms. Cahen and Defrise (1968) found all such vacuum type D metrics, while Ellis (1967) and Stewart and Ellis (1968) gave all locally rotationally symmetric metrics with perfect fluid and electromagnetic field. The approaches used were rather similar, in that tetrads were defined up to the isotropy by properties of the curvature tensor, and the invariance under the isotropy then imposed.

In §§ 4.2. and 5.1. (see Table 5.2) the maximal linear isotropy group was determined for each Petrov and Plebański type. We recall that only in Petrov types D, N, or O was any isotropy possible. In these cases the permitted isotropies were, respectively, the group generated by spatial rotations (3.16) and boosts (3.17), the group of null rotations (3.15), and the Lorentz group; subgroups of these are, of course, permitted.

One can determine every possible subgroup of the Lorentz group; the details are given in many texts. For our purpose, some of these possibilities are irrelevant, because only certain isotropies are consistent with given Weyl and Ricci tensors. To begin with we can combine the information of §§ 4.2., 5.1. and 8.5. to obtain

Theorem 9.2. *The only Einstein spaces* ($R_{ab} = \Lambda g_{ab}$) *with a group of motions* G_r, $r \geqq 7$, *are the spaces of constant curvature* (8.35) *which admit a* G_{10}.

Further we can see that there are no space-times with a G_r ($r \geqq 7$) containing an electromagnetic field with $C = 0$ in (9.1), and that the only possible metrics with exactly a G_7 are a perfect fluid solution with an H_3 of spatial rotations at each point, and a pure radiation solution with an H_3 generated by null rotations (3.15) and spatial rotations (3.16). Both these spaces must be conformally flat. Both in fact exist, being the Einstein static universe and special plane waves (see Chapter 10.). The same energy-momentum tensors would be required for the cases where the maximal group is a G_6 with the same isotropy groups H_3.

Now let us briefly list the less-highly-symmetric cases with isotropy permitted by the algebra of the curvature tensor.

For vacuum, one might have locally isotropic *homogeneous* Petrov type D or type N solutions with an H_2 or H_1. Actually the only case will be shown in the next chapter to be the plane waves of Petrov type N, with a G_6, and an H_2 of null rotations at each point. One may also have inhomogeneous spaces with a multiply-transitive G_3 or G_4, or, with null orbits, G_5 (which is impossible for non-null orbits by Theorem 8.17).

For the Λ-term, the situation is essentially the same as for vacuum, except that there is a Petrov type D solution with a G_6.

With a non-null Einstein-Maxwell field the locally isotropic metrics are either conformally flat or of Petrov type D, and in the latter case the invariant planes of R_{ab}, $C^*{}_{abcd}$ and $F^*{}_{ab}$ all coincide. The only such homogeneous space-time is found in Chapter 10 to admit a G_6; it is the Bertotti-Robinson metric, and is conformally flat. Solutions with a G_4 on V_3 or G_3 on V_2 are admissible.

For a null electromagnetic field, a locally isotropic solution must be conformally flat or Petrov type N, and again the invariant planes of R_{ab}, $C^*{}_{abcd}$ and $F^*{}_{ab}$ agree. The only homogeneous cases turn out to be special plane waves (see Chapter 10.). Cases with lesser symmetry could have a G_5 or G_4 on N_3 or a G_3 on N_2, in principle (but see Chapter 21.). The only conformally flat cases are the special plane waves, cp. § 32.5.

The situation for pure radiation is essentially the same as for null electromagnetic fields (disregarding remarks about $F^*{}_{ab}$).

Finally we come to the case of perfect fluid solutions. The conformally flat perfect fluid solutions (§ 32.5.) admitting an isotropy are the Friedmann-Robertson-Walker universes (§ 12.2.) and the interior Schwarzschild metric (§ 14.1.). The locally isotropic non-conformally-flat perfect fluid admit a H_1 of spatial rotations. Such solutions exist for a G_5 on V_4 (the Gödel solution), G_4 on V_3 and G_3 on V_2, and were all explicitly determined by Ellis (1967) and Stewart and Ellis (1968). Note that no solution with a H_2 of spatial rotations arises, essentially because, as shown in § 8.6., the rotation group has no two-dimensional subgroup. These locally rotationally symmetric solutions were divided by Ellis (1967) into three classes: in his Class *I* there is a timelike congruence with rotation, and the two-planes in which the isotropy acts are non-integrable; in Class *II* these two-planes are integrable, and the timelike congruence is non-rotating; in Class *III* the timelike congruence is non-rotating but the two-planes are not integrable.

The arrangement of the remaining chapters of Part II is given by the following table.

Table 9.1 Metrics with isometries listed by orbit and group action, and where to find them. (3.15) are null rotations, (3.16) spatial rotations, and (3.17) are boosts

Orbit	Maximal group	Isotropy subgroup	Relevant chapters and sections
V_4	G_7	(3.15) and (3.16)	10.5.
		H_3 of space rotations	10.4.
	G_6	(3.15)	10.2., 10.5.
		(3.16) and (3.17)	10.3., 10.5.
	G_5	(3.15), real B	10.5.
		(3.16)	10.4.
	G_4	—	10.
S_3	G_6	H_3 of space rotations	11.1., 12.
	G_4	(3.16)	11., 12., 13.4., 13.6.
	G_3	—	11., 12.
T_3	G_6	3-dim. Lorentz group	11.
	G_4	(3.16) or (3.17)	11., 13.4., 13.6., 14.1.
	G_3	—	11., 20.2.
S_2	G_3	(3.16)	13., 14.
	G_2	—	15., 20.
T_2	G_3	(3.17)	13.
	G_2	—	15., 17.—19.
S_1	G_1	—	15.4.
T_1	G_1	—	15.4., 16.
N_3, N_2, N_1	G_r $(1 \leqq r \leqq 6)$	(3.15) and/or (3.16)	21.

Chapter 10. Homogeneous space-times

10.1. The possible metrics

A homogeneous space-time is one which admits a transitive group of motions. It is quite easy to write down all possible metrics for the case where the group is simply-transitive, or where it contains a simply-transitive subgroup. (See § 8.6. and below.) Difficulties may arise when there is no simply-transitive subgroup, and we will consider this possibility first.

From the remarks in § 9.2. we see that there are only a limited number of cases to consider, and we take each in turn. The metric with constant curvature is (8.35).

If the space-time admits an isotropy group containing the two-parameter group of *null rotations* (3.15), but is not of constant curvature, then it is either of Petrov type N, in which case we can find a complex null tetrad such that (4.10) holds, or it is conformally flat, with a pure radiation energy-momentum tensor, and we can choose a null tetrad such that (5.8) holds with $\Phi^2 = 1$. In either case the tetrad is fixed up to null rotations (together with a spatial rotation in the latter case). The covariant derivative of $\boldsymbol{k}$ in this tetrad must be invariant under the null rotations, which immediately gives

$$\varkappa = \varrho = \sigma = \varepsilon + \bar{\varepsilon} = \tau + \bar{\alpha} + \beta = 0. \tag{10.1}$$

Since τ and σ are invariantly defined for the tetrad described, (7.43) yields

$$\tau(\tau + \beta - \bar{\alpha}) = 0, \tag{10.2}$$

and thus either $\tau = 0$, in which case $\boldsymbol{k}$ is (proportional to) a covariantly constant vector and we arrive at plane waves (§ 21. 5.), or

$$\tau + \beta = 0 = \alpha. \tag{10.3}$$

In the latter case (7.42) yields $\gamma = 0$ and (7.44) shows that

$$\Lambda = -6\tau\bar{\tau} \neq 0. \tag{10.4}$$

It is convenient to alter the tetrad choice so that $\Psi_4 = \Lambda/2$; the equation (7.67) then shows that $\tau = \pm\bar{\tau}$, and, correspondingly, $\Phi_{22} = \mp 5\Lambda/6$. A position-dependent null rotation of the tetrad is still permitted, and may be used to set

$$\bar{\pi} = -\tau, \qquad \lambda = \mu = 0, \qquad \nu = -\tau. \tag{10.5}$$

Since the resulting tetrad has constant spin coefficients, it generates a transformation group whose reciprocal group must be a simply-transitive isometry group. With the choice (10.5) the commutators enable one to introduce coordinates so that the metric is

$$\mathrm{d}s^2 = \frac{3}{|\Lambda|}\left[\frac{\mathrm{d}y^2 + \mathrm{d}z^2}{y^2} - \frac{\mathrm{d}v}{y^2}\left(\mathrm{d}u - \frac{\Lambda\,\mathrm{d}v}{|\Lambda|\,y^2}\right)\right]. \tag{10.6}$$

This is a pure radiation solution of Petrov type N with a cosmological constant. It was first given by Defrise (1969).

In the case with an additional spatial rotation symmetry, this symmetry implies hat for the null tetrad fixed by $\Phi_{22} = 1$, $\tau = 0$, and thus only a special plane wave is possible. From § 21.5, we see the plane wave solutions must be

$$\mathrm{d}s^2 = 2\,\mathrm{d}\zeta\,\mathrm{d}\bar{\zeta} - 2\,\mathrm{d}u\,\mathrm{d}v - 2\left(\bar{A}(u)\,\bar{\zeta}^2 + A(u)\,\zeta^2 + B(u)\,\zeta\bar{\zeta}\right)\mathrm{d}u^2 \tag{10.7}$$

with $R_{ab} = B(u)\,k_a k_b$, and to admit a G_6 or G_7 we require special forms for $A(u)$ and $B(u)$ (see Table 21.1 and § 10.5.).

The other possible cases with a G_6 are those with an isotropy group composed of *boosts* (3.17) and *rotations* (3.16). A short calculation by Schmidt's method (§ 8.6.) reveals that the metric must be that of the product of two two-spaces of constant curvature, i.e.

$$\mathrm{d}s^2 = A^2\left(\mathrm{d}x^2 + \Sigma^2(x, k)\,\mathrm{d}y^2\right) + B^2\left(\mathrm{d}z^2 - \Sigma^2(z, k')\,\mathrm{d}t^2\right), \tag{10.8}$$

where A and B are constants. This space is symmetric, cp. (31.33).

A metric with an isotropy group H_3 consisting of *rotations* must contain a preferred timelike vector field $\boldsymbol{u}$. The isotropy of the covariant derivative of $\boldsymbol{u}$ shows that $\boldsymbol{u}$ is hypersurface-orthogonal and shearfree. The hypersurfaces to which $\boldsymbol{u}$ is orthogonal have constant curvature (by Theorem 8.14), and Theorem 8.16 and (8.38) give the metric as

$$\mathrm{d}s^2 = a^2(t)\left(\mathrm{d}r^2 + \Sigma^2(r, \varepsilon)\,(\mathrm{d}\vartheta^2 + \sin^2\vartheta\,\mathrm{d}\varphi^2)\right) - \mathrm{d}t^2, \tag{10.9}$$

which is the well known Robertson-Walker metric form (Robertson (1935, 1936), Walker (1936)). For a homogeneous space-time we will require $a(t)$ to be constant, since it is clearly an invariant (assuming the H_3 is the maximal isotropy group).

Now we have to consider the possibility of a maximal *isotropy group* H_1.

For the case of an H_1 of spatial rotations, Schmidt's calculations (Schmidt (1968)) show that there is a simply-transitive subgroup G_4, except in the case where the full group of motions is a G_6 and the metric is (10.8). The same holds for the case of an H_1 of Lorentz transformations (3.17). Finally we have the case of an H_1 of null rotations. Here a calculation by Schmidt's method again shows that there is a simply-transitive G_4 subgroup in all cases.

The existence of a *simply-transitive group* enables one to make the solution of the field equations into a purely algebraic problem. To do so one simply chooses a set of reciprocal group generators which form an orthonormal or a complex null tetrad; the connection coefficients (§ 3.3.) or spin coefficients (§ 7.1.) will be constants, and the curvature is easily calculated (cf. § 8.6., equations (8.52)—(8.54)), in terms of the structure constants of the simply-transitive isometry group.

The classification of groups G_4 following Theorem 8.4 (§ 8.2.) permits the further simplification of aligning one of the tetrad vectors with the distinguished vector of the class (A_E, P_C or L^B), except in the trivial case $\boldsymbol{A} = \boldsymbol{L} = \boldsymbol{P} = \boldsymbol{O}$ when the space-time is flat. The resulting Ricci tensors have been given in detail by Hiromoto and Ozsváth (1978). (Essentially the same results can be achieved by taking the reciprocal group generators so that $C^A{}_{BC}$ is in canonical form and algebraically determining the g_{AB}.)

The space-times thus found tend to have rather a large number of different invariant characterizations, and so may be recovered in various ways. As a consequence of Theorem 8.5 all the space-times with a simply-transitive G_4 have a G_3 transitive on hypersurfaces, and so, in principle, recur in Chapters 11. and 21.

Computation of the permissible homogeneous spaces for a given energy-momentum tensor by the systematic treatment of the various possibilities listed above is very laborious, and more elegant proofs are available in some cases, see §§ 10.2. and 10.3.

10.2. Homogeneous vacuum and null Einstein-Maxwell space-times

Consider a homogeneous null electromagnetic field. Taking a complex null tetrad such that $\Phi_2 = 1$, $\Phi_0 = \Phi_1 = 0$, the equations (7.46)–(7.49) yield

$$\varkappa = \sigma = \varrho - 2\varepsilon = \tau - 2\beta = 0 \tag{10.10}$$

and (6.33) then shows that $\varrho = 0$ (since ϱ must be constant, being an invariant). From the Goldberg-Sachs theorem, $\Psi_0 = \Psi_1 = 0$. Now (7.43) and (7.44) give $\Psi_2 = 0$, $\tau(\tau + \beta - \bar{\alpha}) = 0$, (7.39) gives $\alpha - \bar{\beta} = 0$ and thus $\tau = 0$. This leads to plane waves (§ 21.5.), since $\boldsymbol{k}$ must be proportional to a covariantly constant vector.

Homogeneous vacuum spaces with a multiply-transitive group must be type D or N. Taking a geodesic shearfree $\boldsymbol{k}$, the Bianchi identities in the type D case give,

$$\varkappa = \sigma = \lambda = \nu = \varrho = \mu = \tau = \pi = 0 \tag{10.11}$$

and (7.44) then gives $\Psi_2 = 0$. In the type N case, τ is an invariant, and (7.43) and (7.44) yield $\tau = 0$, again giving plane waves. Thus we have

Theorem 10.1. *The plane waves*

$$ds^2 = 2\,d\zeta\,d\bar{\zeta} - 2\,e^{\varepsilon u}\,du\,dv - 2\,du^2[a^2\zeta\bar{\zeta} + b\,\mathrm{Re}\,(\zeta^2\,e^{-2i\gamma u})] \tag{10.12}$$

represent all homogeneous null Einstein-Maxwell fields (with $\pounds_\xi F_{ab} = 0$), and all vacuum homogeneous solutions with a multiply-transitive group.

In (10.12) a, b, γ are constants, $\varepsilon = 0$ or 1, the Petrov type is N if $b \neq 0$ or O if $b = 0$, and the space-time is empty if $a = 0$.

It is possible to interpret the special plane waves (10.35) which have a group

G_7 (see § 10.5.) as electromagnetic null fields whose Maxwell field does not share the space-time's symmetries (cf. § 9.1.).

The group G_6 acting on (10.12) (see § 21.5.) may contain a simply-transitive G_4 and subgroups G_3 of Bianchi types IV, VI_h or VII_h acting on hypersurfaces (see Siklos (1976b, 1978)). One case was found as a Bianchi IV solution (Harvey and Tsoubelis (1977)). The group structure of the vacuum metrics was studied by Klekowska and Osinovsky (1973).

An interesting special case of (10.12) is

$$\begin{aligned} \mathrm{d}s^2 &= 2\,\mathrm{d}\zeta\,\mathrm{d}\bar{\zeta} - 2\,\mathrm{d}u\,\mathrm{d}v - 2H\,\mathrm{d}u^2, \\ 2H &= (x^2 - y^2)\cos 2u - 2xy\sin 2u, \qquad \zeta = (x + \mathrm{i}y)/\sqrt{2}, \end{aligned} \tag{10.13}$$

which is the "anti-Mach" metric of Ozsváth and Schücking (1962). It is geodesically complete and without curvature singularities.

By the methods described in § 10.1., one finds

Theorem 10.2. *The only vacuum solution admitting a simply-transitive G_4 as its maximal group of motions is given by*

$$k^2\,\mathrm{d}s^2 = \mathrm{d}x^2 + \mathrm{e}^{-2x}\,\mathrm{d}y^2 + \mathrm{e}^x\left(\cos\sqrt{3}\,x\,(\mathrm{d}z^2 - \mathrm{d}t^2) - 2\sin\sqrt{3}\,x\,\mathrm{d}z\,\mathrm{d}t\right) \tag{10.14}$$

(Petrov (1962)), *where k is an arbitrary constant.*

The Killing vectors of (10.14) are

$$\partial_t,\ \partial_z,\ \partial_y,\ \partial_x + y\,\partial_y + \frac{1}{2}\left(\sqrt{3}\,t - z\right)\partial_z - \frac{1}{2}\left(t + \sqrt{3}\,z\right)\partial_t \tag{10.15}$$

and the group obeys (8.16) with spacelike $\boldsymbol{P}$. There are subgroups G_3 of Bianchi types I and VII_h acting in timelike hypersurfaces. The solution (10.14) is Petrov type I, and the eigenvalues of the Riemann tensor are the roots of $\lambda^3 = -1$ if $k = 1$.

□ Bonnor (1979a) has pointed out that this is a special case of a cylindrically symmetric vacuum metric (see § 20.2).

Ref.: Debever (1965), Hiromoto and Ozsváth (1978).

10.3. Homogeneous non-null electromagnetic fields

Theorem 10.3. *The only Einstein-Maxwell field which is homogeneous and has a homogeneous non-null Maxwell field is the Bertotti-Robinson solution*

$$\mathrm{d}s^2 = k^2\,(\mathrm{d}\vartheta^2 + \sin^2\vartheta\,\mathrm{d}\varphi^2 + \mathrm{d}x^2 - \sinh^2 x\,\mathrm{d}t^2). \tag{10.16}$$

Proof: From the Rainich conditions (5.21), and the invariance of α, we can obtain

$$\begin{aligned} (\overline{m}^c m_{a;c} - m^c \overline{m}_{a;c})\,k^a &= (\overline{m}^c m_{a;c} - m^c \overline{m}_{a;c})\,l^a = 0, \\ (k^c l_{a;c} - l^c k_{a;c})\,m^a &= 0 \end{aligned} \tag{10.17}$$

by inserting (5.12) in (5.31). (Note that $\Phi_1\overline{\Phi}_1$ must be a constant for homogeneous fields.) (10.17) shows that there are two families of orthogonal two-surfaces, and the space-time is of the form (10.8), the two-surfaces having equal and opposite curvatures.

The group G_6 admitted by (10.16) contains no simply-transitive G_4, in agreement with Ozsváth's (1965c) conclusion that (10.12) includes all electromagnetic solutions with simply-transitive G_4 obeying $F_{ab} \neq 0$, $\pounds_\xi F_{ab} = 0$, $\Lambda = 0$. The group does contain subgroups transitive on S_3, N_3 and T_3, and the metric has therefore been rediscovered as a spherically-symmetric and spatially-homogeneous solution (Lovelock (1967), Dolan (1968)). It was first given by Bertotti (1959) and Robinson (1959).

Some alternative forms of the line element are

$$\mathrm{d}s^2 = \frac{e^2}{r^2}\,(\mathrm{d}r^2 + r^2\,\mathrm{d}\vartheta^2 + \sin^2\vartheta\,d\varphi^2 - \mathrm{d}\tau^2), \tag{10.18}$$

$$\mathrm{d}s^2 = (1 - \lambda y^2)\,\mathrm{d}x^2 + (1 - \lambda y^2)^{-1}\,\mathrm{d}y^2 + (1 + \lambda z^2)^{-1}\,\mathrm{d}z^2 - (1 + \lambda z^2)\,\mathrm{d}t^2. \tag{10.19}$$

In the metric form (10.19) the electromagnetic field is

$$\sqrt{\frac{\varkappa_0}{2}}\,F_{12} = \sqrt{\lambda}\sin\beta, \qquad \sqrt{\frac{\varkappa_0}{2}}\,F_{43} = \sqrt{\lambda}\cos\beta, \qquad \beta = \text{const}. \tag{10.20}$$

The solution is conformally flat; it is the only conformally flat non-null solution of the (source-free) Einstein-Maxwell equations (cf. Theorem 32.16).

McLenaghan and Tariq (1975) and Tupper (1976) presented a homogeneous solution of the Einstein-Maxwell equations whose Maxwell field does not share the space-time symmetry. This metric can be written as

$$\mathrm{d}s^2 = a^2x^{-2}(\mathrm{d}x^2 + \mathrm{d}y^2) + x^2\,\mathrm{d}\varphi^2 - (\mathrm{d}t - 2y\,\mathrm{d}\varphi)^2, \qquad a = \text{const}, \tag{10.21}$$

which admits a simply-transitive group G_4 with Killing vectors ∂_t, ∂_φ, $2\varphi\,\partial_t + \partial_y$, and $\xi = x\,\partial_x - y\,\partial_y - \varphi\,\partial_\varphi$, none of which is hypersurface-orthogonal. In (10.21), $\pounds_\xi F_{ab} \neq 0$, so the condition that the Maxwell field shares the space-time symmetry is essential in Theorem 10.3 (cf. § 9.1.). The solution (10.21) is of Petrov type I and its Maxwell and Weyl tensors have no common null eigendirection (non-aligned case); it is characterized by the existence of a tetrad parallelly propagated along the two geodesic non-expanding null congruences of the non-null Maxwell field. For analogues of the Bertotti-Robinson metric, i.e. a metric form (10.8) with $\Lambda \neq 0$, see, e.g., Cahen and Defrise (1968) and § 31.2.

Ref.: □ Kramer (1978).

10.4. Homogeneous perfect fluid solutions

The only fluid solution with a G_7 is (10.9) with constant a. The field equations give

$$\Lambda = \frac{\varkappa_0}{2}(\mu + 3p) = \frac{3\varepsilon}{a^2} - \mu\varkappa_0. \tag{10.22}$$

For realistic matter we require $\varepsilon = 1$ and $\Lambda > 0$. This is Einstein's static universe. Some alternative metric forms are

$$\mathrm{d}s^2 = \left(1 + \frac{r^2}{4K^2}\right)^{-2} \mathrm{d}x_\alpha\, \mathrm{d}x^\alpha - \mathrm{d}t^2, \qquad r^2 = x_\alpha x^\alpha, \tag{10.23a}$$

$$\Lambda\, \mathrm{d}s^2 = \mathrm{d}\chi^2 + \sin^2\chi(\mathrm{d}\vartheta^2 + \sin^2\vartheta\, \mathrm{d}\varphi^2) - \mathrm{d}t^2, \tag{10.23b}$$

$$\mathrm{d}s^2 = \frac{\mathrm{d}r^2}{(1 - \Lambda r^2)} + r^2(\mathrm{d}\vartheta^2 + \sin^2\vartheta\, \mathrm{d}\varphi^2) - \mathrm{d}t^2. \tag{10.23c}$$

The G_7 includes a simply-transitive G_4, and a G_6 on S_3 with generators

$$\xi_\alpha = \left(1 - \frac{r^2}{4K^2}\right)\partial_\alpha + \frac{1}{2K^2}\, x_\alpha(x^\beta\, \partial_\beta), \qquad \eta^\alpha = \varepsilon^{\alpha\beta\gamma} x_\beta\, \partial_\gamma, \tag{10.24a}$$

with

$$[\xi_\alpha, \xi_\beta] = -\frac{2}{K}\,\varepsilon_{\alpha\beta\gamma}\xi^\gamma; \qquad [\overset{\pm}{\eta}_\alpha, \overset{\pm}{\eta}_\beta] = \mp\frac{2}{K}\,\varepsilon_{\alpha\beta\gamma}\overset{\pm}{\eta}{}^\gamma,\ \overset{\pm}{\eta}_\alpha \equiv \xi_\alpha \pm \frac{1}{K}\,\eta_\alpha, \tag{10.24b}$$

as the non-vanishing commutators. The G_6 is $SO(4) \simeq SO(3) \times SO(3)$, arising from the embedding of the S_3 in R^4 (see § 8.5.). The seventh Killing vector is ∂_t. The fluid velocity is covariantly constant. The G_7 includes G_4 and G_3 transitive on both S_3 and T_3.

There can be no perfect fluid solution with a transitive G_6 (§ 9.2.). The solutions with a G_5 arise as special cases of those with a transitive G_4. In fact there is only one such solution.

Theorem 10.4. *The homogeneous perfect fluid solutions are* (10.23), *and* (10.25)–(10.32) *below. The only such solution with a maximal* G_5 *is the Gödel solution* (10.25).

Proof: This result was proved by Ozsváth (1965d) and Farnsworth and Kerr (1966), by the method outlined in § 10.1. Their results are stated for dust, with a cosmological constant, but by the change of variables $\mu \to \mu + p$, $\Lambda \to \Lambda - \varkappa_0 p$, can be re-interpreted as perfect fluid solutions (the solutions being homogeneous, μ and p are constants, and so this re-interpretation does not affect the arguments in any way). The resulting list of metrics is:

(i) the Gödel solution (Gödel (1949))

$$\mathrm{d}s^2 = a^2\left(\mathrm{d}x^2 + \mathrm{d}y^2 + \frac{1}{2}\,\mathrm{e}^{2x}\, \mathrm{d}z^2 - (\mathrm{d}t + \mathrm{e}^x\, \mathrm{d}z)^2\right) \tag{10.25}$$

(ii) solutions with $A_E = 0$. These must be of case (B), (8.17). If $\boldsymbol{L}$ is null we retrieve the Einstein static and Gödel solutions. If $\boldsymbol{L}$ is timelike one obtains the Farnsworth-Kerr class I solution

$$\mathrm{d}s^2 = a^2\left[(1-k)(\omega^1)^2 + (1+k)(\omega^2)^2 + 2(\omega^3)^2 - \left(\mathrm{d}t + \sqrt{1-2k^2}\,\omega^3\right)^2\right], \tag{10.26}$$

where the ω^a are those for type IX in Table 8.2. This is a rotating solution, extensively discussed by Ozsváth and Schücking (1969). $k = 0$ in (10.26) gives the Einstein static solution. (10.26) admits a G_3 of Bianchi type IX on S_3.

If $\boldsymbol{L}$ is spacelike, there are two metrics, Farnsworth-Kerr classes II and III, given by

$$\mathrm{d}s^2 = a^2\left[(1-k)(\omega^2)^2 + (1+k)(\omega^3)^2 + \left(\mathrm{d}u + \sqrt{1-2k^2}\,\omega^1\right)^2 - 2(\omega^1)^2\right], \tag{10.27}$$

$$\mathrm{d}s^2 = a^2[(1+s)(\omega^2)^2 + (1-s)(\omega^3)^2 + 2\mathrm{d}u^2 - 2(\omega^1)^2] \tag{10.28}$$

with the ω^a of Bianchi type VIII (Table 8.2). These were investigated in detail by Ozsváth (1970). (10.27) and (10.28) both contain (10.25) as a special case. They both admit groups G_3 of Bianchi type VIII on T_3 and type III on S_3.

In these solutions a, k, s are constants; in (10.26) and (10.27) $\Lambda - \varkappa_0 p = \varkappa_0(\mu + p)/2(1 - 4k^2)$, and in (10.28), $\Lambda - \varkappa_0 p = \frac{1}{2}\varkappa_0(\mu + p)$.

(iii) solutions with $A_E \neq 0$. Following Ozsváth (1965b) we use a parameter s^2, $\frac{1}{2} \leqq s^2 \leqq 2$. If $\beta^2 \equiv 1 + 2s^2(1 - s^2)(3 - s^2)$, then we have

$$\beta^2 > 0\colon\ \mathrm{d}s^2 = a^2[(C\,\mathrm{e}^{-Az}\,\mathrm{d}t + D\,\mathrm{e}^{-Bz}\,\mathrm{d}x)^2 + (\mathrm{e}^{-Fz}\,\mathrm{d}y)^2 + \mathrm{d}z^2 - (\mathrm{e}^{-Az}\,\mathrm{d}t + \mathrm{e}^{-Bz}\,\mathrm{d}x)^2], \tag{10.29}$$

$$\beta^2 = 0\colon\ \mathrm{d}s^2 = a^2\left[-\left(\frac{b^2-1}{2b}\right)^2 \mathrm{e}^{-z}\,\mathrm{d}t^2 + (\mathrm{e}^{-Fz}\,\mathrm{d}y)^2 + \mathrm{d}z^2 + (z\,\mathrm{d}t - \mathrm{d}x)^2\,\mathrm{e}^{-z}\right], \tag{10.30}$$

$$\begin{aligned}\beta^2 = -k^2 < 0\colon\ \mathrm{d}s^2 = a^2\big[&\mathrm{e}^{-z}b^{-2}\big((\cos kz + 2k\sin kz)\,\mathrm{d}t \\ &+ (2k\cos kz - \sin kz)\,\mathrm{d}x\big)^2 \\ &- \mathrm{e}^{-z}(\cos kz\,\mathrm{d}t - \sin kz\,\mathrm{d}x)^2 + (\mathrm{e}^{-Fz}\,\mathrm{d}y)^2 + \mathrm{d}z^2\big],\end{aligned} \tag{10.31}$$

where $A = \frac{1}{2}(1-\beta)$, $B = \frac{1}{2}(1+\beta)$, $b = \sqrt{2}\,s(3 - s^2)$, $bC = 2A$, $bD = 2B$, $F = 1 - s^2$, a is constant. These metrics all have

$$\Lambda - \varkappa_0 p = \frac{1}{2}(s^2 - 2)\,a^2; \qquad \varkappa_0(\mu + p) = a^2(2s^2 - 1)(2 - s^2). \tag{10.32}$$

The special cases $s^2 = 2$, $s^2 = 1$, $s^2 = \frac{1}{2}$ give, respectively, the Petrov vacuum solution (10.14), the Gödel solution (10.25), and a type N vacuum solution with a cosmological constant (see § 10.5.). All these metrics have an Abelian G_3 on T_3

and (10.29), (10.30), have G_3 of Bianchi type VI_h on S_3; (10.31) has a G_3 of type VII_h on T_3, (10.30) has a group G_3 of type IV on T_3 and (10.29) a group of type VI on T_3.

Gödel's solution and (10.28) above are of Petrov type D, but the other metrics above are in general of Petrov type I. In Gödel's solution the four-velocity of the fluid is a Killing vector but not hypersurface-orthogonal; the vorticity $\omega = \partial_y$ is covariantly constant. Gödel's solution has interesting global properties (Hawking and Ellis (1973), Ryan and Shepley (1975)).

10.5. Other homogeneous solutions

All homogeneous solutions with a *Λ-term* have been found. Plane waves cannot have a Λ-term, so the type N solutions have a maximal G_5. The type D solutions are either of the form (10.8), in which case they must be composed of two two-spaces of equal curvatures, or have at most a G_5. By the usual arguments, all these spaces have simply-transitive G_4, and G_3 transitive on hypersurfaces. They have been investigated and re-discovered by several authors (Cahen (1964), Ozsváth (1965b), Siklos (1978), Kaigorodov (1962)). The list is as follows.

The only Petrov type N metric is

$$ds^2 = -\frac{12}{\Lambda}\,dz^2 + 10k\,e^{2z}\,dx^2 + e^{-4z}\,dy^2 - 10u\,e^z\,dz\,dx - 2e^z\,du\,dx, \tag{10.33}$$

where $\psi_4 = -k\Lambda$ and $\Lambda < 0$; it has a G_5.

The Petrov type D metric is of the form (10.8), admitting a G_6, as described above: $3\psi_2 = -\Lambda$.

There is a Petrov type III solution, namely

$$\begin{aligned} ds^2 = &-\frac{1}{2}\,dz^2/\Lambda + e^{4z}\,dy^2 + 2e^{-2z}\,dx^2 - 4e^z\,dx\,dy \\ &- 8e^{2z}\,dz\,dy - 2e^{2z}\,du\,dy \end{aligned} \tag{10.34}$$

with $\Lambda < 0$. There are no homogeneous solutions with a Λ-term of Petrov types I or II.

Homogeneous *pure radiation* solutions can also be found. The solutions with G_6 or G_7 include plane waves (10.7). The solutions with a G_7 have $A(u) = 0$, $B(u) = bu^2$ or $B = b$, where b is constant. They can be transformed to the form (Petrov (1966))

$$ds^2 = C^2(u)\,(dx^2 + dy^2) - 2\,du\,dv, \qquad \ddot{C} + 2BC = 0, \tag{10.35}$$

cf. § 21.5. and the metrics (13.18), (21.48).

From § 10.1. we see that the only other pure radiation solution (including Λ) with a G_6 is (10.6). All the remaining solutions must contain a transitive G_4 on V_4. As far as we are aware, they have not been fully enumerated.

Other energy-momentum tensors, beyond the scope of this book, have been considered. Ozsváth (1965a, 1966) found all solutions with dust (or fluid) and a Maxwell field, and (1965c) the unique null Maxwell field with $\Lambda \neq 0$ together with a non-null Maxwell field with $\Lambda \neq 0$.

10.6. Summary

The results of this chapter can be summarized as in Table 10.1.

It is assumed that $\pounds_\xi F_{ab} = 0$ for electromagnetic solutions. The spaces of constant curvature, with a G_{10}, are omitted. The solutions with a G_7 shown are the only possible ones. The symbol A means that all solutions are known; $\nexists$ indicates non-existence.

Table 10.1 Homogeneous solutions

Source	Maximal group			
	G_4	G_5	G_6	G_7
Vacuum	A Petrov (10.14)	$\nexists$	A Plane waves (10.12), $a = 0$	$\nexists$
Einstein-Maxwell non-null field	$\nexists$	$\nexists$	A Bertotti-Robinson (10.16)	$\nexists$
Einstein-Maxwell null field	$\nexists$	$\nexists$	A Plane waves (10.12)	$\nexists$
Perfect fluid	A Ozsváth (10.16)–(10.31)	A Gödel (10.25)	$\nexists$	A Einstein (10.23)
Λ-term	A (10.34)	A (10.33)	A (10.8) special case	$\nexists$
Pure radiation		?	A (10.12)	A (10.35)

Chapter 11. Hypersurface-homogeneous space-times

11.1. The possible metrics

This chapter is concerned with metrics admitting a group of motions transitive on S_3 or T_3. The metrics with a group transitive on N_3 are considered in Chapter 21. As in the case of the homogeneous space-times (Chapter 10.) we first consider the cases with multiply-transitive groups. From Theorems 8.10 and 8.17 we see that only G_6 or G_4 is possible.

G_6 on V_3

By the arguments used in § 10.1., the space-times with a G_6 on S_3 have the metric (10.9), but $a(t)$ is now variable, and the field equations, which necessitate an energy-momentum tensor of perfect fluid type, are (assuming $\dot{a} \neq 0$, since a constant $a(t)$ gives only the Einstein static and de Sitter solutions)

$$3\dot{a}^2 = \varkappa_0 \mu a^2 + \Lambda a^2 - 3\varepsilon, \tag{11.1}$$

$$\dot{\mu} + 3(\mu + p)\,\dot{a}/a = 0. \tag{11.2}$$

If one can find $\mu(a)$ from (11.2), one already has an exact solution, in that one can use a as a new time variable, the coefficient of $\mathrm{d}a^2$ in the line element being given by (11.1). However, in the physical applications it is important to know a as a function of t.

The space-times with a G_6 on T_3 permit only vacuum and Λ-term Ricci tensors (and tachyonic fluid, which we ignore). Thus they will give only the spaces of constant curvature, with a complete G_{10}.

It should be noted that (10.9) always admits G_3 transitive on $t =$ constant; namely, G_3V and G_3VII_h if $\varepsilon = -1$, G_3I and G_3VII_0 if $\varepsilon = 0$, and G_3IX if $\varepsilon = 1$.

G_4 on V_3

The spaces with a G_4 on S_3 or T_3 are easily determined by using Schmidt's method (§ 8.6.) to find the possible G_4, followed by use of Theorems 8.16 and 8.18 to determine the complete metric (MacCallum (1980)). They are of Petrov type D (or conformally flat) unless the isotropy is a null rotation, and they are all among the metrics found by Cahen and Defrise (1968) and Defrise (1969), except for the most general form of one of the Petrov type N metrics. An almost complete list is given by Petrov (1966).

Spatial rotation isotropy. We give first the metrics with a G_4 which are locally rotationally symmetric (*L.R.S.*), i.e. in which the isotropy is a spatial rotation

(Ellis (1967), Stewart and Ellis (1968)). The possible cases are, with $\varepsilon = \pm 1$ and $k = \pm 1$ or 0,

$$ds^2 = \varepsilon\left(dt^2 - A^2(t)\, dx^2\right) + B^2(t)\left(dy^2 + \Sigma^2(y, k)\, dz^2\right), \tag{11.3}$$

$$ds^2 = \varepsilon\left(dt^2 - A^2(t)\, (\sigma^1)^2\right) + B^2(t)\left(dy^2 + \Sigma^2(y, k)\, dz^2\right), \tag{11.4}$$

$$ds^2 = \varepsilon\left(dt^2 - A^2(t)\, dx^2\right) + B^2(t)\, e^{2x}(dy^2 + dz^2), \tag{11.5}$$

where in (11.4) we have

$$k = 1, \quad \sigma^1 = dx + \cos y\, dz, \tag{11.6}$$

$$k = 0, \quad \sigma^1 = dx + y\, dz, \tag{11.7}$$

$$k = -1, \quad \sigma^1 = dx + \cosh y\, dz. \tag{11.8}$$

With the exception of (11.3) with $k = 1$, these metrics can all be written in the form

$$ds^2 = \varepsilon\left(dt^2 - A^2(t)\, (\omega^1)^2\right) + B^2(t)\left((\omega^2)^2 + (\omega^3)^2\right), \tag{11.9}$$

where the ω^α are dual to a basis of reciprocal group generators of a G_3. The possible G_3 are: for (11.3), $k = 0$, $G_3 I$ or $G_3 VII_0$; for (11.3), $k = -1$, $G_3 III$; for (11.4), (11.6), $G_3 IX$; for (11.4), (11.7), $G_3 II$; for (11.4), (11.8), $G_3 VIII$ or $G_3 III$; and for (11.5), $G_3 V$ or $G_3 VII_h$. With the exception of the $G_3 III$, the ω^α of Table 8.2 can be used; for $G_3 III$ one needs instead ω'^α related to the ω^α of Table 8.2 by

$$\omega'^2 = \omega^1, \qquad \omega'^3 = \omega^2 - \omega^3, \qquad \omega'^1 = \omega^2 + \omega^3. \tag{11.10}$$

The exceptional case, (11.3) with $k = 1$, admits no simply-transitive G_3, and, with $\varepsilon = -1$, gives the only spatially-homogeneous solutions with this property (Kantowski and Sachs (1966), Kantowski (1966); cf. Collins (1977)); with $\varepsilon = 1$ it gives the spherically-symmetric static metrics. The metrics (11.3)—(11.5) can also be derived by considering the extensions of Lie algebras of groups G_3 on S_2 or S_3, cf. Shikin (1972), Kantowski (1966).

Ellis' class *I L.R.S.* metrics are (11.4) with $\varepsilon = 1$, class *II* are (11.3) and (11.5), and class *III* are (11.4) with $\varepsilon = -1$.

The metrics (11.3) and (11.4) can jointly be written as

$$ds^2 = Y^2(w)\frac{2\, d\zeta\, d\bar\zeta}{\left(1 + \frac{k}{2}\zeta\bar\zeta\right)^2} + \frac{dw^2}{f(w)} - f(w)\left[dt + il\frac{\zeta\, d\bar\zeta - \bar\zeta\, d\zeta}{\left(1 + \frac{k}{2}\zeta\bar\zeta\right)}\right]^2, \tag{11.11}$$

where $l \neq 0$ corresponds to (11.4), and $f(w)$ may have either sign. The Killing vectors of (11.11) are

$$\begin{aligned} \xi_1 &= i\left(1 - \frac{k}{2}\zeta^2\right)\partial_\zeta - i\left(1 - \frac{k}{2}\bar\zeta^2\right)\partial_{\bar\zeta} + l(\zeta + \bar\zeta)\,\partial_t, \\ \xi_2 &= \left(1 + \frac{k}{2}\zeta^2\right)\partial_\zeta + \left(1 + \frac{k}{2}\bar\zeta^2\right)\partial_{\bar\zeta} + il(\zeta - \bar\zeta)\,\partial_t, \\ \xi_3 &= i(\zeta\partial_\zeta - \bar\zeta\partial_{\bar\zeta}), \quad \xi_4 = \partial_t. \end{aligned} \tag{11.12}$$

A fifth Killing vector, making the space-time homogeneous, can occur, but these metrics either represent a space-time filled with dust and an electromagnetic field (Ozsváth (1966)), or have no known physical interpretation.

In the basis

$$\omega^1 = Y \frac{d\zeta}{\left(1 + \frac{k}{2}\zeta\bar{\zeta}\right)}, \qquad \omega^2 = \overline{\omega}^1, \qquad \omega^3 = \frac{dw}{X},$$
$$\omega^4 = X\left(dt + il\frac{\zeta\, d\bar{\zeta} - \bar{\zeta}\, d\zeta}{\left(1 + \frac{k}{2}\zeta\bar{\zeta}\right)}\right), \qquad f = \varepsilon X^2, \varepsilon = \pm 1, \tag{11.13}$$

the Ricci tensor of (11.11) has, as its only non-zero components,

$$R_{12} = \frac{k}{Y^2} + \frac{2l^2 f}{Y^4} - \frac{f' Y'}{Y} - \frac{f Y'^2}{Y^2} - \frac{f Y''}{Y},$$
$$R_{33} = -\varepsilon\left(\frac{f' Y'}{Y} + \frac{2f Y''}{Y} + \frac{f''}{2}\right), \qquad R_{44} = \varepsilon\left(\frac{f' Y'}{Y} + \frac{2l^2 f}{Y^2} + \frac{f''}{2}\right), \tag{11.14}$$

and is of type [(11) 1, 1] or its specializations. The metric (11.5) has non-zero Ricci tensor components,

$$R_{11} = \frac{\ddot{A}}{A} + \frac{2\dot{A}\dot{B}}{AB} - \frac{2}{A^2},$$
$$R_{22} = R_{33} = \frac{\ddot{B}}{B} + \frac{\dot{B}^2}{B^2} + \frac{\dot{A}\dot{B}}{AB} - \frac{2}{A^2}, \tag{11.15}$$
$$R_{44} = -\frac{\ddot{A}}{A} - \frac{2\ddot{B}}{B}, \qquad R_{14} = \frac{2}{A}\left(\frac{\dot{A}}{A} - \frac{\dot{B}}{B}\right),$$

in the basis ($A\omega^1$, $B\omega^2$, $B\omega^3$, dt). The Ricci tensor has type [(11) 1, 1] or [(11), 2], or their specializations. No G_5 on V_4 is possible, but if $R_{14} = 0$ we recover (10.9), $\varepsilon = -1$.

Boost isotropy. The metrics with a G_4 on T_3 and a boost as the isotropy are

$$ds^2 = dw^2 + A^2(w)\, dx^2 + B^2(w)\left(dy^2 - \Sigma^2(y, k)\, dt^2\right), \tag{11.16}$$
$$ds^2 = dw^2 + A^2(w)\,(\omega^1)^2 - 2B^2(w)\,\omega^2\omega^3, \tag{11.17}$$

where (11.17) covers four cases given by taking the ω^a of Table 8.2 for G_3V or G_3II, or ω'^a related to those given for G_3VIII by

$$\omega'^1 = \omega^1, \quad \omega'^2 = \omega^2 + \omega^3, \qquad \omega'^3 = \pm(\omega^2 - \omega^3). \tag{11.18}$$

The possible simply-transitive G_3 are then: (11.16), $|k| = 1$, G_3III; (11.16), $k = 0$, G_3I or G_3VI_0; (11.17), (11.18), G_3VIII and G_3III; (11.17) with ω^a of G_3II, G_3II; and for (11.17) with ω^a of G_3V, G_3V or G_3VI_h or G_3III. With the exception of the last case (which admits a normal null Killing vector, cp. § 21.4.), these metrics can be combined as

$$ds^2 = Y^2(w) \frac{2\,du\,dv}{\left(1 + \frac{k}{2}uv\right)^2} + \frac{dw^2}{f(w)} + f(w)\left(dy + l\frac{u\,dv - v\,du}{\left(1 + \frac{k}{2}uv\right)}\right)^2, \tag{11.19}$$

with $f(w) > 0$. (11.19) can be derived from (11.11) with $f(w) > 0$ by the complex substitution

$$\zeta \to u, \qquad \bar{\zeta} \to v, \qquad t \to iy, \tag{11.20}$$

which, when applied to (11.12) and (11.14), also yields the Killing vectors and Ricci tensor; the latter must be of type [1 1 (1, 1)] or its specializations.

Null rotation isotropy. The (Petrov type N) metrics with a G_4 on T_3 and a null rotation isotropy are either special cases of the metric (21.1) with a G_3 on N_2 (for which see Petrov (1966), Defrise (1969), Barnes (1979)) or

$$ds^2 = dw^2 + A^2(w)\left[dy^2 - 2e^y\,dv\left(du + B(w)\,e^y\,dv\right)\right]. \tag{11.21}$$

They all have a Ricci tensor of type [1 (1, 2)] or its specialization, so only vacuum, null Einstein-Maxwell and pure radiation energy-momentum tensors are possible.

G_3 on V_3

Finally, there are the metrics with a maximal G_3 on S_3 or T_3, with metrics given by Table 8.2 and

$$ds^2 = -dt^2 + g_{\alpha\beta}(t)\,\omega^\alpha\omega^\beta, \qquad \det(g_{\alpha\beta}) > 0, \tag{11.22}$$

$$ds^2 = dt^2 + g_{\alpha\beta}(t)\,\omega^\alpha\omega^\beta, \qquad \det(g_{\alpha\beta}) < 0. \tag{11.23}$$

A number of the metrics just given recur in other parts of the book, and the solutions of their field equations are accordingly not given in this chapter. The metrics concerned are the following. (11.3) with $k = 1$ or 0 and $\varepsilon = 1$ are spherically and plane symmetric static metrics, and are treated in Chapters 13. and 14. The metrics (11.3) and (11.5) admit a G_3 on S_2, while their counterparts contained in (11.16) and (11.17) admit G_3 on T_2; these metrics are therefore treated in Chapter 13. All solutions with a group of motions transitive on T_3 are stationary or static. The metrics of the form

$$ds^2 = dw^2 + A^2(w)\,dx^2 + B^2(w)\,dy^2 - C^2(w)\,dt^2 \tag{11.24}$$

with a G_3I on T_3 are often interpreted as cylindrically symmetric static metrics (assuming y to be an angular coordinate) and these are treated in Chapter 20.; this includes some metrics with plane symmetry. All solutions with a perfect fluid matter content and a group transitive on S_3 are considered as spatially-homogeneous cosmologies and discussed in Chapter 12.

The metrics (11.16) with $k = 0$, and (11.17), with the ω^α of G_3II and G_3V, admit groups on null orbits and are covered by Chapter 21., as are all the metrics with a null rotation isotropy. Some of the cases of (11.23) also admit groups on null orbits;

for example, nearly all Kellner's metrics with a G_3I on T_3 contain a normal null Killing vector (Kellner (1975)), see (e.g.) (18.23), (21.36), (29.19).

Summarizing the results on the multiply-transitive groups G_4 on V_3 we have

Theorem 11.1. *Apart from the cases with null (sub-)orbits, all metrics with a G_4 on V_3 are covered by the metrics* (11.11), (11.5), *and* (11.19).

11.2. Formulation of the field equations

The Einstein equations for hypersurface-homogeneous space-times reduce to a system of ordinary differential equations. At least for the spatially-homogeneous case, they form a well-posed Cauchy problem (Taub (1951)). In general, they have not been completely integrated, and various approaches have been used to restrict the general case to a more readily integrable system. Nearly all of these have been developed for use in the spatially-homogeneous case, and it is to this that we refer in this section, unless otherwise stated. For general reviews of the qualitative properties of such solutions we refer the reader to Ryan and Shepley (1975), MacCallum (1973, 1979a, b).

The number of degrees of freedom, i.e. the number of arbitrary constants required in a general solution, has been studied by Siklos (1976b), for each Bianchi type (cf. MacCallum (1979b)). Bianchi types I, II and VI_h, $h = -1/9$, (acting on S_3) turn out to be special cases, because the $G^4{}_\alpha$ field equations are not linearly independent constraints on the Cauchy problem.

The solution of the field equations is simplified if the $g_{\alpha\beta}$ in (11.22) can be taken to be a *diagonal matrix*. MacCallum et al. (1970) have shown that if $\boldsymbol{n} = \partial_t$ is a Ricci eigenvector, then in Class G_3A, except types G_3I and G_3II, $g_{\alpha\beta}$ is diagonal, and in Class G_3B, except type VI_h, $h = -1/9$, the vector a^α in (8.50) is an eigenvector of $n^\alpha{}_{;\beta}$ and of the Ricci tensor. The excluded cases require additional assumptions to reach these conclusions, for the reason just mentioned in connection with degrees of freedom. For realistic matter content, the G_3B metrics can only be diagonal if either they are *L.R.S.*, or the $n_{\alpha\beta}$ of (8.50) obey

$$n^\alpha{}_\alpha = 0 \tag{11.25}$$

(MacCallum (1972)). The restriction imposed by (11.25) can apply only in Bianchi types I, III, V, VI, and VIII, and it turns out to give useful new restrictions only in types III and VI (Ellis and MacCallum (1969)). With (11.25), it is convenient to alter the canonical form of the group generators to read

$$[\xi_2, \xi_3] = 0, \qquad [\xi_3, \xi_1] = (1 - A)\,\xi_3, \qquad [\xi_1, \xi_2] = (1 + A)\,\xi_2, \tag{11.26}$$

where $h = -A^2$. The ω^α analogous to those of Table 8.2 are

$$\omega^1 = \mathrm{d}x, \qquad \omega^2 = \mathrm{e}^{(A+1)x}\,\mathrm{d}y, \qquad \omega^3 = \mathrm{e}^{(A-1)x}\,\mathrm{d}z. \tag{11.27}$$

Two standard ways of attacking systems of differential equations have proved fruitful in elucidating qualitative properties of the solutions. A *Lagrangian* or *Hamiltonian formulation* (using only functions of t) proves to be possible only for

Class G_3A in general (MacCallum and Taub (1972), Sneddon (1976)); it can also be achieved for the metrics with $n^\alpha{}_\alpha = 0$. This technique is not well-adapted to searching for exact solutions, though it suggested the idea of "regularizing" the equations. This, in conjunction with *phase plane methods*, provided a second approach leading to a number of new solutions (Collins (1971)).

Alternative sets of variables have been introduced in several ways (compare (11.22) and (8.50)). One of the most widely used is *Misner's* (1968) *parametrization*, which we write in the form

$$S^6 \equiv e^{6\lambda} \equiv \det(g_{\alpha\beta}), \qquad g_{\alpha\gamma} = S^2(\exp 2\beta)_{\alpha\gamma}, \tag{11.28}$$

where β is a symmetric tracefree matrix function of t. If $g_{\alpha\beta}$ is diagonal, one may write

$$\beta_{\alpha\gamma} = \operatorname{diag}\left(\beta_1, -\frac{\beta_1}{2} + \frac{\sqrt{3}\,\beta_2}{2}, -\frac{\beta_1}{2} - \frac{\sqrt{3}\,\beta_2}{2}\right). \tag{11.29}$$

One may now take λ or S to be a new time variable.

An especially simple situation arises in Bianchi type I metrics, and in those G_3V metrics where ∂_t is a Ricci eigenvector, if the Ricci tensor is of the perfect fluid type [(111), 1], because the field equations then yield

$$S^3\dot{\beta}_{\alpha\gamma} \equiv \Sigma_{\alpha\gamma}, \qquad \dot{\Sigma}_{\alpha\gamma} = 0, \tag{11.30a}$$

$$3\dot{S}^2 = \Sigma^2 S^{-4} + \varkappa_0 \mu S^2 + \Lambda S^2 - 3\varepsilon, \tag{11.30b}$$

$$S\dot{\mu} + 3\dot{S}(\mu + p) = 0, \tag{11.30c}$$

where $2\Sigma^2 \equiv \Sigma_{\alpha\beta}\Sigma^{\alpha\beta}$ and $\varepsilon = 0$ for G_3I, $\varepsilon = -1$ for G_3V. One may now take

$$\beta_{\alpha\gamma} = \left(\int \frac{2\Sigma\,dt}{\sqrt{3}\,S^3}\right) \operatorname{diag}\left(\sin\alpha, \sin\left(\alpha + \frac{2\pi}{3}\right), \sin\left(\alpha + \frac{4\pi}{3}\right)\right). \tag{11.31}$$

In G_3I we may take $\Sigma = \sqrt{3}$, while in G_3V the remaining field equation implies $\alpha = 0$. If we can solve (11.30c) for $\mu(S)$, then, by taking S as the time variable, we have found an exact solution up to the quadrature in (11.31). This is similar to the situation arising from (11.1) and (11.2). Another case of this type arises in the metric (11.3), $\varepsilon = -1$, where, for a Ricci tensor of type [(111), 1], the field equations reduce to

$$\frac{2\ddot{B}}{B} + \frac{\dot{B}^2}{B^2} + \frac{k}{B^2} = \Lambda - \varkappa_0 p, \tag{11.32a}$$

$$\frac{\ddot{B}}{B} + \frac{\ddot{A}}{A} + \frac{\dot{A}}{A}\frac{\dot{B}}{B} = \Lambda - \varkappa_0 p, \tag{11.32b}$$

$$\frac{2\dot{A}\dot{B}}{AB} + \frac{\dot{B}^2}{B^2} + \frac{k}{B^2} = \Lambda + \varkappa_0 \mu. \tag{11.32c}$$

For the equation of state (5.36), $p = (\gamma - 1)\,\mu$, with constant γ, the solution of these equations reduces to a series of quadratures; for $k = 0$, $\gamma \neq 2$, one obtains (Stewart and Ellis (1968))

$$A = (B^{3(2-\gamma)/2} + c)^{1/(2-\gamma)} B^{-1/2},$$
$$t = \int (B^{3(2-\gamma)/2} + c)^{(\gamma-1)/(2-\gamma)} B^{1/2} \,\mathrm{d}B, \tag{11.33}$$
$$\varkappa_0\mu = \varkappa_0 p/(\gamma - 1) = 3/(AB^2)^\gamma,$$

where c is a constant.

A quite different approach leading to solutions of the Einstein-Maxwell equations which are homogeneous on S_3 or T_3 arises from an ansatz introduced by Tariq and Tupper (1975) and generalized by Barnes (1978). This is that for a non-null Maxwell field

$$F^*_{ab;c}\, l^c = fF^*_{ab}, \qquad F^*{}_{ab;c}k^c = gF^*_{ab}, \tag{11.34}$$

where $(\boldsymbol{m}, \overline{\boldsymbol{m}}, \boldsymbol{l}, \boldsymbol{k})$ is the principal null tetrad of the Maxwell bivector. (11.34) implies that

$$\varkappa = \nu = \pi = \tau = 0, \tag{11.35a}$$

while if the f and g in (11.34) are zero,

$$\varepsilon = \gamma = 0 \tag{11.35b}$$

also. There is a dual ansatz relating to the vectors $\boldsymbol{m}, \overline{\boldsymbol{m}}$ which gives rise, correspondingly, to the conditions

$$\sigma = \lambda = \varrho = \mu = 0, \tag{11.36a}$$
$$\alpha = \beta = 0. \tag{11.36b}$$

By studying the consistency of the remaining Einstein-Maxwell equations, it can be shown that (11.35a) implies that it is possible to choose a tetrad such that

$$\varrho = e\mu, \qquad \sigma = e\lambda, \qquad \varepsilon = e\gamma, \qquad \alpha = \beta = 0,$$
$$\Psi_0 = \Psi_4, \qquad \Psi_1 = \Psi_3 = 0, \tag{11.37}$$

and such that, for any Newman-Penrose quantity x,

$$\delta x = \overline{\delta} x = (D + e\Delta)\, x = 0, \tag{11.38}$$

where $e = \pm 1$. The similar conditions arising from (11.36a) are

$$\tau = \pi, \qquad \alpha = \beta, \qquad \varkappa = e\nu, \qquad \Psi_0 = e\Psi_4, \qquad \varepsilon = \gamma = 0, \tag{11.39}$$
$$Dx = \Delta x = (\delta + \overline{\delta})\, x = 0. \tag{11.40}$$

The equations (11.38) and (11.40) show that an isometry group acts in the hypersurfaces spanned by $(\boldsymbol{m}, \overline{\boldsymbol{m}}, \boldsymbol{l} + e\boldsymbol{k})$ and $(\boldsymbol{k}, \boldsymbol{l}, \boldsymbol{m} + \overline{\boldsymbol{m}})$ respectively. The solutions in fact all admit a G_3 on S_3 or T_3; an extensive list has been given by Barnes (1978).

The equations (11.38) and (11.40) are analogous to the starting point of Siklos' (1978) treatment of those algebraically special hypersurface-homogeneous space-times in which the normals $\boldsymbol{n}$ to the homogeneous hypersurfaces can be written as

$\sqrt{2}\,\boldsymbol{n} = \mathrm{e}^{-\eta}\boldsymbol{k} + \mathrm{e}^{\eta}\boldsymbol{l}$. The homogeneity implies that

$$\delta x = \bar{\delta} x = (\mathrm{e}^{-2\eta} D - \Delta)\, x = 0. \tag{11.41}$$

Applying the commutators (7.55)—(7.58) to t gives relations between the spin coefficients, and the assumption that $\boldsymbol{k}$ is a repeated principal null direction gives $\varkappa = \sigma = 0$. The remaining tetrad freedom can be used to set $\varepsilon = 0$. Then the field equations for each possible case can be integrated.

It seems likely that each of the above approaches could be generalized.

11.3. Vacuum, Λ-term and Einstein-Maxwell solutions

11.3.1. Solutions with multiply-transitive groups

The case G_6 on V_3 is covered by the Robertson-Walker line element (10.9). No Einstein-Maxwell fields exist. The vacuum and Λ-term solutions are the spaces of constant curvature (8.35); the case $\Lambda > 0$ appears, with different choices of the spatial hypersurfaces, as a solution with each ε, and the $\Lambda = 0$ case occurs with $\varepsilon = 0$ and $\varepsilon = -1$.

The line elements admitting a G_4 on S_3 or T_3 but no null rotation isotropy permit only non-null electromagnetic fields whose principal null directions are aligned with those of the (Petrov type D) Weyl tensor, assuming that the Maxwell field shares the space-time symmetry.

According to Theorem 11.1, the metrics to be considered are (11.5), (11.11), and (11.19).

In the metric (11.5), the condition $R_{14} = 0$ gives only Robertson-Walker solutions.

For the metrics (11.11) and (11.19), the solutions with constant Y are just the Bertotti-Robinson solution and its generalization to non-zero Λ (cf. §§ 10.3. and 10.5.). If Y is not constant in (11.11), the general solution is

$$f(w) = (w^2 + l^2)^{-1} \left[k(w^2 - l^2) - 2mw + e^2 - \Lambda w \left(\frac{w^3}{3} + 2l^2 w - \frac{l^4}{w} \right) \right], \quad Y^2 = w^2 + l^2 \tag{11.42}$$

(Cahen and Defrise (1968)). The solution for (11.19) is the same with the sign of the e^2 term changed. These solutions contain numerous well-known and frequently rediscovered particular cases, e.g. the Schwarzschild and Reissner-Nordström metrics (§ 13.4.), plane symmetric cosmologies (§ 13.6.), and the Taub-NUT metrics, cp. also the metrics (13.11), (13.26), (20.10) and (27.56).

The solutions (11.42) have expanding principal null congruences of the Weyl tensor. For $l = 0 = e = \Lambda$ one obtains the "A-metrics" of Table 16.2. The $l \neq 0$ vacuum cases are the Taub-NUT (Newman, Unti, Tamburino) metrics (Taub (1951), Newman et al. (1963)); for $k = 1$ one obtains the metric

$$\begin{aligned} \mathrm{d}s^2 &= (r^2 + l^2)\,(\mathrm{d}\vartheta^2 + \sin^2\vartheta\,\mathrm{d}\varphi^2) - f(r)\,(\mathrm{d}t + 2l\cos\vartheta\,\mathrm{d}\varphi)^2 + f^{-1}(r)\,\mathrm{d}r^2, \\ f(r) &= (r^2 + l^2)^{-1}\,(r^2 - 2mr - l^2), \qquad m, l \text{ constants,} \end{aligned} \tag{11.43}$$

which has the Killing vectors (see Table 8.2)

$$\xi_1 = \sin\varphi\,\partial_\vartheta + \cos\varphi\left(\cot\vartheta\,\partial_\varphi + 2l\,\frac{1}{\sin\vartheta}\,\partial_t\right),$$

$$\xi_2 = \cos\varphi\,\partial_\vartheta - \sin\varphi\left(\cot\vartheta\,\partial_\varphi + 2l\,\frac{1}{\sin\vartheta}\,\partial_t\right), \tag{11.44}$$

$$\xi_3 = \partial_\varphi, \qquad \xi_4 = \partial_t.$$

The Taub-NUT metrics are stationary in the region $f(w) < 0$ (see § 18.3.) and spatially-homogeneous in the region $f(w) > 0$. The two solutions form part of a single manifold, being joined across a null hypersurface; this manifold has interesting topological properties (see Misner and Taub (1968), Ryan and Shepley (1975), Siklos (1976a)). The "charged" generalization of the Taub metric ($k = 1$, $\Lambda = 0 \neq el$ in (11.42)) was first given by Brill (1964).

The vacuum solutions for (11.19) have non-expanding principal null congruences of the Weyl tensor, and are the only such Petrov type D solutions (Kinnersley (1969b, 1975)); they thus belong to Kundt's class (Chapter 27.). For $l = 0$, they give the "B-metrics" of Table 16.2.

11.3.2. Einstein spaces with a G_3 on V_3

Many of the above solutions can be rediscovered as solutions with a G_3 on V_3, namely: the plane waves (§ 10.2.) with G_3IV, G_3VI_h or G_3VII_h; the homogeneous Petrov type N solution with a Λ-term (10.33), and the Petrov type III solution (10.34), both of which belong to Kundt's class (Chapter 27.) and admit a G_3VI_h, $h = -1/9$; the solutions (11.42), which in the various cases may admit G_3 of types I, II, III, VIII or IX (see § 11.1.); the Bertotti-Robinson-like solutions (10.8), with G_3 of types VIII, VI_0, VI_h and III (in the non-flat cases); and the Petrov solution (10.14) with G_3I and G_3VII_h on T_3.

With the exception of the last, all these solutions are *algebraically special*. Siklos (1978) found two more *algebraically special*. By the method of Siklos (1978), one obtains four more algebraically special hypersurface-homogeneous Einstein spaces, each admitting a maximal G_3VI_h, $h = -1/9$. One is a Robinson-Trautman solution of Petrov type III

$$\begin{aligned} ds^2 = &-2e^{-z}\,du\,dy + u^2\,(dz^2 + e^{4z}\,dx^2)/2 + 2e^{-z}\,dy\,dz \\ &+ (10 + 2\Lambda u^2 + 2/u^2)\,e^{-2z}\,dy^2, \end{aligned} \tag{11.45}$$

i.e. the $\Lambda \neq 0$ generalization (Theorem 24.7) of (24.15), which is itself the only algebraically special vacuum spacetime with diverging rays and a maximal G_3 (Kerr and Debney (1970)).

The second is a non-diverging Petrov type II solution with metric

$$\begin{aligned} ds^2 = &\frac{1}{2}\,e^{2z}\,dx^2 + \frac{1}{8b^2}\,dz^2 + 4u\,e^{-2z}\,dz\,dy \pm \frac{2}{3b}\,e^{-z}\,dy\,dx \\ &-2e^{-2z}\,du\,dy - (8b^2u^2 - 1/18b^2)\,e^{-4z}\,dy^2, \end{aligned} \tag{11.46}$$

where $\Lambda = -8b^2$. It has no (non-trivial) vacuum limit.

The other two solutions are twisting solutions, one of Petrov type N and the other of Petrov type III, which can be jointly written as

$$k\,\mathrm{d}s^2 = -2\left(x^a\,\mathrm{d}u - \frac{\mathrm{d}y}{(a+1)\,x}\right)\left[\mathrm{d}t + a(t\,\mathrm{d}x + \mathrm{d}y)/x + f(t)\left(x^a\,\mathrm{d}u - \frac{\mathrm{d}y}{(a+1)\,x}\right)\right]$$
$$+ (\mathrm{d}x^2 + \mathrm{d}y^2)\,(t^2+1)/2x^2. \tag{11.47}$$

Type N: $a = 2$, $k\Lambda = -3$, $f(t) = (t^2 - 1)/2$.

Type III: $a = 1/2$, $78k\Lambda = -32$, $f(t) = (13t^2 + 17)/32$.

The type N solution was first found by Leroy (1970), and the type III solution by □ Siklos and MacCallum (1980).

The Killing vectors for (11.45) and (11.46) are

$$\partial_x,\ \partial_y,\ \partial_z + x\,\partial_x - 2y\,\partial_y, \tag{11.48}$$

and for (11.47)

$$\partial_u,\ \partial_y,\ x\partial_x + y\,\partial_y - au\,\partial_u. \tag{11.49}$$

A modification of Siklos' method can be used to study those algebraically special hypersurface-homogeneous Einstein spaces in which the repeated principal null direction does not lie in the surfaces of transitivity of the group. All such metrics belong to Kundt's class.

We now come to the *algebraically general* vacuum solutions with a G_3 on V_3 as their maximal group. In many cases there are solutions with homogeneus S_3 and T_3 related by a substitution of the form $t \leftrightarrow \mathrm{i}x$. The first example of this is the well-known general solution for a vacuum metric with a G_3I on S_3,

$$\begin{gathered}\mathrm{d}s^2 = t^{2p_1}\,\mathrm{d}x^2 + t^{2p_2}\,\mathrm{d}y^2 + t^{2p_3}\,\mathrm{d}z^2 - \mathrm{d}t^2,\\ p_1 + p_2 + p_3 = 1 = {p_1}^2 + {p_2}^2 + {p_3}^2; \qquad p_1, p_2, p_3 \text{ constants.}\end{gathered} \tag{11.50}$$

This form of the metric is usually associated with the name of Kasner (1921), who gave the related metric

$$\begin{gathered}\mathrm{d}s^2 = x^{2a_1}\,\mathrm{d}x^2 + x^{2a_2}\,\mathrm{d}y^2 + x^{2a_3}\,\mathrm{d}z^2 - x^{2a_4}\,\mathrm{d}t^2,\\ a_2 + a_3 + a_4 = a_1 + 1, \qquad {a_2}^2 + {a_3}^2 + {a_4}^2 = (a_1 + 1)^2,\end{gathered} \tag{11.51}$$

and stated he had found all similar solutions. Another form of the metric which is frequently used was given by Narlikar and Karmarkar (1946). The solution (11.50) is easily found from (11.30), (11.31), and can be generalized to non-zero Λ as (Saunders (1967))

$$\Lambda > 0, \qquad S^3 = a\sinh\omega t, \qquad \exp\int S^{-3}\,\mathrm{d}t = S^{-3}\,(\cosh\omega t - 1), \tag{11.52a}$$

$$\Lambda < 0, \qquad S^3 = a\sin\omega t, \qquad \exp\int S^{-3}\,\mathrm{d}t = S^{-3}\,(1 - \cos\omega t), \tag{11.52b}$$

$$\omega^2 = |\Lambda|/3 \qquad a^2 = 3/|\Lambda|, \tag{11.52c}$$

cp. also (13.32). It should be noted that for $p_1 = 1$, $p_2 = p_3 = 0$, (11.50) is flat space-time. The related solution (11.51) can be transformed to the Levi-Civita cylindrically symmetric vacuum solution (20.8). The non-flat plane symmetric case ($a_2 = a_3$) was given by Taub (1951), cp. (13.30).

The Kasner solutions play an important role in the discussion of certain cosmological questions (see the reviews cited in § 11.1.).

Kellner (1975) gives, as the only Einstein spaces with a G_3I on T_3 without a null Killing vector, a (stationary cylindrically symmetric) metric of the form (11.51) and its generalization to non-zero Λ.

The general vacuum solution with a G_3II on S_3 (Taub (1951)) is, using the ω^α of Table 8.2,

$$\begin{aligned} \mathrm{d}s^2 &= X^{-2}(\omega^1)^2 + X^2[\mathrm{e}^{2A\tau}(\omega^2)^2 + \mathrm{e}^{2B\tau}(\omega^3)^2 - \mathrm{e}^{2(A+B)\tau}\,\mathrm{d}\tau^2], \\ kX^2 &= \cosh k\tau, \qquad 4AB = k^2; \qquad k, A, B \text{ constant.} \end{aligned} \tag{11.53}$$

The general vacuum solution for G_3VI_h on S_3 obeying (11.25), $h \neq -1/9$, is (Ellis and MacCallum (1969), MacCallum (1971))

$$\begin{aligned} \mathrm{d}s^2 &= X^2\big((\omega^1)^2 - \mathrm{d}u^2/A^2\big) + Y^2(\omega^2)^2 + Z^2(\omega^3)^2, \\ X^2 &= (\sinh 2u)^{1-h^{-1}}\,(\tanh u)^{e\sqrt{1-3h}/h}, \\ Y^2 &= (\sinh 2u)^{1+A^{-1}}\,(\tanh u)^{e\sqrt{(3h-1)/h}}, \\ Z^2 &= (\sinh 2u)^{1-A^{-1}}\,(\tanh u)^{-e\sqrt{(3h-1)/h}}, \quad e = \pm 1, \ A - \sqrt{-h}, \end{aligned} \tag{11.54}$$

using (11.27). There is also a special case (Collins (1971), Evans (1978))

$$\begin{aligned} \mathrm{d}s^2 &= X^2[(\omega^1)^2 + X^{2(A-1)/(A^2+1)}(\omega^2)^2 + X^{-2(1+A)/(A^2+1)}(\omega^3)^2] - \mathrm{d}t^2, \\ X &= (1 + A^2)\,t/A. \end{aligned} \tag{11.55}$$

By taking appropriate limits of (11.54) one can derive the G_3VI_0 on S_3 vacuum solution (Ellis and MacCallum (1969))

$$\mathrm{d}s^2 = u^{-1/2}\,\mathrm{e}^{u^2/2}\big((\omega^1)^2 - \mathrm{d}u^2\big) + 2u\big((\omega^2)^2 + (\omega^3)^2\big), \tag{11.56}$$

and the general G_3V on S_3 vacuum solution (Joseph (1966))

$$\mathrm{d}s^2 = \sinh 2a\tau\big((\omega^1)^2 - \mathrm{d}\tau^2\big) + (\tanh a\tau)^{\sqrt{3}}\,(\omega^2)^2 + (\tanh a\tau)^{-\sqrt{3}}\,(\omega^3)^2. \tag{11.57}$$

For $h = -1/9$, (11.54) is only a special solution.

Apart from the spaces with higher symmetries, no vacuum solutions with G_3 of types III, IV, VIII or IX are known, and the only known type VII solutions, apart from the algebraically special ones listed earlier, are a solution due to Lukash (1974), with G_3VII_h, $h = 4/11$, on S_3, i.e., with ω^α from Table 8.2,

$$\begin{aligned} \mathrm{d}s^2 &= -\mathrm{d}t^2 + A^2(\omega^1)^2 + B^2\big(\mathrm{e}^{2\mu}\,(\sin\Phi\,\omega^2 + \cos\Phi\,\omega^3)^2 \\ &\qquad + \mathrm{e}^{-2\mu}\,(\cos\Phi\,\omega^2 - \sin\Phi\,\omega^3)^2\big), \\ A^2 &= a^2\,\mathrm{e}^{11/4}\,\sinh^{-3/8} 2\xi, \qquad B^2 = b^2 \sinh 2\xi, \\ \Phi &= c - \sqrt{11}\,\xi, \qquad \mu = \frac{1}{2}\ln\coth\xi, \qquad \mathrm{d}t = A\,\mathrm{d}\xi, \end{aligned} \tag{11.58}$$

and the vacuum solution (Barnes (1978))

$$\begin{aligned}
\mathrm{d}s^2 &= A^2\Sigma\,(2\sqrt{2}\,kx, e)\,(\mathrm{d}u^2 - e\,\mathrm{d}v^2) - 2A^2\,\Sigma'\,(2\sqrt{2}\,kx, e)\,\mathrm{d}u\,\mathrm{d}v \\
&\quad + P^2\,\mathrm{d}x^2 + U^2\,\mathrm{d}z^2; \qquad k \text{ constant},\ e = \pm 1, \\
P &= \exp\left(\frac{1}{2z}\right)\left(\frac{e(z-4)}{z}\right)^{1/8}, \qquad A = \left(\frac{e(z-4)}{z}\right)^{1/4}, \\
U^{-1} &= -\sqrt{2}\,ekz\exp\left(\frac{1}{2z}\right)[ez^3(z-4)^5]^{1/8},
\end{aligned} \tag{11.59}$$

with $G_3VI_0(VII_0)$ when $e = -1$ $(e = 1)$. Σ is defined as in (8.38).

11.3.3. Einstein-Maxwell solutions with a G_3 on V_3

As in the previous sub-section, the plane waves and Bertotti-Robinson-like metrics admit subgroups G_3 of the full group of symmetry. Apart from these and the other metrics which admit a multiply transitive group (see § 11.3.1.), there are a few known cases.

The Einstein-Maxwell solutions admitting a G_3I on T_3 which can be interpreted as cylindrically symmetric static metrics are treated in § 20.2. We note that the distinction between longitudinal and azimuthal fields depends on which of the coordinates is considered to be the angular coordinate; the solutions are otherwise identical. The corresponding solutions with G_3I on S_3 can be derived from (20.9) by $t \leftrightarrow \mathrm{i}\varrho$ and have been explicitly given by Datta (1965). They include solutions of interest as cosmologies with a magnetic field; solutions of this type were given by Rosen (1962) and Jacobs (1969), and can be generalized to include fluids (see §§ 12.3., 12.4.). They are related to the vacuum G_3II solution (11.53) (Collins (1972)) and include plane symmetric solutions included also in (11.42). The solutions are

$$\mathrm{d}s^2 = -A^{-1}\,\mathrm{d}t^2 + A\,\mathrm{d}x^2 + B\,\mathrm{d}y^2 + C\,\mathrm{d}z^2, \tag{11.60}$$

$$B = C = t^2, \qquad A = bt^{-1} - at^{-2} \tag{11.61}$$

or

$$AB = t^\lambda, \quad AC = t^{2-\lambda}, \quad A = C_1t^\mu + C_2t^{-\mu}, \quad 4\mu^2 = \lambda(2-\lambda). \tag{11.62}$$

Griffiths (1976a) found a metric with G_3I on T_3 representing interacting electromagnetic waves. After a coordinate transformation it can be written as

$$\begin{aligned}
\mathrm{d}s^2 &= c^2r^2(r^2\,\mathrm{d}r^2 - \mathrm{d}t^2) + r^4(r-a)^2\,\mathrm{d}\Theta^2 \\
&\quad + r^2\big(\mathrm{d}z - 2c(r-b)\,(r-a)\,\mathrm{d}\Theta\big)^2.
\end{aligned} \tag{11.63}$$

The other Einstein-Maxwell hypersurface-homogeneous solutions known to us all arise from the ansatz (11.35a) or (11.36a). The case (11.35) leads to (10.21) when

$\varrho + \bar{\varrho} = 0$, and to the solution

$$\mathrm{d}s^2 = \frac{4t^2}{3}\,\mathrm{d}x^2 + t(\mathrm{e}^{-2x}\,\mathrm{d}y^2 + \mathrm{e}^{2x}\,\mathrm{d}z^2) - \mathrm{d}t^2 \tag{11.64}$$

with a G_3VI_0 ($n^\alpha{}_\alpha = 0$) on S_3 (Tariq and Tupper (1975)), in the case of twistfree rays ($\varrho - \bar{\varrho} = 0$). The dual ansatz leads to the solution

$$\mathrm{d}s^2 = \mathrm{d}r^2 + \frac{4r^2}{3}\,(\omega^1)^2 + r\big((\omega^2)^2 - (\omega^3)^2\big) \tag{11.65}$$

with the ω^a of G_3VII_0 from Table 8.2. (11.64) and (11.65) form two real slices of a complex manifold (Barnes (1977)). A number of solutions satisfying the weaker ansatz (11.35a) have been found, both twisting and twistfree (Barnes (1978)): the twisting solutions admit a G_3I or G_3II on S_3 or T_3; for the twistfree case only a vacuum solution with G_3VI_0 on S_3 or T_3 has been found. Among these, the only solutions not discussed earlier are the electromagnetic G_3II solutions given by

$$\begin{gathered}\mathrm{d}s^2 = P^2\,\mathrm{d}x^2 + R^2\,\mathrm{d}y^2 - 2eQ^2(\mathrm{d}v + 2efx\,\mathrm{d}y)^2 + eU^2\,\mathrm{d}t^2,\\ e = \pm 1; \qquad a, b, c, f, p \text{ constant}; \qquad a^2 + b^2 = c^2 + 1,\end{gathered} \tag{11.66}$$

where

$$\begin{aligned}&QP = F^{-(a+1)/2a}, \quad QR = F^{-(a-1)/2a}, \quad \Phi_1 = \Phi\mathrm{e}^{\mathrm{i}p}\,\frac{(1 - t^2 - 2\mathrm{i}t)}{(1 + t^2)},\\ &\sqrt{2}\,f(1 + t^2)\,Q^3FU = 1, \quad e\Phi^2 = 2bf^2F^2Q^4,\\ &Q^2 = \frac{\big(2ct + b(1 + t^2)\big)}{(1 + t^2)},\\ &F = \begin{cases}\exp\left(\dfrac{2a}{\sqrt{1 - a^2}}\tan^{-1}\dfrac{bt + c}{\sqrt{1 - a^2}}\right) & |a| < 1,\\ \exp\left(\dfrac{-2a}{b(1 + t)}\right), & \text{if} \quad |a| = 1,\\ \left(\dfrac{bt + c - \sqrt{a^2 - 1}}{bt + c + \sqrt{a^2 - 1}}\right)^{a/\sqrt{a^2-1}}, & |a| > 1,\end{cases}\end{aligned} \tag{11.67}$$

or, for $e = 1$,

$$Q = 1, \quad P = t^{(E-1)/2}, \quad R = t^{(E+1)/2}, \quad \Phi_1 = \sqrt{2}\,f\,\mathrm{e}^{\mathrm{i}p}t^{-1+2\mathrm{i}f}. \tag{11.68}$$

The dual ansatz (11.36a), with $\tau + \bar{\pi} \neq 0$, similarly gives solutions with G_3I and G_3II on T_3, together with (10.14), while the case $\tau + \bar{\pi} = 0$ gives the vacuum metric (11.59) with G_3VI_0 or G_3VII_0 on T_3. The G_3II electromagnetic solutions are

$$
\begin{aligned}
\mathrm{d}s^2 &= FP^{-2}\,\Sigma(B,-e)\,(\mathrm{d}u^2 + e\,\mathrm{d}v^2) - 2FP^{-2}\Sigma'(B,-e)\,\mathrm{d}u\,\mathrm{d}v \\
&\quad + P^2\left(\mathrm{d}x - 2\sqrt{2}\,fu\,\mathrm{d}v\right)^2 + U^2\,\mathrm{d}z^2; \qquad e = \pm 1, \qquad f \text{ constant}, \\
P^2 &= \frac{2cz + b(1+z^2)}{1+z^2}, \quad B = \frac{1}{a}\ln F; \quad a, b, c \text{ constants}, \\
a^2 + b^2 &= c^2 + e, \qquad U^{-1} = \sqrt{2}\,fPF\left(2cz + b(1+z^2)\right), \\
\Phi_1 &= \sqrt{-2b}\,fP^2F\,\mathrm{e}^{\mathrm{i}p}\,(1 - z^2 - 2\mathrm{i}z)/(1+z^2), \quad p \text{ constant}, \\
F &= \left(\frac{bz + c - \sqrt{a^2 - e}}{bz + c + \sqrt{a^2 - e}}\right)^{a/\sqrt{(a^2-e)}},
\end{aligned}
\tag{11.69}
$$

with Σ as in (8.38).

11.4. Perfect fluid solutions homogeneous on T_3

Apart from the stationary or static solutions with spherical, plane and cylindrical symmetry (see Chapters 13., 14. and 20.) and the examples provided by the solutions given in § 10.4., few T_3-homogeneous fluid solutions are known. The S_3-homogeneous perfect fluid cosmologies will be treated in Chapter 12.

Of the solutions with G_4 on T_3 only the *L.R.S.* cases need be considered, because the four-velocity is invariant under the isotropy. The *L.R.S.* cases must give algebraically special solutions (cf. Chapter 29.). Solutions of Ellis' class *III* cannot occur for G_4 on T_3.

The metric (11.11), $f(w) < 0$, permits a perfect fluid matter content only if the four-velocity is $\boldsymbol{u} = u\,\partial_w$ (this follows from the symmetry and $R_{34} = 0$). Any pair of functions f and Y obeying

$$\frac{1}{2}f'' + \frac{k}{Y^2} + f\left((3-e)\frac{l^2}{Y^4} + e\frac{Y''}{Y} - \frac{Y'^2}{Y^2}\right) = 0, \quad e \equiv \frac{f(w)}{|f(w)|}, \tag{11.70}$$

gives a perfect fluid solution.

The dust solutions of Ellis' class *I* are completely known. They have $\omega \neq 0 = \sigma = \Theta$ and are stationary. Apart from the Gödel solution, (10.25), they are (Ellis (1967))

$$
\begin{aligned}
&\Lambda > 0,\ Y^2 = a\cos\beta w + b\sin\beta w + k/2\Lambda, \quad f = 1, \\
&\qquad \beta^2 = 4\Lambda, \quad \Lambda(a^2 + b^2) = l^2 + k^2/4\Lambda;
\end{aligned}
\tag{11.71}
$$

$$\Lambda = 0,\ Y^2 = kw^2 + 2aw + b; \quad a^2 = l^2 + kb, \quad f = 1; \tag{11.72}$$

$$
\begin{aligned}
&\Lambda < 0, \quad Y^2 = a\,\mathrm{e}^{\beta w} + b\,\mathrm{e}^{-\beta w} + k/2\Lambda, \quad f = 1, \\
&\qquad \beta^2 = -4\Lambda,\ 4\Lambda ab = l^2 + k^2/4\Lambda.
\end{aligned}
\tag{11.73}
$$

The Ellis class II solutions are included in § 13.5. Solutions of class I containing perfect fluids and Maxwell fields are given up to quadratures in Stewart and Ellis (1968); the only explicit cases with fluid involve unrealistic equations of state and we therefore omit them, except for (29.26).

The fluid solutions for (11.5), which have a T_3-homogeneous part, have been qualitatively studied by Collins (1974) and Shikin (1975).

An example of a solution with a G_3I on T_3 is the metric (16.67) (Barnes (1972)). In his study of shearfree dust, Ellis (1967) found a solution with a G_3II on T_3,

$$\begin{aligned} ds^2 &= dx^2 + Y^2F^{-2}\,dy^2 + Y^2F^2\,dz^2 - (dt + 2YFay\,dz)^2, \\ F &= \exp\left(b\int Y^{-2}\,dx\right); \qquad a, b, c = \sqrt{a^2+b^2} \quad \text{constants}, \\ Y^2 &= \frac{c}{\sqrt{\Lambda}}\sin 2\sqrt{\Lambda}\,x \quad (\Lambda > 0), \quad Y^2 = 1 + 2cx \quad (\Lambda = 0), \\ Y^2 &= \frac{c}{\sqrt{|\Lambda|}}\sinh 2\sqrt{|\Lambda|}\,x \quad (\Lambda < 0). \end{aligned} \tag{11.74}$$

11.5 Summary of all metrics with G_r on V_3

To help the reader in identifying any metric with a G_r on V_3 which he may have found, we append here three tables. The first lists the places where line elements with larger maximal symmetry may be found, the second summarizes the actual solutions with a maximal G_4 on V_3 for each energy-momentum type considered in this book, and the third similarly lists the solutions with a maximal G_3 on V_3.

Table 11.1 Subgroups G_3 on V_3 occurring in solutions with multiply-transitive groups. The spaces (8.35) of constant curvature are omitted here

Bianchi type of G_3	Maximal group				
	G_7 or G_6 on V_4	G_5 or G_4 on V_4	G_6 on S_3	G_4 on S_3 or T_3 (L.R.S.)	G_4 on T_3 (non-L.R.S.)
I	(10.23)	(10.14) (10.29)–(10.31)	§§ 11.1., 12.2.	(11.3) $k = 0$	(11.16)
II	—	—	—	(11.4) & (11.7)	(11.17)
III	(10.8)	—	—	(11.3) $k = -1$, (11.4) & (11.8)	(11.16), (11.17)
IV	§ 10.2.	(10.29)	—	—	—
V	—	—	§§ 11.1., 12.2.	(11.5)	(11.17)
VI	§ 10.2. (10.8)	(10.29)–(10.30) (10.33), (10.34)	—	—	(11.16), (11.17)
VII	§ 10.2.	(10.14) (10.31)	§§ 11.1., 12.2.	(11.3) $k = 0$ (11.5)	—
VIII	(10.8)	(10.25) (10.27)–(10.28)	—	(11.4) & (11.8)	(11.17)
IX	(10.23)	(10.26)	§§ 11.1., 12.2.	(11.4) & (11.6)	—

Table 11.2 Solutions given in this book with a maximal G_4 on V_3

Energy-momentum	Metric				
	(11.11)		(11.5)	(11.19)	
	(11.3)	(11.4)		(11.16)	(11.17)
Dust	(11.71) –(11.73) (12.10) –(12.14)		(12.15)	$\nexists$	$\nexists$
Perfect fluid	(12.10) –(12.14) (29.26)	(12.17) –(12.19)	(12.16)	$\nexists$	$\nexists$
Vacuum, and Einstein-Maxwell (with $\Lambda \neq 0$)	(11.42) (includes, e.g., (13.26))		$\nexists$	counterpart of (11.42) (includes (13.11), (13.32), (27.56))	

Table 11.3 Solutions given in this book with a maximal G_3 on V_3
(A = all solutions are known)

Bianchi type	Vacuum	Λ-term	Einstein-Maxwell	Dust	Perfect fluid
I	A (11.50) (11.51) ≡ (20.8) (18.23)	(11.52)	(10.60)–(11.62) (11.63), (20.9) (20.11)–(20.12)	For S_3: A (12.20)	For S_3: A (12.21)–(12.22) (16.67), (29.19)
II	For S_3: A (11.53)	—	(11.66)–(11.69)	(11.74) (12.23)	(12.23) (12.24)
V	For S_3: A (11.57)	—	—	—	—
VI_0	(11.56) (11.59)	—	(11.64)	(12.25)	(12.25) (12.27)
VI_h	(11.54)–(11.55) (24.15)	(11.45) –(11.47)	—	—	(12.30)–(12.32)
VII_0	(11.59)	—	(11.65)	—	§ 12.4.
VII_h	(11.58)	—	—	—	(12.28)–(12.29)

Chapter 12. Spatially-homogeneous perfect fluid cosmologies

12.1 Introduction

In this chapter we give solutions containing a perfect fluid and admitting an isometry group transitive on spacelike orbits S_3. The relevant metrics are (11.22) and (11.3), $k = 1$.

The implications of these metrics as cosmological models are beyond the scope of this book, and we refer the reader to standard texts (e.g. McVittie (1956), Peebles (1971), Weinberg (1972)), which deal principally with the Robertson-Walker metrics, and to the reviews cited in § 11.1. Solutions containing both fluid and a magnetic field are of cosmological interest, and exact solutions have been given by (e.g.) Doroshkevich (1965), Shikin (1966), Thorne (1967), Jacobs (1969), Melvin (1975), Dunn and Tupper (1976), and Ozsváth (1977); details of these solutions are omitted here, but they frequently contain, as special cases, solutions for fluid without a Maxwell field. Similarly, they and the fluid solutions may contain as special cases the Einstein-Maxwell and vacuum fields given in Chapter 11.

Perfect fluids with bulk viscosity have been investigated by a number of authors (see e.g. Murphy (1973), Nightingale (1973), and references therein) because they give solutions without singularities.

There is an especially close connection between vacuum solutions and solutions with a perfect fluid with equation of state $p = \mu$. Wainwright et al. (1979) have given a generation procedure for cases admitting an Abelian G_2 (see § 15.1.). Solutions for the further cases of (11.4) with $k \neq 0$ have been found by Maartens and Nel (1978), who found them as illustrations of a general method in which the differential operator in an ordinary differential equation is factorized. The resulting solutions, and the previously discovered examples, are given below.

In general, the known solutions assume an equation of state (5.36), $p = (\gamma - 1)\,\mu$, or its specializations $\gamma = 1, 4/3, 2$, and we take this to be so, in this chapter, unless otherwise stated. If the four-velocity is aligned with the ∂_t of (11.3) or (11.22), which together cover all possible metrics, then, using the Bianchi identities (5.5) and the definition (11.28), we find

$$\varkappa_0 \mu = M S^{-3\gamma}, \qquad \dot{M} = 0. \tag{12.1}$$

The general dynamics of the solutions in which the four-velocity is not aligned with ∂_t ("*tilted*" solutions) have been studied by King and Ellis (1973), and Ellis and King (1974). Rather few exact tilted solutions are known, because the diagonalization theorems discussed in § 11.2. do not apply.

12.2. Robertson-Walker cosmologies

The problem of integrating (11.1), (11.2), has frequently been attacked. Apart from the choice $a(t)$, the most commonly used new time variable is

$$\Psi = \int \frac{dt}{a}. \tag{12.2}$$

The general behaviour for fluids with (12.1), $S \equiv a$, has been investigated by Harrison (1967), who gives many references to the earlier literature. Apart from the de Sitter metrics (3.35) and the Einstein static universe (10.23), some of the best known solutions are the dust solutions with $\Lambda = 0$ (Friedmann (1922, 1924), Einstein and de Sitter (1932)), which are (with $3m = M$)

$$\varepsilon = 1, \qquad a = m \sin^2 \Psi/2, \qquad 2t = m(\Psi - \sin \Psi), \tag{12.3a}$$

$$\varepsilon = 0, \qquad a = (3\sqrt{m}\, t/2)^{2/3}, \tag{12.3b}$$

$$\varepsilon = -1, \qquad a = m \sinh^2 \Psi/2, \qquad 2t = m(\sinh \Psi - \Psi), \tag{12.3c}$$

and the Tolman (1934b) radiation solution

$$\Lambda = \varepsilon = 0, \qquad a = (2\sqrt{m}\, t)^{1/2}. \tag{12.4}$$

The general solutions for $\Lambda\varepsilon = 0$ with (12.1) are

$$\varepsilon = 0 < \Lambda, \qquad a^{3\gamma} = \frac{3m}{\Lambda} \sinh^2 \left(\frac{3\gamma}{2}\left(\frac{\Lambda}{3}\right)^{1/2} t\right), \tag{12.5a}$$

$$\varepsilon = 0 = \Lambda, \qquad a^{3\gamma} = \left(\frac{3\gamma}{2}\sqrt{m}\, t\right)^2, \tag{12.5b}$$

$$\varepsilon = 0 > \Lambda, \qquad a^{3\gamma} = -\frac{3m}{\Lambda} \sin^2 \left(\frac{3\gamma}{2}\left(\frac{-\Lambda}{3}\right)^{1/2} t\right); \tag{12.5c}$$

$$\varepsilon = 1, \quad \Lambda = 0, \quad a^{3\gamma-2} = m \sin^2 \eta, \qquad t = \frac{2}{|3\gamma - 2|} \int a \, d\eta, \tag{12.6a}$$

$$\varepsilon = -1, \quad \Lambda = 0, \quad a^{3\gamma-2} = m \sinh^2 \eta, \qquad t = \frac{2}{|3\gamma - 2|} \int a \, d\eta. \tag{12.6b}$$

The solutions (12.6a and b), which include (12.3a and c) and models due to Tolman and Whittaker, were given by Harrison (1967). The solutions for $\Lambda = 0$ and $\gamma = 1, 4/3, 2$, were found by Tauber (1967), using the manifestly conformally flat form of the metric (the form using (12.2) has become known as the conformal form). The solutions were also given, with a useful compendium of other cases, by Vajk (1969).

For $\Lambda\varepsilon \neq 0$ and $\gamma = 1$ or $4/3$, solutions can be given in terms of elliptic functions (Lemaître (1927), Edwards (1972), Kharbediya (1976)). For $\gamma = 4/3$ there are some solutions in terms of elementary functions (Harrison (1967)), e.g.

$$\frac{2\Lambda}{3} a^2 = 1 - \cosh 2 \left(\frac{\Lambda}{3}\right)^{1/2} t + \left(\frac{\Lambda}{\Lambda_c}\right)^{1/2} \sinh 2 \left(\frac{\Lambda}{3}\right)^{1/2} t, \tag{12.7}$$

where the critical value of Λ, Λ_c, satisfies the Einstein static equation $2\Lambda_c = \varkappa_0(\mu + 3p)$.

It is of cosmological interest to consider a fluid composed of a combination of dust and incoherent radiation. The solutions with a non-interacting mixture have been given by a number of authors (see McIntosh (1968), Sistero (1972), May (1975), Sapar (1970), and references therein). They are, if $\Lambda = 0$,

$$\varkappa_0 \mu = \frac{3m}{a^3} + \frac{3N}{a^4}, \tag{12.8}$$

$$t = \frac{m}{2} \sin^{-1} \left(\frac{2a - m}{(m^2 + 4N)^{1/2}}\right) - (ma + N - a^2)^{1/2}, \quad \varepsilon = 1, \tag{12.9a}$$

$$t = \frac{2(ma - 2N)(ma + N)^{1/2}}{3m^2}, \quad \varepsilon = 0, \tag{12.9b}$$

$$t = (ma + N + a^2)^{1/2} - \frac{m}{2} \ln \left(a + \frac{m}{2} + (ma + N + a^2)^{1/2}\right), \quad \varepsilon = -1. \tag{12.9c}$$

The expressions (12.9) are perhaps no more useful than the form with the time variable a, and for practical calculation parametric forms of the relations (12.9) have been developed. Some of these extend to the case of interacting matter and radiation (McIntosh (1968), May (1975) and references therein).

Other combinations of fluids, formed like (12.8), can be integrated, at least in terms of elliptic or other special functions; for details see Vajk (1969), McIntosh (1972), McIntosh and Foyster (1972), Zaikov (1971), Sistero (1972).
Ref.: Knight and Bergmann (1974).

12.3. Cosmologies with a G_4 on S_3

The metrics (11.3) with $\varepsilon = -1$ have been extensively investigated as cosmological models. The field equations for perfect fluids are (11.32). The known exact solutions were collected by Vajk and Eltgroth (1970). For the *case* $k = 0$ these (*L.R.S.*) solutions are all special cases of the solutions with a G_3I on S_3 (see § 12.4.). Particular solutions were found by Doroshkevich (1965) and Jacobs (1968), while the general solution (for (11.3) with $k = 0$, $1 < \gamma < 2$) can be given either as (11.33), or as

$$\mathrm{d}t = C\,\mathrm{d}\Psi, \qquad C = (-c)^{(\gamma-1)/(2-\gamma)} [3(\gamma - 1)]^{1/(2-\gamma)} \times \left|\frac{1}{2} a\right|^{\gamma/(2-\gamma)} [2/3(2 - \gamma)] (\sinh \Psi)^{\gamma/(2-\gamma)}, \tag{12.10}$$

$$A = M^{1/\gamma}(-c)^{(\gamma-1)/(2-\gamma)} (\gamma - 1)^{2(3-2\gamma)/3\gamma(2-\gamma)}$$
$$\times \left(\frac{1}{2} a\right)^{2/3(2-\gamma)} (\cosh \Psi + 1)^{1/(2-\gamma)} [3(\cosh \Psi - 1)]^{-1/3(2-\gamma)}; \quad a > 0,$$

$$A = M^{1/\gamma}(-c)^{(\gamma-1)/(2-\gamma)} (\gamma - 1)^{2(3-2\gamma)/3\gamma(2-\gamma)}$$
$$\times \left|\frac{1}{2} a\right|^{2/3(2-\gamma)} (\cosh \Psi - 1)^{1/(2-\gamma)} [3(\cosh \Psi + 1)]^{-1/3(2-\gamma)}; \quad a < 0, \tag{12.10}$$

$$B = \left[\frac{3}{2} (\gamma - 1)\, a \,(\cosh \Psi - 1)\right]^{2/3(2-\gamma)}; \quad a > 0,$$

$$B = \left[\frac{3}{2} (\gamma - 1)\, |a| \,(\cosh \Psi + 1)\right]^{2/3(2-\gamma)}; \quad a < 0,$$

(Vajk and Eltgroth (1970)). The *L.R.S.* solution for $\gamma = 2$ is given by

$$\mathrm{d}s^2 = -\mathrm{d}t^2 + t^{2/(1+2\lambda)}\, \mathrm{d}x^2 + t^{2\lambda/(1+2\lambda)}\, (\mathrm{d}y^2 + \mathrm{d}z^2),$$
$$\varkappa_0\mu = \frac{(2 + \lambda)\, \lambda}{(1 + 2\lambda)^2\, t^2}, \qquad \lambda > 0 \quad \text{or} \quad \lambda < -2. \tag{12.11}$$

For the *cases* $k = \pm 1$, the solutions for $\gamma = 1$, 4/3 and 2 can be given in parametric form as below.

For $\gamma = 1$,

$$k = 1, \qquad AB = \frac{bM}{2} \big(\Psi \sin \Psi + 2(1 + \cos \Psi) + E \sin \Psi\big),$$
$$B = b(1 + \cos \Psi), \qquad \mathrm{d}t = B\, \mathrm{d}\Psi; \qquad b, E \text{ constants}; \tag{12.12a}$$

$$k = -1, \qquad AB = \frac{bM}{2} \big(\Psi \sinh \Psi - 2\,(\cosh \Psi - 1) + E \sinh \Psi\big),$$
$$B = b(\cosh \Psi - 1), \qquad \mathrm{d}t = B\, \mathrm{d}\Psi; \qquad b, E \text{ constants}. \tag{12.12b}$$

For $\gamma = 4/3$,

$$\mathrm{d}t = C\, \mathrm{d}\Psi,$$

$$C = \frac{9k}{4} \cosh^2 \Psi [a(b \sinh \Psi + 3kc)^{-1/2} + b^{-1/2}\, (b \sinh \Psi - 6kc)],$$

$$B = \left[\left(-\frac{M}{3c}\right) (b \sinh \Psi + 3kc)\right]^{3/4} A^{-1/2}; \qquad a, b, c \text{ constants}, \tag{12.13}$$

$$A = a + b^{-2}\, (b \sinh \Psi - 6kc)\, (b \sinh \Psi + 3\, kc)^{1/2}.$$

For $\gamma = 2$,

$$\begin{aligned} &\mathrm{d}t = C\,\mathrm{d}\Psi, \qquad C^{-1} = -cA/M = B^{-1}\left(k(y^2+by+c)\right)^{1/2}, \\ &B = a\,|y^2+by+c|^{1/2}\exp\left[\frac{b}{(b^2-4c)^{1/2}}\,g\left(\frac{2\Psi+b}{(b^2-4c)^{1/2}}\right)\right], \\ &g(x) = \tanh^{-1}x \quad (k=1), \qquad g(x) = \coth^{-1}x \quad (k=-1). \end{aligned} \tag{12.14}$$

The solutions (12.12) were found by Doroshkevich (1965), Kantowski and Sachs (1966), Thorne (1967), and Shikin (1966), (12.12)—(12.14) by Kantowski (1966), (12.12b) and (12.13) by Kompaneets and Chernov (1964); □ Ftaclas and Cohen (1979) have found the special case $B = t$ of (12.12b). The forms quoted here are given by Vajk and Eltgroth (1970). The qualitative properties for (11.3) have been studied by Collins (1971, 1977).

The field equations for the metric (11.5) have been integrated for dust by Farnsworth (1967) in the form

$$\begin{aligned} &\mathrm{d}s^2 = a^2(g'-g)^2\,\mathrm{d}x^2 + (g\,\mathrm{e}^{-x})^2\,(\mathrm{d}y^2+\mathrm{d}z^2) - \mathrm{d}t^2, \\ &g = g(t+bx), \qquad \varkappa_0\mu = -6\left(\ddot{g} - \frac{1}{3}\Lambda g\right)\Big/(g-b\dot{g}), \\ &2g\ddot{g} + \dot{g}^2 - \Lambda g^2 - 1/a^2 = 0; \qquad a, b \text{ constants}. \end{aligned} \tag{12.15}$$

Farnsworth also gives, as a special case, a Tolman-Bondi solution (cf. § 13.5.). The $\gamma = 2$ solution has been found (Maartens and Nel (1978), Wainwright et al. (1979)); it is

$$\begin{aligned} &B^2 = a\sinh 2\xi + b\cosh 2\xi; \qquad a, b, c, d \text{ constants}, \\ &A\ = cB^2\exp\left(-d\int B^{-2}\,\mathrm{d}\xi\right), \qquad t = \int A\,\mathrm{d}\xi. \end{aligned} \tag{12.16}$$

The general behaviour of (11.5) for various γ has been studied by Collins (1974) and Shikin (1975).

The field equations for perfect fluid solutions of the *L.R.S.* G_3II metric, (11.4) with $k = 0$, have been reduced to a series of quadratures and a second-order differential equation by Maartens and Nel (1978),

$$\begin{aligned} &A = a\mu^{1/\gamma}\exp\left[2\int\mu^{(2-\gamma)/\gamma}(\eta-\eta_0)\,\mathrm{d}\eta\right], \\ &B = b\mu^{-1/\gamma}\exp\left[-\int\mu^{(2-\gamma)/\gamma}(\eta-\eta_0)\,\mathrm{d}\eta\right], \\ &\eta\ = [(2-\gamma)\,d^{2(3-\gamma)}]^{1/2}\int(AB^2)^{1-\gamma}\,\mathrm{d}t, \\ &\mu\mu'' - \frac{4}{\gamma}\mu'^2 - 8(\eta-\eta_0)\,\mu^{2/\gamma}\mu' - 6\gamma(\eta-\eta_0)^2\,\mu^{4/\gamma} \\ &\qquad + [\gamma - \gamma(\gamma+2)/2(2-\gamma)\,d^{2(3-\gamma)}]\,\mu^{(2+\gamma)/\gamma} = 0, \end{aligned} \tag{12.17}$$

for $\gamma \neq 2$. Some explicit solutions have been found by Collins (1971); they arise from the non-*L.R.S.* G_3II solutions given in the next section. Similarly, the $\gamma = 2$ solutions can be found from the general solution for G_3II in the next section.

The metric (11.4) with $k \neq 1$ has some perfect fluid solutions not obeying (5.36), found by Collins et al. (1980). For the case $k = -1$, this solution can be written as

$$\begin{aligned} ds^2 = k^2 \Bigg(\sinh^{10} \xi \left[-d\xi^2 + \frac{1}{9} (dy^2 + \sinh^2 y \, dz^2) \right] \\ + \frac{8}{81} \sinh^8 \xi \, (dx + \cosh y \, dz)^2 \Bigg), \end{aligned} \tag{12.18}$$

where k is an arbitrary constant. (12.18) was found (in another form) by complexifying the coordinates in a stationary axisymmetric solution. The hypersurface normal $\boldsymbol{n}$ has the properties that Θ and σ (as defined in § 6.1.) are proportional, and σ_{ab} has two equal eigenvalues: Collins et al. (1980) generalized the properties of (12.18) to give ansätze which led to new solutions in the other cases of (11.4), and also for Bianchi type VI_0 and II metrics without rotational symmetry.

Solutions with matter and a magnetic field have been found by Batakis and Cohen (1972) and Damião Soares and Assad (1978). The $\gamma = 2$ solutions have been found by Maartens and Nel (1978) and (with a magnetic field) by Ruban (1978); one has

$$dt = AB^2 \, d\eta, \qquad A^2 = \frac{2\alpha \, e^{-\alpha\eta}}{e^{-2\alpha\eta} + 4l^2}, \qquad \gamma^2 = \alpha^2 + 4M,$$

$$AB = \frac{1}{2}\gamma \begin{cases} \cosh^{-1} \frac{1}{2} \gamma(\eta + \eta_0), & k = 1, \\ \exp\left(\frac{1}{2} \gamma(\eta + \eta_0)\right), & k = 0, \\ \sinh^{-1} \frac{1}{2} \gamma(\eta + \eta_0), & k = -1. \end{cases} \tag{12.19}$$

The $k = 1$ case is also given by Barrow (1978).

12.4. Cosmologies with a G_3 on S_3

Once again, because of the occurrence of G_3 as subgroups of larger symmetry groups, we have to refer to §§ 10.4., 11.1. and 12.2., 12.3. We now list the other known solutions, taking first those obeying $p = (\gamma - 1)\,\mu$.

The most easily solved case is that with a G_3I. Jacobs (1968) gives an extensive list of solutions. For dust, with $\Lambda \neq 0$, we have (Saunders (1967))

$$\begin{aligned} \Lambda > 0, \quad & S^3 = a \sinh \omega t + \frac{M}{2\Lambda} (\cosh \omega t - 1), \\ & S^3 F = \cosh \omega t - 1, \\ \Lambda = 0, \quad & S^3 = 3\left(\frac{Mt^2}{4} + t\right), \quad S^3 F = t^2, \\ \Lambda < 0, \quad & S^3 = a \sin \omega t + \frac{M}{2\Lambda} (\cos \omega t - 1), \\ & S^3 F = (1 - \cos \omega t), \end{aligned} \tag{12.20}$$

where (11.52c) holds and $F = \exp\left(\int S^{-3}\,\mathrm{d}t\right)$. With $M = 0$, (12.20) gives (11.52). The case $\Lambda = 0$ was earlier given by Robinson (1961) and Raychaudhuri (1958). The *L.R.S.* case ($\alpha = 0, \frac{\pi}{2}$ in (11.32)) has been rediscovered several times.

The solutions with $\gamma = 2$ are

$$\Lambda > 0, \quad S^3 = \sqrt{\frac{3+M}{\Lambda}}\sinh \omega t, \quad F = (\tanh \omega t/2)^b, \tag{12.21a}$$

$$\Lambda = 0, \quad S^3 = \sqrt{3(3+M)}\,t, \quad F = t^b, \tag{12.21b}$$

$$\Lambda < 0, \quad S^3 = \sqrt{(3+M)/|\Lambda|}\sin \omega t, \quad F = (\tan \omega t/2)^b, \tag{12.21c}$$

$$b \equiv \sqrt{3/(3+M)}.$$

Jacobs reformulated the problem as

$$\begin{gathered} y^2 \equiv \frac{1}{S^{3\gamma}} + \frac{3}{M}, \qquad F = \left(\frac{y - \sqrt{3/M}}{y + \sqrt{3/M}}\right)^{1/(\gamma-2)}, \\ \frac{\sqrt{3}\,M(2-\gamma)}{2}\,t = \int \left(y^2 - \frac{3}{M}\right)^{(\gamma+1)/(2-\gamma)} \mathrm{d}y, \end{gathered} \tag{12.22}$$

and was able to integrate these equations for $\gamma = 4/3$, $\gamma = 1 + n/(n+1)$, and $\gamma = 1 + (2n+1)/(2n+3)$ with integer n. The solutions for combinations of fluids with $\gamma = 1$, 4/3, 5/3 and 2 can be expressed in terms of elliptic functions (Ellis and MacCallum (1969), Jacobs (1968)).

The solutions with electromagnetic fields, given by Jacobs (1969), can be used to solve the G_3II case (Collins (1972)). Collins (1971) has found a special G_3II solution, which can be written, using (11.28) and (11.29), as

$$\begin{gathered} \beta_1 = \frac{1}{4}(2-3\gamma)\,\lambda, \qquad \beta_2 = \int \left[\frac{3}{16}\,\frac{(2+\gamma)(10-3\gamma)\,E^2\,\mathrm{e}^{-3(2-\gamma)\lambda}}{1 + E^2\,\mathrm{e}^{-3(2-\gamma)\lambda}}\right]^{1/2} \mathrm{d}\lambda, \\ t = \int \frac{3}{4}\left(\frac{\mathrm{e}^{-3\gamma\lambda} + E^2\,\mathrm{e}^{-6\lambda}}{(2-\gamma)(3\gamma-2)}\right)^{-1/2} \mathrm{d}\lambda, \qquad M = \frac{6-\gamma}{(2-\gamma)(3\gamma-2)}, \end{gathered} \tag{12.23}$$

and the general G_3II non-tilted $\gamma = 2$ solution,

$$\begin{gathered} \mathrm{d}s^2 = A^2(-\mathrm{d}t^2 + \mathrm{d}x^2) + t\left(B(\mathrm{d}y + 2nbx\,\mathrm{d}z)^2 + B\,\mathrm{d}z^2\right), \\ A^2 = t^{\alpha^2 + \frac{1}{2}n^2 - n}(1 + b^2t^{2n}), \qquad B = t^{n-1}(1 + b^2t^{2n})^{-1}, \end{gathered} \tag{12.24}$$

given here in the form due to Wainwright et al. (1979). The tilted G_3II solution with $\gamma = 2$ has been reduced to an ordinary differential equation defining the third Painlevé transcendent, followed by quadratures (Maartens and Nel (1978)). Special cases of (12.23) have been integrated by Ruban (1978), while (12.24) was also investigated by Send (1972).

A class of perfect fluid G_3VI_0 space-times was found by Dunn and Tupper (1976) by generalizing the metric (11.64) to the form

$$\mathrm{d}s^2 = -\mathrm{d}t^2 + (m-n)^{-2}\,t^2\,\mathrm{d}x^2 + t^{-2(m+n)}(\mathrm{e}^{-2x}\,\mathrm{d}y^2 + \mathrm{e}^{2x}\,\mathrm{d}z^2), \tag{12.25}$$

where m and n are constants. One obtains

$$\varkappa_0\mu = (m^2 + mn + n^2)/t^2, \qquad \varkappa_0 p = -4mn/t^2, \tag{12.26}$$

which includes the dust solution found by Ellis and MacCallum (1969). The metric (12.25) is in fact identical with the G_3VI_0 solution for $n^\alpha{}_\alpha = 0$ given by Collins (1971), and the dust solution was also found by Evans (1978).

The vacuum solution (11.56) can be used to generate the $p = \mu$ solution (Wainwright et al. (1979))

$$\begin{aligned}
&\mathrm{d}s^2 = A^2(-\mathrm{d}t^2 + \mathrm{d}x^2) + t(B\,\mathrm{d}y^2 + B^{-1}\,\mathrm{d}z^2),\\
&B = t^m\,\mathrm{e}^{2nx}, \qquad A^2 = t^{\alpha^2+\frac{1}{2}(m^2-1)}\exp\left(\left(n^2+\frac{1}{2}\beta^2\right)t^2\right),\\
&nm = \alpha\beta,
\end{aligned} \tag{12.27}$$

which is tilted if $\beta \neq 0$. The non-tilted case is the G_3VI_0 $(n^\alpha{}_\alpha = 0)$ solution found by Ellis and MacCallum (1969).

For Bianchi type VII_h there is a $p = \mu$ solution found from the vacuum solution (11.58) by Wainwright et al. (1979), which is given by replacing $A(\xi)$ in (11.58) with

$$A^2 = a^2(\sinh 2\xi)^{\alpha^2+\beta^2-3/8}\,(\tanh\xi)^{2\alpha\beta}\exp(m^2\xi), \tag{12.28}$$

h by $1/m^2$, and Φ by $c - 2m\xi$. One needs

$$m^2 - \frac{11}{4} + 2(\alpha^2 - \beta^2) = 0. \tag{12.29}$$

If $\beta \neq 0$, this solution is tilted; the non-tilted case was found by Barrow (1978).

For Bianchi type V, solutions of (11.30) could be given in terms of elliptic functions for $\gamma = 2$ or $4/3$.

The only other solutions known to us with G_3 on S_3 containing a perfect fluid obeying (5.36) are special cases of the Bianchi type VI_h metrics. Collins (1971) has found special solutions with $n^\alpha{}_\alpha = 0$, which can be put in the form

$$\begin{aligned}
&\mathrm{d}s^2 = a^2t^2(\omega^1)^2 + t^{b+c}(\omega^2)^2 + t^{b-c}(\omega^3)^2 - \mathrm{d}t^2,\\
&b = \left(\frac{2}{3\gamma}+1\right), \qquad c = \sqrt{-h}\left(\frac{2}{\gamma}-3\right), \qquad (3\gamma-2)(1-3h) < 4,
\end{aligned} \tag{12.30}$$

using (11.27).

Collins also gave the solution for the G_3VI_h ($n^\alpha{}_\alpha = 0$) metric with $4 = (1 - 3h) \times (3\gamma - 2)$, in the form, using (11.27) and (11.29),

$$\beta_1 = \frac{\beta_2}{\sqrt{-3h}} = -\frac{\beta}{\sqrt{1-3h}}, \qquad X \equiv e^{\lambda}, \qquad Y^2 \equiv e^{\beta+2\lambda},$$
$$t = \pm\left(\frac{12E}{M}\right)^{1/2} \int X^{3(\gamma-1)}\, Y^{-(3\gamma-4)/2}\, \mathrm{d}Y, \tag{12.31}$$
$$Y^{-3(2-\gamma)/2} X^{3(2-\gamma)} - 4E\, Y^{3(2-\gamma)/2} - \frac{12E}{M}\frac{(2-\gamma)}{(3\gamma+2)}\, Y^{(3\gamma+2)/2} = F,$$

where E, F are constants. (12.31) generalizes the radiation solution (12.13).

Wainwright et al. (1979) have used the vacuum solution (11.54) to generate the $\gamma = 2$ G_3VI_h solution, which takes the same form with the powers of $\tanh u$ in Y^2 and Z^2 altered to m and $-m$, and X^2 altered to

$$X^2 = (\sinh 2u)^{\alpha^2+\beta^2+(m^2+n^2-1)/2}\, (\tanh u)^{mn+2\alpha\beta},$$
$$hn^2 = -1, \qquad m^2 - n^2 - 3 + 2(\alpha^2 - \beta^2) = 0. \tag{12.32}$$

This solution is tilting if $\beta \neq 0$; the non-tilting case was also found by Ruban (1978), and, in another form, by Collins (1971). It includes G_3III solutions, and (12.16).

The fluid solutions other than those obeying (5.36) are relatively few. For G_3I with *L.R.S.*, metric (11.3), $k = 0$, the cases

$B = \sqrt{t}$ (Singh and Singh (1968)); $B = t$ (Singh and Abdussattar (1974)); $B = \sinh^{2/3}(at)$ (Horský and Novotný (1972)), $B = t^{2/3}$ (Patel and Vaidya (1969)); $B = T$, $bT\, e^{a/T}\, \mathrm{d}T = \mathrm{d}t$ (Singh and Abdussattar (1973))

have been discussed.

The other case known to us is the solution with a G_3VII_0 found by Demiański and Grishchuk (1972), up to a differential equation, by imposing the condition that the orbits of the group are flat three-spaces. The model is tilted, and the matter is rotating, shearing and expanding. The other metrics in this section all contain non-rotating fluid.

Chapter 13. Groups G_3 on non-null orbits V_2. Spherical and plane symmetry

13.1. Metric, Killing vectors, and Ricci tensor

A Riemannian space V_q admitting a group G_r, $r = q(q+1)/2$, is a space of constant curvature (§ 8.5.). Hence the orbits V_2 of a group G_3 of motions must have constant Gaussian curvature K, and the 2-metric $\mathrm{d}\sigma^2$ of the spacelike (S_2) or timelike (T_2) orbits can be written in the form (8.41):

$$\mathrm{d}\sigma^2 = Y^2[(\mathrm{d}x^1)^2 \pm \Sigma^2(x^1, k)\,(\mathrm{d}x^2)^2] \tag{13.1}$$

with $k = KY^2$, and $\Sigma(x^1, k) = \sin x^1$, x^1, $\sinh x^1$ respectively for $k = 1, 0, -1$ as in § 8.5. In (13.1) and in the following formulae the upper and lower signs refer to spacelike and timelike orbits, respectively. The function Y in (13.1) is independent of the coordinates x^1 and x^2 in the orbits V_2. However, Y in general depends on coordinates x^3 and x^4 because the orbits V_2 are subspaces ($x^3 = \text{const}$, $x^4 = \text{const}$) of space-time V_4.

From Theorem 8.18 it follows that the orbits V_2 admit *orthogonal surfaces* in V_4. By performing a coordinate transformation in the 2-spaces orthogonal to the orbits we can put the space-time metric into diagonal form (Goenner and Stachel (1970))

$$\begin{gathered}\mathrm{d}s^2 = Y^2[(\mathrm{d}x^1)^2 \pm \Sigma^2(x^1, k)\,(\mathrm{d}x^2)^2] + \mathrm{e}^{2\lambda}(\mathrm{d}x^3)^2 \mp \mathrm{e}^{2\nu}(\mathrm{d}x^4)^2,\\ Y = Y(x^3, x^4), \qquad \lambda = \lambda(x^3, x^4), \qquad \nu = \nu(x^3, x^4).\end{gathered} \tag{13.2}$$

Note that for timelike orbits T_2 the coordinate x^2 is timelike, while x^4 is spacelike. For spacelike orbits S_2, it is sometimes more convenient to use the space-time metric

$$\mathrm{d}s^2 = Y^2(u, v)\,[(\mathrm{d}x^1)^2 + \Sigma^2(x^1, k)\,(\mathrm{d}x^2)^2] - 2G(u, v)\,\mathrm{d}u\,\mathrm{d}v \tag{13.3}$$

(cp. (13.18) and (13.24) below).

In the S_2 case, we have to distinguish between $Y_{,m}Y^{,m} > 0$ (R-region), $Y_{,m}Y^{,m} < 0$ (T-region) (see McVittie and Wiltshire (1975) and references cited therein), and $Y_{,m}Y^{,m} = 0$. The case $Y_{,m}Y^{,m} < 0$ cannot occur for timelike orbits T_2, because of the Lorentzian signature of the metric.

Coordinate transformations which preserve the form (13.2) of the metric can be used to reduce the number of functions in the line element (13.2). For instance, we can set $Y = x^3$ (canonical coordinates) or $Y = x^3\,\mathrm{e}^{\lambda}$ (isotropic coordinates) provided that $Y_{,m}Y^{,m} > 0$.

In the coordinate system of (13.2), the Killing vectors ξ_A are given by

$$\begin{aligned}\xi_1 &= \cos x^2\,\partial_1 - \sin x^2\,\Sigma_{,1}\Sigma^{-1}\,\partial_2, \qquad \xi_2 = \partial_2,\\ \xi_3 &= \sin x^2\,\partial_1 + \cos x^2\,\Sigma_{,1}\Sigma^{-1}\,\partial_2\end{aligned} \tag{13.4a}$$

for spacelike orbits and

$$\xi_1 = \cosh x^2\, \partial_1 - \sinh x^2\, \Sigma_{,1}\Sigma^{-1}\, \partial_2, \qquad \xi_2 = \partial_2,$$
$$\xi_3 = -\sinh x^2\, \partial_1 + \cosh x^2\, \Sigma_{,1}\Sigma^{-1}\, \partial_2 \tag{13.4b}$$

for timelike orbits, with $\Sigma = \Sigma(x^1, k) = \sin x^1, x^1, \sinh x^1$ respectively for $k = 1, 0, -1$. The group types are: for S_2, IX $(k = 1)$, VII_0 $(k = 0)$, VIII $(k = -1)$, and for T_2, VIII $(k = 1)$, VI_0 $(k = 0)$, and VIII $(k = -1)$. These Bianchi types belong to class G_3A (§ 8.2.). The spacelike and timelike metrics on V_2, and their corresponding groups, are related by complex transformations.

The existence of a higher-dimensional group of motions, G_r, $r > 3$, imposes further restrictions on the functions ν, λ and Y in the metric (13.2) (Takeno (1966)). For certain Ricci tensor types, a group G_3 on V_2 *implies* a G_4 on V_3 (§ 13.4.). The de Sitter universe (8.36), the Friedmann models (§ 12.2.), the Kantowski-Sachs solutions (12.12), and the static spherically symmetric perfect fluid solutions (§ 14.1.) are solutions admitting a G_r, $r > 3$, with a subgroup G_3 on V_2.

For the metric (13.2), a natural choice of a tetrad metric g_{ab} and of a basis $\{\omega^a\}$ of 1-forms is

$$g_{ab} = \mathrm{diag}\,(1, \pm 1, 1, \mp 1), \tag{13.5}$$
$$\omega^1 = Y\,\mathrm{d}x^1, \qquad \omega^2 = Y\Sigma(x^1, k)\,\mathrm{d}x^2, \qquad \omega^3 = e^\lambda\,\mathrm{d}x^3, \qquad \omega^4 = \mathrm{e}^\nu\,\mathrm{d}x^4.$$

The non-zero tetrad components G^b_a of the Einstein tensor, with respect to this basis, are

$$G^4_4 = -\frac{k}{Y^2} + \frac{2}{Y}\,\mathrm{e}^{-2\lambda}\left(Y'' - Y'\lambda' + \frac{Y'^2}{2Y}\right) \mp \frac{2}{Y}\,\mathrm{e}^{-2\nu}\left(\dot{Y}\dot{\lambda} + \frac{\dot{Y}^2}{2Y}\right), \tag{13.6a}$$

$$G^3_3 = -\frac{k}{Y^2} \mp \frac{2}{Y}\,\mathrm{e}^{-2\nu}\left(\ddot{Y} - \dot{Y}\dot{\nu} + \frac{\dot{Y}^2}{2Y}\right) + \frac{2}{Y}\,\mathrm{e}^{-2\lambda}\left(Y'\nu' + \frac{Y'^2}{2Y}\right), \tag{13.6b}$$

$$G^1_1 = G^2_2 = \mathrm{e}^{-2\lambda}\left(\nu'' + \nu'^2 - \nu'\lambda' + \frac{Y''}{Y} + \frac{Y'}{Y}\nu' - \frac{Y'}{Y}\lambda'\right)$$
$$\mp\, \mathrm{e}^{-2\nu}\left(\ddot{\lambda} + \dot{\lambda}^2 - \dot{\lambda}\dot{\nu} + \frac{\ddot{Y}}{Y} + \frac{\dot{Y}}{Y}\dot{\lambda} - \frac{\dot{Y}}{Y}\dot{\nu}\right), \tag{13.6c}$$

$$G^4_3 = \pm\frac{2}{Y}\,\mathrm{e}^{-\nu-\lambda}(\dot{Y}' - \dot{Y}\nu' - Y'\dot{\lambda}). \tag{13.6d}$$

Dash and dot denote differentiation with respect to the coordinates x^3 and x^4 respectively.

Ref.: Israel (1958), Takeno and Kitamura (1969), Collins (1977).

13.2. Some implications of the existence of an isotropy group H_1

The group G_3 on V_2 implies an isotropy group H_1 (§ 9.2.) and this, in turn, implies a (one-dimensional) linear isotropy subgroup of the Lorentz group $L_+^\uparrow$ in the tangent space T_p. Therefore the principal tetrad (§ 4.2.) cannot be determined uniquely; only degenerate Petrov types (N, D, O) are possible. As G_3 acts on *non-null* orbits, H_1 describes spatial rotations (boosts) for spacelike (timelike) orbits. For Petrov type N, the invariance subgroup of $L_+^\uparrow$ consists of null rotations (3.15). Therefore type N cannot occur and we have proved

Theorem 13.1. *Space-times admitting a group G_3 of motions acting on non-null orbits V_2 are of Petrov type D or O.*

Similarly, the existence of an isotropy group H_1 leads to

Theorem 13.2. *The Ricci tensor of a space-time admitting a group G_3 on V_2 has at least two equal eigenvalues.*

Any invariant timelike vector field in V_4 (with zero Lie derivative with respect to ξ_A) necessarily lies in the 2-spaces orthogonal to the group orbits S_2, because otherwise the preferred vector field would not be invariant under the isotropies. Hence we can formulate

Theorem 13.3. *Perfect fluid and dust solutions cannot admit a group G_3 on timelike orbits T_2.*

In the coordinate system (13.2) (upper sign) the 4-velocity $\boldsymbol{u}$ of a perfect fluid has the form $\boldsymbol{u} = u^3 \partial_3 + u^4 \partial_4$ and from the invariance of $\boldsymbol{u}$ under isometries one infers that the components u^3, u^4 cannot depend on x^1 and x^2. Therefore $\boldsymbol{u}$ is hypersurface-orthogonal and there is a transformation of x^3 and x^4 which takes $\boldsymbol{u}$ into

$$\boldsymbol{u} = \mathrm{e}^{-\nu} \partial_4 \tag{13.7}$$

(comoving coordinates) and, simultaneously, preserves the form of the metric (13.2).

Ref.: Goenner and Stachel (1970).

13.3. Spherical and plane symmetry

The group G_3 on V_2 contains two special cases of particular physical interest: spherical and plane symmetry.

In the first four decades of research on General Relativity the overwhelming majority of exact solutions were obtained by solving the field equations under the assumption of spherical symmetry. The exterior and interior Schwarzschild solutions and the Friedmann model of relativistic cosmology are well-known examples.

Originally, problems with spherical and plane symmetry were treated more or less intuitively. In the modern literature the group theoretical approach is preferred and spherical and plane symmetry are invariantly defined by the

Definition: *A space-time V_4 is said to be spherically (plane) symmetric if it admits a group G_3 IX (G_3 VII$_0$) of motions acting on spacelike 2-spaces S_2 and if the non-metric fields inherit the same symmetry.*

Each orbit S_2 has constant positive (zero) Gaussian curvature, $k = 1$ ($k = 0$). The isotropy group H_1 represents a spatial rotation in the tangent space of S_2. According to Theorem 13.1, spherically and plane symmetric space-times are of Petrov type D or O.

In the expressions (13.6) for the components of the Einstein tensor we have to choose the upper signs, and $k = 1$ ($k = 0$). We specialize the metric (13.2) to *spherical symmetry* ($k = 1$),

$$\begin{aligned} ds^2 &= Y^2(d\vartheta^2 + \sin^2\vartheta\, d\varphi^2) + e^{2\lambda}\, dr^2 - e^{2\nu}\, dt^2, \\ Y &= Y(r, t), \qquad \lambda = \lambda(r, t), \qquad \nu = \nu(r, t). \end{aligned} \tag{13.8}$$

For *plane symmetry* ($k = 0$), we use Cartesian coordinates in the orbits:

$$\begin{aligned} ds^2 &= Y^2(dx^2 + dy^2) + e^{2\lambda}\, dz^2 - e^{2\nu}\, dt^2, \\ Y &= Y(z, t), \qquad \lambda = \lambda(z, t), \qquad \nu = \nu(z, t), \\ \xi_1 &= \partial_x, \qquad \xi_2 = \partial_y, \qquad \xi_3 = x\,\partial_y - y\,\partial_x. \end{aligned} \tag{13.9}$$

For vacuum, Einstein-Maxwell and pure radiation fields, and for dust, the general solutions admitting a G_3 on V_2 are known (§§ 13.4., 13.5.). Some plane symmetric perfect fluid solutions are listed in § 13.6. Chapter 14. gives a survey of spherically symmetric perfect fluids.

13.4. Vacuum, Einstein-Maxwell and pure radiation fields

For vacuum and Einstein-Maxwell fields (including the cosmological constant Λ), and pure radiation fields, the algebraic types of the energy-momentum tensor are [(111, 1)], [(11) (1, 1)], and [(11, 2)], cp. § 5.2.

The eigenvalues λ_i of the Einstein tensor are

$$\begin{aligned} &\lambda_1 = \lambda_2 = G^1_1 = G^2_2, \\ &\lambda_{3,4} = \frac{1}{2}\,(G^3_3 + G^4_4) \pm \Delta^{1/2}, \qquad \Delta \equiv \frac{1}{4}\,(G^3_3 - G^4_4)^2 \mp (G^4_3)^2. \end{aligned} \tag{13.10}$$

The symmetry (G_3 on V_2) implies that a double eigenvalue ($\lambda_1 = \lambda_2$) exists (cf. Theorem 13.2). For the algebraic types under consideration, there are *two* double eigenvalues (which might coincide); $\lambda_3 = \lambda_4$ implies $\Delta = 0$.

13.4.1. Time-like orbits

For time-like orbits T_2, Table 5.2 shows immediately that the algebraic type is [11(1,1)]. Since the type [(1 1, 2)] is impossible we have

Theorem 13.4. *Einstein-Maxwell fields with an electromagnetic null field, and pure radiation fields, cannot admit a G_3 on T_2.*

For vacuum and non-null Einstein-Maxwell fields, we have to distinguish between $Y_{,a} \neq 0$ and $Y_{,a} = 0$. If $Y_{,a} \neq 0$, one can put $Y = x^3$ (canonical coordinates), because of $Y_{,a}Y^{,a} > 0$, and the evaluation of $G^3_3 = G^4_4$ and $G^4_3 = 0$ in the metric (13.2) (lower signs) gives $\lambda' + \nu' = 0$ and $\dot{\lambda} = 0$. In ν an additive term dependent on x^4 can be made zero by transforming x^4. Thus we have $\nu = -\lambda$, $\dot{\lambda} = 0$. In the coordinates of (13.2), the general solution of the Einstein-Maxwell equations (including Λ) reads (cp. (11.42))

$$\mathrm{e}^{2\nu} = k - 2m/x^3 - e^2/(x^3)^2 - \Lambda(x^3)^2/3 > 0, \tag{13.11}$$
$$\lambda = -\nu, \qquad Y = x^3, \qquad m, e \text{ constants}, \qquad k = 0, \pm 1,$$

the only non-vanishing tetrad component of the (non-null) electromagnetic field tensor being (up to a constant duality rotation, see § 5.4.)

$$\sqrt{\varkappa_0/2}\, F_{34} = e/(x^3)^2. \tag{13.12}$$

These type D Einstein-Maxwell fields belong to Kundt's class (Chapter 27).

For $Y = Y_0 = \text{const}$, two double eigenvalues exist,

$$\lambda_1 = \lambda_2 = K_\perp, \qquad \lambda_3 = \lambda_4 = -kY_0^{-2}, \tag{13.13}$$

where $K_\perp$ denotes the Gaussian curvature of the 2-spaces with

$$\mathrm{d}\sigma_\perp{}^2 = (\mathrm{e}^\lambda\, \mathrm{d}x^3)^2 + (\mathrm{e}^\nu\, \mathrm{d}x^4)^2 \tag{13.14}$$

orthogonal to the orbits T_2. There is only one Einstein-Maxwell field (for $K_\perp = kY_0^{-2} < 0$), namely the Bertotti-Robinson solution (10.16).

13.4.2. Space-like orbits

For space-like orbits S_2, and $Y_{,a}Y^{,a} > 0$, $\lambda_3 = \lambda_4$ implies that

$$\Delta = 0 \Leftrightarrow \partial(\mathrm{e}^{\nu+\lambda})/\partial r + \partial(\mathrm{e}^{2\lambda})/\partial t = 0 \tag{13.15}$$

($Y = x^3 = r$, $x^4 = t$) is satisfied (Plebański and Stachel (1968), Goenner and Stachel (1970)). Then, introducing a null coordinate u according to

$$\mathrm{e}^{\nu+\lambda} = \partial u/\partial t, \qquad \mathrm{e}^{2\lambda} = -\partial u/\partial r, \tag{13.16}$$

one can transform the line element (13.2) (upper signs) into the simpler form

$$\mathrm{d}s^2 = r^2\, \mathrm{d}\sigma^2 - 2\, \mathrm{d}u\, \mathrm{d}r - 2H(u, r)\, \mathrm{d}u^2 \tag{13.17}$$

($\mathrm{d}\sigma^2$ as in (13.1)). For the Ricci tensor types under consideration, the corresponding metrics can be completely determined. The results are listed in Table 13.1.

In the coordinate frame (13.2), the general solution of the Einstein-Maxwell equations (including Λ) differs from its counterpart (13.11)–(13.12) in the case of timelike orbits only by a sign ($e^2 \to -e^2$ in the expression for $\mathrm{e}^{2\nu}$), see § 11.3.

The assumption $Y_{,a}Y^{,a} > 0$ enabled us to put $Y = x^3$. For $Y_{,a}Y^{,a} < 0$, an analogous treatment with $Y = x^4$ leads again to this solution, but now with $x^3 = r$

and $x^4 = t$ interchanged, and with $(k - 2m/t + e^2/t^2 - \Lambda t^2/3) < 0$. The cases $Y_{,a}Y^{,a} > 0$ and $Y_{,a}Y^{,a} < 0$ correspond to the R- and T-regions of the same solutions. There is an additional timelike or spacelike Killing vector $\xi = \partial_t$ (resp. $\xi = \partial_r$) which is hypersurface-orthogonal and commutes with the three generators of G_3 on V_2.

Table 13.1 The vacuum, Einstein-Maxwell and pure radiation solutions with G_3 on S_2 $(Y_{,a}Y^{,a} > 0)$

Plebański type	$2H(u, r)$	Solution for $k = 1$
[(111, 1)]	$k - \frac{2m}{r} - \frac{1}{3}\Lambda r^2$	$\Lambda = 0$: Schwarzschild (1916a) $\Lambda \neq 0$: Kottler (1918)
[(11) (1 1)]	$k - \frac{2m}{r} + \frac{e^2}{r^2} - \frac{1}{3}\Lambda r^2$	$\Lambda = 0$: Reissner (1916), Nordström (1918),
[(11, 2)]	$k - \frac{2m(u)}{r} - \frac{1}{3}\Lambda r^2$	$\Lambda = 0$: Vaidya (1951)

The case of spacelike orbits with $Y_{,a}Y^{,a} = 0$ must be treated separately (Foyster and McIntosh (1972)). It is advisable to start with the coordinate system (13.3). It turns out that no vacuum solutions exist (for $\Lambda \neq 0$ see Nariai (1951)) and that the only Einstein-Maxwell fields are the Bertotti-Robinson solution ($Y = \text{const}$), and the special *pp* wave (§§ 21.5. and 10.5.)

$$\mathrm{d}s^2 = Y^2(u)\,(\mathrm{d}x^2 + \mathrm{d}y^2) - 2\,\mathrm{d}u\,\mathrm{d}v \tag{13.18}$$

(which is flat for $Y_{,uu} = 0$).

13.4.3. Generalized Birkhoff theorem

From the results obtained hitherto in this section we see that all the Einstein-Maxwell fields (including Λ-terms) admitting a group G_3 on V_2 have (at least) one additional Killing vector.

Theorem 13.5 (Cahen and Debever (1965), Barnes (1973b), Goenner (1970)). *Metrics with a group G_3 of motions on non-null orbits V_2 and with Ricci tensors of types* [(11) (1, 1)] *and* [(111, 1)] *admit a group G_4, provided that $Y_{,a}Y^{,a} \neq 0$.*

Obviously, this theorem generalizes Birkhoff's Theorem (Birkhoff (1923)): the only vacuum solution with spherical symmetry is the Schwarzschild solution

$$\mathrm{d}s^2 = r^2(\mathrm{d}\vartheta^2 + \sin^2\vartheta\,\mathrm{d}\varphi^2) + (1 - 2m/r)^{-1}\,\mathrm{d}r^2 - (1 - 2m/r)\,\mathrm{d}t^2. \tag{13.19}$$

Note that the additional Killing vector $\xi = \partial_t$ of the Schwarzschild solution is spacelike in the T-region ($r < 2m$). The original formulation of Birkhoff's Theorem (the only vacuum solution with spherical symmetry is *static*) was criticized by Petrov (1963b, c). Petrov's contribution to Birkhoff's Theorem is discussed in Bergmann et al. (1965), and Hamoui (1969).

For the Ricci tensor type [(11, 2)] the group G_3 on V_2 does not imply the existence of a G_4: The Vaidya metric (Table 13.1)

$$\mathrm{d}s^2 = r^2(\mathrm{d}\vartheta^2 + \sin^2\vartheta\,\mathrm{d}\varphi^2) - 2\,\mathrm{d}u\,\mathrm{d}r - (1 - 2m(u)/r)\,\mathrm{d}u^2, \tag{13.20}$$

$m(u)$ being a disposable function of the null coordinate u, has G_3 on V_2 as the maximal group of motions (unless $m = \text{const}$).

13.4.4. Spherically and plane symmetric fields

The *spherically symmetric* Einstein-Maxwell field with $\Lambda = 0$ is the Reissner-Nordström solution

$$\begin{aligned}\mathrm{d}s^2 &= r^2(\mathrm{d}\vartheta^2 + \sin^2\vartheta\,\mathrm{d}\varphi^2)\\ &\quad + (1 - 2m/r + e^2/r^2)^{-1}\,\mathrm{d}r^2 - (1 - 2m/r + e^2/r^2)\,\mathrm{d}t^2,\end{aligned} \tag{13.21}$$

which describes the exterior field of a spherically symmetric charged body. For $e = 0$, we obtain (13.19). We give the Schwarzschild solution (13.19) in various other coordinate systems which are frequently used:

$$\mathrm{d}s^2 = \left(1 + \frac{m}{2\bar{r}}\right)^4 (\mathrm{d}x^2 + \mathrm{d}y^2 + \mathrm{d}z^2) - \frac{\left(1 - \dfrac{m}{2\bar{r}}\right)^2}{\left(1 + \dfrac{m}{2\bar{r}}\right)^2}\,\mathrm{d}t^2, \quad r = \bar{r}\left(1 + \frac{m}{2\bar{r}}\right)^2 \tag{13.22}$$

(isotropic coordinates);

$$\begin{aligned}&\mathrm{d}s^2 = r^2(\mathrm{d}\vartheta^2 + \sin^2\vartheta\,\mathrm{d}\varphi^2) - 2\,\mathrm{d}u\,\mathrm{d}r - (1 - 2m/r)\,\mathrm{d}u^2,\\ &u = t - \int \frac{\mathrm{d}r}{1 - 2m/r}\end{aligned} \tag{13.23}$$

(Eddington (1924), Finkelstein (1958));

$$\begin{aligned}&\mathrm{d}s^2 = r^2(\mathrm{d}\vartheta^2 + \sin^2\vartheta\,\mathrm{d}\varphi^2) - \frac{32m^3}{r}\,\mathrm{e}^{-r/2m}\,\mathrm{d}u\,\mathrm{d}v,\\ &u = -\left(\frac{r}{2m} - 1\right)^{1/2} \mathrm{e}^{r/4m}\,\mathrm{e}^{-t/4m}, \qquad v = \left(\frac{r}{2m} - 1\right)^{1/2} \mathrm{e}^{r/4m}\,\mathrm{e}^{t/4m}\end{aligned} \tag{13.24}$$

(Kruskal (1960), Szekeres (1960));

$$\begin{aligned}&\mathrm{d}s^2 = Y^2(\mathrm{d}\vartheta^2 + \sin^2\vartheta\,\mathrm{d}\varphi^2) + \frac{(Y'\,\mathrm{d}r)^2}{1 - \varepsilon f^2(r)} - \mathrm{d}\tau^2,\\ &\dot{Y}^2 - \frac{2m}{Y} = -\varepsilon f^2(r)\end{aligned} \tag{13.25}$$

($\varepsilon = 0$: Lemaître (1933), $\varepsilon = 1$, $f^2 = (1 + r^2)^{-1}$: Novikov (1963)).

The *plane symmetric* Einstein-Maxwell field with $\Lambda = 0$,

$$\mathrm{d}s^2 = z^2(\mathrm{d}x^2 + \mathrm{d}y^2) + (m/z + e^2/z^2)^{-1}\,\mathrm{d}z^2 - (m/z + e^2/z^2)\,\mathrm{d}t^2 \tag{13.26}$$

(McVittie (1929)), is either static (z being a spacelike coordinate, $m/z + e^2/z^2 > 0$, $Y_{,a}Y^{,a} > 0$), or a spatially homogeneous solution ($m/z + e^2/z^2 < 0$, $Y_{,a}Y^{,a} < 0$), cp. (11.42). With the aid of a transformation of the z-coordinate, the solution (13.26) can be transformed into the form given by Patnaik (1970), and Letelier and Tabensky (1974):

$$\mathrm{d}s^2 = Y^2(z)\,(\mathrm{d}x^2 + \mathrm{d}y^2) + \frac{1}{2}\,Y'(z)\,(\mathrm{d}z^2 - \mathrm{d}t^2), \tag{13.27}$$

where $Y(z)$ is determined implicitly by the equation

$$(Y - A)^2 + 2A^2 \ln (Y + A) = -Cz, \qquad A,\, C \text{ constants}. \tag{13.28}$$

In the metric (13.27), the (non-null) electromagnetic field is given by

$$F_{12} = C_1, \qquad F_{34} = \frac{1}{2}\,C_2 Y' Y^{-2}, \qquad A = \frac{\varkappa_0}{2C}\,(C_1^2 + C_2^2). \tag{13.29}$$

If $Y_{,a}Y^{,a} > 0$, the general plane symmetric *vacuum* solution (Taub (1951)) is the static metric (see (11.51))

$$\mathrm{d}s^2 = z^{-1/2}(\mathrm{d}z^2 - \mathrm{d}t^2) + z(\mathrm{d}x^2 + \mathrm{d}y^2), \qquad z > 0. \tag{13.30}$$

The case $Y_{,a}Y^{,a} < 0$ leads to the Kasner metric (11.50) with $p_1 = p_2 = 2/3$, $p_3 = -1/3$,

$$\mathrm{d}s^2 = t^{-1/2}(\mathrm{d}z^2 - \mathrm{d}t^2) + t(\mathrm{d}x^2 + \mathrm{d}y^2), \qquad t > 0. \tag{13.31}$$

The plane symmetric vacuum solution with a Λ-term can be written in the form (Novotný and Horský (1974))

$$\begin{aligned} \mathrm{d}s^2 &= \sin^{4/3}(az)\,(\mathrm{d}x^2 + \mathrm{d}y^2) + \mathrm{d}z^2 - \cos^2(az)\sin^{-2/3}(az)\,\mathrm{d}t^2, \\ a &\equiv \sqrt{3\Lambda}/2, \qquad \Lambda > 0. \end{aligned} \tag{13.32}$$

This solution belongs to the class given by Carter (1968b).

The plane symmetric pure radiation field

$$\mathrm{d}s^2 = z^2(\mathrm{d}x^2 + \mathrm{d}y^2) - 2\mathrm{d}u\,\mathrm{d}z + 2m(u)\,z^{-1}\,\mathrm{d}u^2. \tag{13.33}$$

resembles the (spherically symmetric) Vaidya solution (13.20).

Ref.: Datta (1967), Carlson and Safko (1978).

13.5. Dust solutions

For dust, the group orbits cannot be timelike (Theorem 13.3). We will take a comoving system of reference (13.7), and, as $T^{ab}{}_{;b} = (\mu u^a u^b)_{;a} = 0$ implies $\nu' = 0$, we start with

$$\mathrm{d}s^2 = Y^2(r, t)\,[\mathrm{d}\vartheta^2 + \Sigma^2(\vartheta, k)\,\mathrm{d}\varphi^2] + \mathrm{e}^{2\lambda(r,t)}\,\mathrm{d}r^2 - \mathrm{e}^{2\nu(t)}\,\mathrm{d}t^2, \tag{13.34}$$

$\Sigma(\vartheta, k)$ being defined as in (13.1). Of the components (13.6) of the Einstein tensor (to be taken with the upper signs), only $G^4_4 = -\varkappa_0\mu$ is non-zero.

For $Y' \neq 0$, we choose $\nu = 0$ and integrate the field equation $G^4_3 = 0$ by

$$e^{2\lambda} = Y'^2/[k - \varepsilon f^2(r)], \qquad \varepsilon = 0, \pm 1; \tag{13.35}$$

$f(r)$ is an arbitrary function, and ε is to be chosen such that $e^{2\lambda}$ becomes positive. With (13.35), we obtain a first integral of $G^3_3 = 0$ by

$$\dot{Y}^2 - 2m(r)/Y = -\varepsilon f^2(r). \tag{13.36}$$

$G^1_1 = G^2_2 = 0$ are satisfied identically, and $G^4_4 = -\varkappa_0\mu$ yields

$$\varkappa_0\mu(r, t) = 2m'/Y'Y^2. \tag{13.37}$$

The differential equation (13.36) can be completely integrated. The solution for $\varepsilon = 0$ is

$$t - t_0(r) = \pm \frac{2}{3} Y^{3/2}[2m(r)]^{1/2}, \qquad \varepsilon = 0, \tag{13.38a}$$

and for $\varepsilon \neq 0$

$$\begin{aligned} t - t_0(r) &= \pm h(\eta)\, m(r)\, f^{-3}(r), \\ Y &= h'(\eta)\, m(r)\, f^{-2}(r), \\ h(\eta) &= \begin{cases} \eta - \sin\eta & \varepsilon = +1 \\ \sinh\eta - \eta & \varepsilon = -1. \end{cases} \end{aligned} \tag{13.38b}$$

Theorem 13.6. *The general dust solution admitting a G_3 on V_2 with $Y' \neq 0$ is given in a comoving system of reference by*

$$ds^2 = Y^2(r, t)\,[d\vartheta^2 + \Sigma^2(\vartheta, k)\,d\varphi^2] + \frac{Y'^2\,dr^2}{k - \varepsilon f^2(r)} - dt^2, \tag{13.39}$$

where Y is given by (13.38), *and m, f and t_0 are arbitrary functions of r* (Lemaître (1933), Tolman (1934a), Datt (1938), Bondi (1947), Horský et al. (1977)). Note that the radial coordinate is defined only up to a scale transformation.

Originally, dust solutions were considered only for the case of spherical symmetry ($k = +1$), and $Y' \neq 0$ was tacitly assumed. Special solutions contained here are (i) the Schwarzschild solution ($m = \text{const}$; neither t nor r are uniquely defined; the choice $\varepsilon = 0$, $t_0 = r$ leads to the Lemaître form (13.25)), and (ii) the Friedmann dust universes (12.3).

Spherically symmetric charged dust configurations have been discussed by several authors; for references see e.g. Vickers (1973), Misra and Srivastava (1974), and Raychaudhuri (1975).

For $Y' = 0$, $Y = Y(t)$, we choose $Y = t$ ($Y = \text{const}$ leads to $\mu = 0$). All these solutions have the property $Y_{,a}Y^{,a} < 0$. Now $G^3_3 = 0$ is integrated by

$$e^{-2\nu} = at^{-1} - k \tag{13.40}$$

and, writing $e^\lambda = V e^{-\nu}$, $G^1_1 = G^2_2 = 0$ gives the differential equation

$$\ddot{V} + \dot{V}\left(\frac{1}{t} - 3\dot{\nu}\right) = 0 \tag{13.41}$$

for $V(r, t)$, which is solved by

$$V(r, t) = B(r) \int t^{1/2}(a - kt)^{-3/2} \, \mathrm{d}t + A(r). \tag{13.42}$$

If $B(r)$ is zero, we regain the vacuum case. Assuming $B(r) \neq 0$, we can transform B to unity by a scale transformation of r. Introducing $x = \mathrm{e}^{\nu}$ as a new variable of integration in (13.42), the final result is the

Theorem 13.7. *The general dust solution admitting a G_3 on V_2, with $Y' = 0$, is given in comoving coordinates by*

$$\mathrm{d}s^2 = t^2[\mathrm{d}\vartheta^2 + \Sigma^2(\vartheta, k) \, \mathrm{d}\varphi^2] + \mathrm{e}^{2\lambda(r,t)} \, \mathrm{d}r^2 - \mathrm{e}^{2\nu(t)} \, \mathrm{d}t^2,$$

$$\mathrm{e}^{\lambda} = \mathrm{e}^{-\nu} \left[\int^{e^{\nu}} \frac{2x^2 \, \mathrm{d}x}{1 + kx^2} + A(r) \right], \quad \mathrm{e}^{2\nu} = \frac{t}{a - kt}, \tag{13.43}$$

$$\varkappa_0 \mu(r, t) = 2(\dot{\lambda} + \dot{\nu})/t \, \mathrm{e}^{2\nu}.$$

These solutions are generalizations of the Kantowski-Sachs ($k \neq 0$) and Bianchi type I ($k = 0$) solutions given in Chapter 12. (in rescaled time-coordinates) and specialize to them for $A = \text{const}$ (Ellis (1967)).

13.6. Plane symmetric perfect fluids

The plane symmetric *static* perfect fluids with an equation of state $\mu = \mu(p)$ are described by (Taub (1972))

$$\mathrm{d}s^2 = z^2(\mathrm{d}x^2 + \mathrm{d}y^2) + \frac{z}{F(z)} \, \mathrm{d}z^2 - \mathrm{e}^{2\nu} \, \mathrm{d}t^2, \tag{13.44}$$

$$2z \frac{p'}{\mu(p) + p} = 1 - \varkappa_0 p z^3 F^{-1}, \qquad F' = -\varkappa_0 \mu(p) \, z^2, \tag{13.45}$$

$$\big(\mu(p) + p\big) \, \nu' = -p'. \tag{13.46}$$

For a given function $\mu = \mu(p)$, the differential equations (13.45) determine $p = p(z)$ and $F = F(z)$. The solution for $p = \mu/3$ was obtained by Teixeira et al. (1977b). In this case the functions p and F in (13.45) are given by

$$p = p_0(36z^2 - 12z^7 + z^{12}), \quad F = \frac{\varkappa_0 p_0}{5} (216 - 108z^5 + 18z^{10} - z^{15}). \tag{13.47}$$

The solution for $\mu = \text{const}$ was given by Taub (1956) and, in terms of hypergeometric functions, by Avakyan and Horský (1975), and Horský (1975); the solution (13.32) ($\varkappa_0 \mu = -\varkappa_0 p = \Lambda$) is a special case thereof.

Non-static incompressible fluids ($\mu = \text{const}$) are investigated in Taub (1956). Tabensky and Taub (1973) obtained the general plane symmetric solution for an irrotational perfect fluid with the equation of state $p = \mu$ (see also § 20.5. and

Theorem 15.1); it is

$$\begin{aligned}
&ds^2 = t^{-1/2} e^{\Omega}(dz^2 - dt^2) + t(dx^2 + dy^2), \qquad t > 0, \\
&\Omega = 2 \int t[(\sigma_{,t}{}^2 + \sigma_{,z}{}^2)\, dt + 2\sigma_{,t}\sigma_{,z}\, dz], \\
&\sigma_{,tt} + t^{-1}\sigma_{,t} - \sigma_{,zz} = 0, \\
&\varkappa_0 p = \varkappa_0 \mu = t^{1/2} e^{-\Omega}(\sigma_{,t}{}^2 - \sigma_{,z}{}^2), \qquad (\varkappa_0 p)^{1/2} u_i = \sigma_{,i}.
\end{aligned} \tag{13.48}$$

For $\sigma = \text{const}$, the line element (13.48) goes over into (13.31).

If the metric depends only on the ratio z/t, the field equations for perfect fluids reduce to ordinary differential equations (Taub (1972)) and the equation of state necessarily has the form $p = (\gamma - 1)\,\mu$.

The metric

$$ds^2 = -dt^2 + e^{2\beta(t)}(dx^2 + dy^2) + e^{2\alpha(t)}\, dz^2, \tag{13.49}$$

dependent only on time, describes plane symmetric Bianchi type I models (§ 12.4.).

Ref.: Roy and Singh (1977).

Chapter 14. Spherically symmetric perfect fluid solutions

14.1. Static solutions

14.1.1. Field equations and first integrals

Static spherically symmetric perfect fluid solutions have been widely discussed as relativistic models of a star in mechanical and thermodynamical equilibrium. One usually takes Schwarzschild (or canonical) coordinates defined by

$$\mathrm{d}s^2 = r^2\,\mathrm{d}\Omega^2 + \mathrm{e}^{2\lambda(r)}\,\mathrm{d}r^2 - \mathrm{e}^{2\nu(r)}\,\mathrm{d}t^2, \qquad \mathrm{d}\Omega^2 \equiv \mathrm{d}\vartheta^2 + \sin^2\vartheta\,\mathrm{d}\varphi^2, \tag{14.1}$$

and the field equations then read

$$\varkappa_0 \mu r^2 = -G^4_4 r^2 = [r(1 - \mathrm{e}^{-2\lambda})]', \tag{14.2a}$$

$$\varkappa_0 p r^2 = G^3_3 r^2 = -1 + \mathrm{e}^{-2\lambda}(1 + 2r\nu'), \tag{14.2b}$$

$$\varkappa_0 p = G^1_1 = G^2_2 = \mathrm{e}^{-2\lambda}[\nu'' + \nu'^2 - \nu'\lambda' + (\nu' - \lambda')/r]. \tag{14.2c}$$

These field equations should be supplemented by an equation of state

$$f(\mu, p) = 0. \tag{14.3}$$

From the four equations (14.2)—(14.3), the four unknown functions μ, p, λ, and ν can be determined. Physically, and to get a realistic stellar model, one should start with a reasonable equation of state and impose some regularity conditions, see e.g. Glass and Goldmann (1978). In practice, to get analytic expressions for the solutions, the field equations are often solved by making an ad hoc assumption for one of the metric functions or for the energy density, the equation of state being computed from the resulting line element.

The field equations can be cast in various mathematical forms, each of them admitting different tricks for finding solutions. Often the starting point for constructing exact solutions is the condition of isotropy (of pressure) $G^3_3 = G^1_1$, which in full reads

$$\nu'' + \nu'^2 - \nu'\lambda' - (\nu' + \lambda')/r + (\mathrm{e}^{2\lambda} - 1)/r^2 = 0. \tag{14.4}$$

Once a solution (ν, λ) of this equation has been found, one can compute μ and p from (14.2).

An obvious first integral of (14.2a) is

$$\mathrm{e}^{-2\lambda} = 1 - 2m(r)/r, \qquad 2m(r) \equiv \varkappa_0 \int^r \mu(r)\, r^2\,\mathrm{d}r. \tag{14.5}$$

Inserting this into (14.2b) one obtains

$$2r(r - 2m)\,\nu' = \varkappa_0 r^3 p + 2m. \tag{14.6}$$

Eliminating ν' by means of

$$(\mu + p)\,\nu' = -p', \tag{14.7}$$

which immediately follows from $T^{ab}{}_{;b} = 0$, one gets

$$2r(r - 2m)\,p' = -(\mu + p)\,(\varkappa_0 r^3 p + 2m). \tag{14.8}$$

The form (14.5)—(14.8) can be useful if $\lambda(r)$ or $\mu(r)$ or an equation of state is prescribed.

A different form of the field equations has been obtained by Buchdahl (1959) by introducing the new variables

$$x \equiv r^2, \qquad \zeta \equiv e^{\nu}, \qquad w \equiv mr^{-3}. \tag{14.9}$$

The equations (14.5)—(14.8) yield

$$\begin{aligned} e^{-2\lambda} &= 1 - 2xw, \qquad \varkappa_0\mu = 6w + 4xw_{,x}, \\ \varkappa_0 p &= -2w + (4 - 8xw)\,\zeta_{,x}/\zeta, \end{aligned} \tag{14.10}$$

and

$$(2 - 4xw)\,\zeta_{,xx} - (2w + 2xw_{,x})\,\zeta_{,x} - w_{,x}\zeta = 0. \tag{14.11}$$

The last equation, (14.11), is a differential equation *linear* in both ζ and w, and an analytic expression for one of them may be found if the other is prescribed suitably. Moreover, if (by some other method) a solution (ζ, w) is known, a (possibly) new solution $(\hat{\zeta}, \hat{w})$ can be generated by

$$\hat{\zeta} = \zeta, \qquad \hat{w} = w + C(\zeta + 2x\zeta_{,x})^{-2} \exp\left[4 \int \zeta_{,x}(\zeta + 2x\zeta_{,x})^{-1}\,dx\right] \tag{14.12}$$

(Heintzmann (1969)). Once a solution (ζ, w) of (14.11) is known, λ, μ and p can be computed from (14.10).

Sometimes isotropic coordinates,

$$ds^2 = e^{2\lambda}(r^2\,d\Omega^2 + dr^2) - e^{2\nu}\,dt^2, \tag{14.13}$$

prove useful (Kuchowicz (1972)), and even the general form (13.8) of the line element may make the analytic expressions for the solution simple (Buchdahl (1967)).

14.1.2. Solutions

The best known of the spherically symmetric static perfect fluid solutions is the interior Schwarzschild solution (Schwarzschild (1916b))

$$\begin{aligned} &\varkappa_0\mu = 3R^{-2} = \text{const}, \qquad \varkappa_0 p = \frac{3b\sqrt{1 - r^2/R^2} - a}{R^2\left(a - b\sqrt{1 - r^2/R^2}\right)}, \\ &ds^2 = r^2\,d\Omega^2 + dr^2/(1 - r^2/R^2) - \left(a - b\sqrt{1 - r^2/R^2}\right)^2 dt^2. \end{aligned} \tag{14.14}$$

To fit the exterior Schwarzschild solution (13.19) at some boundary $r = r_0$, the constants a and b must satisfy $2a = 3\sqrt{1 - r_0{}^2/R^2}$, $2b = 1$.

Solutions with $\mu = \text{const}$, but possessing a singularity at $r = 0$ ($e^{-2\lambda} = 1 - c_1 r^2 + c_2/r$), have been discussed by Volkoff (1939) and Wyman (1948).

Solutions with a simple equation of state have been found in various cases, e.g. for $\mu + 3p = \text{const}$ (Whittaker (1968)), for $\mu = 3p$ (Klein (1947)), for $p = \mu + \text{const}$ (Buchdahl and Land (1968)), and for $\mu = (1 + a)\sqrt{p} - ap$ (Buchdahl (1967)). None of these equations of state is very realistic. But if one takes e.g. polytropic fluid spheres $p = a\mu^{1+1/n}$ (Klein (1953), Tooper (1964), Buchdahl (1964)) or a mixture of an ideal gas and radiation (Suhonen (1968)), one soon has to use numerical methods.

Several other classes of explicit perfect fluid solutions are known, most of them being unphysical. For some of them, we will list the key assumption and give the references for further details. The solutions may have a singularity at $r = 0$, but can then be used for the outer regions of composite spheres.

Table 14.1 Some static spherically symmetric perfect fluid solutions

μ:	ar^b $a(1 - br^2)$	Wyman (1948), Kuchowicz (1966) Mehra (1966), Kuchowicz (1967)*
$e^{-2\lambda}$:	$a + br^{2a/m}$, $a^{-1} = 4m - 2 - m^2$; ar^b; $a - 2\ln r$ $a + br^2$; $a + br^{-2a}$; $(1 + ar^2)/(1 + cr^2)$	Tolman (1939), Wyman (1948) Kuchowicz (1968) Kuchowicz (1968) Buchdahl (1959)
$e^{2\nu}$:	$(ar^{1-\alpha} - br^{1+\alpha})^2$ $a + br^2$ $a(r + b)^2$; ar^b; ae^{br^2} ar^2e^{br}	Tolman (1939), Kuchowicz (1970) Tolman (1939) Kuchowicz (1968) Kuchowicz (1970)

Some solutions take a simple form when written in isotropic coordinates (14.13). For examples see Nariai (1950), Kuchowicz (1972), □ Bayin (1978), and Buchdahl (1964). In these coordinates, the condition of isotropy takes the simple form (Kustaanheimo and Qvist (1948))

$$LG'' = 2L''G, \qquad L \equiv e^{-\lambda}, \qquad G \equiv L\,e^{\nu}. \tag{14.15}$$

14.2. Non-static solutions

14.2.1. The basic equations

In dealing with time-dependent, spherically symmetric perfect fluid solutions, most of the authors prefer a comoving frame of reference,

$$ds^2 = Y^2(r, t)\,d\Omega^2 + e^{2\lambda(r,t)}\,dr^2 - e^{2\nu(r,t)}\,dt^2, \quad u^i = (0, 0, 0, e^{-\nu}). \tag{14.16}$$

In this coordinate system, the field equations read (cf. 13.6)

$$\varkappa_0\mu = \frac{1}{Y^2} - \frac{2}{Y}\,e^{-2\lambda}\left(Y'' - Y'\lambda' + \frac{Y'^2}{2Y}\right) + \frac{2}{Y}\,e^{-2\nu}\left(\dot{Y}\dot{\lambda} + \frac{\dot{Y}^2}{2Y}\right), \tag{14.17a}$$

$$\varkappa_0 p = -\frac{1}{Y^2} + \frac{2}{Y} e^{-2\lambda} \left(Y'\nu' + \frac{Y'^2}{2Y} \right) - \frac{2}{Y} e^{-2\nu} \left(\ddot{Y} - \dot{Y}\dot{\nu} + \frac{\dot{Y}^2}{2Y} \right), \tag{14.17b}$$

$$\begin{aligned} \varkappa_0 p Y = {} & e^{-2\lambda}[(\nu'' + \nu'^2 - \nu'\lambda') Y + Y'' + Y'\nu' - Y'\lambda'] \\ & - e^{-2\nu}[(\ddot{\lambda} + \dot{\lambda}^2 - \dot{\lambda}\dot{\nu}) Y + \ddot{Y} + \dot{Y}\dot{\lambda} - \dot{Y}\dot{\nu})], \end{aligned} \tag{14.17c}$$

$$0 = \dot{Y}' - \dot{Y}\nu' - Y'\dot{\lambda}. \tag{14.17d}$$

Two simple consequences of $T^{ab}{}_{;b} = 0$ are the relations

$$p' = -(\mu + p)\,\nu', \qquad \dot{\mu} = -(\mu + p)\,(\dot{\lambda} + 2\dot{Y}/Y). \tag{14.18}$$

Like all perfect fluid solutions, the spherically symmetric solutions can be classified according to their kinematical properties, i.e. the 4-velocity's rotation, acceleration, expansion, and shear, cp. § 6.1. Spherical symmetry implies that $\omega_{ab} = 0$, so the velocity field must be hypersurface-orthogonal, and in the coordinate system (14.16) the other quantities in question are given by

$$\begin{aligned} & \dot{u}_i = (0, 0, \nu', 0), \qquad \Theta = e^{-\nu}(\dot{\lambda} + 2\dot{Y}/Y), \\ & \sigma^1_1 = \sigma^2_2 = -\frac{1}{2}\,\sigma^3_3 = \frac{1}{3}\,e^{-\nu}(\dot{Y}/Y - \dot{\lambda}). \end{aligned} \tag{14.19}$$

Most of the known solutions have vanishing shear. In this case, (14.19) implies the relation $\dot{Y}/Y = \dot{\lambda}$, whose integral is $Y = e^{\lambda}g(r)$. Thus, by a coordinate transformation $\hat{r} = \hat{r}(r)$, we can transform (14.16) into

$$ds^2 = e^{2\lambda(r,t)}(r^2\, d\Omega^2 + dr^2) - e^{2\nu(r,t)}\, dt^2, \tag{14.20}$$

i.e. we can introduce a coordinate system which is *simultaneously* comoving and isotropic. In (14.20), the r-coordinate is defined up to a transformation

$$\hat{r} = 1/r, \qquad e^{2\hat{\lambda}} = e^{2\lambda} r^4. \tag{14.21}$$

If the shear does not vanish, isotropic coordinates (14.20) can again be introduced, but they cannot be comoving (and the equations (14.17) no longer hold).

14.2.2. Solutions without shear and expansion

Because of (14.19), vanishing shear and expansion implies $\dot{\lambda} = \dot{Y} = 0$; ν is the only metric function which may depend on t. But as $\dot{\nu}$ enters the field equations (14.17) only as a coefficient of $\dot{Y}$ and $\dot{\lambda}$, and $\ddot{\nu}$ does not appear at all, no time derivative is contained in the field equations, which are the same as in the static case and are given in canonical coordinates by (14.2). These equations show that the energy density μ is a function of r alone, whereas the pressure p may depend on t if ν' does.

The solutions in question are either static (all static solutions are shear- and expansionfree) or can be generated from static solutions as follows. Take any static solution

$$ds^2 = r^2\, d\Omega^2 + e^{2\lambda(r)}\, dr^2 - e^{2\nu}\, dt^2 \tag{14.22}$$

and replace $\nu(r)$ by the general time-dependent solution $\nu(r, t)$ of the condition of isotropy (14.4). Introducing $N \equiv e^\nu$, this condition reads

$$N'' - N'(\lambda' + 1/r) + N(e^{2\lambda} - 1 + r\lambda')/r^2 = 0, \qquad N \equiv e^\nu. \tag{14.23}$$

The function $\lambda(r)$ being taken from (14.22), this is a linear differential equation for N, with coefficients independent of t, and so its general solution can be written as

$$e^\nu = N = f_1(t)\, N_1(r) + f_2(t)\, N_2(r). \tag{14.24}$$

The functions f_1 and f_2 are disposable, and N_1 and N_2 are any two linearly independent solutions of (14.23). Equations (14.22) and (14.24) give all non-static, expansion- and shearfree solutions (Kustaanheimo and Qvist (1948), Leibovitz (1971)).

14.2.3. Expanding solutions without shear

Field equations and first integrals

For $\sigma_{\alpha\beta} = 0$, but $\Theta \neq 0$, it follows from (14.19) that $\dot\lambda$ must be non-zero. The field equation (14.17d), i.e. the equation $\dot\lambda' = \dot\lambda\nu'$ (Y being $r\, e^\lambda$), is then integrated by

$$\nu = \ln \dot\lambda - f(t). \tag{14.25}$$

The metric now takes the form

$$ds^2 = e^{2\lambda(r,t)}\,(r^2\, d\Omega^2 + dr^2) - \dot\lambda^2\, e^{-2f(t)}\, dt^2, \tag{14.26}$$

the expansion being given by $\Theta(t) = 3\, e^{f(t)}$. The remaining field equations can be written as

$$\varkappa_0\mu = 3e^{2f} - e^{-2\lambda}(2\lambda'' + \lambda'^2 + 4\lambda'/r), \tag{14.27a}$$

$$\varkappa_0 p\dot\lambda = e^{-3\lambda}\, \partial_t[e^\lambda\, (\lambda'^2 + 2\lambda'/r) - e^{3\lambda+2f}], \tag{14.27b}$$

$$\varkappa_0 p\dot\lambda = e^{-3\lambda}\, \partial_t[e^\lambda(\lambda'' + \lambda'/r) - e^{3\lambda+2f}]. \tag{14.27c}$$

Elimination of p from (14.27b–c) gives the condition of isotropy

$$e^\lambda(\lambda'' - \lambda'^2 - \lambda'/r) = -\tilde F(r), \tag{14.28}$$

$\tilde F(r)$ being an arbitrary function of integration (Wyman (1946), Narlikar (1947), Kustaanheimo and Qvist (1948)). Introduction of

$$L \equiv e^{-\lambda}, \qquad x \equiv r^2, \qquad F(x) \equiv 4r^2\tilde F, \tag{14.29}$$

(Kustaanheimo and Qvist (1948)) transforms (14.28) into

$$L_{,xx} = L^2 F(x). \tag{14.30}$$

L^2 is proportional to the Weyl tensor invariant Ψ_2 (□ Glass (1979)).

To get a metric, one has to prescribe $f(t)$ (which fixes the t-coordinate) and $F(r^2)$, and find a solution $L(r^2, t)$ of (14.30), which in general will contain two arbitrary functions of time. Energy density and pressure can then be computed from (14.27).

A different form of the field equations has been given by McVittie (1967) for metrics satisfying

$$e^{\lambda(r,t)} = P(r)\, S(t)\, e^{\eta(z)}, \qquad z = \ln\,[Q(r)/S(t)], \qquad e^{-f} = \dot{S}/S. \tag{14.31}$$

A general class of solutions

A rather large class of solutions of (14.30) was found by Kustaanheimo and Qvist (1948). They chose

$$F(x) = (ax^2 + 2bx + c)^{-5/2}. \tag{14.32}$$

The perhaps surprising power $-5/2$ can be understood by observing that equation (14.30) must be invariant under the coordinate transformation (14.21), which implies the transformation law $\hat{F} = F(\hat{x}^{-1})\,\hat{x}^5$.

With (14.32), equation (14.30) for $F \neq 0$ leads to

$$(ax^2 + 2bx + c)^2\, u_{,xx} + 2ax(ax^2 + 2bx + c)\, u_{,x} + (ac - b^2)\, u = u^2, \tag{14.33}$$

$$u \equiv L(ax^2 + 2bx + c)^{-1/2}, \tag{14.34}$$

which is integrated by

$$\int^{u} \frac{du}{\sqrt{\frac{2}{3}\, u^3 + (b^2 - ac)\, u^2 + A(t)}} = \int^{x} \frac{dx}{ax^2 + 2bx + c} + B(t). \tag{14.35}$$

The function $u = u(x) = u(r^2)$ (which may be expressed in terms of elliptic functions) gives $e^{-\lambda} = L$ via (14.34), and, choosing $f(t)$, we can compute the full metric (14.26). For $F = 0$, (14.30) gives

$$e^{-\lambda} = L = A(t)\, r^2 + B(t). \tag{14.36}$$

Among the solutions (14.35)—(14.36) are contained several which were either known before or rediscovered later (cf. Wagh (1955), Wyman (1978) and the references given in Chakravarty et al. (1976)). To the authors' knowledge, all explicitly known shearfree and expanding spherically symmetric perfect fluid solutions are contained in the class (14.32). In Table 14.2, we list a few interesting subcases, which correspond

Table 14.2 Some subcases of the $F = (ax^2 + 2bx + c)^{-5/2}$ class of solutions, compare (14.35)—(14.36)

		F	A, B
McVittie solution (1933)		$[x(x + 4R^2\ ]^{-5/2}$	$A = 0$
$\mu = \mu(t)$ Kustaanheimo (1947)		$(2bx)^{-5/2},\ b \neq 0$	$6A = b(3e^{2f} - \varkappa_0\mu)$
		0	$12AB = 3e^{2f} - \varkappa_0\mu$
Equation of state $p = p(\mu)$ Wyman (1946)	$\mu = \mu(t)$ (Friedmann)	0	$B = \varepsilon A,\ \varepsilon = 0, \pm 1$
	$\mu = \mu(r, t)$	1	$A = \text{const},\ B = t$ $e^{-2f} = -4At$

to special choices of the functions $A(t)$, $B(t)$ and/or of the real constants a, b, and c. We will discuss these subcases in the following paragraphs.

The subclass $A(t) = 0$

For $A(t) = 0$, $a \neq 0$, $ac - b^2 \neq 0$, and with the notation

$$ax^2 + 2bx + c = a(x - x_1)(x - x_2), \qquad B(t) = 2 \ln C(t)/a(x_1 - x_2), \tag{14.37}$$

we obtain from (14.35) the solution

$$e^{-\lambda(r,t)} = \frac{3}{2} a^{5/2}(x_1 - x_2)^2 (r^2 - x_1)(r^2 - x_2) C(t) \left[1 - \sqrt{\frac{r^2 - x_1}{r^2 - x_2}} C(t)\right]^{-2} \tag{14.38}$$

(Kustaanheimo and Qvist (1948)). It contains, for $x_1 = 0$, $x_2 = -4R^2$, the metric of McVittie (1933),

$$\begin{aligned} ds^2 = {} & \frac{\left[1 + \dfrac{m}{2r\, e^{g(t)/2}}\left(1 + \dfrac{r^2}{4R^2}\right)^{1/2}\right]^4}{[1 + r/4R^2]^2} e^{g(t)} (r^2 \, d\Omega^2 + dr^2) \\ & - \left[\frac{1 - \dfrac{m}{2r\, e^{g(t)/2}}\left(1 + \dfrac{r^2}{4R^2}\right)^{1/2}}{1 + \dfrac{m}{2r\, e^{g(t)/2}}\left(1 + \dfrac{r^2}{4R^2}\right)^{1/2}}\right]^2 dt^2 . \end{aligned} \tag{14.39}$$

For $A(t) = 0$ and $a = 0$, one obtains the solution

$$\begin{aligned} & ds^2 = S^2(t)(1 + h)^4 (r^2 \, d\Omega^2 + dr^2) - (1 - h)^2 (1 + h)^{-2} \, dt^2, \\ & h \equiv S^{-1}(t)(\alpha r^2 + \beta)^{-1/2}, \end{aligned} \tag{14.40}$$

rediscovered by Nariai (1967).

Solutions with a homogeneous distribution of matter

If we assume a homogeneous distribution of matter, i.e. $\mu = \mu(t)$, then we have the two field equations (14.27a) and (14.28) to consider; they read

$$e^{-2\lambda}(2\lambda'' + \lambda'^2 + 4\lambda'/r) = 3e^{2f(t)} - \varkappa_0\mu(t) \equiv C(t), \tag{14.41}$$

$$e^{\lambda}(\lambda'' - \lambda'^2 - \lambda'/r) = -\tilde{F}(r). \tag{14.42}$$

Differentiating the first equation with respect to r, and eliminating λ'' and λ''' by means of the second equation, we obtain $3\tilde{F} + r\tilde{F}' = 0$, i.e. because of (14.29)

$$F(x) = (2bx)^{-5/2}. \tag{14.43}$$

To satisfy (14.41), which requires that the left-hand side depends only on t, we have to make an appropriate choice of $A(t)$ and $B(t)$ in the general formula (14.35). The final result is: all shearfree, spherically symmetric, expanding perfect fluid solutions with the energy density μ depending only on t are given by

$$ds^2 = e^{\lambda(r,t)} (r^2 \, d\Omega^2 + dr^2) - \dot{\lambda}^2 e^{-2f(t)} \, dt^2 \tag{14.44}$$

with

$$\mathrm{e}^{-\lambda} = \sqrt{2b}\, ur, \qquad b = \text{const} \neq 0,$$
$$\int \frac{\mathrm{d}u}{\sqrt{\frac{2}{3}u^3 + b^2u^2 + b(3\mathrm{e}^{2f} - \varkappa_0\mu)/6}} = \frac{1}{b}\ln r + B(t), \tag{14.45}$$

or

$$\mathrm{e}^{-\lambda} = A(t) + B(t)\, r^2, \qquad 12AB = 3\mathrm{e}^{2f} - \varkappa_0\mu \quad (b = 0) \tag{14.46}$$

(Kustaanheimo (1947)). Subcases have been discovered and rediscovered by several authors, see Gupta (1959) and the references given in Cook (1975). In general, the pressure p will depend on both t and r; the subcase $p = p(t)$ is contained in the solutions considered in the following paragraphs.

Solutions with an equation of state $p = p(\mu)$

Solutions which obey an equation of state $p = p(\mu)$ have been discussed by Wyman (1946), and special cases by Taub (1968). Because of (14.18), $\dot{Y}/Y = \dot{\lambda}$ and (14.25), they satisfy (as do the shearfree perfect fluid cases)

$$p' = -(\mu + p)\,\dot{\lambda}'/\dot{\lambda}, \qquad \dot{\mu} = -3(\mu + p)\,\dot{\lambda}. \tag{14.47}$$

If $\mu + p$ vanishes, then μ and p are constant, and the solution in question is the static vacuum solution of Kottler, see Theorem 13.5 and Table 13.1.

If $\mu + p$ is different from zero, but μ' vanishes $\big(\mu = \mu(t),\, p = p(t)\big)$, then we have $\mu' = p' = \nu' = \dot{\lambda}' = 0$. We choose the time coordinate t so that ν is zero, and infer from (14.26) that

$$\lambda = \lambda_1(r) + \lambda_2(t), \qquad \dot{\lambda} = \dot{\lambda}_2 = \mathrm{e}^{f} \tag{14.48}$$

holds. This special time dependence of λ is compatible with (14.42) only if $\tilde{F} = 0 = F$. So the solutions with $\mu = \mu(t)$, $p = p(t)$ are the subcase

$$\mathrm{e}^{-\lambda} = A(t)\left[1 + \frac{\varepsilon}{4}r^2\right], \qquad \varepsilon = 0, \pm 1, \qquad \mathrm{e}^{\nu} = 1 \tag{14.49}$$

of the solutions (14.46). These are exactly the Friedmann-like universes (§ 12.2.).

If neither $\mu + p$ nor μ' vanishes, then $\dot{\lambda}'$ is not zero and can be eliminated from (14.47), which leads to

$$\dot{\mu}\mu' = (\mu + p)\,\dot{\mu}'. \tag{14.50}$$

Equation (14.50) is integrated by

$$\ln \dot{\mu} = \ln M(\mu) + \ln \dot{\alpha}(t), \qquad \ln M(\mu) \equiv \int \frac{\mathrm{d}\mu}{\mu + p(\mu)}, \tag{14.51}$$

and in a further step by

$$H(\mu) = \alpha(t) + \beta(r), \qquad H(\mu) \equiv \int M^{-1}(\mu)\, \mathrm{d}\mu. \tag{14.52}$$

Choosing the time coordinate so that $\alpha = t$ ($\dot{\alpha} = 0$ is prohibited by $\mu + p \neq 0$, $\dot{\lambda} \neq 0 \Rightarrow \dot{\mu} \neq 0$, cp. (14.47)), we see that because of (14.52) and (14.47) the functions μ and λ must have the special time-dependence

$$\mu(r, t) = \mu(v), \qquad \lambda(r, t) = \lambda_1(v) + \lambda_2(r^2), \qquad v \equiv t + G(r^2). \tag{14.53}$$

To determine the functions λ_1, λ_2, G, and the function F occurring in the first integral (14.30), we insert (14.53) into (14.30). Writing

$$e^{-\lambda} = L = u(v)\, l(x), \qquad x = r^2, \tag{14.54}$$

we obtain

$$\dot{u}(G_{,xx}l + 2G_{,x}l_{,x}) + ul_{,xx} + \ddot{u}lG_{,x}^2 = u^2l^2F. \tag{14.55}$$

In this equation, only u and its derivatives depend on t. Since by assumption, $\dot{u}$, l, and $G_{,x}$ are non-zero, either u, $\dot{u}^2$ and $\ddot{u}$ are proportional to each other (which is impossible), or at least one of the coefficients of these functions vanishes. From this reasoning, we conclude that

$$G_{,xx}l + 2G_{,x}l_{,x} = 0, \qquad l_{,xx} = 0, \qquad \ddot{u} = \mathrm{const} \cdot u^2 = l^2Fu^2G_{,x}^{-2} \tag{14.56}$$

holds. One can show that the two cases $l = ax + b$ and $l = b$ are equivalent (i.e. connected by a transformation (14.21)), and that $F = 0$ gives $\mu = \mu(t)$. So we need only consider the case $F = 1$, $l = \mathrm{const}$. We now have to choose the arbitrary functions of integration in the solution with $F = 1$ so that μ and λ have the functional form (14.53). This is done by specializing (14.35) to

$$\int \frac{du}{\sqrt{\frac{2}{3}u^3 + A}} = t + r^2, \qquad A = \mathrm{const},$$

$$e^{-\lambda} = u(t + r^2), \qquad e^{-2f(t)} = -4At, \tag{14.57}$$

$$ds^2 = e^{2\lambda}(r^2\, d\Omega^2 + dr^2) - \dot{\lambda}^2\, e^{2f}\, dt^2$$

(Wyman (1946)). One can easily check by computing μ and p from (14.27) that $\mu = \mu(t + r^2)$ and $p = p(t + r^2)$ hold.

14.2.4. Solutions with non-vanishing shear

As shown in the preceding section, solutions without shear are fairly well known. In contrast, only a few solutions with shear have been found and discussed so far. For a physical approach to the basic quantities and equations, see Misner and Sharp (1964).

A negative result concerning solutions with shear is due to Thompson and Whitrow (1967) and Misra and Srivastava (1973): if in the comoving frame of reference (14.16) the mass density μ is a function only of t, and if the metric is regular at $r = 0$ (i.e. $Y = 0$, $Y' = e^{\lambda}$), then the 4-velocity is necessarily shearfree.

Solutions with shear but without acceleration

For $\dot{u}^a = 0$, ν is a function only of t, in the comoving frame of reference (14.16), and so is p. The field equation (14.17d) therefore implies

$$\dot{Y}' = \dot{\lambda} Y'. \tag{14.58}$$

If $Y' = 0$ ($Y \neq \text{const}$), then we choose $Y = t$, and the remaining field equations (14.17) read

$$\varkappa_0 \mu t^2 = 2\dot{\lambda} t\, e^{-2\nu} + 1 + e^{-2\nu}, \tag{14.59a}$$

$$\varkappa_0 p t^2 = 2\dot{\nu} t\, e^{-2\nu} - 1 - e^{-2\nu}, \tag{14.59b}$$

$$\varkappa_0 p = -e^{-2\nu}\, [\ddot{\lambda} + \dot{\lambda}^2 - \dot{\lambda}\dot{\nu} + (\dot{\lambda} - \dot{\nu})/t]. \tag{14.59c}$$

These equations (and their solutions) closely resemble the static case, cf. § 14.1 and equation (14.2). Solutions with $e^\lambda = t^n$ have been considered by McVittie and Wiltshire (1975). Solutions admitting an additional spacelike Killing vector are discussed in Chapter 11.

If $Y' \neq 0$, then we choose $\nu = 0$ and integrate (14.58) by

$$e^{2\lambda} = Y'^2/\big(1 - \varepsilon f^2(r)\big), \qquad \varepsilon = 0, \pm 1. \tag{14.60}$$

From the field equation (14.17b) we obtain

$$\varkappa_0 p(t)\, Y^2 = -2 Y\ddot{Y} - \dot{Y}^2 - \varepsilon f^2(r). \tag{14.61}$$

Equation (14.17c) follows from (14.61) by differentiation with respect to r, and μ can be computed from

$$\varkappa_0 \mu = -3p(t) - 2\ddot{Y}'/Y'. \tag{14.62}$$

In (14.61), $p(t)$ can be prescribed, but to avoid zero shear, $\dot{Y}'Y = \dot{Y}Y'$ is forbidden. For $p = \text{const}$, equation (14.61) can be solved by quadratures.

Solutions with shear but without expansion

Because of (14.19), solutions with zero expansion but non-zero shear have to obey $\dot{\lambda} = -2\dot{Y}/Y$, $\dot{Y} \neq 0$. Together with the field equation (14.17d), this leads to

$$ds^2 = Y^2\, d\Omega^2 + Y^{-4}\, dr^2 - Y^4 \dot{Y}^2 f^2(t)\, dt^2. \tag{14.63}$$

Equation (14.18b) shows that the energy density μ is a function only of r, so that the field equations (14.17a, b) read

$$2Y''Y^5 + 5Y'^2Y^4 + 3Y^{-4}f^{-2} + \varkappa_0\mu(r)\, Y^2 - 1 = 0, \tag{14.64}$$

$$\varkappa_0 p = -Y^{-2} + 5Y'^2Y^2 + 2\dot{Y}'Y'Y^3\dot{Y}^{-1} + 3Y^{-6}f^{-2} + 2\dot{f}f^{-3}\dot{Y}^{-1}Y^{-5}. \tag{14.65}$$

The condition of isotropy is satisfied if equations (14.64)–(14.65) are. In (14.64), $\mu(r)$ and $f(t)$ can be prescribed, and p can then be computed from (14.65), once Y is known. For constant μ, equation (14.64) is integrated by

$$Y'^2Y^5 - Y^{-3}f^{-2}(t) + \varkappa_0\mu Y^3/3 - Y = A(t), \tag{14.66}$$

which can be solved by quadratures (Skripkin (1960)).

Solutions with shear, acceleration and expansion

A large variety of spherically symmetric perfect fluid solutions is to be expected in this most general class, but only a few special cases have been treated so far.

In the comoving frame of reference (14.16), the solution

$$\mathrm{d}s^2 = r^2 f(t)\,\mathrm{d}\Omega^2 + \mathrm{d}r^2 - \frac{r^2}{4}\,\mathrm{d}t^2, \qquad 2f = 1 + A\,\mathrm{e}^t + B\,\mathrm{e}^{-t}, \tag{14.67}$$

was found by Gutman et al. (1967). A solution with a different $f(t)$ and equation of state $p = \mu$ was given by Wesson (1978).

Solutions in a non-comoving frame of reference have been studied by Vaidya (1968), and McVittie and Wiltshire (1977) (not all their solutions have non-zero shear!).

Chapter 15. Groups G_2 and G_1 on non-null orbits

In this chapter, we give in full only those known solutions admitting a G_2 or G_1 which are not included in Chapters 16.—20. We confine ourselves to non-null orbits, the case of G_2 and G_1 on null orbits being treated separately in Chapter 21.

15.1. Group structures G_2 and group orbits V_2

The normal forms of the space-time metrics for the two inequivalent structures of groups G_2 described in § 8.2. are (Petrov (1966), p. 150)

$$G_2\,I\colon\; g_{ij} = g_{ij}(x^3, x^4), \qquad \xi = \partial_1, \qquad \eta = \partial_2,$$

$$G_2\,II\colon\; g_{ij} = \begin{pmatrix} e^{-2x^2}a_{11} & e^{-x^2}a_{12} & e^{-x^2}a_{13} & 0 \\ e^{-x^2}a_{12} & a_{22} & a_{23} & 0 \\ e^{-x^2}a_{13} & a_{23} & a_{33} & 0 \\ 0 & 0 & 0 & e_4 \end{pmatrix} \tag{15.1}$$

$$a_{ij} = a_{ij}(x^3, x^4), \qquad e_4 = \pm 1, \qquad \xi = \partial_1, \qquad \eta = x^1\,\partial_1 + \partial_2.$$

The 2-surface of transitivity (group orbit) spanned by the two Killing vectors ξ and η is spacelike or timelike respectively when the square of the simple bivector $\xi_{[a}\eta_{b]}$ is positive or negative.

Chapters 17.—19. are devoted to the stationary axisymmetric fields; they have *timelike* group orbits T_2. Cylindrically symmetric fields (Chapter 20.) have *spacelike* group orbits S_2. In different regions of the same space-time the orbits may have different character (timelike or spacelike); the field of uniformly accelerated particles (§ 18.2.) provides an example.

In this chapter we deal with some physically interesting solutions (colliding plane waves and inhomogeneous cosmological models) which admit an Abelian group G_2 on spacelike orbits S_2, but are not cylindrically symmetric.

The field equations are very complicated to solve and no exact solutions have been obtained for either of the metrics (15.1) without additional simplifications. The further restrictions imposed may be either degeneracy of the Weyl tensor or special properties of the Killing vector fields. The symmetry groups of the known algebraically special solutions have not yet been systematically investigated and we cannot completely answer the question of which of the algebraically special solutions

(Part III) admit a G_2 or G_1 on non-null orbits. Some remarks concerning the link between groups of motions and Petrov types are contained in Chapter 33.

A restriction which may be imposed on commuting Killing vector fields is the existence of 2-surfaces orthogonal to the group orbits (orthogonally transitive group). The Killing vectors then obey the relations

$$\xi_{[a;b}\xi_c\eta_{d]} = 0 = \eta_{[a;b}\eta_c\xi_{d]} \tag{15.2}$$

(see (6.11) and § 17.2.). Hypersurface-orthogonal Killing vectors satisfy the more stringent condition $\xi_{[a;b}\xi_{c]} = 0 = \eta_{[a;b}\eta_{c]}$ which, of course, implies (15.2).

If an orthogonally transitive group G_2 acts on spacelike orbits S_2, the space-time metric can be written in the form

$$\mathrm{d}s^2 = \mathrm{e}^M(\mathrm{d}z^2 - \mathrm{d}t^2) + W[\mathrm{e}^\Psi(\mathrm{d}x + A\,\mathrm{d}y)^2 + \mathrm{e}^{-\Psi}\,\mathrm{d}y^2], \qquad W > 0, \tag{15.3}$$

where all functions are independent of x and y. The function W is invariantly defined by $W^2 = 2\xi_{[a}\eta_{b]}\xi^a\eta^b$. If the Killing vectors are hypersurface-orthogonal, the function A in (15.3) can be made zero. For $A = 0 = \Psi$ one regains from (15.3) the plane symmetric line element (13.9).

For the Ricci tensor of the metric (15.3), see e.g. Wainwright et al. (1979). For later use we give only

$$R^1_1 + R^2_2 = \mathrm{e}^{-M}W^{-1}(W_{,44} - W_{,33}) = -\varkappa_0(T^3_3 + T^4_4). \tag{15.4}$$

In some cases (vacuum fields, Einstein-Maxwell fields with $F^*_{ab}\xi^a\eta^b = 0$, perfect fluids with $p = \mu$), the energy-momentum tensor obeys the condition

$$T^3_3 + T^4_4 = 0. \tag{15.5}$$

Then, from (15.4), we obtain the general solution

$$W = f(u) + g(v), \qquad \sqrt{2}\,u = t - z, \qquad \sqrt{2}\,v = t + z, \tag{15.6}$$

with arbitrary functions $f(u)$ and $g(v)$.

In the vacuum case with $A = 0$ in (15.3), the function Ψ obeys the linear differential equation

$$(W\Psi_{,3})_{,3} - (W\Psi_{,4})_{,4} = 0 \tag{15.7}$$

(see also § 20.3.), and the rest of the field equations determine M in terms of a line integral.

From the Ricci tensor of (15.3) one infers

Theorem 15.1. *If the metric* (15.3) *satisfies* $R_{ab} = 0$, *then the field equations*

$$\hat{R}_{ab} = 2\sigma_{,a}\sigma_{,b}, \qquad \sigma_{,a}\xi^a = 0 = \sigma_{,a}\eta^a, \tag{15.8}$$

are satisfied by the metric $\mathrm{d}\hat{s}^2$ *which differs from* (15.3) *by the substitution*

$$\hat{M} = M + \Omega, \qquad \Omega_{,a} = 2W(W_{,b}W^{,b})^{-1}\,\sigma^{,c}(2W_{,c}\sigma_{,a} - \sigma_{,c}W_{,a}) \tag{15.9}$$

(for a proof, see Wainwright et al. (1979)).

For perfect fluids with the equation of state $p = \mu$ and with irrotational 4-velocity

$$u_a = (-\sigma_{,b}\sigma^{,b})^{-1/2}\,\sigma_{,a}, \tag{15.10}$$

the field equations read (Tabensky and Taub (1973))

$$R_{ab} = 2\sigma_{,a}\sigma_{,b}, \qquad \varkappa_0 p = \varkappa_0 \mu = -\sigma_{,c}\sigma^{,c} \tag{15.11}$$

and the Bianchi identities imply $\sigma^{,a}{}_{;a} = 0$. Theorem 15.1 enables one to generate such perfect fluid solutions from vacuum solutions. An example is given in § 13.6. Wainwright et al. (1979) derived a variety of homogeneous and inhomogeneous cosmological models by Theorem 15.1 (see also § 12.4.).

Most of the inhomogenous models admit a three-parameter homothety group (see § 31.4.) and are thus self-similar (□ McIntosh (1978)). Some solutions are interpreted as gravitational pulses in a homogeneous (Bianchi type I) model (□ Wainwright and Marshman (1979)).

The gradient of W provides an invariant characterization of the corresponding space-times and determines essentially the character and global properties of a solution: $W_{,a}$ is a null vector for the plane waves (§ 21.5.); it is a spacelike vector for the Einstein-Rosen waves (§ 20.3.); and it is spacelike, timelike and null in different regions of the gravitational waves forming closed universes (§ 15.3.).

15.2. Colliding plane waves

Fig. 15.1 illustrates the following situation: two plane waves (regions II and III) propagate in opposite directions, and collide. The incoming waves determine the data on the null surfaces $u = 0$, $v > 0$ and $v = 0$, $u > 0$, and, therefore, (via a characteristic initial value problem) the field in the interaction region IV (hatched region in Fig. 15.1).

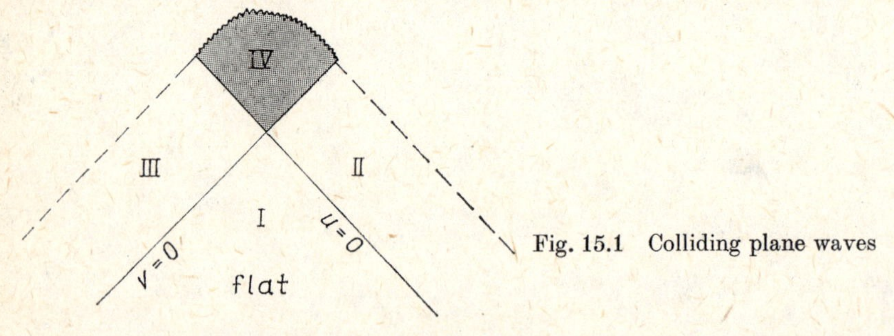

Fig. 15.1 Colliding plane waves

Exact solutions representing a collision of two plane gravitational waves were first given by Szekeres (1970), and by Khan and Penrose (1971). To treat this problem exactly, two simplifying assumptions are made: (i) there exists an orthogonally transitive *Abelian group* G_2 acting on spacelike 2-surfaces, even in the interaction region; and (ii) both incoming waves have constant polarization (§ 21.5.). The line elements in the regions II and III of Fig. 15.1 may be put in the form (see (15.3))

$$ds^2 = -2e^M \, du \, dv + W(e^\Psi \, dx^2 + e^{-\Psi} \, dy^2), \tag{15.12}$$

$$\text{region II}: \; M = M(u), \qquad W = W(u), \qquad \Psi = \Psi(u),$$

$$\text{region III}: \; M = M(v), \qquad W = W(v), \qquad \Psi = \Psi(v).$$

These suppositions, together with the field equations $R_{ab} = 0$ and the junction conditions (continuity of the metric and its first derivatives) at $u = 0$ and $v = 0$, imply that one can use the common form (15.12) of the metric in all regions of Fig. 15.1, the functions M, W and Ψ now depending on the null coordinates u and v. The null hypersurfaces $u = \text{const}$ and $v = \text{const}$ (wave fronts) intersect in spacelike 2-surfaces (group orbits).

An exact solution of (15.7) satisfying the junction conditions is given by Szekeres (1972). Studying the superposition of gravitational waves, Griffiths (1975) rediscovered the vacuum solution (10.14) admitting a simply transitive group G_4. Bell and Szekeres (1974) give the metric

$$ds^2 = -2 \, du \, dv + \cos^2 (au - bv) \, dx^2 + \cos^2 (au + bv) \, dy^2 \tag{15.13}$$

for the interaction zone of colliding plane waves in the Einstein-Maxwell theory. (15.13) is just the Bertotti-Robinson solution (10.16) in a slightly changed coordinate system.

Ref.: Griffiths (1976a, b), Sbytov (1976), □ Halil (1979).

15.3. Closed universes built from gravitational waves

The class of inhomogeneous cosmological models given by Gowdy (1971) admits an Abelian group G_2 acting on spacelike 2-surfaces, whereas homogeneous universes (Chapter 12.) admit a group G_3 transitive on spacelike hypersurfaces. The Gowdy

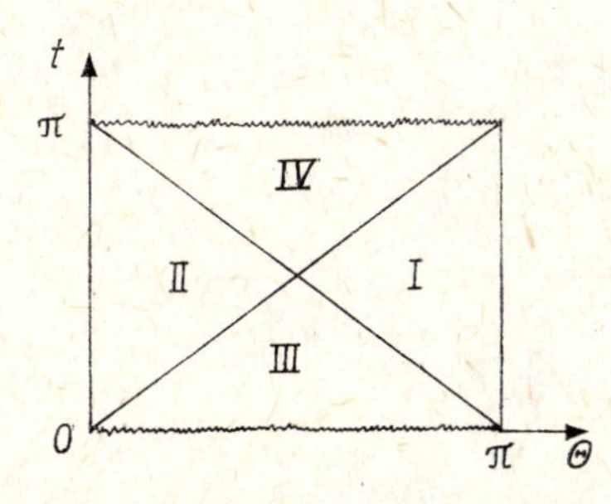

Fig. 15.2 Gowdy universe: coordinate patch $0 < t < \pi$, $0 \leqq \Theta \leqq \pi$ ($\Theta = 0$ and $\Theta = \pi$ are identified), $W_{,a}W^{,a} > 0$ in I and II, $W_{,a}W^{,a} < 0$ in III, IV

vacuum solutions represent closed universes built from gravitational waves. Unlike plane waves these gravitational waves have wave fronts of finite extent.

The metric representing these models is the same as (15.3), $A = 0$, namely

$$ds^2 = e^M (d\Theta^2 - dt^2) + W(e^\Psi \, d\chi^2 + e^{-\Psi} \, d\varphi^2). \tag{15.14}$$

The function W (see (15.6)) is chosen to be $W = \sin \Theta \sin t$. In the case of spherical topology of the space sections, Θ, χ, and φ are interpreted as generalized Euler angle coordinates. The resulting space-time metrics satisfying the regularity conditions at $\Theta = 0$ and $\Theta = \pi$ are inhomogeneous generalizations of the Friedmann models and have initial and final collapse singularities at $t = 0$ and $t = \pi$ (Fig. 15.2). The 3-spaces $t = \text{const}$ are distorted 3-spheres (Gowdy (1975)). For an electromagnetic generalization, see □ Charach (1979).

15.4. Group G_1 on non-null orbits

One can always transform a non-null Killing vector into the form $\xi = \partial_n$, the metric is independent of x^n in this coordinate system.

Stationary gravitational fields (Chapter 16.) admit a timelike Killing vector. The reduction formulae for the Ricci tensor as derived in § 16.2., hold also in the case of a spacelike Killing vector.

Without any further assumption, it is hardly possible to solve Einstein's field equations, in contrast to the case of a null Killing vector (§ 21.4.). In this section we consider special assumptions which lead to solutions admitting a group G_1 (on non-null orbits).

Harrison (1959) started with the metrical ansatz ("linked pairs" form)

$$\begin{aligned} g_{ij} &= e_i \delta_{ij} A_i^2(x^1, x^4)\, B_i^2(x^3, x^4) \qquad \text{(no summation)}, \\ e_1 &= e_2 = e_3 = 1, \qquad e_4 = -1 \end{aligned} \tag{15.15}$$

(or a diagonal metric derived therefrom by an interchange of coordinates) and obtained a series of exact vacuum solutions admitting a group G_1. The spacelike (or timelike) Killing vector $\xi = \partial_2$ (or $\xi = \partial_4$) is hypersurface-orthogonal. The separation of variables in the particular form (15.15) either leads to solutions in closed form or reduces the problem to an ordinary differential equation. We give one example from each of these cases for illustration. Solution in closed form:

$$\mathrm{d}s^2 = \sum_{i=1}^{4} e_i a_i^2 [(x^4)^2 + x^1]^{n_i} (x^3)^{2k_i} (\mathrm{d}x^i)^2 \tag{15.16}$$

i	1	2	3	4
n_i	$1 + \sqrt{2}$	$-\sqrt{2}$	$2 + \sqrt{2}$	$1 + \sqrt{2}$
k_i	$(1 + 2\sqrt{2})/7$	$(2 - 3\sqrt{2})/7$	0	1
a_i	$c = \text{const}$	1	$(3 - \sqrt{2})/7$	1

Solution in non-closed form:

$$\mathrm{d}s^2 = \sum_{i=1}^{4} e_i \left[\frac{x^1}{\sqrt{3}} - \frac{(x^4)^2}{12}\right]^{n_i} \left[\exp \int \frac{v\,\mathrm{d}z}{z}\right]^{k_i} (-x^3)^{l_i} \left(\frac{v^2 - 1}{x^3 - 1}\right)^{m_i} (\mathrm{d}x^i)^2 \tag{15.17}$$

i	1	2	3	4
n_i	$1+\sqrt{3}$	$-\sqrt{3}$	$3+\sqrt{3}$	$2+\sqrt{3}$
k_i	$1+2/\sqrt{3}$	-1	$1+2/\sqrt{3}$	$1+2/\sqrt{3}$
l_i	$-(1+1/\sqrt{3})$	$1/\sqrt{3}$	$-(2+1/\sqrt{3})$	$-1/\sqrt{3}$
m_i	0	0	0	1

$$\frac{\mathrm{d}v}{\mathrm{d}z} = \frac{v^2-1}{4z}\left(\frac{2v}{z-1} + \frac{4}{\sqrt{3}}\right), \qquad z = x^3.$$

The complex transformation $x^2 \to \mathrm{i}x^4$, $x^4 \to \mathrm{i}x^2$ takes (15.16) and (15.17) into static solutions. However, not every Harrison solution with a spacelike Killing vector has a (real) static counterpart.

Because of their analytical complexity, the metrics given by Harrison have been checked by coumputer (d'Inverno and Russell Clark (1971)). Altogether there are 17 vacuum solutions — including (15.16), (15.17) — which are of non-degenerate Petrov type I and do not admit a G_r, $r > 1$ (Collinson (1964)). The ansatz (15.15) also leads to solutions, mostly of type D, which belong to Weyl's class of static fields (§ 18.1.) or to the class of Einstein-Rosen waves (§ 20.3.).

By means of a separation ansatz which is similar to (but not identical with) the ansatz (15.15), Harris and Zund (1978) obtained a class of (type I) vacuum solutions admitting (at least) a G_1.

Starting with the ansatz

$$\begin{aligned} &\mathrm{d}s^2 = h(t)\,\{\mathrm{e}^{2k}[(\mathrm{d}x^1)^2 + (\mathrm{d}x^2)^2] + W^2\,\mathrm{d}\varphi^2\} - \mathrm{e}^{2U}\,\mathrm{d}t^2, \\ &k = k(x^1, x^2), \qquad W = W(x^1, x^2), \qquad U = U(x^1, x^2), \qquad \mathrm{d}h/\mathrm{d}t \neq 0, \end{aligned} \tag{15.18}$$

Rastall (1960) solved the vacuum equations, but the resulting solutions are flat (□ Koppel (1967)).

Ref.: Hoenselaers (1978c).

Chapter 16. Stationary gravitational fields

Stationary gravitational fields are characterized by the existence of a timelike Killing vector, i.e. one can choose coordinates so that the metric is independent of a timelike coordinate. A stationary space-time is said to be *static* if the Killing vector is hypersurface-orthogonal.

In §§ 16.1—16.3. we derive the field equations from a projection formalism using differential geometric concepts. Some methods outlined for stationary fields, e.g. the projection formalism (§ 16.1.) or geodesic eigenrays (§ 16.5.), also apply, with slight changes, to the case with a spacelike Killing vector.

Some classes of solutions are given in §§ 16.5.—16.7. Only in a few cases are exact stationary solutions without an additional symmetry known. The stationary fields admitting a second Killing vector describing axial symmetry will be treated in the subsequent chapters.

16.1. The projection formalism

A stationary space-time invariantly determines a differentiable 3-manifold Σ_3 defined by the smooth map (§ 2.2) $\Psi : \mathcal{M} \to \Sigma_3$ (Geroch (1971)), where $\mathcal{M}$ is the space-time V_4 and $\Psi = \Psi(p)$ denotes the trajectory of the timelike Killing vector ξ passing through the point p of V_4. The elements of Σ_3 are the orbits of the 1-dimensional group of motions generated by ξ. The 3-space Σ_3 is called the factor space V_4/G_1.

Only in the case of static gravitational fields is there a natural way of introducing subspaces V_3 (orthogonal to the Killing trajectories). The factor space Σ_3 provides a generalization applicable to stationary, as well as static, space-times; it must be regarded as the image of a map rather than a hypersurface in V_4.

In Geroch (1971) it is shown in detail that there is a one-to-one correspondence between tensor fields on Σ_3 and tensor fields $T_{a..}^{\;\cdot\cdot b}$ on V_4 satisfying

$$\xi^a T_{a..}^{\;\cdot\cdot b} = 0, \qquad \xi_b T_{a..}^{\;\cdot\cdot b} = 0, \qquad \pounds_\xi T_{a..}^{\;\cdot\cdot b} = 0. \tag{16.1}$$

Tensors on V_4 subject to these conditions are simply called tensors on Σ_3; the algebra of the space-time tensors satisfying (16.1) is completely and uniquely mapped to the tensor algebra on Σ_3. Examples of tensors on Σ_3 are the projection tensor h_{ab} (the metric tensor on Σ_3) and the Levi-Civita tensor ε_{abc} on Σ_3, where

$$h_{ab} \equiv g_{ab} + u_a u_b, \qquad u^a \equiv (-F)^{-1/2}\, \xi^a, \qquad F \equiv \xi_a \xi^a, \tag{16.2}$$

$$\varepsilon_{abc} \equiv \varepsilon_{dabc} u^d. \tag{16.3}$$

The derivative on Σ_3 defined by

$$T_{a..\|e}^{..b} \equiv h_a^c \dots h_d^b h_e^f T_{c..;f}^{..d} \tag{16.4}$$

satisfies all the axioms for the covariant derivative associated with the metric h_{ab}. In particular, it is symmetric (torsion-free) and the metric tensor h_{ab} is covariantly constant, $h_{ab\|c} = 0$.

The Riemann curvature tensor on Σ_3 can be calculated from the identity (cp. (2.78))

$$v_{a\|bc} - v_{a\|cb} = v^d \overset{3}{R}_{dabc}, \tag{16.5}$$

$\boldsymbol{v}$ being an arbitrary vector field on Σ_3. The curvature tensors on Σ_3 and V_4 are related by the equation

$$\overset{3}{R}_{abcd} = h_{[a}^p h_{b]}^q h_{[c}^r h_{d]}^s \left\{ R_{pqrs} + \frac{2}{F} (\xi_{q;p}\xi_{s;r} + \xi_{r;p}\xi_{s;q}) \right\} \tag{16.6}$$

(Lichnerowicz (1955), Jordan et al. (1960)). In order to simplify our expressions we define the *twist vector* $\boldsymbol{\omega}$,

$$\omega^a \equiv \varepsilon^{abcd}\xi_b\xi_{c;d}, \qquad \omega^a\xi_a = 0, \qquad \pounds_\xi \omega = 0. \tag{16.7}$$

Applying the equation (3.37) for complex self-dual bivectors to $\xi^*_{a;b} = \xi_{a;b} + \mathrm{i}\tilde{\xi}_{a;b}$, we obtain the simple relation

$$2\xi_{a;b} = F^{-1}(\varepsilon_{abcd}\xi^c\omega^d + 2\xi_{[a}F_{,b]}) \tag{16.8}$$

for the covariant derivative of the Killing vector with respect to the space-time metric g_{ab}, so that such terms can be eliminated from (16.6). The resulting expression will be used in calculating the Ricci tensor.

16.2. The Ricci tensor on Σ_3

First we introduce a complex vector $\boldsymbol{\Gamma}$ on Σ_3,

$$\Gamma_a = -2\xi^c\xi^*_{c;a} = -F_{,a} + \mathrm{i}\omega_a, \qquad \Gamma^a\xi_a = 0, \qquad \pounds_\xi \Gamma = 0. \tag{16.9}$$

With the aid of equation (8.24) and the symmetry relations (2.80) of the curvature tensor it can easily be verified that the equations

$$\xi^{*\ \ ;b}_{a;b} = -R_{ad}\xi^d, \qquad \xi^*_{a;b} = -F^{-1}(\xi_{[a}\Gamma_{b]})^* \tag{16.10}$$

hold. Taking the divergence of the complex vector $\boldsymbol{\Gamma}$ gives

$$\Gamma^a{}_{;a} = \Gamma^a{}_{\|a} + \frac{1}{2} F^{-1}F_{,a}\Gamma^a = -F^{-1}\Gamma^a\Gamma_a + 2\xi^a\xi^b R_{ab}. \tag{16.11}$$

The real and imaginary parts of this equation are

$$F^{,a}{}_{\|a} = \frac{1}{2} F^{-1}F_{,a}F^{,a} - F^{-1}\omega_a\omega^a - 2\xi^a\xi^b R_{ab}, \tag{16.12}$$

$$\omega^a{}_{\|a} = \frac{3}{2} F^{-1}F_{,a}\omega^a. \tag{16.13}$$

Equation (16.12) expresses the component $\xi^a\xi^b R_{ab}$ of the Ricci tensor in V_4 in terms of tensors and their covariant derivatives on Σ_3. In order to derive an analogous formula for the components $h^a_c\xi^b R_{ab}$, we calculate the curl of $\boldsymbol{\omega}$:

$$\omega_{[b\|a]} = \varepsilon_{smnr}(\xi^m\xi^{n;r})_{;c}h^c_{[a}h^s_{b]} = 2\xi^m\xi^d R_{dcms}^{\sim} h^c_{[a}h^s_{b]}$$
$$= -2\xi^d \tilde{R}_{d[ab]m}\xi^m = \varepsilon_{abmn}\xi^m R^n_d\xi^d. \tag{16.14}$$

The result of this short calculation is the formula

$$(-F)^{-1/2}\,\varepsilon^{abc}\omega_{c\|b} = 2h^a_b R^b_c\xi^c. \tag{16.15}$$

From this equation we conclude that the vanishing of the components $h^a_b R^b_c\xi^c$ implies that, at least locally, the twist vector $\boldsymbol{\omega}$ is a gradient, $\omega_a = \omega_{,a}$ (see Theorem 2.1).

Finally, we can derive the formula

$$\overset{3}{R}_{ab} = \frac{1}{2}F^{-1}F_{,a\|b} - \frac{1}{4}F^{-2}F_{,a}F_{,b} + \frac{1}{2}F^{-2}(\omega_a\omega_b - h_{ab}\omega_c\omega^c) + h^m_a h^n_b R_{mn} \tag{16.16}$$

by a straightforward calculation in which we insert (16.8) into (16.6), contract the equation (16.6) for the curvature tensor with the projection tensor and apply some of the previous relations of the present section.

The equations (16.12), (16.15) and (16.16) express the Ricci tensor of a stationary space-time in terms of tensors and their covariant derivatives on the 3-space Σ_3. All the equations are written in a 4-dimensionally covariant manner and refer to a general basis $\{\boldsymbol{e}_a\}$ of V_4. Nevertheless, they are in fact 3-dimensional relations. If we take a basis such that

$$\{\boldsymbol{e}_a\} = \{\boldsymbol{e}_\alpha, \boldsymbol{\xi}\}, \qquad \boldsymbol{e}_\alpha \cdot \boldsymbol{\xi} = 0, \tag{16.17}$$

$\boldsymbol{\xi}$ being the Killing vector, then the equations (16.15) and (16.16) give just the tetrad components with respect to the 3-basis $\{\boldsymbol{e}_\alpha\}$ in Σ_3.

For a specified Ricci tensor, $R_{ab} = \varkappa_0\left(T_{ab} - \frac{1}{2}Tg_{ab}\right)$, the Einstein field equations are equivalent to the system of differential equations (16.12), (16.13), (16.15), (16.16).

Given a four-dimensional manifold $\mathscr{M}$, a vector field $\boldsymbol{\xi}$ on $\mathscr{M}$ (with prescribed *contravariant* components), and a solution (h_{ab}, F, ω^a) of these field equations, we can find the corresponding stationary *metric*

$$g_{ab} = h_{ab} + F^{-1}\xi_a\xi_b, \tag{16.18}$$

for which $\boldsymbol{\xi}$ is the timelike Killing vector field.

That the quantities ξ_a (covariant components!) can be calulated from (h_{ab}, F, ω^a) is proved as follows. The 1-form $F^{-1}\xi_a\,\mathrm{d}x^a$ can be determined up to a gradient,

$$F^{-1}\xi_a \to F^{-1}\xi_a + \chi_{,a}, \qquad \chi_{,a}\xi^a = 0, \tag{16.19}$$

from the equation (16.8),

$$2(F^{-1}\xi_{[a})_{;b]} = F^{-2}\varepsilon_{abcd}\xi^c\omega^d, \tag{16.20}$$

provided that the exterior derivative of the bivector $(F^{-1}\xi_{[a})_{;b]}$ vanishes, i.e. if

$$\xi^d(\omega^a{}_{;a} - 2F^{-1}F_{,a}\omega^a) + \omega^a\xi^d{}_{;a} - \xi^a\omega^d{}_{;a} = 0. \tag{16.21}$$

Since $\pounds_\xi\omega$ is zero, we have, as an integrability condition for (16.20), precisely the equation (16.13).

For practical calculations, it is convenient to take a *coordinate system* adapted to the congruence $\boldsymbol{\xi} = \partial_t$,

$$\begin{gathered} \mathrm{d}s^2 = h_{\mu\nu}\,\mathrm{d}x^\mu\,\mathrm{d}x^\nu + F(\mathrm{d}t + A_\mu\,\mathrm{d}x^\mu)^2, \\ g_{\mu\nu} = h_{\mu\nu} + F^{-1}\xi_\mu\xi_\nu, \qquad g_{\mu 4} = \xi_\mu = FA_\mu, \qquad g_{44} = \xi_4 = F. \end{gathered} \tag{16.22}$$

If the equation (16.13) is satisfied, the quantities ξ_μ (covariant components of $\boldsymbol{\xi}$) can be obtained from the equation (16.20), which, in the special coordinate system (16.22), takes the form

$$2A_{[\mu,\nu]} = (-F)^{-3/2}\,\varepsilon_{\mu\nu\varrho}\omega^\varrho \tag{16.23}$$

(metric $h_{\mu\nu}$). The remaining freedom in the choice of the gauge function in (16.19) is irrelevant; it corresponds simply to a transformation of the time coordinate, $t \to t + \chi(x^\mu)$.

16.3. Conformal transformation of Σ_3 and the field equations

The equation (16.16) containing second derivatives of F can be simplified considerably by applying a conformal transformation of Σ_3,

$$\Sigma_3 \to \hat{\Sigma}_3\colon\quad \gamma_{ab} \equiv \hat{h}_{ab} = -Fh_{ab}. \tag{16.24}$$

According to equation (3.85), the Ricci tensor $\overset{3}{R}_{ab}$ of Σ_3 is connected with the Ricci tensor $\hat{R}_{ab}$ of $\hat{\Sigma}_3$ by the formula

$$\begin{aligned} \hat{R}_{ab} = \overset{3}{R}_{ab} - \frac{1}{2}\,F^{-1}F_{,a\|b} + \frac{3}{4}\,F^{-2}F_{,a}F_{,b} \\ - \frac{1}{2}\,F^{-1}h_{ab}\left(F_{,c}{}^{\|c} - \frac{1}{2}\,F^{-1}F_{,c}F^{,c}\right). \end{aligned} \tag{16.25}$$

We insert this equation into (16.16) and denote covariant derivatives with respect to γ_{ab} by a double point in front of the index. Then we arrive at

Theorem 16.1. *The complete set of Einstein equations for stationary fields takes the form*

$$\hat{R}_{ab} = \frac{1}{2}\,F^{-2}(F_{,a}F_{,b} + \omega_a\omega_b) + \varkappa_0(h^c_a h^d_b - F^{-2}\gamma_{ab}\xi^c\xi^d)\left(T_{cd} - \frac{1}{2}\,Tg_{cd}\right), \tag{16.26}$$

$$F_{,a}{}^{;a} = F^{-1}\gamma^{ab}(F_{,a}F_{,b} - \omega_a\omega_b) - 2\varkappa_0 F^{-1}\xi^a\xi^b\left(T_{ab} - \frac{1}{2}\,Tg_{ab}\right), \tag{16.27}$$

$$\omega_a^{;a} = 2F^{-1}\gamma^{ab}F_{,a}\omega_b, \tag{16.28}$$

$$F\varepsilon^{abc}\omega_{c,b} = -2\varkappa_0 h^a_b T^b_c \xi^c. \tag{16.29}$$

16.4. Vacuum and Einstein-Maxwell equations for stationary fields

For stationary *vacuum fields*, (16.29) implies $\omega_a = \omega_{,a}$. The existence of the twist potential ω was proved by Papapetrou (1963). The real functions F and ω can be combined to form a complex scalar potential (cp. (16.9))

$$\Gamma = -F + \mathrm{i}\omega. \tag{16.30}$$

Scalar potentials play an important role in a procedure for generating solutions. This question will be discussed in Chapter 30.

In terms of Γ, the equations (16.26)—(16.28) reduce to

$$\hat{R}_{ab} = \frac{1}{2}\,F^{-2}\Gamma_{,(a}\overline{\Gamma}_{,b)}, \tag{16.31 a}$$

$$\Gamma_{,a}{}^{;a} + F^{-1}\gamma^{ab}\Gamma_{,a}\Gamma_{,b} = 0. \tag{16.31 b}$$

For stationary *Einstein-Maxwell fields*, we assume that the Lie derivative of the Maxwell field tensor vanishes. Then we can introduce a complex potential Φ by

$$\sqrt{\varkappa_0/2}\;\xi^a F^*_{ab} = \Phi_{,b}, \qquad \Phi_{,a}\xi^a = 0, \qquad F^{*ab}{}_{;b} = 0. \tag{16.32}$$

With the energy-momentum tensor (5.7),

$$T_{ab} = \frac{1}{2}\,F^{*c}_{a}\overline{F^*_{bc}}, \qquad \sqrt{\varkappa_0/2}\;F^*_{ab} = 2F^{-1}(\xi_{[a}\Phi_{,b]})^*, \tag{16.33}$$

the right-hand side of the field equation (16.29) takes the form

$$\begin{aligned}&-\varkappa_0 h^a_b\xi^c F^{*bd}\overline{F^*_{cd}} = -\sqrt{2\varkappa_0}\;h^a_b F^{*bd}\overline{\Phi}_{,d}\\ &= -2\mathrm{i}F^{-1}\varepsilon^{adbc}\xi_b\Phi_{,c}\overline{\Phi}_{,d} = -2\mathrm{i}F\varepsilon^{abc}(\overline{\Phi}\Phi_{,c})_{,b}.\end{aligned} \tag{16.34}$$

Thus we conclude from (16.29) that the combination $\omega_c + 2\mathrm{i}\overline{\Phi}\Phi_{,c}$ is a gradient (Harrison (1968)). It is convenient to introduce a complex scalar potential, the Ernst potential $\mathscr{E}$, by the equation (Ernst (1968b))

$$\mathscr{E}_{,a} \equiv -F_{,a} + \mathrm{i}\omega_a - 2\overline{\Phi}\Phi_{,a} = \Gamma_a - 2\overline{\Phi}\Phi_{,a}, \qquad \mathrm{Re}\,\mathscr{E} = -F - \overline{\Phi}\Phi. \tag{16.35}$$

The existence of the potential $\mathscr{E}$ is guaranteed, in a similar manner to that of the potential Φ in (16.32), by the equations

$$\xi^a K^*_{ab} = \mathscr{E}_{,b}, \quad \mathscr{E}_{,a}\xi^a = 0, \quad K^{*ab}{}_{;b} = 0, \quad K^*_{ab} \equiv -2\xi^*_{a;b} - \sqrt{2\varkappa_0}\;\overline{\Phi}F^*_{ab}. \tag{16.36}$$

Now we insert the expressions

$$h^c_a h^d_b R_{cd} = 2F^{-1}\left[\Phi_{,(a}\bar{\Phi}_{,b)} - \frac{1}{2}\gamma_{ab}\Phi^{,c}\bar{\Phi}_{,c}\right], \qquad \xi^a\xi^b R_{ab} = F\Phi_{,c}\bar{\Phi}^{,c}, \tag{16.37}$$

into the field equations and substitute for ω_a with the aid of equation (16.35).

Theorem 16.2. *The field equations for stationary Einstein-Maxwell fields outside the sources reduce to the following system of equations, referred to the metric* γ_{ab}:

$$\hat{R}_{ab} = \frac{1}{2}F^{-2}(\mathscr{E}_{,(a} + 2\bar{\Phi}\Phi_{,(a)})(\bar{\mathscr{E}}_{,b)} + 2\Phi\bar{\Phi}_{,b)}) + 2F^{-1}\Phi_{,(a}\bar{\Phi}_{,b)}, \tag{16.38}$$

$$\mathscr{E}_{,a}{}^{;a} + F^{-1}\gamma^{ab}\mathscr{E}_{,a}\Gamma_b = 0, \tag{16.39}$$

$$\Phi_{,a}{}^{;a} + F^{-1}\gamma^{ab}\Phi_{,a}\Gamma_b = 0 \tag{16.40}$$

(Harrison (1968), Neugebauer and Kramer (1969)).

The source-free Maxwell equations are equivalent to (16.40), the field equations (16.26)–(16.28) have been written in terms of the complex potential $\mathscr{E}$ in the complex form (16.38)–(16.39), and the field equation (16.29) is automatically satisfied by introducing the complex potential $\mathscr{E}$.

By specialization of the potentials $\mathscr{E}$ and Φ we can describe the physical situations given in Table 16.1.

Table 16.1 The complex potentials $\mathscr{E}$ and Φ for some physical problems

Physical problem	Potentials	
	$\mathscr{E}$	Φ
Stationary Einstein-Maxwell fields	complex	complex
Electrostatic Einstein-Maxwell fields	real	real
Magnetostatic Einstein-Maxwell fields	real	imaginary
Stationary vacuum fields	complex	0
Static vacuum fields	real	0
Conformastationary Einstein-Maxwell fields (§ 16.7.)	0	complex

The potentials Φ and $\mathscr{E}$ can also be introduced when electromagnetic and material *currents* which are everywhere parallel to the Killing vector,

$$j_{[a}\xi_{b]} = 0 = u_{[a}\xi_{b]}, \tag{16.41}$$

are present. In this case, the (generalized Poisson) equations for $\mathscr{E}$ and Φ read

$$\begin{aligned} &\mathscr{E}_{,a}{}^{;a} + F^{-1}\mathscr{E}_{,a}\Gamma^a = \xi_a T^a, \qquad \Phi_{,a}{}^{;a} + F^{-1}\Phi_{,a}\Gamma^a = \sqrt{\frac{\varkappa_0}{2}}\,\xi_a j^a, \\ &j^a = \sigma u^a, \qquad T^a = -\varkappa_0\xi^a\left[3p + \mu + \sqrt{\frac{2}{\varkappa_0}}(-F)^{-1/2}\sigma\bar{\Phi}\right]. \end{aligned} \tag{16.42}$$

Theorem 16.3. *Stationary asymptotically flat and asymptotically source-free Einstein-Maxwell fields are static provided that the currents satisfy the condition* (16.41) *everywhere.* (The proof uses (16.42) and Stokes' Theorem, see Carter (1972).)

16.5. Geodesic eigenrays

Assuming the existence of a timelike Killing vector, the Einstein field equations have been written as equations in a 3-dimensional space. It is also possible to develop a (3-dimensional) triad formalism and a corresponding spinor technique (Perjés (1970)) in the 3-dimensional Riemannian space Σ_3 of the Killing trajectories.

In a stationary space-time, a null congruence $\boldsymbol{k}$ normalized by $k_a \xi^a = 1$ determines a spacelike unit vector $\boldsymbol{n}$ by

$$n^a = k^a - F^{-1}\xi^a, \qquad n^a n_a = 1, \qquad n^a \xi_a = 0, \qquad \pounds_\xi \boldsymbol{n} = 0. \tag{16.43}$$

One can introduce a triad $\{\boldsymbol{e}_\alpha\} = (\boldsymbol{n}, \boldsymbol{m}, \overline{\boldsymbol{m}})$ orthogonal to $\boldsymbol{\xi}$. Applying the conformal transformation (16.24), we rewrite the geodesic condition on $\boldsymbol{k}$ in the form of an equation over $\hat{\Sigma}_3$ (metric γ_{ab}),

$$k^b k_{a;b} = 0 \Leftrightarrow F n^b n_{a:b} + F_{,a} - n_a n^b F_{,b} + \hat{\varepsilon}_{abc} \omega^b n^c = 0. \tag{16.44}$$

An *eigenray* $\boldsymbol{n}$ is defined by the equation

$$F_{,a} - n_a n^b F_{,b} + \hat{\varepsilon}_{abc} \omega^b n^c = 0. \tag{16.45}$$

In the case of static gravitational fields (§ 16.6.), the two different eigenrays are given by

$$n_a = \pm (F_{,b} F^{,b})^{-1/2} F_{,a}. \tag{16.46}$$

From (16.44) we infer that an eigenray $\boldsymbol{n}$ is geodesic in $\hat{\Sigma}_3$ if and only if the null congruence $\boldsymbol{k}$ is geodesic in V_4,

$$k^b k_{a;b} = 0 \Leftrightarrow n^b n_{a:b} = 0. \tag{16.47}$$

The shear σ of an eigenray is defined by

$$\sigma = n_{a:b} m^a m^b, \qquad n_a m^a = 0, \qquad m_a \overline{m}^a = 1. \tag{16.48}$$

One can show that the existence of a geodesic and shearfree eigenray implies that space-time admits a geodesic and shearfree null congruence.

The stationary *vacuum* solutions admitting geodesic (but not necessarily shearfree) eigenrays are completely known (Kóta and Perjés (1972)). The geodesic eigenray conditions (16.45), (16.47) allow the integration of the field equations in a 3-dimensional version of the Newman-Penrose formalism. The explicit forms of the vacuum metrics with *shearing geodesic eigenrays* are:

$$\begin{aligned} ds^2 = {} & f^{-1}\left[f^0\left(r^{1-1/\sqrt{2}}\, dx^2 + r^{1+1/\sqrt{2}}\, dy^2\right) + dr^2\right] \\ & - f\left(dt - \sqrt{2}\, Qy\, dx + f^{-1}\, dr\right)^2, \end{aligned} \tag{16.49}$$

$f = P(x + Qy)\left(r^{1/\sqrt{2}} + Q^2 r^{-1/\sqrt{2}}\right)^{-1}, \qquad f^0 = P(x + Qy), \qquad P, Q$ real constants, and

$$\mathrm{d}s^2 = f^{-1}\left[f^0\left(r^{1-1/\sqrt{2}}\,\mathrm{d}x^2 + r^{1+1/\sqrt{2}}\,\mathrm{d}y^2\right) + \mathrm{d}r^2\right] - f\left(\mathrm{d}t - \frac{1}{\sqrt{2}}\frac{x^2}{y}\,\mathrm{d}y + f^{-1}\,\mathrm{d}r\right)^2, \tag{16.50}$$

$f = (ax + by)\left(y^2 r^{1/\sqrt{2}} + x^2 r^{-1/\sqrt{2}}\right)^{-1}$, $f^0 = (ax + by)\,y^{-2}$, a, b real constants.

Lukács (1973) gave all vacuum metrics with a *spacelike* Killing vector and shearing geodesic eigenrays. These metrics are very similar to (16.49) and (16.50). The stationary Einstein-Maxwell fields for which the shearing geodesic eigenrays of the Maxwell and gravitational fields coincide are given by Lukács and Perjés (1973). For dust solutions see Lukács (1974).

16.6. Static fields

16.6.1. Definitions

A stationary solution is called *static* if the timelike Killing vector is hypersurface-orthogonal,

$$\xi_{(a;b)} = 0, \qquad \xi_{[a}\xi_{b;c]} = 0, \qquad \xi^a\xi_a < 0. \tag{16.51}$$

We mention an *equivalent characterization* of static fields: in a static space-time there is a vector field $\boldsymbol{u}$ with the properties (Ehlers and Kundt (1962))

$$u_{a;b} = -\dot{u}_a u_b, \qquad \ddot{u}_{[a}u_{b]} = 0, \qquad u^a u_a = -1. \tag{16.52}$$

(a dot means $\nabla_{\boldsymbol{u}}$). From (6.24), (6.25) we obtain the equation

$$\dot{u}_a\dot{u}_c + \dot{u}_{a;c} + \ddot{u}_a u_c = u^b u^d R_{dabc}, \tag{16.53}$$

the antisymmetric part of which tells us that $\dot{\boldsymbol{u}}$ is a gradient, $\dot{u}_a = U_{,a}$. Therefore, $\boldsymbol{\xi} = e^U\boldsymbol{u}$ is a hypersurface-orthogonal Killing vector field.

In § 6.2. we have shown that the Weyl tensor of a static space-time is of Petrov type *I*, *D* or *O*. The vector $\boldsymbol{u}$ satisfying (16.52) is a principal vector of the Weyl tensor and an eigenvector of the Ricci tensor. The tensor Q_{ac} defined in (3.62) is purely real, and the traceless symmetric matrix $\mathbf{Q}$ can be transformed to a diagonal form with *real eigenvalues* with the aid of orthogonal tetrad transformations preserving the principal vector $\boldsymbol{u}$.

The equations (16.26)—(16.28) reduce to

$$\begin{aligned} \hat{R}_{ab} &= 2U_{,a}U_{,b} + h_a^c h_b^d R_{cd} - \mathrm{e}^{-4U}\gamma_{ab}\xi^c\xi^d R_{cd}, \\ U_{,a}{}^{;a} &= -\mathrm{e}^{-4U}\xi^a\xi^b R_{ab}, \qquad F = -\mathrm{e}^{2U}. \end{aligned} \tag{16.54}$$

In particular, the *vacuum field equations* have the remarkably simple form

$$\hat{R}_{ab} = 2U_{,a}U_{,b}. \tag{16.55}$$

The potential equation $U_{,a}{}^{;a} = 0$ follows from (16.55) in virtue of the contracted Bianchi identities for $\hat{R}_{ab}$. It can easily be verified that $\hat{R} = 0$ implies $R_{abcd} = 0$.

The spacelike hypersurfaces (space sections) orthogonal to the Killing vector are totally geodesic, i.e., geodesics in the space sections are simultaneously geodesics of the space-time. In general there exists only one timelike Killing vector in the static space-time; the space sections are uniquely determined by the metric.

In a coordinate frame with $\xi = \partial_t$ the line element has the structure

$$ds^2 = d\sigma^2 - e^{2U}\, dt^2, \qquad d\sigma^2 = h_{\mu\nu}\, dx^\mu\, dx^\nu = e^{-2U}\gamma_{\mu\nu}\, dx^\mu\, dx^\nu, \tag{16.56}$$

the metric functions being independent of the time coordinate t. The preferred coordinate system (16.56) is unique up to purely spatial transformations, $x^{\nu'} = x^{\nu'}(x^\mu)$, and linear transformations of t with constant coefficients, $t' = at + b$.

In the case of a *conformastat* metric (see Synge (1960), p. 339)

$$ds^2 = \Psi^4(x, y, z)\,(dx^2 + dy^2 + dz^2) - e^{2U(x,y,z)}\, dt^2, \tag{16.57}$$

the curvature scalar of the conformally flat spaces $t = \text{const}$ is

$$\overset{3}{R} = -8\Psi^{-5}\Delta\Psi. \tag{16.58}$$

16.6.2. Vacuum solutions

All *degenerate* (type D) static vacuum fields are known. They are given in Table 16.2, together with the simple eigenvalue λ of the Weyl tensor. All these metrics admit at least an Abelian group G_2 and belong to the class of solutions investigated by Weyl (§ 18.1.).

The degenerate static vacuum solutions were originally found by Levi-Civita (1917—19). The invariant classification into the subclasses of Table 16.2 is given in Ehlers and Kundt (1962). The classes A and B are connected by the complex substitution $t \to i\varphi$, $\varphi \to it$. The fields of classes A and B admit an isometry group G_4 and an isotropy group H_1. The "C-metric" admits an Abelian group G_2 of motions The spacelike (class A) or timelike (class B) surfaces determined by the eigen-

Table 16.2 The degenerate static vacuum solutions

Class	Metric	Eigenvalue λ
A		
AI	$ds^2 = r^2(d\vartheta^2 + \sin^2\vartheta\, d\varphi^2) + (1 - b/r)^{-1}\, dr^2 - (1 - b/r)\, dt^2$	$-br^{-3}$
AII	$ds^2 = z^2(dr^2 + \sinh^2 r\, d\varphi^2) + (b/z - 1)^{-1}\, dz^2 - (b/z - 1)\, dt^2$	bz^{-3}
$AIII$	$ds^2 = z^2(dr^2 + r^2\, d\varphi^2) + z\, dz^2 - z^{-1}\, dt^2$	z^{-3}
B		
BI	$ds^2 = (1 - b/r)^{-1}\, dr^2 + (1 - b/r)\, d\varphi^2 + r^2(d\vartheta^2 - \sin^2\vartheta\, dt^2)$	$-br^{-3}$
BII	$ds^2 = (b/z - 1)^{-1}\, dz^2 + (b/z - 1)\, d\varphi^2 + z^2(dr^2 - \sinh^2 r\, dt^2)$	bz^{-3}
$BIII$	$ds^2 = z\, dz^2 + z^{-1}\, d\varphi^2 + z^2(dr^2 - r^2\, dt^2)$	z^{-3}
C ("C-metric")	$ds^2 = \dfrac{1}{(x + y)^2}\left[\dfrac{dx^2}{f(x)} + \dfrac{dy^2}{h(y)} + f(x)\, d\varphi^2 - h(y)\, dt^2\right]$ $f(x) = \pm(x^3 + ax + b) > 0$ $h(y) = -f(-y)$	$\pm(x + y)^3$

bivectors of the curvature tensor (§ 4.2.) have constant Gaussian curvature. *AI* is the Schwarzschild solution (13.19). Classes *A* and *B* are included in (11.42) and its timelike counterpart.

The Harrison metrics (§ 15.4.) include some static *non-degenerate* (type *I*) vacuum solutions. Another type *I* solution is that given by Das (1973). Here we present the latter solution in a somewhat different coordinate system:

$$ds^2 = a^2 z^{-2b}[|z\, d\zeta + c\,\bar{\zeta}\, dz|^2 + (z\, dz)^2] - z^{2b}\, dt^2, \qquad b^2 + c^2 = 1 \tag{16.59}$$

(a, b, c real constants). For $b = 1$, the space-time is flat.

Ref.: Trümper (1962).

16.6.3. Electrostatic and magnetostatic Einstein-Maxwell fields

Restriction to *electrostatic* fields,

$$\bar{\Phi} = \Phi = \chi, \qquad \bar{\mathscr{E}} = \mathscr{E} = e^{2U} - \chi^2 \tag{16.60}$$

(χ = electrostatic potential), or *magnetostatic* fields,

$$\bar{\Phi} = -\Phi = \psi, \qquad \bar{\mathscr{E}} = \mathscr{E} = e^{2U} - \psi^2 \tag{16.61}$$

(ψ = magnetostatic potential), simplifies the differential equations (16.38)–(16.40) for stationary Einstein-Maxwell fields outside the sources. In the electrostatic case, the equations read

$$\hat{R}_{ab} = 2(U_{,a}U_{,b} - e^{-2U}\chi_{,a}\chi_{,b}), \tag{16.62a}$$

$$U_{,a}^{\ ;a} = e^{-2U}\gamma^{ab}\chi_{,a}\chi_{,b}, \qquad \chi_{,a}^{\ ;a} = 2\gamma^{ab}U_{,a}\chi_{,b}. \tag{16.62b}$$

The equations governing magnetostatic fields follow from (16.62) by the substitution $\chi \to \psi$.

In order to simplify these differential equations, we assume a relationship $U = U(\chi)$ between the potentials U and χ. This ansatz and (16.62b) imply

$$e^{2U} = 1 - 2c\chi + \chi^2, \qquad c = \text{const}, \tag{16.63}$$

or, in parametric representation:

$$\begin{aligned} c^2 > 1:&\quad \chi = -(c^2 - 1)^{1/2} \coth Y + c, \qquad e^{2U} = (c^2 - 1) \sinh^{-2} Y,\\ c^2 < 1:&\quad \chi = -(1 - c^2)^{1/2} \cot Y + c, \qquad e^{2U} = (1 - c^2) \sin^{-2} Y,\\ c^2 = 1:&\quad e^{2U} = Y^{-2}, \qquad \chi = -Y^{-1} + c. \end{aligned} \tag{16.64}$$

From (16.64) it follows that the field equations (16.62) reduce to $\hat{R}_{ab} = \pm 2 Y_{,a} Y_{,b}$ for $c^2 \neq 1$, and to $\hat{R}_{ab} = 0$, $Y_{,a}^{\ ;a} = 0$ for $c^2 = 1$. The class $c^2 = 1$ of static Einstein-Maxwell fields (without spatial symmetry) will be given in § 16.7.

Inserting the relation (16.63) into the field equations with charged perfect fluid sources (see (16.42)) one obtains (Gautreau and Hoffman (1973))

$$c\sigma = \sqrt{\frac{\varkappa_0}{2}}\, \varepsilon; \tag{16.65}$$

the charge density σ divided by the active gravitational mass density $\varepsilon = (3p + \mu)\, e^U + \sigma\chi$ is a constant.

We conclude this section with two theorems.

Theorem 16.4 (Esposito and Glass (1976)). *Stationary electrovac space-times are static if and only if the Weyl tensor and the Maxwell tensor are both of pure electric type,* $B_{ac} = \tilde{C}_{abcd} u^b u^d = 0$, $B_a = \tilde{F}_{ab} u^b = 0$.

Theorem 16.5 (Banerjee 1970a)). *There are no static Einstein-Maxwell fields with an electromagnetic null field.*

16.6.4. Perfect fluid solutions

Barnes (1972) determined *all static degenerate* (type D or O) perfect fluid solutions by a method closely analogous to that for vacuum fields. The metrics admitting an isotropy group H_1 are already contained in Chapters 11., 13. and 14. The metrics without an isotropy group are

$$\begin{gathered} ds^2 = (x + y)^{-2} \left\{ \frac{dx^2}{f(x)} + \frac{dy^2}{h(y)} + f(x)\, d\varphi^2 - h(y) \left[A \int h^{-3/2}\, dy + B \right]^2 dt^2 \right\}, \\ f(x) = \pm x^3 + ax + b, \qquad h(x) = -f(-x) - \varkappa_0 \mu/3, \qquad \mu = \text{const}; \end{gathered} \tag{16.66}$$

$$\begin{gathered} ds^2 = (n + mx)^{-2} \{F^{-1}(x)\, dx^2 + F(x)\, d\varphi^2 + dz^2 - x^2\, dt^2\}, \\ F(x) = a(n^2 \ln x + 2mnx + m^2x^2/2) + b, \qquad m = \pm 1, 0, \qquad x > 0; \end{gathered} \tag{16.67}$$

$$\begin{gathered} ds^2 = N^{-2}(z) \{G^{-1}(x)\, dx^2 + G(x)\, d\varphi^2 + dz^2 - x^2\, dt^2\}, \\ G(x) = ax^2 + b \ln x + c, \qquad x > 0, \\ N(z) = A \sin\left(\sqrt{a}z\right),\ Az,\ A \sinh\left(\sqrt{-a}\, z\right) \text{ for } a > 0,\ a = 0,\ a < 0, \text{ resp.} \end{gathered} \tag{16.68}$$

(a, b, c, m, n, A, B real constants). The metrics (16.66) and (16.68) admit an Abelian group G_2; the metric (16.67) admits a G_3. For vanishing pressure and energy density, (16.66) goes over into the "C-metric" of Table 16.2.

The class of *conformastat* (in the sense of (16.57)) perfect fluid solutions was given by Melnick and Tabensky (1975). Spatial symmetry is not assumed. To construct a solution explicitly, one has to specify a real function $\gamma = \gamma(s)$. The functions U and Ψ in the metric (16.57) then follow from

$$2\gamma U'^2 = \gamma'', \qquad \Psi^{-2} = e^U \gamma(z + m), \qquad s = \frac{1}{2}(r^2 + q)(z + m)^{-1} \tag{16.69}$$

(q, m constants). Pressure and energy density can be evaluated from U and Ψ. The choice $\gamma = s^2$ leads to the metric

$$ds^2 = \frac{(z + m)^4}{(r^2 + q)^6} (dx^2 + dy^2 + dz^2) - \left(\frac{r^2 + q}{z + m}\right)^2 dt^2, \tag{16.70}$$

which for $q = m = 0$ is a vacuum solution of Weyl's class (§ 18.1).

Theorem 16.6 (Glass (1975)). *The necessary and sufficient condition for a shearfree perfect fluid to be rotationfree is that the Weyl tensor is of pure electric type.*

16.7. The conformastationary class of Einstein-Maxwell fields

An interesting class of stationary Einstein-Maxwell fields without spatial symmetry is characterized by the 3-space $\hat{\Sigma}_3$ being flat,

$$ds^2 = e^{-2U}(dx^2 + dy^2 + dz^2) - e^{2U}(dt + A_\mu\, dx^\mu)^2. \tag{16.71}$$

We call this metric *conformastationary*, by analogy with the definition (16.57) of a conformastat metric.

The field equations (16.38)—(16.40) are satisfied by

$$\hat{R}_{ab} = 0, \qquad \mathscr{E} = 0, \qquad \Phi^{-1}{}_{,a}{}^{;a} = 0. \tag{16.72}$$

The functions U and A_μ in the line element (16.71) can be determined from a solution $V = \Phi^{-1}$ of the potential equation $\Delta V = 0$ in the flat 3-space $\hat{\Sigma}_3$ via the relations (16.35) and (16.23), i.e., in 3-dimensional vector notation,

$$e^{2U} = (V\overline{V})^{-1}, \qquad \operatorname{curl} \boldsymbol{A} = i(V \operatorname{grad} \overline{V} - \overline{V} \operatorname{grad} V), \qquad \Delta V = 0. \tag{16.73}$$

This class of solutions was discovered by Neugebauer (1969), Perjés (1971), and Israel and Wilson (1972). The fields are static if $\operatorname{curl} \boldsymbol{A} = 0$. In particular, purely electric fields ($\overline{\Phi} = \Phi$) are static. Papapetrou (1947) and Majumdar (1947) have given this special class of static Einstein-Maxwell fields without spatial symmetry.

The asymptotic form of the electromagnetic potentials and the space-time metric shows that the source of the class under consideration satisfies $|Q| = \sqrt{\varkappa_0/2}\, M$, $\mu = \pm\sqrt{\varkappa_0/2}\, J$ (M = mass, Q = charge, μ = magnetic moment, J = angular momentum). The conformastationary solutions are the exterior fields of charged spinning sources in equilibrium under their mutual electromagnetic and gravitational forces.

The linearity of the differential equation for V allows a *superposition* of solutions. An example of an exterior field of two isolated sources will be given as (19.23).

If the geometry is regular outside the sources, the condition

$$\int_S (\overline{V} \operatorname{grad} V - V \operatorname{grad} \overline{V})\, d\boldsymbol{f} = 0 \tag{16.74}$$

is satisfied for every exterior closed 2-surface S. From this regularity condition one can derive restrictions on the source parameters which guarantee that no stresses between the spinning sources occur (Israel and Spanos (1973)). For a discussion of the regularity conditions in stationary fields, and further references, see Ward (1976).

Chapter 17. Stationary axisymmetric fields: basic concepts and field equations

17.1. The Killing vectors

We now consider physical systems which, in addition to being stationary (with Killing vector ξ) possess a further symmetry: axial symmetry. This symmetry is described by a *spacelike* Killing vector field η, the trajectories of which are *closed* (compact) curves. The Killing vector η vanishes on the *rotation axis* at all times.

The Killing vectors ξ and η generate a group G_2. The general metric for gravitational fields admitting a group G_2 was given in (15.1). Here we demand that ξ and η commute

$$\xi^a{}_{;b}\eta^b - \eta^a{}_{;b}\xi^b = 0, \qquad \xi^a\xi_a < 0, \qquad \eta^a\eta_a > 0, \tag{17.1}$$

i.e. that ξ and η generate an *Abelian group* G_2. Carter (1970) has shown that the cases of the greatest physical importance, *asymptotically flat* stationary axisymmetric gravitational fields, necessarily admit an Abelian group G_2, so that for them (17.1) does not impose an additional restriction.

The group parameter φ along the trajectories of η has the standard periodicity 2π only if the regularity condition,

$$X_{,a}X^{,a}/4X \to 1, \qquad (X = \eta^a\eta_a,\ \eta = \partial_\varphi), \tag{17.2}$$

on the limit at the rotation axis is satisfied. This normalization of the parameter φ is always possible if we assume Lorentzian geometry ("elementary flatness") in the vicinity of the rotation axis. Otherwise the regularity condition (17.2) is violated and there are singularities (rods or struts) on the axis.

The Killing vectors ξ and η are *uniquely* determined if we demand that (i) η has compact trajectories (in general other subgroups G_1 will have non-compact trajectories), (ii) η is normalized by (17.2), and (iii) space-time is asymptotically flat and ξ is normalized so that $F = \xi^a\xi_a \to -1$ holds in the limit at large distances.

17.2. Orthogonal surfaces

The Killing trajectories form 2-dimensional orbits T_2; the simple bivector

$$v_{ab} = 2\xi_{[a}\eta_{b]}, \qquad v_{ab}v^{ab} < 0, \tag{17.3}$$

is surface-forming (see equations (17.1) and (6.11)). The surface element orthogonal to the group orbits is spanned by the dual bivector

$$\tilde{v}_{ab} = \frac{1}{2}\,\varepsilon_{abcd}v^{cd} = 2m_{[a}\overline{m}_{b]}, \qquad m_a\xi^a = 0 = m_a\eta^a. \tag{17.4}$$

The simple bivector $\tilde{v}_{ab}$ is surface-forming if and only if the condition

$$v_{ab}\tilde{v}^{bc}{}_{;c} = 2\xi_{[a}\eta_{b]}(\overline{m}^c m^b{}_{;c} - m^c\overline{m}^b{}_{;c}) = 0$$
$$\Leftrightarrow \varepsilon_{abcd}\eta^a\xi^b\xi^{c;d} = 0 = \varepsilon_{abcd}\xi^a\eta^b\eta^{c;d} \tag{17.5}$$

is satisfied (cp. (6.11)). In general, the Killing vectors of an arbitrary stationary axisymmetric space-time do not obey (17.5), but only solutions obeying (17.5) are explicitly known.

The condition (17.5) can be reformulated as a restriction on the algebraic form of the Ricci tensor:

Theorem 17.1 (Kundt and Trümper (1966)). *Stationary axisymmetric fields admit 2-spaces orthogonal to the group orbits if and only if the conditions*

$$\xi^d R_{d[a}\xi_b\eta_{c]} = 0 = \eta^d R_{d[a}\xi_b\eta_{c]} \tag{17.6}$$

are satisfied.

Proof (see, e.g., also Carter (1972)): For a Killing vector ξ the identity (8.24),

$$\xi_{a;bc} = R_{abcd}\xi^d, \tag{17.7}$$

holds. From (17.7) (and an analogous relation for η) we get

$$(\xi_{[a;b}\xi_{c]})^{;c} = \frac{2}{3}\,\xi^d R_{d[a}\xi_{b]},$$
$$(\eta_{[a;b}\eta_{c]})^{;c} = \frac{2}{3}\,\eta^d R_{d[a}\eta_{b]}. \tag{17.8}$$

With the aid of these equations, together with the cyclic symmetry relation of the curvature tensor and the commutativity of ξ and η, we obtain

$$(\xi_{[a;b}\xi_c\eta_{d]})^{;d} = -\frac{1}{2}\,\xi^d R_{d[a}\xi_b\eta_{c]},$$
$$(\eta_{[a;b}\eta_c\xi_{d]})^{;d} = -\frac{1}{2}\,\eta^d R_{d[a}\eta_b\xi_{c]}. \tag{17.9}$$

Using the fact that a completely antisymmetric tensor is proportional to the ε-tensor, we convert (17.9) into the equivalent dual equations

$$(\varepsilon^{abcd}\xi_{a;b}\xi_c\eta_d)^{;e} = -2\varepsilon^{abce}\xi^d R_{da}\xi_b\eta_c,$$
$$(\varepsilon^{abcd}\eta_{a;b}\eta_c\xi_d)^{;e} = -2\varepsilon^{abce}\eta^d R_{da}\eta_b\xi_c. \tag{17.10}$$

From these equations we conclude that the conditions (17.6) are a necessary consequence of the original criterion (17.5). Conversely, if the conditions (17.6) are satisfied, the twist scalars $\varepsilon_{abcd}\eta^a\xi^b\xi^{c;d}$ and $\varepsilon_{abcd}\xi^a\eta^b\eta^{c;d}$ are constants. Both twist scalars vanish on the rotation axis ($\eta = 0$). Hence they vanish in a connected domain which intersects the rotation axis if the conditions (17.6) hold in that domain. Then the geometry admits 2-spaces orthogonal to the orbits of the Abelian group of motions G_2.

For vacuum fields ($R_{ab} = 0$), the existence of orthogonal 2-spaces was first demonstrated by Papapetrou (1966) in his pioneering work on this subject.

The conditions (17.6) hold for a wide class of energy-momentum tensors. For example, decomposition of the metric into orthogonal 2-spaces is possible for perfect fluid solutions, and for Einstein-Maxwell fields, provided that the 4-velocity of the fluid, and the electromagnetic 4-current density vector, respectively, satisfy the so-called circularity condition

$$u_{[a}\xi_b\eta_{c]} = 0 = j_{[a}\xi_b\eta_{c]}, \tag{17.11}$$

i.e., the trajectories of $\boldsymbol{u}$ and $\boldsymbol{j}$ lie in the surfaces of transitivity of the group G_2. In this terminology, the conditions (17.6) mean that $R^a_b\xi^b$ and $R^a_b\eta^b$ are circular vector fields. The circularity condition (17.11) is a natural generalization of the condition (16.41) (4-velocity and current parallel to the Killing vector) to the case of stationary axisymmetric fields. Obviously, for perfect fluids the condition $u_{[a}\xi_b\eta_{c]} = 0$ implies (17.6).

Now we show that $j_{[a}\xi_b\eta_{c]} = 0$ for Einstein-Maxwell fields also implies (17.6). The Maxwell equations

$$F^{*ab}{}_{;b} = j^a, \tag{17.12}$$

and the assumption that the Lie derivatives of F^*_{ab} with respect to both the commuting Killing vectors $\boldsymbol{\xi}$ and $\boldsymbol{\eta}$ vanish, lead to

$$(F^*_{ab}\xi^a\eta^b)_{;d} = \mathrm{i}\varepsilon_{abcd}\xi^a\eta^b j^c = 0. \tag{17.13}$$

$F^*_{ab}\xi^a\eta^b$ is constant, and it is zero on the axis of rotation. Hence,

$$F^*_{ab}\xi^a\eta^b = 0 = \xi_{[a}\eta_b F^*_{cd]}. \tag{17.14}$$

(In flat space-time electrodynamics this relation says that the azimuthal components of the electric and magnetic field vectors are zero.) The relation (17.14) in turn implies (17.6).

If in an Einstein-Maxwell geometry, outside the sources, two Killing vectors $\boldsymbol{\xi}$, $\boldsymbol{\eta}$ satisfy the conditions (17.1), (17.3), (17.5), then the Maxwell field shares the space-time symmetry: $\pounds_\xi F_{ab} = 0 = \pounds_\eta F_{ab}$ (Michalski and Wainwright (1975)).

17.3. The metric and the projection formalism

Theorem 17.1 is of fundamental importance in the construction of exact solutions of the Einstein equations. If we adapt the coordinate system to the Killing vectors, the line element of a stationary axisymmetric field admitting 2-spaces orthogonal to the Killing vectors $\boldsymbol{\xi} = \partial_t$ and $\boldsymbol{\eta} = \partial_\varphi$ can be written in the form (Lewis (1932), Papapetrou (1966))

$$\mathrm{d}s^2 = \mathrm{e}^{-2U}(\gamma_{MN}\,\mathrm{d}x^M\,\mathrm{d}x^N + W^2\,\mathrm{d}\varphi^2) - \mathrm{e}^{2U}(\mathrm{d}t + A\,\mathrm{d}\varphi)^2, \tag{17.15}$$

where the metric functions U, γ_{MN}, W, A depend only on the coordinates $x^M = (x^1, x^2)$ which label the points on the 2-surfaces S_2 orthogonal to the orbits. The form (17.15) is preserved under the transformations

$$x^{M'} = x^{M'}(x^N), \qquad t' = at + b\varphi, \qquad \varphi' = c\varphi + dt, \tag{17.16}$$

where $a, \ldots, d$ are constants. The functions W, U, A behave like scalars under the coordinate transformations $x^{M'} = x^{M'}(x^N)$. The metric (17.15) exhibits the reflection symmetry $(t, \varphi) \to (-t, -\varphi)$.

(17.15) is the specialization of (16.22) (including the conformal transformation (16.24)) to the stationary axisymmetric fields obeying the conditions (17.5). In the coordinate system (17.15), the scalar products of the Killing vectors are

$$\xi^a\xi_a = -e^{2U}, \qquad \eta^a\eta_a = e^{-2U}W^2 - e^{2U}A^2, \qquad \xi^a\eta_a = -e^{2U}A, \qquad W^2 = -2\xi_{[a}\eta_{b]}\xi^a\eta^b. \tag{17.17}$$

In general, $\boldsymbol{\xi}$ and $\boldsymbol{\eta}$ are not hypersurface-orthogonal. However, it can easily be verified (Bardeen (1970)) that the timelike vector

$$\boldsymbol{\zeta} = \boldsymbol{\xi} - \xi^a\eta_a(\eta^b\eta_b)^{-1}\,\boldsymbol{\eta} \tag{17.18}$$

is orthogonal to the hypersurfaces $t = \text{const}$. In general, $\boldsymbol{\zeta}$ is not a Killing vector.

Without loss of generality we can introduce *isotropic coordinates* in V_2,

$$\gamma_{MN} = e^{2k}\delta_{MN}. \tag{17.19}$$

Moreover, if W is a potential function, it can be transformed either to $W = x^1$ or to $W = 1$ with the aid of a coordinate transformation $\zeta' = f(\zeta)$, $\sqrt{2}\,\zeta = x^1 + ix^2$, preserving the isotropic coordinate condition. (The case $W = 1$ is not of interest here; metrics of this type will be treated in § 21.5.)

The isotropic coordinates with $W = x^1 = \varrho$ are called *Weyl's canonical coordinates* $\left(x^1 = \varrho,\ x^2 = z;\ \sqrt{2}\,\zeta = \varrho + iz\right)$. In these coordinates, the space-time metric (17.15) is given by

$$ds^2 = e^{-2U}[e^{2k}(d\varrho^2 + dz^2) + \varrho^2\, d\varphi^2] - e^{2U}(dt + A\, d\varphi)^2. \tag{17.20}$$

For stationary axisymmetric fields, one can develop a projection formalism (Geroch (1972)) similar to that for stationary fields given in § 16.1. We introduce the notation

$$\lambda_{AB} \equiv \xi_A{}^a\xi_{Ba}, \qquad A, B = 1, 2, \tag{17.21}$$

for the scalar products of the Killing vectors $\boldsymbol{\xi}_1 = \boldsymbol{\xi}$, $\boldsymbol{\xi}_2 = \boldsymbol{\eta}$, and identify the projection tensor

$$H_{ab} \equiv g_{ab} + W^{-2}\lambda^{AB}\xi_{Aa}\xi_{Bb}, \qquad W^2 \equiv -\frac{1}{2}\,\lambda^{AB}\lambda_{AB}, \tag{17.22}$$

with the metric tensor on a differentiable manifold Σ_2. Capital Latin indices are raised and lowered with the aid of the alternating symbols (3.65), as in (3.66). Covariant derivatives (denoted by D_a) and the Riemann curvature tensor on Σ_2 are defined in analogy to (16.4) and (16.6) ($h_{ab} \to H_{ab}$).

If 2-spaces V_2 orthogonal to the group orbits exist, Σ_2 can be identified with these

2-spaces and the two scalars $C_A = \varepsilon^{MN}\varepsilon^{abcd}\xi_{Ma}\xi_{Nb}\xi_{Ac;d}$ vanish. Then the Ricci tensor R_{ab} in V_4 can be written in terms of tensors and metric operations on V_2 as follows,

$$\overset{2}{R}_{ab} = (2W)^{-2}\lambda^{AB}{}_{,a}\lambda_{AB,b} + W^{-1}D_aW_{,b} + H^c_aH^d_bR_{cd}, \tag{17.23}$$

$$D^a(W^{-1}\lambda_{AB,a}) = \frac{1}{2}W^{-3}\lambda_{AB}H^{ab}\lambda^{CD}{}_{,a}\lambda_{CD,b} - 2W^{-1}R_{ab}\xi^a_A\xi^b_B, \tag{17.24}$$

$$H^c_a\xi^d_AR_{cd} = 0. \tag{17.25}$$

Ref.: □ Kundu (1978).

17.4. The field equations for stationary axisymmetric Einstein-Maxwell fields

We specialize the field equations (16.38)—(16.40) for stationary Einstein-Maxwell fields outside the sources to the axisymmetric case. Assuming that the electromagnetic field satisfies the condition (17.14), $F^*_{ab}\xi^a\eta^b = 0$, we can start with the line element (17.15) (see Theorem 17.1), and the potentials $\mathscr{E}$ and Φ depend only on the spatial coordinates in the 2-spaces orthogonal to the group orbits. Exceptional cases (e.g. the field of an axial current) where the condition (17.14) is violated are given in § 20.2.

All field equations can be written covariantly with respect to the metric γ_{MN}. After the substitution $-F = \mathrm{Re}\,\mathscr{E} + \Phi\overline{\Phi}$ (see (16.35)), the field equations (16.39) and (16.40) read

$$(\mathrm{Re}\,\mathscr{E} + \Phi\overline{\Phi})\,W^{-1}(W\mathscr{E}_{,M})^{;M} = \mathscr{E}_{,M}(\mathscr{E}^{,M} + 2\overline{\Phi}\Phi^{,M}), \tag{17.26}$$

$$(\mathrm{Re}\,\mathscr{E} + \Phi\overline{\Phi})\,W^{-1}(W\Phi_{,M})^{;M} = \Phi_{,M}(\mathscr{E}^{,M} + 2\overline{\Phi}\Phi^{,M}). \tag{17.27}$$

The field equations (16.38) reduce to

$$W_{,M}{}^{;M} = 0, \tag{17.28}$$

$$K\gamma_{MN} - \frac{W_{,M;N}}{W} = \frac{(\mathscr{E}_{,M} + 2\overline{\Phi}\Phi_{,M})\,(\overline{\mathscr{E}}_{,N} + 2\Phi\overline{\Phi}_{,N})_{(M,N)}}{2(\mathrm{Re}\,\mathscr{E} + \overline{\Phi}\Phi)^2} - 2\,\frac{\overline{\Phi}_{,(M}\Phi_{,N)}}{\mathrm{Re}\,\mathscr{E} + \overline{\Phi}\Phi}, \tag{17.29}$$

where K denotes the Gaussian curvature of the 2-space $\hat{V}_2$ with metric γ_{MN}. According to the formulae (16.23) and (16.35), the metric coefficients A and $F = -e^{2U}$ in the line element (17.15) can be determined from the complex potentials $\mathscr{E}$ and Φ via

$$A_{,M} = W\,e^{-4U}\varepsilon_{MN}\omega^N, \qquad \omega_N = \mathrm{Im}\,\mathscr{E}_{,N} - i(\overline{\Phi}\Phi_{,N} - \Phi\overline{\Phi}_{,N}), \tag{17.30}$$

$$e^{2U} = \mathrm{Re}\,\mathscr{E} + \overline{\Phi}\Phi, \tag{17.31}$$

where ε_{MN} is the Levi-Civita tensor in $\hat{V}_2$.

Because of (17.28) we can introduce Weyl's canonical coordinates, ϱ $(= W)$ and z;

the functions U, k and A in the metric (17.20) are determined by (17.31) and

$$k,_{\zeta} = \sqrt{2}\varrho\left[\frac{(\mathscr{E},_{\zeta} + 2\bar{\Phi}\Phi,_{\zeta})\,(\bar{\mathscr{E}},_{\zeta} + 2\Phi\bar{\Phi},_{\zeta})}{4(\mathrm{Re}\,\mathscr{E} + \bar{\Phi}\Phi)^2} - \frac{\bar{\Phi},_{\zeta}\Phi,_{\zeta}}{\mathrm{Re}\,\mathscr{E} + \bar{\Phi}\Phi}\right], \tag{17.32}$$

$$A,_{\zeta} = \varrho\,\frac{\mathrm{i}(\mathrm{Im}\,\mathscr{E}),_{\zeta} + \bar{\Phi}\Phi,_{\zeta} - \Phi\bar{\Phi},_{\zeta}}{(\mathrm{Re}\,\mathscr{E} + \bar{\Phi}\Phi)^2}, \quad \sqrt{2}\,\partial_{\zeta} = \partial_{\varrho} - \mathrm{i}\partial_{z}. \tag{17.33}$$

Because of the field equations (17.26), (17.27), the integrability conditions of (17.32) and (17.33) are satisfied identically, and k and A can be calculated simply by line integration. Two of the three equations (17.29) are just the equations (17.32) for determining k. The remaining field equation of (17.29) contains the Gaussian curvature $K = -2\mathrm{e}^{-2k}k,_{\zeta\bar{\zeta}}$ of $\hat{V}_2$ and has, in Weyl's canonical coordinates, the form

$$k,_{MM} + \frac{(\mathscr{E},_M + 2\bar{\Phi}\Phi,_M)\,(\bar{\mathscr{E}},_M + 2\Phi\bar{\Phi},_M)}{4(\mathrm{Re}\,\mathscr{E} + \bar{\Phi}\Phi)^2} - \frac{\bar{\Phi},_M\Phi,_M}{\mathrm{Re}\,\mathscr{E} + \bar{\Phi}\Phi} = 0. \tag{17.34}$$

It is identically satisfied because of the other field equations.

Thus we have proved

Theorem 17.2. *The Einstein-Maxwell equations for stationary axisymmetric fields can be completely reduced to the simultaneous system* (17.26), (17.27) *of elliptic differential equations of the second order for the two complex potentials* $\mathscr{E}$ *and* Φ. *Every solution of these equations determines a metric* (17.20) *via* (17.31)–(17.33).

For exterior electrostatic fields ($\Phi = \bar{\Phi} = \chi$), the field equations (17.26)–(17.27) read

$$W^{-1}(WU,_M)^{;M} = \mathrm{e}^{-2U}\chi,_M\chi^{,M}, \qquad (W\,\mathrm{e}^{-2U}\chi,_M)^{;M} = 0. \tag{17.35}$$

In Weyl's canonical coordinates, (17.35) can be replaced by the equations

$$\begin{aligned} \Delta U &= \varrho^{-1}(\varrho U,_A),_A = U,_AU,_A + k,_{A,A}, \\ (\Delta U)^2 &= (2U,_1U,_2 - k,_2/\varrho)^2 + (U,_1{}^2 - U,_2{}^2 - k,_1/\varrho)^2 \end{aligned} \tag{17.36}$$

for the metric functions U and k. Two special classes of solutions are those with $\Delta U = 0$ (Weyl's *vacuum* solutions, § 18.1.), and $k = 0$ (Papapetrou-Majumdar solutions (§ 16.7.) with axial symmetry).

17.5. Various forms of the field equations for stationary axisymmetric vacuum fields

All stationary axisymmetric vacuum fields ($\Phi = 0$, $\mathscr{E} = \Gamma$) are described by the metric (17.15). The field equation (17.26), specialized to the vacuum case, is the Ernst equation (Ernst (1968a))

$$(\Gamma + \bar{\Gamma})\,W^{-1}(W\Gamma,_M)^{;M} = 2\Gamma,_M\Gamma^{,M}. \tag{17.37}$$

Splitting Γ into its real and imaginary parts, we have to solve a simultaneous system of two elliptic differential equations of the second order,

$$W^{-1}(WU_{,M})^{;M} = -\frac{1}{2}\,\mathrm{e}^{-4U}\omega_{,M}\omega^{,M}, \qquad (W\,\mathrm{e}^{-4U}\omega_{,M})^{;M} = 0. \tag{17.38}$$

These equations are covariant with respect to transformations in the 2-space $\hat{V}_2$ with the metric γ_{MN}. In Weyl's canonical coordinates (ϱ, z), (17.37) reads

$$(\Gamma + \bar{\Gamma})\,(\Gamma_{,\varrho\varrho} + \varrho^{-1}\Gamma_{,\varrho} + \Gamma_{,zz}) = 2(\Gamma_{,\varrho}{}^2 + \Gamma_{,z}{}^2). \tag{17.39}$$

Once a solution $\Gamma = \mathrm{e}^{2U} + \mathrm{i}\omega$ of this differential equation is given, we find the full metric (17.20) with the aid of line integrals for k and A,

$$k_{,\zeta} = \sqrt{2}\,\varrho\,\frac{\Gamma_{,\zeta}\bar{\Gamma}_{,\zeta}}{(\Gamma + \bar{\Gamma})^2}, \quad A_{,\zeta} = 2\varrho\,\frac{(\Gamma - \bar{\Gamma})_{,\zeta}}{(\Gamma + \bar{\Gamma})^2}, \quad \sqrt{2}\,\partial_\zeta = \partial_\varrho - \mathrm{i}\,\partial_z, \tag{17.40}$$

the integrability conditions being automatically satisfied. Thus the problem has been reduced essentially to the equation (17.39) for the complex potential Γ (see Theorem 17.2). At first glance, the simple structure of this non-linear differential equation is encouraging, but actually only some special solutions and restricted classes of solutions have been found as yet. To attack the problem under consideration, at least partially, reformulated field equations might be of some help. Therefore we collect together here some other formulations.

(i) Introducing a new function $S \equiv -U + \frac{1}{2}\cdot \ln W$, we obtain from (17.38) the system of differential equations

$$W^{-1}(WS_{,M})^{;M} = \frac{1}{2}\,\mathrm{e}^{-4S}A_{,M}A^{,M}, \qquad (W\,\mathrm{e}^{-4S}A_{,M})^{;M} = 0 \tag{17.41}$$

for the unknown functions S and A. Up to a sign the equations (17.41) and (17.38) have exactly the same form.

(ii) We retain Weyl's canonical coordinates $\left(\sqrt{2}\,\zeta = \varrho + \mathrm{i}z\right)$, but go over to new variables M and N defined by

$$M = 2(\zeta + \bar{\zeta})\,(\Gamma + \bar{\Gamma})^{-1}\,(\Gamma + \bar{\Gamma})_{,\zeta}, \qquad N = 2(\zeta + \bar{\zeta})\,(\Gamma + \bar{\Gamma})^{-1}\,(\Gamma - \bar{\Gamma})_{,\zeta}. \tag{17.42}$$

Inserting these expressions into the field equation (17.39), and taking into account the integrability condition, we obtain a simultaneous system of partial differential equations of the *first* order for the two complex functions M and N,

$$\begin{aligned} 2(\zeta + \bar{\zeta})\,M_{,\bar{\zeta}} &= M - \bar{M} - \bar{N}N, \\ 2(\zeta + \bar{\zeta})\,N_{,\bar{\zeta}} &= N + \bar{N} - \bar{N}M. \end{aligned} \tag{17.43}$$

Once a solution to these equations is known, we get the potential Γ by a line integral.

(iii) The Ernst equation (17.37) can be reformulated by introducing a new potential ξ by

$$\xi \equiv (1 - \mathscr{E})/(1 + \mathscr{E}). \tag{17.44}$$

ξ satisfies the differential equation

$$(\xi\bar{\xi} - 1)\, W^{-1}(W\xi_{,M})^{;M} = 2\bar{\xi}\xi_{,M}\xi^{,M}. \tag{17.45}$$

This last version of the field equations has proved to be especially useful in constructing new solutions (see § 18.5.).

(iv) The function k in the metric (17.20) satisfies a differential equation of the fourth order, which can be derived from the fact that the right-hand side of equation (17.29) (with $\Phi = 0$) multiplied by $(-2)\,\mathrm{d}x^M\,\mathrm{d}x^N$, is the metric of a space of constant negative curvature. From the left-hand side of (17.29) we infer that the line element $\varrho^{-1}(\varrho k_{,A})_{,A}(\mathrm{d}\varrho^2 + \mathrm{d}z^2) - 2k_{,A}\,\mathrm{d}x^A\,\mathrm{d}\varrho/\varrho$ is associated with this space of constant curvature so that we obtain a differential equation for k alone:

$$\begin{aligned} 2D(A_{,zz} + C_{,\varrho\varrho} - 2B_{,\varrho z}) - D_{,z}A_{,z} - D_{,\varrho}C_{,\varrho} - BA_{,\varrho}C_{,z} \\ + BA_{,z}C_{,\varrho} + 2CA_{,\varrho}B_{,z} + 2AC_{,z}B_{,\varrho} - 4BB_{,\varrho}B_{,z} = 4D^2 \end{aligned} \tag{17.46}$$

$$A \equiv -2k_{,AA} + k_{,\varrho}/\varrho, \qquad B \equiv 2k_{,z}/\varrho$$
$$C \equiv -2k_{,AA} - k_{,\varrho}/\varrho, \qquad D \equiv AC - B^2.$$

From a given (non-constant) solution k of (17.46), the potential ξ can be constructed up to a phase factor, $\xi \to \mathrm{e}^{\mathrm{i}\alpha}\xi$ (ignoring pure coordinate transformations).

Other forms of the field equations are given, e.g., in Chandrasekhar (1978), Kramer and Neugebauer (1968b) and □ Cox and Kinnersley (1979).

The comparison of (17.38) with the equations (17.41), and with the very similar equations (17.35) for electrostatic fields, leads to

Theorem 17.3. *Given a stationary axisymmetric vacuum solution* (U, ω), *the substitution* (Kramer and Neugebauer (1968b))

$$S' = -U' + \frac{1}{2}\ln W = U, \qquad A' = \mathrm{i}\omega, \tag{17.47}$$

yields another vacuum solution (U', A'), *and the substitution* (Bonnor (1961))

$$U' = 2U, \qquad \chi = \mathrm{i}\omega, \tag{17.48}$$

yields a static solution (U', χ) *of the Einstein-Maxwell equations.*

These new solutions will be *real* if one can analytically continue the parameters in the solutions to compensate for the imaginary unit in the substitutions.

Applying the Newman-Penrose formalism (Chapter 7.) to stationary axisymmetric vacuum solutions, and choosing the real null directions $\boldsymbol{k}$ and $\boldsymbol{l}$ as linear combinations of the Killing vectors ξ and η, we obtain the result: solutions of Petrov type *III* cannot occur. Type *II* solutions are members of the class $A^2 = W^2\,\mathrm{e}^{-4U}$ (§ 18.4.) (Collinson and Dodd (1969)).

Chapter 18. Stationary axisymmetric vacuum solutions

This chapter gives a survey of the known stationary axisymmetric vacuum solutions. In Chapter 30., the reader can find some solutions not given explicitly here. The stationary cylindrically symmetric vacuum field is given in § 20.2.

18.1. Static axisymmetric vacuum solutions (Weyl's class)

If we assume that one of the Killing vectors in the metric (17.15), say ξ, is hypersurface-orthogonal, then A can be put equal to zero and the second Killing vector is also hypersurface-orthogonal. The *static* axisymmetric vacuum solutions (Weyl (1917)) are invariantly characterized by the existence of two Killing vectors ξ, η satisfying

$$\begin{aligned} &\xi_{(a;b)} = 0, \qquad \xi_{[a;b}\xi_{c]} = 0, \qquad \xi^a\xi_a < 0, \\ &\eta_{(a;b)} = 0, \qquad \eta_{[a;b}\eta_{c]} = 0, \qquad \eta^a\eta_a > 0, \\ &\qquad\qquad \xi^c\eta^a{}_{;c} - \eta^c\xi^a{}_{;c} = 0. \end{aligned} \tag{18.1}$$

Because in this case the twist potential ω vanishes, $\Gamma = -e^{2U}$ is real, and the Ernst equation (17.39) is simply the potential equation $\Delta U = 0$ for a real function U independent of the azimuthal coordinate φ. In Weyl's canonical coordinates (17.20), the *field equations* for static axisymmetric vacuum solutions read

$$\begin{aligned} &\Delta U = \varrho^{-1}(\varrho U_{,M})_{,M} = 0, \\ &k_{,\varrho} = \varrho(U_{,\varrho}{}^2 - U_{,z}{}^2), \qquad k_{,z} = 2\varrho U_{,\varrho}U_{,z}. \end{aligned} \tag{18.2}$$

The function k can be calculated by means of a line integral. Although $\Delta U = 0$ is a linear differential equation, the equations for k manifest the non-linearity of the Einstein field equations. For static axisymmetric vacuum solutions the *regularity condition* (17.2) on the axis of symmetry means $\lim_{\varrho\to 0} k = 0$.

Two solutions of Weyl's class are completely equivalent if the functions U and k of one solution differ only by additive constants from those of the other solution.

The potential equation $\Delta U = 0$ may be solved using various coordinates in the Euclidean 3-space. In *spherical* coordinates (r, ϑ), the asymptotically flat solutions are

$$U = \sum_{n=0}^{\infty} a_n r^{-(n+1)} P_n(\cos\vartheta),$$
$$k = -\sum_{l,m=0}^{\infty} a_l a_m (l+1)(m+1)(l+m+2)^{-1} r^{-(l+m+2)} (P_l P_m - P_{l+1} P_{m+1}), \tag{18.3}$$

$P_n = P_n(\cos\vartheta)$ being the Legendre polynomials. The simplest case is the solution

$$U = -m/r, \qquad 2k = -m^2 \sin^2\vartheta / r^2 \tag{18.4}$$

(Chazy (1924), Curzon (1924)). Though the potential U is the spherically symmetric solution of $\Delta U = 0$, the solution (18.4) is not spherically symmetric. *Prolate spheroidal* coordinates (x, y) have been used recently in discovering new exact stationary solutions (see § 18.5.). These coordinates are connected with Weyl's canonical coordinates (ϱ, z) by the relations (Zipoy (1966))

$$\varrho = \sigma(x^2-1)^{1/2}(1-y^2)^{1/2}, \qquad z = \sigma x y, \qquad \sigma = \text{const}$$
$$2\sigma x = r_+ + r_-, \qquad 2\sigma y = r_+ - r_-, \qquad {r_\pm}^2 \equiv \varrho^2 + (z \pm \sigma)^2, \tag{18.5}$$

$$\mathrm{d}\varrho^2 + \mathrm{d}z^2 = \sigma^2(x^2-y^2)\,[(x^2-1)^{-1}\,\mathrm{d}x^2 + (1-y^2)^{-1}\,\mathrm{d}y^2]. \tag{18.6}$$

The surfaces $x = \text{const}$ and $y = \text{const}$ are orthogonal families of, respectively, ellipsoids and hyperboloids. The special case of Weyl solutions with

$$\mathrm{e}^{2U} = \left(\frac{x-1}{x+1}\right)^{\delta} \tag{18.7}$$

has been investigated by Voorhees (1970) and Zipoy (1966). For $\delta = 1$, (18.7) represents the Schwarzschild solution; it is expressed in terms of Weyl's canonical coordinates (ϱ, z) by

$$U = \frac{1}{2}\ln\left(\frac{r_+ + r_- - 2m}{r_+ + r_- + 2m}\right), \qquad k = \frac{1}{2}\ln\left(\frac{(r_+ + r_-)^2 - 4m^2}{4r_+ r_-}\right),$$
$$r_\pm^2 = \varrho^2 + (z \pm m)^2 \tag{18.8}$$

(put $\sigma = m$ in (18.5)). For $\delta = 2$, (18.7) represents the Darmois (1927) solution with the potential

$$U = \ln\left(\frac{r_1 + r_2 - m}{r_1 + r_2 + m}\right), \qquad r_{1,2}^2 \equiv \varrho^2 + (z \pm m/2)^2 \tag{18.9}$$

(put $\sigma = m/2$ in (18.5)), m being again the mass parameter.

The linear *superposition* of N collinear particles with masses m_A and positions b_A on the z-axis yields the value (Israel and Khan (1964))

$$\lim_{\varrho\to 0} k = \frac{1}{4}\sum_{A,B=1}^{N} \ln\left[\frac{(r_{A+}r_{B-} + l_{A+}l_{B-})(r_{A-}r_{B+} + l_{A-}l_{B+})}{(r_{A-}r_{B-} + l_{A-}l_{B-})(r_{A+}r_{B+} + l_{A+}l_{B+})}\right],$$
$$l_{A\pm} \equiv z - b_A \pm m_A, \qquad r_{A\pm} \equiv |l_{A\pm}|, \tag{18.10}$$

for k on the axis $\varrho = 0$. The regularity condition $\lim_{\varrho \to 0} k = 0$ cannot be satisfied everywhere on the axis outside the sources: the singularities prevent the masses moving towards each other by virtue of their mutual gravitational attraction. In Einstein's theory, a two-body system in static equilibrium is impossible without such singularities — a very satisfactory feature of this non-linear theory.

Finally we remark that a Weyl solution is flat if the potential U is (up to an additive constant) equal to 0, $\ln \varrho$ or $\frac{1}{2} \ln \left(\sqrt{\varrho^2 + z^2} + z\right)$.

Ref.: Szekeres (1968).

18.2. Fields of uniformly accelerated particles

The part of flat space-time

$$ds^2 = dR^2 + R^2 d\Phi^2 + dZ^2 - dT^2 \tag{18.11}$$

for which $Z^2 > T^2$, $Z > 0$ (the hatched region in Fig. 18.1) can be transformed into Weyl's canonical form with the aid of the coordinate transformation

$$\begin{aligned} R &= (r - z)^{1/2}, & Z &= (r + z)^{1/2} \cosh t, \\ T &= (r + z)^{1/2} \sinh t, & \Phi &= \varphi. \end{aligned} \tag{18.12}$$

Then the metric coefficients U and k are given by

$$e^{2U} = r + z, \qquad e^{2k} = (r + z)/2r, \qquad r = (\varrho^2 + z^2)^{1/2} \tag{18.13}$$

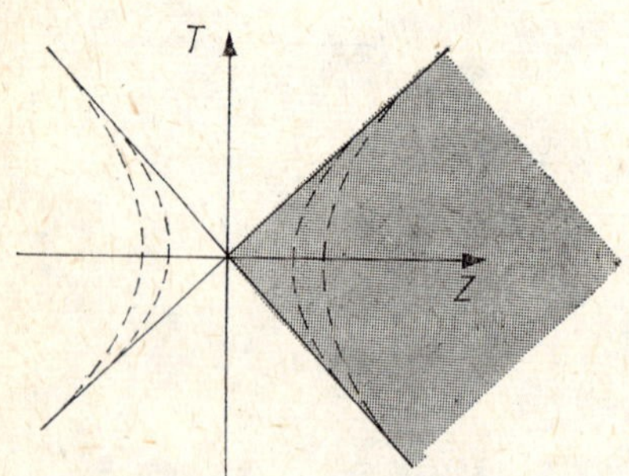

Fig. 18.1 Accelerated particles. The world lines of particles are dotted

The linear superposition of the potential $U = \frac{1}{2} \ln (r + z)$ and a potential U for a particle solution yields a gravitational field of particles moving with uniform acceleration along the axis of symmetry. Taking as the particle solution the field of two Curzon-Chazy particles, the result of the superposition is the metric (Bonnor and

Swaminarayan (1964))

$$ds^2 = e^{2\lambda} \cdot \frac{d\varrho^2 + dz^2}{2r} + e^{-2U} \frac{\varrho^2}{r+z} d\varphi^2 - e^{2U}(r+z)\, dt^2,$$

$$U = -\frac{m_1}{r_1} - \frac{m_2}{r_2} + \text{const}, \quad r_A{}^2 = \varrho^2 + (z - b_A)^2, \quad A = 1, 2,$$

$$2\lambda = -\frac{m_1{}^2\varrho^2}{r_1{}^4} - \frac{m_2{}^2\varrho^2}{r_2{}^4} + 4m_1m_2 \frac{\varrho^2 + (z-b_1)(z-b_2)}{r_1r_2(b_1-b_2)^2} \tag{18.14}$$

$$+ \frac{2m_1r}{b_1r_1} + \frac{2m_2r}{b_2r_2} + \text{const}.$$

This line element may be transformed with the inverse transformation of (18.12) into

$$ds^2 = e^{2\lambda}\, dR^2 + R^2 e^{-2U}\, d\Phi^2 + (Z^2 - T^2)^{-1} [e^{2\lambda} (Z\, dZ - T\, dT)^2$$
$$- e^{2U} (Z\, dT - T\, dZ)^2], \tag{18.15}$$

with U and λ from (18.14). The Killing vectors $\xi = T\,\partial_Z + Z\,\partial_T$ and $\eta = \partial_\Phi$ generate an Abelian group G_2; ξ is timelike for $Z^2 > T^2$, but spacelike for $Z^2 < T^2$.

In the region $Z^2 > T^2$, $Z < 0$, two additional (mirror) particles appear. The additive constants in U and k may be chosen so that two particles of positive masses are moving freely, whereas the other two particles are connected by stress singularities on the axis. The world lines of the point singularities are $Z = \pm(T^2 + 2b_A)^{1/2}$, $R = 0$. The sources do not remain in a finite region of space for all time. Consequently, the solution (18.14) cannot serve as a model describing the emission of gravitational waves from an isolated source.

18.3. The class of solutions $U = U(\omega)$ (Papapetrou's class)

The assumption $U = U(\omega)$ considerably simplifies the problem of finding stationary axisymmetric vacuum solutions. From the equations (17.38) one learns that

$$e^{4U} = -\omega^2 + C_1\omega + C_2, \tag{18.16}$$

and that there is a function $V(\omega)$ satisfying the potential equation in flat 3-space

$$\Delta V = 0, \qquad V = V(\omega) = \int \frac{d\omega}{-\omega^2 + C_1\omega + C_2} = \int e^{-4U}\, d\omega. \tag{18.17}$$

The right-hand side of (18.16) can assume positive values only if $s^2 = C_2 + C_1^2/4 > 0$. Given a solution V of $\Delta V = 0$, one obtains the associated complex potential $\Gamma = e^{2U} + i\omega$ from

$$\Gamma = s[\operatorname{sech}(sV) + i \tanh(sV)]. \tag{18.18}$$

The class $U = U(\omega)$ was discovered by Papapetrou (1953) and given originally in the equivalent form

$$\begin{aligned} e^{-2U} &= \alpha \cosh(\Omega_{,z}) + \beta \sinh(\Omega_{,z}), \\ A &= (\alpha^2 - \beta^2)^{1/2} \varrho \Omega_{,\varrho}, \qquad \Delta\Omega = 0 \end{aligned} \tag{18.19}$$

(α, β constants), in which the metric function A can be obtained from the potential function Ω by pure differentiation. In general, this class of solutions is Petrov type I. The solutions (18.19) are asymptotically well-behaved only if the mass term $\sim r^{-1}$ does not occur in e^{2U}. This is an obvious defect of these solutions; whereas the isolated sources have angular momentum, they are without mass.

One well-known solution of the Papapetrou class $U = U(\omega)$ is the NUT-solution (11.43). In terms of Weyl's canonical coordinates the NUT-solution is given by

$$\begin{aligned} e^{2U} &= \frac{(r_+ + r_-)^2 - 4(m^2 + l^2)}{(r_+ + r_- + 2m)^2 + 4l^2}, \quad e^{2k} = \frac{(r_+ + r_-)^2 - 4(m^2 + l^2)}{4r_+r_-}, \\ A &= \frac{l}{\sqrt{m^2 + l^2}}(r_+ - r_-), \qquad r_\pm{}^2 \equiv \varrho^2 + \left(z \pm \sqrt{m^2 + l^2}\right)^2 \end{aligned} \tag{18.20}$$

(Gautreau and Hoffman (1972)). For $l = 0$, these metric functions go over into the corresponding expressions (18.8) of the Schwarzschild solution. Many papers (e.g. Sackfield (1971), Bonnor (1969a)) are devoted to a physical interpretation of the NUT-solution, for which asymptotically Minkowskian coordinates do not exist.

Ref.: Bonnor and Swaminarayan (1965).

18.4. The class of solutions $S = S(A)$

The substitution (17.47) generates from the Papapetrou class $U = U(\omega)$ another class of stationary vacuum solutions characterized by the functional relationship

$$e^{4S} = \varrho^2 e^{-4U} = A^2 + C_1 A + C_2. \tag{18.21}$$

The function $\Psi = \int e^{-4S}\, dA$ satisfies the potential equation $\Delta\Psi = 0$. One has three distinct cases,

$$\varrho^2 e^{-4U} = \begin{cases} A^2 - 1, & h < 0, \\ A^2 + 1, & h > 0, \qquad h = 4C_2 - C_1^2, \\ A^2, & h = 0, \end{cases} \tag{18.22}$$

to consider. The subclass $h < 0$ is just Weyl's class: with the aid of a real linear transformation $t' = at + b$, $\varphi' = c\varphi + dt$, the function A can be made equal to zero. The subclass $h > 0$ was discovered by Lewis (1932) and is transformed to static solutions ($A = 0$) by *complex* linear transformations of the coordinates φ and t (Hoffman (1969a)).

The last subclass ($h = 0$), due to van Stockum (1937), has a line element of the simple form

$$ds^2 = \varrho^{-1/2}(d\varrho^2 + dz^2) - 2\varrho \, d\varphi \, dt + \varrho\Omega \, dt^2, \qquad \Delta\Omega = 0. \tag{18.23}$$

These solutions are of Petrov type *II*. They admit at most a group G_3. There exists a null Killing vector, ∂_φ (Chapter 21.).

The *cylindrically symmetric* stationary vacuum solutions form a subcase of the class $S = S(A)$ and are given in § 20.2.

Ref.: Levy (1968).

18.5. The Kerr solution and the Tomimatsu-Sato class

The Kerr and Tomimatsu-Sato solutions possibly describe exterior gravitational fields of stationarily rotating axisymmetric isolated sources. However, no satisfactory interior solutions are known.

The Kerr solution was found by a systematic study of algebraically special vacuum solutions. From its original form (28.49) (Kerr (1963a)) the metric can be transformed to Boyer-Lindquist coordinates (r, ϑ) which are related to Weyl's canonical coordinates (ϱ, z) and to prolate spheroidal coordinates (x, y) by

$$\varrho = \sqrt{r^2 - 2mr + a^2} \sin\vartheta, \qquad z = (r - m)\cos\vartheta, \tag{18.24a}$$

$$\sigma x = r - m, \qquad y = \cos\vartheta, \qquad \sigma = \text{const} \tag{18.24b}$$

(Boyer and Lindquist (1967)). In these coordinates, the Kerr solution reads

$$\begin{aligned} ds^2 = {} & \left(1 - \frac{2mr}{r^2 + a^2\cos^2\vartheta}\right)^{-1} \left[(r^2 - 2mr + a^2\cos^2\vartheta)\right. \\ & \times \left.\left(d\vartheta^2 + \frac{dr^2}{r^2 - 2mr + a^2}\right) + (r^2 - 2mr + a^2)\sin^2\vartheta \, d\varphi^2\right] \\ & - \left(1 - \frac{2mr}{r^2 + a^2\cos^2\vartheta}\right)\left(dt + \frac{2mar\sin^2\vartheta}{r^2 - 2mr + a^2\cos^2\vartheta} d\varphi\right)^2. \end{aligned} \tag{18.25}$$

Special cases are: $a = 0$ (Schwarzschild solution) and $a = m$ ("extreme" Kerr solution). The form (18.25) exhibits the existence of 2-surfaces orthogonal to the trajectories of the two Killing vectors ∂_t and ∂_φ. The Kerr solution is of Petrov type D and admits a non-trivial Killing tensor (§ 31.3.). For a discussion of the Kerr solution and for further references the reader is referred to Stewart and Walker (1973), Carter (1972), and Misner et al. (1973).

Carter (1968b), Demiański (1973), and Frolov (1974) generalized the Kerr solution (18.25) to include the Λ-term. Demiański and Newman (1966) constructed the solution (30.55) which contains the Kerr and NUT-metrics as special cases.

For the Kerr solution, the complex potential ξ defined in (17.44) takes the very simple form (Ernst (1968a))

$$\xi^{-1} = px - iqy, \qquad p^2 + q^2 = 1, \tag{18.26}$$

if one uses prolate spheroidal coordinates. The full metric can be obtained from ξ:

$$e^{2U} = \frac{p^2x^2 + q^2y^2 - 1}{(px+1)^2 + q^2y^2}, \qquad e^{2k} = \frac{p^2x^2 + q^2y^2 - 1}{p^2(x^2 - y^2)},$$
$$A = \frac{2mq}{p^2x^2 + q^2y^2 - 1} \cdot (1 - y^2)(px + 1) \tag{18.27}$$

($mq = a$, $mp = \sigma$). The potential ξ in (18.26) is a special solution of the differential equation (17.45),

$$(\xi\bar{\xi} - 1)\{[(x^2 - 1)\,\xi_{,x}]_{,x} + [(1 - y^2)\,\xi_{,y}]_{,y}\} = 2\bar{\xi}[(x^2 - 1)\,\xi_{,x}{}^2 + (1 - y^2)\,\xi_{,y}{}^2]. \tag{18.28}$$

Exploiting the very symmetric form of this equation in the variables x and y, Tomimatsu and Sato (1972, 1973) succeeded in constructing a series of new solutions (TS-solutions) containing an integer parameter δ. The potential ξ of these solutions is a quotient $\xi = \beta/\alpha$, α and β being polynomials in the coordinates x and y. For $\delta = 1, 2, 3$, these polynomials are ($p^2 + q^2 = 1$)

$\underline{\delta = 1}$: $\alpha = px - iqy, \quad \beta = 1.$ (Kerr solution)

$\underline{\delta = 2}$: $\alpha = p^2(x^4 - 1) - 2ipqxy(x^2 - y^2) - q^2(1 - y^4),$
$\beta = 2px(x^2 - 1) - 2iqy(1 - y^2).$

$\underline{\delta = 3}$: $\alpha = p(x^2 - 1)^3(x^3 + 3x) + iq(1 - y^2)^3(y^3 + 3y)$
$\quad - pq^2(x^2 - y^2)^3(x^3 + 3xy^2) - ip^2q(x^2 - y^2)^3(y^3 + 3x^2y),$
$\beta = p^2(x^2 - 1)^3(3x^2 + 1) - q^2(1 - y^2)^3(3y^2 + 1)$
$\quad - 12ipqxy(x^2 - y^2)(x^2 - 1)(1 - y^2).$ (18.29)

The constant σ in the relation (18.5) between the coordinates (ϱ, z) and (x, y), the angular momentum J, and the quadrupole moment Q are given by

$$\sigma = \frac{mp}{\delta}, \qquad J = m^2q, \qquad Q = m^3\left(\frac{\delta^2 - 1}{3\delta^2}\,p^2 + q^2\right), \tag{18.30}$$

m being the mass parameter of the TS-solutions. The TS-solutions are asymptotically flat. The corresponding Weyl solutions ($q = 0$) are the solutions (18.7) for integer parameter δ.

A class closely related to the TS-solutions was given by Chandrasekhar (1978).

Ref.: Gibbons and Russell-Clark (1973), Tanabe (1974), Ernst (1976b), Yamazaki (1977a, b), Hori (1978).

18.6. Other solutions

The function k in the Kerr metric (18.27) depends only on the coordinate $\eta = (x^2 - 1)/(1 - y^2)$. The fact that $k = k(\eta)$ is a property of the whole class of TS-solutions leads to a generalization of this class to arbitrary continuous parameter δ. Cosgrove

(1977, 1978) made the ansatz

$$(1 - \nu^2)\, k_{,\nu} = 2h, \quad 2\eta(1 + \eta)\, k_{,\eta} = l(\eta), \quad \eta = (x^2 - 1)/(1 - y^2),$$
$$\nu = y/x, \qquad h \text{ constant}, \tag{18.31}$$

for the function k obeying the fourth-order differential equation (17.46) and obtained the ordinary differential equation (see also Dale (1978))

$$\eta^2(1 + \eta)^2\, l''^2 = 4[\eta l'^2 - ll' - h^2]\,[-(1 + \eta)\, l' + l - \delta^2] \tag{18.32}$$

(δ constant). For $h = 0$, and with the boundary condition $l(\eta) = \delta^2 p^{-2} + O(\eta^{-1})$ as $\eta \to \infty$, the equation (18.32) defines asymptotically flat solutions which are regular on the axis outside a finite region. This class contains 3 parameters, $\sigma = mp/\delta$, q and δ ($p^2 + q^2 = 1$), related to mass m, angular momentum J and quadrupole moment Q according to (18.30), and goes over into the TS-class for integer values of δ.

For $h \neq 0$, Cosgrove (1978) obtained special solutions in closed form, e.g., using the parametrization $\delta^2 = n^2 + 2bn + 2b^2$, $h = b(n + b)$, one finds for $n = 1$ that

$$\begin{aligned} e^{2k} &= C^2(x^2 - 1)^{b^2}\,(1 - y^2)^{b^2} \cdot (x - y)^{-(2b+1)^2}\,(x + y)^{-1} \\ &\quad \times [p^2(x^2 - 1)^{2b+1} - q^2(1 - y^2)^{2b+1}], \\ \Gamma &= \left[\frac{(x - 1)\,(1 + y)}{(x + 1)\,(1 - y)}\right]^b \\ &\quad \times \frac{p^2(x^2 - 1)^{2b+1} - q^2(1 - y^2)^{2b+1} - 2ipq(x - 1)^{2b}\,(1 + y)^{2b}\,(x + y)}{p^2(x^2 - 1)^{2b}\,(x + 1)^2 + q^2(1 - y^2)^{2b}\,(1 - y)^2}. \end{aligned} \tag{18.33}$$

This solution is the $h \neq 0$ generalization of the Kerr metric and is, in general, not asymptotically flat.

Limiting procedures give rise to new solutions. There are several ways of performing the limit $q \to 1$ in the TS-solutions. Assuming that $p \cdot x$ remains finite, one always obtains the extreme Kerr metric ($m = a$) regardless of which value of δ we start with. Kinnersley and Kelley (1974) took the limit such that px^{2S-1} remains finite and introduced this quantity as a new coordinate. In this way they obtained a new class of solutions,

$$\xi = \frac{(1 - \Theta^2)^{S-1}\,[(1 + \Theta)^S - (1 - \Theta)^S] + i(r/m)^{2S-1}\,[(1 + \Theta)^{S-1} + (1 - \Theta)^{S-1}]}{(r/m)^{2S-1}\,[(1 + \Theta)^{S-1} - (1 - \Theta)^{S-1}] - i(1 - \Theta^2)^{S-1}\,[(1 + \Theta)^S + (1 - \Theta)^S]} \tag{18.34}$$

($\Theta = \cos\vartheta$; r, ϑ are spherical coordinates), for all real values of the parameter S. These solutions are not asymptotically flat, except for $S = 1$.

Other limiting procedures lead to the closed-form solution determined by the complex potential

$$\Gamma = r^{2+2c}\,\frac{\eta^c[p^2\eta^{2c+3} - q^2 + ipq(2c + 3)\,(1 + \eta)\,\eta^{c+1}]}{(1 + \eta)^{2c+2}\,[p^2\eta^{2c+1} - q^2 - ipq(2c + 1)\,(1 + \eta)\,\eta^c]},$$
$$\eta = (1 + \cos\vartheta)/(1 - \cos\vartheta), \tag{18.35}$$

(Cosgrove (1978)), and to a rotating version of the Chazy-Curzon solution (18.4) (Cosgrove (1977)).

The Ernst equation (17.37) may be separated in some coordinate systems. The ansatz (Ernst (1977))

$$\Gamma = r^k Y_k(\cos\vartheta) \tag{18.36}$$

(no summation) leads to the ordinary differential equation

$$\frac{1}{2}(Y_k + \overline{Y}_k)\left[(\sin\vartheta)^{-1}(\sin\vartheta\, Y_{k,\vartheta})_{,\vartheta} + k(k+1)\, Y_k\right] = k^2 Y_k^2 + (Y_{k,\vartheta})^2 \tag{18.37}$$

for the complex function $Y_k = Y_k(\cos\vartheta)$. For $k = 2$, the solution is the special case $c = 0$ of (18.35).

Marek (1968) found stationary axisymmetric vacuum solutions under the assumption

$$a_{,\zeta} = \varrho\, e^{-2U} h(\varrho), \qquad \sqrt{2}\,\partial_\zeta = \partial_\varrho - i\,\partial_z, \tag{18.38}$$

which corresponds to a separation of the Ernst equation (17.37) in Weyl's canonical coordinates,

$$\Gamma = e^{-mz} R_m(\varrho). \tag{18.39}$$

Non-trivial solutions of the types (18.36) and (18.39) do not have the desired asymptotic behaviour ($\Gamma \to 1$) for $r \to \infty$.

Ref.: Hoenselaers (1978a).

Chapter 19. Non-empty stationary axisymmetric solutions

In this chapter we continue the survey of stationary axisymmetric solutions. We will give all known Einstein-Maxwell and perfect fluid solutions except the stationary cylindrically symmetric fields (see § 20.2.) and some Einstein-Maxwell fields generated by special methods (see Chapter 30.).

19.1. Einstein-Maxwell fields

19.1.1. Electrostatic solutions

Substituting the magnetostatic potential ψ for the potential χ of an electrostatic solution one obtains its magnetostatic counterpart; we need not consider the magnetostatic case separately.

Assuming the functional dependence $e^{2U} = 1 - 2c\chi + \chi^2$ (cp. (16.63)) of the potentials U and χ plus axial symmetry, Weyl (1917) has given a class of solutions (*Weyl's electrovac class*). (The sources of these solutions have a constant specific charge density, see § 16.6.3.). For axisymmetric electrostatic Einstein-Maxwell fields the formula (17.32) yields

$$k_{,\zeta} = (\zeta + \bar{\zeta})\,(U_{,\zeta}{}^2 - e^{-2U}\chi_{,\zeta}{}^2), \tag{19.1}$$

so that for Weyl solutions the function k can easily be constructed from a solution Y of the potential equation $\Delta Y = 0$, according to

$$k_{,\zeta} = \pm(\zeta + \bar{\zeta})\,Y_{,\zeta}{}^2 \quad (c^2 \neq 1); \qquad k_{,\zeta} = 0 \quad (c^2 = 1). \tag{19.2}$$

The Weyl solutions with $c^2 = 1$ are just the Papapetrou-Majumdar solutions (see § 16.7.) with axial symmetry. In the sense of (17.48), Weyl's electrovac class corresponds to Papapetrou's class (§ 18.3.). The spherically symmetric solution in Weyl's electrovac class is the well-known Reissner-Nordström solution (13.21). In Weyl's canonical coordinates, it is given by

$$\begin{gathered} ds^2 = \frac{(R+m)^2}{r_+ r_-}(d\varrho^2 + dz^2) + \frac{(R+m)^2}{R^2 - d^2}\varrho^2\,d\varphi^2 - \frac{R^2 - d^2}{(R+m)^2}dt^2, \\ \chi = \frac{e}{R+m}, \quad R \equiv \frac{1}{2}(r_+ + r_-), \quad r_\pm{}^2 \equiv \varrho^2 + (z \pm d)^2, \quad d \equiv (m^2 - e^2)^{1/2} \end{gathered} \tag{19.3}$$

(see e.g. Gautreau et al. (1972)).

Herlt (1978) found a new class of solutions which differs from Weyl's class and contains a subclass of asymptotically flat solutions. From every real solution Ω of the linear differential equation

$$\Omega_{,\varrho\varrho} - \varrho^{-1}\Omega_{,\varrho} + \Omega_{,zz} = 0 \tag{19.4a}$$

the gravitational and electrostatic potentials are calculated according to the relations

$$\begin{aligned} e^{2U} &= (\Omega^{-1} + G)^2, \qquad \chi = \Omega^{-1} - G, \\ G &= \Omega_{,\varrho}[\varrho(\Omega_{,\varrho}{}^2 + \Omega_{,z}{}^2) - \Omega\Omega_{,\varrho}]^{-1}. \end{aligned} \tag{19.4b}$$

The metric function k can be obtained from (19.1). The solutions of *Herlt's class* (19.4) are in general of non-degenerate Petrov type *I*. This class was constructed by an application of generation techniques (§ 30.4.) to the complex van Stockum class (18.23). Some special solutions of the class (19.4) are known in the literature. For instance, to obtain the Bonnor (1966) solution (30.57) for a mass endowed with a magnetic dipole, the function Ω has to be chosen as

$$\Omega = A(e^{\alpha}r_1 + e^{-\alpha}r_2 + 2m) \tag{19.5}$$

(r_1, r_2 as in (18.9); A, α = real constants). A generalization of this solution to the exterior field of a rotating charged source will be given explicitly in § 30.5.2. The asymptotically flat vacuum solution in the class (19.4) is the Darmois solution (18.9). Recently Herlt (1979a) generalized his class to include asymptotically flat electrovac solutions which go over to the Schwarzschild solution for vanishing electrostatic field. An asymptotically flat three-parameter solution was given by □ Bonnor (1979b).

Ref.: Bonnor (1954), Tauber (1957), Misra (1962a), Perjés (1968), Misra and Pandey (1973).

19.1.2. A general class and its limits

A class of solutions to the Einstein-Maxwell equations (including the cosmological constant Λ) is given by the line element (Plebański and Demiański (1976), see also Debever (1971))

$$\begin{aligned} ds^2 = (1 - pq)^{-2}\Bigg[&\frac{p^2 + q^2}{X}\,dp^2 + \frac{p^2 + q^2}{Y}\,dq^2 + \frac{X}{p^2 + q^2}\,(d\tau + q^2\,d\sigma)^2 \\ &- \frac{Y}{p^2 + q^2}\,(d\tau - p^2\,d\sigma)^2\Bigg] \end{aligned} \tag{19.6}$$

$$X = X(p) = (-g^2 + \gamma - \Lambda/6) + 2lp - \varepsilon p^2 + 2mp^3 - (e^2 + \gamma + \Lambda/6)\,p^4,$$

$$Y = Y(q) = (e^2 + \gamma - \Lambda/6) - 2mq + \varepsilon q^2 - 2lq^3 + (g^2 - \gamma - \Lambda/6)\,q^4.$$

These solutions admit (at least) a group G_2 with commuting Killing vectors ∂_τ and ∂_σ. In certain ranges of the coordinates the vector ∂_τ is timelike, and σ may be interpreted as an azimuthal coordinate: the solutions are stationary and axisymmetric.

Besides the cosmological constant, (19.6) contains six real parameters: m and l are the mass and NUT-parameter (see (18.20)); γ and ε are related to the angular momentum per unit mass, a, and the acceleration b; e and g are the electric and magnetic charges. With respect to a complex null tetrad associated with the preferred null directions

$$Y^{-1}(q^2\,\partial_\tau - \partial_\sigma) \pm \partial_q, \tag{19.7}$$

the only non-vanishing tetrad component of the Weyl tensor is

$$\Psi_2 = -(m + il)\left(\frac{1 - pq}{q + ip}\right)^3 + (e^2 + g^2)\left(\frac{1 - qp}{q + ip}\right)^3 \cdot \frac{1 + pq}{q - ip}. \tag{19.8}$$

Consequently, the solutions are of Petrov type D (or O), see (4.25). The space-time is flat if $m = l = 0$, $e = g = 0$, $\Lambda = 0$. The complex potential Φ (formed with respect to the Killing vector ∂_τ) and the complex invariant of the non-null electromagnetic field are given by

$$\Phi = \frac{e + ig}{q + ip}, \qquad \varkappa_0 F^*_{ab} F^{*ab} = 8F^{-1}\Phi_{,a}\Phi^{,a} = -8(e + ig)^2 \left(\frac{1 - pq}{q + ip}\right)^4. \tag{19.9}$$

Some well-known classes of solutions can be obtained from the general metric (19.6) by appropriate *limiting procedures* (§ 30.6.). We give two examples of such "contractions" consisting of both a coordinate transformation and a simultaneous redefinition of constants.

Case I:

We consider the scale transformation

$$\begin{aligned} &p \to n^{-1}p, \qquad q \to n^{-1}q, \qquad \sigma \to n^3\sigma, \qquad \tau \to n\tau, \\ &m + il \to n^{-3}(m + il), \qquad e + ig \to n^{-2}(e + ig), \qquad \varepsilon \to n^{-2}\varepsilon, \\ &\gamma - \Lambda/6 \to n^{-4}\gamma, \qquad \Lambda \to \Lambda, \end{aligned} \tag{19.10}$$

perform the limit $n \to \infty$, and obtain from (19.6) the metric

$$\begin{aligned} &ds^2 = \frac{p^2 + q^2}{X}\,dp^2 + \frac{p^2 + q^2}{Y}\,dq^2 + \frac{X}{p^2 + q^2}(d\tau + q^2\,d\sigma)^2 - \frac{Y}{p^2 + q^2}(d\tau - p^2\,d\sigma)^2, \\ &X = \gamma - g^2 + 2lp - \varepsilon p^2 - \Lambda p^4/3, \\ &Y = \gamma + e^2 - 2mq + \varepsilon q^2 - \Lambda q^4/3. \end{aligned} \tag{19.11}$$

This metric has been studied in detail by Plebański (1975). It includes the family of Einstein-Maxwell fields found by Carter (1968a, b), who investigated space-times with an Abelian group of motions G_2 in which the Hamilton-Jacobi equation is separable. In addition to (19.11) (with $g = 0$), Carter gives other families which are also limiting cases of the general form (19.6). We remark that, for instance, the Bertotti-Robinson solution (10.16) can be obtained from (19.6) by a contraction procedure (Plebański (1975)). Besides Λ, the metric (19.11) contains five real parameters (ε can be reduced to $\varepsilon = 0, \pm 1$ by the scale transformation (19.10), n finite). For $\Lambda = 0$, $\varepsilon = 1$ we obtain from (19.11) the metric

$$\begin{aligned} \mathrm{d}s^2 &= (p^2+q^2)\left(\mathrm{d}\vartheta^2 + \frac{\mathrm{d}q^2}{Y}\right) + \frac{a^2 \sin^2\vartheta}{p^2+q^2}(\mathrm{d}\tau + q^2\,\mathrm{d}\sigma)^2 - \frac{Y}{p^2+q^2}(\mathrm{d}\tau - p^2\,\mathrm{d}\sigma)^2 \\ p &= l - a\cos\vartheta, \qquad a^2 = \gamma - g^2 + l^2, \\ Y &= a^2 + e^2 + g^2 - l^2 - 2mq + q^2, \end{aligned} \tag{19.12}$$

which in turn contains as an important special case ($l = g = 0$) the well-known Kerr-Newman solution, which will be dealt with separately in § 19.1.3. The coordinate transformation

$$q = r, \qquad p = -a\cos\vartheta, \qquad \sigma = -\varphi/a, \qquad \tau = t + a\varphi, \tag{19.13}$$

brings the metric (19.12) into the form (19.19) and the multiple principal null directions (19.7) take the form given in (19.18) (up to normalization).

Case II:

We make the substitution

$$\begin{aligned} &p \to n^{-1}p, \qquad q \to -nq^{-1}, \qquad \sigma \to n^{-1}\sigma, \qquad \tau \to n^{-1}\tau, \\ &l \to nl, \qquad \varepsilon \to n^2\varepsilon, \qquad m \to n^3 m, \qquad e + \mathrm{i}g \to n^2(e + \mathrm{i}g), \\ &\gamma \to \gamma + n^4 g^2, \qquad \Lambda \to \Lambda, \end{aligned} \tag{19.14}$$

in the metric (19.6). Taking the limit $n \to \infty$, we obtain the line element

$$\begin{aligned} \mathrm{d}s^2 &= (p+q)^{-2}\left(\frac{\mathrm{d}p^2}{X} + \frac{\mathrm{d}q^2}{Y} + X\,\mathrm{d}\sigma^2 - Y\,\mathrm{d}\tau^2\right), \\ X &= (\gamma - \Lambda/6) + 2lp - \varepsilon p^2 + 2mp^3 - (e^2+g^2)\,p^4, \\ Y &= -(\gamma + \Lambda/6) + 2lq + \varepsilon q^2 + 2mq^3 + (e^2+g^2)\,q^4. \end{aligned} \tag{19.15}$$

These space-times are generalizations of the degenerate static vacuum field denoted as the C-metric in Table 16.2. We now consider the case $\Lambda = 0$, $\varepsilon = 1$, $g = 0$. The parameter l can be put equal to zero by a simple coordinate transformation. Then the transformation ($\gamma = b^2$)

$$q = r^{-1} + bx, \qquad p = -bx, \qquad \sigma = z/b, \qquad \tau = u - \int Q^{-1}\,\mathrm{d}q, \tag{19.16}$$

turns the metric (19.15) into the form

$$\begin{aligned} \mathrm{d}s^2 &= r^2(G^{-1}\,\mathrm{d}x^2 + G\,\mathrm{d}z^2) - 2\,\mathrm{d}u\,\mathrm{d}r - 2br^2\,\mathrm{d}u\,\mathrm{d}x - 2H\,\mathrm{d}u^2, \\ G &= 1 - x^2 - 2mbx^3 - e^2b^2x^4, \\ 2H &= 1 - 2m/r + e^2/r^2 - b^2r^2G + brG' + 6mbx + 6e^2b^2x^2 - 4be^2x/r. \end{aligned} \tag{19.17}$$

This is the gravitational field generated by two uniformly accelerated charged mass points, the parameter b being the acceleration parameter (Kinnersley and Walker (1970), Walker and Kinnersley (1972), Plebański and Demiański (1976)). Note that

b may be put equal to zero in (19.17) but not in (19.15), because the transformation (19.16) involves b explicitly. Putting $b = 0$ in (19.17) we get the Reissner-Nordström solution in terms of the retarded time coordinate u (cp. (13.17) and Table 13.1). The singularity between the sources is removed in a more general metric (Ernst (1976a)), which contains an additional parameter describing an electric field which causes the uniform acceleration of the charges.

In Table 19.1, taken from Plebański and Demiański (1976), the special cases of the metric (19.6) which had been given earlier in the literature are listed.

Table 19.1 Stationary axisymmetric Einstein-Maxwell fields. Only the parameters marked by a cross (×) are different from zero in the corresponding solution

m	l	a	b	e	g	Λ	References
×	×	×	×	×	×	×	Debever (1971), Plebański and Demiański (1976)
×	×	×	×	×	×		Kinnersley (1969b)
×	×	×		×	×	×	Plebański (1975)
×	×	×		×	×		Demiański and Newman (1966)
×	×	×		×		×	Carter (1968a, b)
×	×		×	×		×	Carter (1968a, b)
×	×		×	×			Carter (1968a, b)
×	×			×			Brill (1964)
×		×		×			Newman et al. (1965)
				×	×		Bertotti (1959), Robinson (1959)
×				×			Reissner (1916), Nordström (1918)

Ref.: Weir and Kerr (1977), Debever (1969, 1976).

19.1.3. The Kerr-Newman solution

Newman et al. (1965) applied a complex substitution to the preferred complex null tetrad of the Kerr solution (see § 30.6.) and obtained the complex null tetrad

$$k^i = (1, 0, a/\Delta, (r^2 + a^2)/\Delta), \qquad l^i = \frac{1}{2}(r^2 + a^2\cos^2\vartheta)^{-1}(-\Delta, 0, a, r^2 + a^2),$$
$$m^i = 2^{-1/2}(r + \mathrm{i}a\cos\vartheta)^{-1}(0, \; 1, \mathrm{i}/\sin\vartheta, \mathrm{i}a\sin\vartheta), \quad \Delta \equiv r^2 + a^2 + e^2 - 2mr \tag{19.18}$$

($x^1 = r$, $x^2 = \vartheta$, $x^3 = \varphi$, $x^4 = t$), which represents a solution of the Einstein-Maxwell equations. The corresponding metric is given by

$$\mathrm{d}s^2 = (r^2 + a^2\cos^2\vartheta)\left(\frac{\mathrm{d}r^2}{\Delta} + \mathrm{d}\vartheta^2\right) + \sin^2\vartheta\left[r^2 + a^2 + \frac{a^2\sin^2\vartheta}{r^2 + a^2\cos^2\vartheta}(2mr - e^2)\right]\mathrm{d}\varphi^2$$
$$- \left(1 - \frac{2mr - e^2}{r^2 + a^2\cos^2\vartheta}\right)\mathrm{d}t^2 - \frac{2a\sin^2\vartheta}{r^2 + a^2\cos^2\vartheta}(2mr - e^2)\,\mathrm{d}t\,\mathrm{d}\varphi. \tag{19.19}$$

For $e = 0$, it goes over into the Kerr metric (18.25). The Kerr-Newman solution (19.19) possibly describes the exterior gravitational field of a rotating charged source and contains three real parameters: m (mass), e (charge), and a (angular momentum per unit mass). For a discussion of the Kerr-Newman geometry we refer to Misner et al. (1973), and Carter (1972).

The only non-vanishing components of the Weyl and Maxwell tensors, with respect to the null tetrad (19.18), are

$$\Psi_2 = -\frac{m(r + \mathrm{i}a \cos\vartheta) - e^2}{(r - \mathrm{i}a\cos\vartheta)^3\,(r + \mathrm{i}a\cos\vartheta)},$$
$$\sqrt{\frac{\varkappa_0}{2}}\,\Phi_1 = \frac{1}{2}\,\frac{e}{(r - \mathrm{i}a\cos\vartheta)^2}. \tag{19.20}$$

The Kerr-Newman solution is of Petrov type D; the electromagnetic field is non-null. The spin coefficients are listed in Bose (1975). In the metric (19.19) the complex scalar potentials Φ and $\mathscr{E}$, with respect to the Killing vector $\xi = \partial_t$, are given by

$$\Phi = \frac{e}{r - \mathrm{i}a\cos\vartheta}, \qquad \mathscr{E} = 1 - \frac{2m}{r - \mathrm{i}a\cos\vartheta}. \tag{19.21}$$

$\mathscr{E}$ is a linear function of Φ. As can be seen from Φ, the magnetic dipole moment vanishes if the source is either non-rotating ($a = 0$, Reissner-Nordström solution) or uncharged ($e = 0$, Kerr solution).

The Kerr-Newman solution with $|e| = m$ is conformastationary (see § 16.7.) and can be derived from the solution

$$V = \left(1 - \frac{m}{r - \mathrm{i}a\cos\vartheta}\right)^{-1} = 1 + \frac{m}{R}, \qquad R^2 = \varrho^2 + (z - \mathrm{i}a)^2,$$
$$\varrho = \sqrt{(r-m)^2 + a^2}\,\sin\vartheta, \qquad z = (r - m)\cos\vartheta, \tag{19.22}$$

of the potential equation $\Delta V = 0$, see (16.73). The *superposition* of N collinear Kerr-Newman sources for which the gravitational attraction and the electrostatic repulsion are balanced ($|e_A| = m_A$, $A = 1, \ldots, N$) and for which the spins are parallel or antiparallel along the axis of symmetry leads to the expression

$$V = 1 + \sum_{A=1}^{N} \frac{m_A}{R_A}, \qquad R_A{}^2 \equiv \varrho^2 + (z - l_A)^2. \tag{19.23}$$

The real and imaginary parts of l_A give respectively the position of the source A on the z-axis and the angular momentum per unit mass. The explicit metric for $N = 2$ has been obtained by Parker et al. (1973), and Kobiske and Parker (1974). The regularity condition (16.74) implies $\mathrm{Im}\,[m_1 + m_1\overline{m}_2(l_1 - \bar{l}_2)^{-1}] = 0$ $= \mathrm{Im}(m_1 + m_2)$. For two equal sources with oppositely directed spins ($m_1 = m_2$ real, $l_1 = -l_2$), no singularities along the axis between the particles occur. In general, the superposition (19.23) of Kerr-Newman solutions with $|e| = m$ gives rise to naked singularities (Hartle and Hawking (1972)); black hole metrics necessarily belong to the *static* subclass (l_A real) discovered by Papapetrou (1947) and Majumdar (1947).

19.1.4. Null fields and pure radiation fields

All the Einstein-Maxwell fields considered in this chapter have a non-null electromagnetic field. In the null case, and for pure radiation fields, the stationary axisymmetric solutions admit a null Killing vector (§ 21.4.) (Gürses (1977)).

19.2. Perfect fluid solutions

In all known rotating perfect fluid solutions the 4-velocity of the fluid obeys the circularity condition (17.11),

$$u^{[a}\xi_A^b\xi_B^{c]} = 0, \qquad u^a = (-H)^{-1/2}(\xi^a + \Omega\eta^a) = (-H)^{-1/2} S^A\xi_A^a,$$
$$H \equiv \lambda_{AB}S^AS^B, \qquad \lambda_{AB} \equiv \xi_A^a\xi_{Ba}, \qquad S^A \equiv (1, \Omega). \tag{19.24}$$

Then there exist 2-surfaces orthogonal to the group orbits (§ 17.2.) and the metric can be written as

$$ds^2 = e^{-2U}[e^{2k}(d\varrho^2 + dz^2) + W^2\, d\varphi^2] - e^{2U}(dt + A\, d\varphi)^2. \tag{19.25}$$

If the angular velocity Ω of the system (with respect to infinity) is a constant, then the system rotates rigidly, i.e. the fluid is in shearfree and expansionfree motion,

$$\sigma = 0 = \Theta \Leftrightarrow u_{(a;b)} + u_{(a}\dot{u}_{b)} = 0. \tag{19.26}$$

The general case ($\Omega_{,a} \neq 0$) is that of differential rotation.

19.2.1. The general dust metric

All stationary rotating dust metrics satisfying the condition (19.24) are known up to quadratures (Winicour (1975)). Following this paper, we substitute $R_{ab} = \varkappa_0\mu(u_au_b + \frac{1}{2}g_{ab})$ into the equations (17.24),

$$D^a(W^{-1}\lambda_{AB,a}) = \frac{1}{2} W^{-3}\lambda_{AB}\lambda^{CD,a}\lambda_{CD,a} + \varkappa_0\mu\, W^{-1}(\lambda_{AB} + 2W^2H^{-1}S_AS_B),$$
$$2W^2 \equiv -\lambda^{AB}\lambda_{AB}, \tag{19.27}$$

and take into account the relation

$$T^{ab}{}_{;b} = 0 \rightarrow S^AS^B\lambda_{AB,a} = 0. \tag{19.28}$$

The last equation, if rewritten in the form

$$H_{,a} = 2\eta\Omega_{,a}, \qquad \eta \equiv \lambda_{12} + \Omega\lambda_{22}, \tag{19.29}$$

shows that for differential rotation both η and Ω are functions of H. The arbitrary function $\eta = \eta(H)$ determines an auxiliary function $\beta = \beta(H)$ defined by

$$\beta_{,a} \equiv H_{,a}/(H\eta). \tag{19.30}$$

From the field equations (19.27) we obtain the relations

$$W^{-1}[(\beta W^2)_{,a} + (H/\eta)(\eta^2/H)_{,a}] = \varepsilon_{ab}\gamma^{,b} \to \Delta\gamma = 0, \tag{19.31}$$

$$D^a W_{,a} = 0, \tag{19.32}$$

and an expression for the mass density μ. Because of the 2-dimensional Laplace equation (19.32) for W we choose $W = \varrho$ in the metric (19.25). (The case $W = 1$ does not allow us to interpret φ as an azimuthal coordinate.) The remaining field equations obtained from (17.23)—(17.25) either determine the conformal factor e^{2k} in the line element (19.25) or are satisfied in consequence of (19.27), (19.28).

In order to construct dust metrics, we have to choose (i) a function $\eta = \eta(H)$ and (ii) an axisymmetric solution γ of the potential equation $\Delta\gamma = 0$ in flat 3-space. Once $\gamma = \gamma(\varrho, z)$ and $\eta = \eta(H)$ are given, one obtains $\beta = \beta(H)$ from (19.30), the function $\beta W^2 + 2\eta - \int \eta H^{-1}\, dH$ and consequently $H = H(\varrho, z)$ from (19.31), and finally the angular velocity Ω from (19.29). Hence, the scalar products of the Killing vectors are completely known,

$$\begin{aligned} \lambda_{11} &= \xi^a \xi_a = g_{44} = H^{-1}[(H - \eta\Omega)^2 - \Omega^2\varrho^2] = -e^{2U}, \\ \lambda_{12} &= \xi^a \eta_a = g_{34} = H^{-1}\Omega(\varrho^2 - \eta^2) + \eta = -A\, e^{2U}, \\ \lambda_{22} &= \eta^a \eta_a = g_{33} = -H^{-1}(\varrho^2 - \eta^2) = e^{-2U}\varrho^2 - A^2\, e^{2U}. \end{aligned} \tag{19.33}$$

The conformal factor e^{2k} can be determined by means of a line integral and the mass density is given by the formula

$$4\varkappa_0\mu = \eta^{-2}[H^2\varrho^{-2}(\eta^2/H)_{,a}(\eta^2/H)^{,a} - \varrho^2 H^{-2} H_{,a} H^{,a}]. \tag{19.34}$$

The general class of dust solutions described above contains some special cases given earlier in the literature (e.g. Maitra (1966)).

Ref.: Vishveshwara and Winicour (1977), Caporali (1978), Hoenselaers and Vishveshwara (1979).

19.2.2. Rigidly rotating dust and perfect fluid solutions

For rigid rotation ($\Omega = \text{const}$ in (19.24)), there exists a (timelike) Killing vector (linear combination of ξ and η with constant coefficients) parallel to the 4-velocity of the fluid. We can identify this Killing vector with $\xi = \partial_t$ (comoving system).

In a comoving coordinate system (19.25), the field equations (16.26)—(16.29) for the energy-momentum tensor of a rigidly rotating perfect fluid read

$$W_{,\zeta\bar{\zeta}} = \varkappa_0 p W\, e^{2k-2U}, \tag{19.35a}$$

$$\begin{aligned} &U_{,\zeta\bar{\zeta}} + (2W)^{-1}(U_{,\zeta}W_{,\bar{\zeta}} + U_{,\bar{\zeta}}W_{,\zeta}) + \frac{1}{2} W^{-2}\, e^{4U} A_{,\zeta} A_{,\bar{\zeta}} \\ &\quad = \varkappa_0(\mu + 3p)\, e^{2k-2U}/4, \end{aligned} \tag{19.35b}$$

$$A_{,\zeta\bar{\zeta}} - (2W)^{-1}(A_{,\zeta}W_{,\bar{\zeta}} + A_{,\bar{\zeta}}W_{,\zeta}) + 2(A_{,\zeta}U_{,\bar{\zeta}} + A_{,\bar{\zeta}}U_{,\zeta}) = 0, \tag{19.35c}$$

$$W,_{\zeta\bar{\zeta}} - 2W,_{\zeta}k,_{\zeta} + 2W(U,_{\zeta})^2 - (2W)^{-1}\,\mathrm{e}^{4U}(A,_{\zeta})^2 = 0, \tag{19.35d}$$

$$k,_{\zeta\bar{\zeta}} + U,_{\zeta}U,_{\bar{\zeta}} + (2W)^{-2}\,\mathrm{e}^{4U}A,_{\zeta}A,_{\bar{\zeta}} = \varkappa_0 p\,\mathrm{e}^{2k-2U}/2. \tag{19.35e}$$

The conservation law

$$T^a_{b;a} = 0 \Rightarrow p,_{\zeta} + (\mu + p)\,U,_{\zeta} = 0 \tag{19.36}$$

is a consequence of the field equations (19.35). Conversely, (19.35e) follows from (19.36) and the other equations in (19.35) (Trümper (1967)). In the metric (19.25), the *regularity condition* (17.2) takes the form

$$\lim_{\varrho\to 0}\,[\varrho^{-1}\,\mathrm{e}^{U-k}(\mathrm{e}^{-2U}W^2 - \mathrm{e}^{2U}A^2)^{1/2}] = 1. \tag{19.37}$$

For *dust metrics* ($p = 0$), we can put $\sqrt{2}W = \zeta + \bar{\zeta}$, $U = 0$, and the field equations are

$$\begin{aligned} &2(\zeta + \bar{\zeta})\,A,_{\zeta\bar{\zeta}} - A,_{\zeta} - A,_{\bar{\zeta}} = 0, \qquad 2k,_{\zeta} = -(\zeta+\bar{\zeta})^{-1}\,(A,_{\zeta})^2,\\ &\varkappa_0\mu = 2\varrho^{-2}A,_{\zeta}A,_{\bar{\zeta}}\,\mathrm{e}^{-2k} = 2u_{[a;b]}u^{a;b}. \end{aligned} \tag{19.38}$$

The mass density is positive definite. This class of dust solutions is due to van Stockum (1937). A special member of this class was analysed by Bonnor (1977) with the surprising result that, contrary to Newtonian mechanics, a density gradient in the z-direction (axis of rotation) occurs. (For $r \to \infty$, the mass density tends very rapidly to zero.)

All stationary rigidly rotating dust metrics with geodesic and/or shearfree eigenrays (§ 16.5.) are given in Lukacs (1974). They are axially symmetric and belong to the van Stockum class (19.38).

Comparing the equations (19.38) with the equations (18.2) for Weyl's vacuum solutions, we conclude that there exists a one-to-one correspondence between static axisymmetric vacuum solutions and stationary axisymmetric dust solutions. This statement is a special case of the more general

Theorem 19.1 (Ehlers (1962)). *To every static vacuum solution* (g_{ij}, ξ_k) *a (rigidly rotating) stationary dust solution* $(\hat{g}_{ij}, u_i, \mu)$ *can be assigned by*

$$\begin{aligned} &\hat{g}_{ij} = -\xi_m\xi^m g_{ij} + \xi_i\xi_j - u_iu_j, \qquad u_i\xi^i = -1, \qquad \mathrm{e}^{2U} = -\xi^i\xi_i,\\ &u_{[i;j]} = \varepsilon_{ijkl}U^{,k}\xi^l, \qquad u_{(i;j)} = 0, \qquad \varkappa_0\mu = 4U,_iU^{,i}. \end{aligned} \tag{19.39}$$

For *non-zero pressure,* we know one axially symmetric perfect fluid solution without a higher symmetry: the Wahlquist (1968) solution. It is given in terms of generalized oblate-spheroidal coordinates (ξ, η, φ) by

$$\begin{aligned} \mathrm{d}s^2 = r_0^2(\xi^2+\eta^2)\left[\frac{\mathrm{d}\xi^2}{(1-k^2\xi^2)\,h_1} + \frac{\mathrm{d}\eta^2}{(1+k^2\eta^2)\,h_2} + \frac{c^2h_1h_2}{h_1-h_2}\,\mathrm{d}\varphi^2\right]&\\ - \mathrm{e}^{2U}(\mathrm{d}t + A\,\mathrm{d}\varphi)^2,& \end{aligned} \tag{19.40}$$

$$e^{2U} = \frac{h_1 - h_2}{\xi^2 + \eta^2}, \qquad A = c r_0 \left(\frac{\xi^2 h_2 + \eta^2 h_1}{h_1 - h_2} - \eta_0^2 \right),$$

$$\begin{aligned} h_1 &= h_1(\xi) = 1 + \xi^2 - 2m r_0^{-1} \xi (1 - k^2 \xi^2)^{1/2} \\ &\quad + \xi b^{-2} [\xi - k^{-1} (1 - k^2 \xi^2)^{1/2} \operatorname{arc\,sin} (k\xi)] \\ h_2 &= h_2(\eta) = 1 - \eta^2 - 2a r_0^{-1} \eta (1 + k^2 \eta^2)^{1/2} \\ &\quad - \eta b^{-2} [\eta - k^{-1} (1 + k^2 \eta^2)^{1/2} \operatorname{arc\,sinh} (k\eta)] \\ h_2(\eta_0) &= 0, \qquad c^{-1} = \frac{1}{2} (1 + k^2 {\eta_0}^2)^{1/2} \, \mathrm{d}h_2/\mathrm{d}\eta|_{\eta=\eta_0}, \\ p &= \frac{1}{2} \mu_0 (1 - b^2 e^{2U}), \qquad \mu = \frac{1}{2} \mu_0 (3b^2 e^{2U} - 1), \qquad \varkappa_0 \mu_0 = 2k^2 (b r_0)^{-2}. \end{aligned} \tag{19.40}$$

The equation of state is

$$\mu + 3p = \mu_0 = \text{const}. \tag{19.41}$$

Wahlquist's solution (19.40) is of Petrov type D and, in general, admits a group G_2. The real constants r_0, k, b, m, a are arbitrary and the constants η_0, c are adjusted so that the solution behaves properly on the axis, i.e. that the metric satisfies the regularity condition (19.37). For $m = a = 0$, the solution (19.40) is a singularity-free interior solution for a rigidly rotating fluid body bounded by a finite surface of zero pressure. For $m = a = 0$, $\xi = r/r_0$, $\eta = \cos\vartheta$ and in the limit $r_0 \to 0$ we obtain from (19.40) a spherically symmetric static solution (§ 14.1.2.) with $\mu + 3p = \text{const}$.

Vaidya (1977) published the perfect fluid solution (with $\mu + 3p = 0$)

$$\begin{aligned} \mathrm{d}s^2 &= M^2[(1 - a^2 \sin^2\alpha/R^2)^{-1} \mathrm{d}\alpha^2 + \sin^2\alpha \, \mathrm{d}\beta^2] \\ &\quad -2(\mathrm{d}u + a \sin^2\alpha \, \mathrm{d}\beta) \, \mathrm{d}\tau + (1 + 2\hat{m}N)(\mathrm{d}u + a \sin^2\alpha \, \mathrm{d}\beta)^2, \\ M^2 &= (R^2 - a^2) \sin^2 (r/R) + a^2 \cos^2\alpha, \\ N &= R \sin (r/R) \cos (r/R)/M^2, \qquad u = \tau - r, \qquad \hat{m}, a, R \text{ constants}, \end{aligned} \tag{19.42}$$

and interpreted it as the Kerr metric in the cosmological background of the Einstein universe. The metric (19.42) is the limiting case $\mu_0 \to 0$ ($\mu_0 b^2$ finite and non-zero) of the Wahlquist solution (19.40). The coordinate transformation leading from this special Wahlquist solution to (19.42) reads (☐ Herlt and Herrmann (1980))

$$\begin{aligned} \eta &= \cos\alpha, \qquad \xi = \frac{R}{r_0} \sin\frac{r}{R}, \\ \mathrm{d}t &= \mathrm{d}u + (M^2 + a^2 \sin^2\alpha)(M^2 + a^2 \sin^2\alpha - 2\hat{m}NM^2)^{-1} \, \mathrm{d}r, \\ \mathrm{d}\varphi &= \mathrm{d}\beta - a(M^2 + a^2 \sin^2\alpha - 2\hat{m}NM^2)^{-1} \, \mathrm{d}r, \end{aligned} \tag{19.43}$$

and the parameters are related by

$$r_0^2 = a^2 \frac{R^2}{R^2 - a^2}, \qquad k^2 = \frac{a^2}{R^2 - a^2}, \qquad m = \frac{R^2}{R^2 - a^2} \hat{m}. \tag{19.44}$$

Other known stationary axisymmetric perfect fluid solutions belong to the locally rotationally symmetric space-times (§§ 11.4., 12.3.) or to the homogeneous space-times (§ 10.4.) or they are cylindrically symmetric (§ 20.2.). For instance, the field equations (19.35) can be solved under the condition $\omega = \omega(U)$, $\omega_{,\zeta} = -\mathrm{i}W^{-1}\,\mathrm{e}^{4U}A_{,\zeta}$ (Herlt (1972)). This class corresponds to Papapetrou's class (§ 18.3.) for vacuum fields and is locally rotationally symmetric.

The solutions due to Wahlquist and Herlt, and the interior Schwarzschild solution (14.14) are the only perfect fluid solutions (with rigid rotation) for which the Hamilton-Jacobi equation for null geodesics is separable (Bonanos (1976)).

No stationary axisymmetric perfect fluid solutions with non-zero pressure and *non-rigid* motion are known to us. For the problem of matching a perfect fluid source to the Kerr solution, see Roos (1976, 1977).

The only stationary, axisymmetric, and conformally flat perfect fluid solution is the interior Schwarzschild solution (14.14) (Collinson (1976a)).

Ref.: □ Hoenselaers and Vishveshwara (1979).

Chapter 20. Cylindrical symmetry

20.1. General remarks

Cylindrically symmetric fields are axisymmetric about an infinite axis (Killing vector $\boldsymbol{\eta} = \partial_\varphi$) and translationally symmetric along that axis (Killing vector $\boldsymbol{\zeta} = \partial_z$). The two Killing vectors $\boldsymbol{\eta}$ and $\boldsymbol{\zeta}$ are spacelike and generate an Abelian group G_2.

We assume the existence of 2-surfaces orthogonal to the group orbits and start with the line element

$$\mathrm{d}s^2 = \mathrm{e}^{-2U}[\gamma_{MN}\,\mathrm{d}x^M\,\mathrm{d}x^N + W^2\,\mathrm{d}\varphi^2] + \mathrm{e}^{2U}(\mathrm{d}z + A\,\mathrm{d}\varphi)^2, \tag{20.1}$$

which is independent of φ and z. (For fields regular on the axis, which satisfy

$$\zeta^a T_{a[b}\zeta_c\eta_{d]} = 0 = \eta^a T_{a[b}\eta_c\zeta_{d]}, \tag{20.2}$$

these 2-surfaces must exist, cp. § 17.1.). Note that plane symmetry (§§ 13.4., 13.5.) can be treated as a special case of cylindrical symmetry (put $A = 0$ and $\mathrm{e}^{-2U}W^2 = \mathrm{e}^{2U} \to Y^2$, $\varphi \to x$, $z \to y$ in (20.1) to obtain the metric (13.9)).

The cylindrically symmetric line element (20.1) can formally be obtained from the corresponding stationary axisymmetric line element (17.15) by the complex substitution

$$t \to \mathrm{i}z, \qquad z \to \mathrm{i}t, \qquad A \to \mathrm{i}A. \tag{20.3}$$

The indefinite 2-metric γ_{MN} in (20.1) can always be chosen in the form

$$\gamma_{MN}\,\mathrm{d}x^M\,\mathrm{d}x^N = \mathrm{e}^{2k}(\mathrm{d}\varrho^2 - \mathrm{d}t^2). \tag{20.4}$$

Throughout this chapter we assume that the cylindrically symmetric fields obey the condition (15.5), i.e.

$$T^\varrho_\varrho + T^t_t = 0 \qquad W_{,\varrho\varrho} - W_{,tt} = 0, \tag{20.5}$$

and that the gradient of W is a spacelike vector,

$$W_{,\varrho}{}^2 - W_{,t}{}^2 > 0, \tag{20.6}$$

so that we may choose $W = \varrho$. (Similar solutions can be obtained for $W_{,a}W^{,a} < 0$, $W = t$.) If the field admits the reflectional symmetry $\varphi \to -\varphi$, $z \to -z$, one can put $A = 0$ in (20.1) and the two commuting Killing vectors are hypersurface-orthogonal.

The *stationary* cylindrically symmetric fields are treated in § 20.2., the non-stationary fields will be the subject of the subsequent sections.

20.2. Stationary cylindrically symmetric fields

The stationary cylindrically symmetric fields are hypersurface-homogeneous space-times (Chapter 11.); they admit an Abelian group G_3 acting on timelike orbits T_3, the three Killing vectors being $\xi = \partial_t$, $\eta = \partial_\varphi$, $\zeta = \partial_z$. For solutions with G_3 on T_3 see also Chapter 11.

The stationary cylindrically symmetric solutions can be obtained either as special cases of stationary axisymmetric fields (with Killing vectors ξ, η) or of cylindrically symmetric fields (with Killing vectors η, ζ) by demanding a third symmetry.

Vacuum solutions

The general stationary cylindrically symmetric vacuum solution is given by

$$\begin{aligned} \mathrm{d}s^2 &= \mathrm{e}^{-2U}[\mathrm{e}^{2k}(\mathrm{d}\varrho^2 + \mathrm{d}z^2) + \varrho^2\,\mathrm{d}\varphi^2] - \mathrm{e}^{2U}(\mathrm{d}t + A\,\mathrm{d}\varphi)^2, \\ \mathrm{e}^{2U} &= \varrho(a_1\varrho^n + a_2\varrho^{-n}), \qquad n^2 a_1 a_2 = -C^2, \\ A &= \frac{C}{na_2}\cdot\frac{\varrho^n}{a_1\varrho^n + a_2\varrho^{-n}} + B, \qquad \mathrm{e}^{2k-2U} = \varrho^{(n^2-1)/2}. \end{aligned} \tag{20.7}$$

The complex constants have to be so chosen that the metric becomes real. The constant n can be either real or imaginary, the corresponding solutions belong to the Weyl class (§ 18.1.) or Lewis class (§ 18.4.), respectively. The cylindrically symmetric gravitational field inside a rotating hollow cylinder was treated by Davies and Caplan (1971) and Frehland (1972). It turns out that the regularity condition (17.2) on the axis of symmetry allows only flat space-time. The static vacuum solution (Levi-Civita (1917–19))

$$\mathrm{d}s^2 = \varrho^{-2m}[\varrho^{2m^2}(\mathrm{d}\varrho^2 + \mathrm{d}z^2) + \varrho^2\,\mathrm{d}\varphi^2] - \varrho^{2m}\,\mathrm{d}t^2 \tag{20.8}$$

is a special case of (20.7). This metric is flat for $m = 0, 1$, and Petrov type D for $m = 1/2, 2, -1$. (20.8) is identical with the Kasner solution (11.51).

Einstein-Maxwell fields

There are three different types of *static* cylindrically symmetric Einstein-Maxwell fields: (i) angular magnetic field (caused by an axial current), (ii) longitudinal magnetic field (caused by an angular current), and (iii) radial electric field (caused by an axial charge distribution). The corresponding metrics are

(i) $$\mathrm{d}s^2 = \varrho^{2m^2}G^2(\mathrm{d}\varrho^2 - \mathrm{d}t^2) + \varrho^2 G^2\,\mathrm{d}\varphi^2 + G^{-2}\,\mathrm{d}z^2, \tag{20.9a}$$

(ii) $$\mathrm{d}s^2 = \varrho^{2m^2}G^2(\mathrm{d}\varrho^2 - \mathrm{d}t^2) + G^{-2}\,\mathrm{d}\varphi^2 + \varrho^2 G^2\,\mathrm{d}z^2, \tag{20.9b}$$

(iii) $$\mathrm{d}s^2 = \varrho^{2m^2}G^2(\mathrm{d}\varrho^2 + \mathrm{d}z^2) + \varrho^2 G^2\,\mathrm{d}\varphi^2 - G^{-2}\,\mathrm{d}t^2 \tag{20.9c}$$

($G = C_1\varrho^m + C_2\varrho^{-m}$; C_1, C_2, m = real constants), cp. (11.60)–(11.62). Starting with the Rainich conditions (5.20), (5.21), Witten (1962) constructed and interpreted the solutions (20.9a, b). Superpositions of cylindrically symmetric Einstein-Maxwell fields were considered by Safko (1977).

The Melvin solution (Bonnor (1954), Melvin (1964))

$$\mathrm{d}s^2 = (1 + B_0^2\varrho^2/4)^2 (\mathrm{d}\varrho^2 + \mathrm{d}z^2 - \mathrm{d}t^2) + (1 + B_0^2\varrho^2/4)^{-2} \varrho^2 \mathrm{d}\varphi^2 \tag{20.10}$$

is a special case of (20.9b) and of (27.55). The gravitational field (20.10) originates in a "uniform" magnetic field B_0 along the z-axis. The Mukherji (1938) solution describing the gravitational field of a charged line-mass is contained in (20.9c).

Chitre et al. (1975) gave the particular static solution

$$\mathrm{d}s^2 = c^2\varrho^{-4/9} \exp (a^2\varrho^{2/3}) (\mathrm{d}\varrho^2 - \mathrm{d}t^2) + \varrho^{4/3} \mathrm{d}\varphi^2 + \varrho^{2/3} (\mathrm{d}z + a\varrho^{2/3} \mathrm{d}\varphi)^2 \tag{20.11}$$

(a, c constants) with $A = a\varrho^{2/3} \neq 0$ in (20.1), which admits a G_3 on T_3. The metric (Wilson (1968))

$$\mathrm{d}s^2 = (\mathrm{d}x^1)^2 + (1/x^1) (\mathrm{d}x^2)^2 + (\mathrm{d}x^3)^2 - \left(\mathrm{d}x^4 - \sqrt{3} \ln x^1 \, \mathrm{d}x^3\right)^2 \tag{20.12}$$

is a *stationary* Einstein-Maxwell field admitting an Abelian group G_3 on T_3.

Dust solutions

The stationary cylindrically symmetric dust solutions are contained in the general classes treated in § 19.2. The specialization of the Winicour solution to a cylindrically symmetric model can be found in Vishveshwara and Winicour (1977). Maitra (1966) gave a solution for non-rigidly rotating dust (with $\Theta = 0$, but $\sigma \neq 0$). The general stationary cylindrically symmetric dust solution was investigated by King (1974). Specializing the van Stockum class (19.38) to cylindrical symmetry we get the metric

$$\mathrm{d}s^2 = \mathrm{e}^{-a^2\varrho^2} (\mathrm{d}\varrho^2 + \mathrm{d}z^2) + \varrho^2 \mathrm{d}\varphi^2 - (\mathrm{d}t + a\varrho^2 \mathrm{d}\varphi)^2, \tag{20.13}$$

which can be matched to the vacuum solution (20.7), for certain values of the parameters. It is remarkable that the external field is static, though the dust rotates.

Perfect fluid solutions

Starting with the general cylindrically symmetric *static* line element

$$\mathrm{d}s^2 = \mathrm{d}\varrho^2 + \mathrm{e}^{2\beta(\varrho)} \mathrm{d}\varphi^2 + \mathrm{e}^{2\gamma(\varrho)} \mathrm{d}z^2 - \mathrm{e}^{2\delta(\varrho)} \mathrm{d}t^2 \tag{20.14}$$

and introducing the new functions $\zeta \equiv \beta + \gamma + \delta$, and $u \equiv \exp(3\delta)$, Evans (1977) reduced the perfect fluid problem to a second-order linear differential equation for u,

$$u'' - \zeta'u' + 3(\zeta'' + \mathrm{e}^{-2\zeta}) u = 0. \tag{20.15}$$

The arbitrary function ζ and the constants of integration should be assigned in agreement with the regularity conditions on the axis (e.g. $\mathrm{e}^\beta/\varrho \to 1$ as $\varrho \to 0$). The metric, and p and μ, are determined from (20.15) and from

$$\begin{aligned} &(\beta - \gamma)' = \mathrm{e}^{-\zeta}, \qquad \varkappa_0(5p - \mu) = 2(\mathrm{e}^\zeta)''/\mathrm{e}^\zeta, \\ &4\varkappa_0 p = (\zeta' + 3\delta')(\zeta' - \delta') - \mathrm{e}^{-2\zeta}. \end{aligned} \tag{20.16}$$

Evans (1977) investigated models with an equation of state $p = \text{const} \cdot \mu$. A very simple solution (with $5p = \mu$) results from the choice $e^{\zeta} = \varrho$. Teixeira et al. (1977a) studied static perfect fluids with $3p = \mu$. See also □ Bronnikov (1979).

Krasiński (1978) considered stationary cylindrically symmetric perfect fluids with *rigid rotation* (see § 19.2.2.) under the restriction that the Killing vector ζ parallel to the axis is proportional to the vorticity vector $\boldsymbol{\omega}$,

$$\zeta_{[a}\omega_{b]} = 0, \qquad \omega^a \equiv \varepsilon^{abcd}u_b u_{c;d}. \tag{20.17}$$

The solution is given by

$$\begin{aligned} ds^2 &= (E\varrho h)^{-1}\, dx^2 + b^{-1}Eh^{-2}\, dy^2 + b\varrho^{-1}h^3\, dz^2 - h^{-2}(dt + x\, dy)^2, \\ E &= bx^2 - F(x), \qquad F'' = 2\varkappa_0 f(x), \qquad \varrho = ah^5E^{-1}\exp\left(b\int E^{-1}x\, dx\right), \\ h &= u^{1/3} \end{aligned} \tag{20.18}$$

(a, b constants), where $u = u(x)$ satisfies the differential equation

$$Eu'' - (E' - bx)\, u' - \frac{3}{4}\, u(E'' - E'^2E^{-1} + bxE'E^{-1} - b) = 0. \tag{20.19}$$

The pressure and mass density can be calculated from

$$p = \int \varrho h'\, dx, \qquad \mu = \varrho h - p. \tag{20.20}$$

The function $f(x)$ in (20.18) is arbitrary; there is no limitation concerning the equation of state. For $f(x) = 1$, (20.19) can be solved in terms of hypergeometric functions (Krasiński (1974, 1975)).

20.3. Vacuum fields

With the aid of the complex substitution (20.3) one obtains the field equations for cylindrically symmetric vacuum fields,

$$2W(WU_{,M})^{;M} = e^{4U}A_{,M}A^{,M}, \qquad \left(\frac{1}{W}\, e^{4U}A_{,M}\right)^{;M} = 0, \tag{20.21}$$

from the corresponding equations (17.41) for stationary axisymmetric vacuum fields. The metric γ_{MN} now has the signature $(+-)$. With the choice (20.4) and $W = \varrho$, the field equations (20.21) read

$$\begin{aligned} U_{,\varrho\varrho} + \varrho^{-1}U_{,\varrho} - U_{,tt} &= \frac{1}{2}\, \varrho^{-2}e^{4U}(A_{,\varrho}{}^2 - A_{,t}{}^2), \\ A_{,\varrho\varrho} - \varrho^{-1}A_{,\varrho} - A_{,tt} &= 4(A_{,t}U_{,t} - A_{,\varrho}U_{,\varrho}) \end{aligned} \tag{20.22}$$

and k is determined by

$$\begin{aligned} k_{,\varrho} &= \varrho(U_{,\varrho}{}^2 + U_{,t}{}^2) + \frac{1}{4}\, \varrho^{-1}\, e^{4U}(A_{,t}{}^2 + A_{,\varrho}{}^2), \\ k_{,t} &= 2\varrho U_{,\varrho}U_{,t} + \frac{1}{2}\, \varrho^{-1}\, e^{4U}A_{,\varrho}A_{,t}. \end{aligned} \tag{20.23}$$

The integrability condition for (20.23) is satisfied by virtue of (20.22).

The substitution (20.3) takes (real) stationary axisymmetric solutions into (in general complex) cylindrically symmetric solutions. In some cases one can obtain real counterparts by analytic continuation of the parameters in the solution. No systematic investigation of this problem has yet been made.

A real solution of (20.21) which seems not to have a real stationary axisymmetric counterpart was given by Papapetrou (1966). It can be written in terms of a null coordinate u as

$$\begin{aligned} \gamma_{MN}\,\mathrm{d}x^M\,\mathrm{d}x^N &= \mathrm{e}^{2k}[-\mathrm{d}u\,\mathrm{d}v + G(u)\,v\,\mathrm{d}u^2], \qquad G(u) = \dot{F}(u)/F(u), \\ W &= (u - v^{-1})\,F(u), \qquad \mathrm{e}^{4U} = WF(u), \qquad A^2 = v^{-1}. \end{aligned} \tag{20.24}$$

This vacuum solution contains an arbitrary function $F(u)$. The conformal factor e^{2k} was calculated by Reuss (1968).

The cylindrically symmetric counterparts of the static axisymmetric solutions (Weyl's class, § 18.1.) are characterized by the existence of two hypersurface-normal spacelike Killing vectors. In this case one can put $A = 0$ and all solutions of this class can be obtained from the cylindrical wave equation and a line integral:

$$\begin{aligned} &\varrho^{-1}(\varrho U_{,\varrho})_{,\varrho} - U_{,tt} = 0, \\ &k = \int [\varrho(U_{,\varrho}{}^2 + U_{,t}{}^2)\,\mathrm{d}\varrho + 2\varrho U_{,\varrho} U_{,t}\,\mathrm{d}t]. \end{aligned} \tag{20.25}$$

The complex substitution $t \to \mathrm{i}z$, $z \to \mathrm{i}t$ leading from (18.2) (Weyl's class) to (20.25) was first mentioned by Beck (1925). The interpretation of the solutions of (20.25) as cylindrical gravitational waves is due to Einstein and Rosen (1937). Because of the cylindrical symmetry, the Einstein-Rosen waves do not describe the exterior fields of bounded radiating sources. The solutions of the class (20.25) are of Petrov type *I* or *II* (Petrov (1966), p. 447). The Einstein-Rosen waves were investigated by Marder (1958a, b). Superposing a cylindrical wave and flat space-time, Marder (1969) constructed a spherical-fronted pulse wave. For an analogous method for static Weyl solutions, see § 18.2.

Ref.: Stachel (1966), Kompaneets (1958), □ Cox and Kinnersley (1979).

20.4. Einstein-Maxwell fields and pure radiation fields

The metric (20.1) is very similar to that in the stationary axisymmetric case, and many results of Chapter 17. apply here. This justifies us in confining ourselves to hypersurface-orthogonal Killing vectors (reflectional symmetry $z \to -z$, $\varphi \to -\varphi$),

$$\mathrm{d}s^2 = \mathrm{e}^{-2U}[\mathrm{e}^{2k}(\mathrm{d}\varrho^2 - \mathrm{d}t^2) + W^2\,\mathrm{d}\varphi^2] + \mathrm{e}^{2U}\,\mathrm{d}z^2. \tag{20.26}$$

The existence of the Killing vector $\zeta = \partial_z$ enables us to introduce (real) scalar potentials Θ and η (cf. § 16.4.),

$$\sqrt{\frac{\varkappa_0}{2}}\,\zeta^a F^*_{ab} = \Theta_{,b} - \mathrm{i}\eta_{,b}. \tag{20.27}$$

If Θ and η are functions of ϱ and t, then the non-zero components of the field tensor F_{mn} with respect to the metric (20.26) are

$$\sqrt{\frac{\varkappa_0}{2}}\,F_{z\varrho} = \Theta_{,\varrho}, \quad \sqrt{\frac{\varkappa_0}{2}}\,F_{zt} = \Theta_{,t}, \quad \sqrt{\frac{\varkappa_0}{2}}\,F_{\varphi t} = W\,\mathrm{e}^{-2U}\eta_{,\varrho}, \qquad (20.28)$$
$$\sqrt{\frac{\varkappa_0}{2}}\,F_{\varphi\varrho} = W\,\mathrm{e}^{-2U}\eta_{,t}.$$

It follows from (20.28) that the condition $T^\varrho_\varrho + T^t_t = 0$ is satisfied. Choosing $W = \varrho$, we get the field equations from (17.26)—(17.33) by means of the substitution $t \to \mathrm{i}z$, $z \to \mathrm{i}t$:

$$U_{,\varrho\varrho} + \varrho^{-1}U_{,\varrho} - U_{,tt} = -\mathrm{e}^{-2U}(\Theta_{,\varrho}{}^2 - \Theta_{,t}{}^2 + \eta_{,\varrho}{}^2 - \eta_{,t}{}^2), \qquad (20.29\text{a})$$

$$\Theta_{,\varrho\varrho} + \varrho^{-1}\Theta_{,\varrho} - \Theta_{,tt} = 2(U_{,\varrho}\Theta_{,\varrho} - U_{,t}\Theta_{,t}), \qquad (20.29\text{b})$$

$$\eta_{,\varrho\varrho} + \varrho^{-1}\eta_{,\varrho} - \eta_{,tt} = 2(U_{,\varrho}\eta_{,\varrho} - U_{,t}\eta_{,t}), \qquad (20.29\text{c})$$

$$k_{,\varrho} = \varrho(U_{,\varrho}{}^2 + U_{,t}{}^2) + \varrho\,\mathrm{e}^{-2U}(\Theta_{,\varrho}{}^2 + \Theta_{,t}{}^2 + \eta_{,\varrho}{}^2 + \eta_{,t}{}^2), \qquad (20.29\text{d})$$

$$k_{,t} = 2\varrho U_{,\varrho}U_{,t} + 2\varrho\,\mathrm{e}^{-2U}(\Theta_{,\varrho}\Theta_{,t} + \eta_{,\varrho}\eta_{,t}), \qquad (20.29\text{e})$$

$$\eta_{,t}\Theta_{,\varrho} - \eta_{,\varrho}\Theta_{,t} = 0. \qquad (20.29\text{f})$$

From the last equation we conclude that either (i) one of the potentials Θ or η is a constant (and can be made zero by a gauge transformation) or (ii) the potentials are mutually dependent, $\eta = \eta(\Theta)$. The first case is contained as a special case in the equations (20.32) and (20.33) below. For $\eta = \eta(\Theta)$, the equations (20.29b) and (20.29c) are consistent only if

$$(\Theta_{,\varrho}{}^2 - \Theta_{,t}{}^2)\,\mathrm{d}^2\eta/\mathrm{d}\Theta^2 = 0. \qquad (20.30)$$

For electromagnetic *null fields*, the first factor in (20.30) vanishes,

$$F_{ab}F^{*ab} = 0 \Rightarrow \Theta_{,\varrho}{}^2 - \Theta_{,t}{}^2 = 0. \qquad (20.31)$$

In this case, the potentials Θ and η are arbitrary functions of $u = (t - \varrho)/\sqrt{2}$ or $v = (t + \varrho)/\sqrt{2}$. The general solutions of the field equations are (Misra and Radhakrishna (1962))

$$\mathrm{d}s^2 = \varrho^{-1/2}\exp\left[-2^{3/2}\int(\dot{\Theta}^2 + \dot{\eta}^2)\,\mathrm{d}u\right](\mathrm{d}\varrho^2 - \mathrm{d}t^2) + \varrho(\mathrm{d}\varphi^2 + \mathrm{d}z^2) \qquad (20.32)$$

for $\Theta = \Theta(u)$, $\eta = \eta(u)$ (dot denotes derivative with respect to u), and a similar solution for $\Theta = \Theta(v)$, $\eta = \eta(v)$.

For electromagnetic *non-null fields*, the relation (20.30) demands that η is a linear function of Θ,

$$\Theta = \Psi\cos\alpha, \quad \eta = \Psi\sin\alpha, \quad \alpha = \text{const}. \qquad (20.33)$$

In terms of the potential Ψ introduced in (20.33), the field equations (20.29a—c) read

$$U_{,\varrho\varrho} + \varrho^{-1}U_{,\varrho} - U_{,tt} = \mathrm{e}^{-2U}(\Psi_{,t}{}^2 - \Psi_{,\varrho}{}^2), \qquad (20.34)$$
$$\Psi_{,\varrho\varrho} + \varrho^{-1}\Psi_{,\varrho} - \Psi_{,tt} = 2(U_{,\varrho}\Psi_{,\varrho} - U_{,t}\Psi_{,t}).$$

Up to the complex replacement $t \to \mathrm{i}z$, $z \to \mathrm{i}t$, these equations have exactly the same structure as the equations (17.38) for stationary axisymmetric vacuum fields.

Theorem 20.1. *From a stationary axisymmetric vacuum solution* (U, ω) *(in Weyl's canonical coordinates) one obtains a corresponding cylindrically symmetric Einstein-Maxwell field* (U, Ψ) *by the substitution*

$$t \to \mathrm{i}z, \qquad z \to \mathrm{i}t, \qquad 2U \to U, \qquad \omega \to \Psi. \tag{20.35}$$

For instance, the solution (Radhakrishna (1963))

$$W = \varrho, \qquad \Psi = at, \qquad \mathrm{e}^U = ab^{-1}\varrho \cosh(b \ln \varrho), \tag{20.36}$$

belongs to the class related to the Lewis class (§ 18.4.) by (20.35).

The generation method outlined in § 30.5. applies also when two spacelike Killing vectors exist. Harrison (1965), Misra (1962b, 1966), and Singatullin (1973) generated cylindrically symmetric wave solutions of the Einstein-Maxwell equations from vacuum solutions.

The ansatz (Harrison (1965))

$$U = U(\eta), \qquad \Psi = \ln \varrho - \int Q(\eta)\, \mathrm{d}\eta, \qquad \cosh \eta = t/\varrho, \tag{20.37}$$

reduces the equations (20.34) to a system of two ordinary first order differential equations,

$$\begin{aligned} &Q' = 2U'(Q + \cosh \eta), \\ &Q^2 + \mathrm{e}^{2U}(U'^2 + C) = 1, \qquad C = \text{const}. \end{aligned} \tag{20.38}$$

Harrison (1965) discussed the special solution $C = 0$, $U = 0$, $Q = -1$.

Pure radiation fields obey, in the metric (20.26) ($W = \varrho$), the equations (Krishna Rao (1964))

$$\begin{aligned} &U_{,\varrho\varrho} + \varrho^{-1}U_{,\varrho} - U_{,tt} = 0, \\ &k_{,\varrho} + k_{,t} = \varrho(U_{,\varrho} + U_{,t})^2. \end{aligned} \tag{20.39}$$

The other field equations are then automatically satisfied. Hence for every Einstein-Rosen wave (U_0, k_0) one can generate a pure radiation field metric (U, k) by

$$\begin{aligned} &U = U_0, \qquad k = k_0 + f(u), \qquad \sqrt{2}\, u = t - \varrho, \\ &\text{or} \\ &U = U_0, \qquad k = k_0 + g(v), \qquad \sqrt{2}\, v = t + \varrho. \end{aligned} \tag{20.40}$$

This result can be generalized to the metric (20.1) with $A \neq 0$ (Krishna Rao (1970)). The pure radiation field

$$\mathrm{d}s^2 = \mathrm{e}^{2k(u)}(\mathrm{d}\varrho^2 - \mathrm{d}t^2) + \varrho^2\, \mathrm{d}\varphi^2 + \mathrm{d}z^2 \tag{20.41}$$

generated from flat space-time ($U_0 = k_0 = 0$) has a Weyl tensor of type N (Krishna Rao (1963)). Looking at the solution (20.32) we find that a field generated by (20.40) satisfies the Maxwell equations only if $\mathrm{e}^{2U} = \varrho$.

Ref.: Kompaneets (1959), Singh et al. (1965), Lal and Prasad (1969), Roy and Tripathi (1972).

20.5. Perfect fluid solutions

For the special equation of state $p = \mu$ and for irrotational 4-velocity the field equations are given by (15.10) and (15.11). Letelier (1975) applied these relations to radially moving, cylindrically symmetric perfect fluids with $p = \mu$. It is convenient to choose the coordinate system (20.26). When the motion of the fluid is purely radial, i.e. $\sigma = \sigma(\varrho, t)$, the condition $T^\varrho_\varrho + T^t_t = 0$ is satisfied and the field equation $R^\varphi_\varphi + R^z_z = 0$ can easily be integrated as

$$W = f(t - \varrho) + g(t + \varrho). \tag{20.42}$$

The equations $\sigma_{,a}{}^{;a} = 0$ and $R^\varphi_\varphi - R^z_z = 0$ lead to two identical linear differential equations for σ and U,

$$(W\sigma_{,t})_{,t} - (W\sigma_{,\varrho})_{,\varrho} = 0, \qquad (WU_{,t})_{,t} - (WU_{,\varrho})_{,\varrho} = 0 \tag{20.43}$$

and the remaining field equations merely determine k in the metric (20.26) in terms of a line integral. Theorem 15.1 can be used to generate perfect fluid solutions with the equation of state $p = \mu$ from vacuum solutions.

□ Allnutt (1980) recently found a solution (of Petrov type D) which reads

$$\begin{aligned} ds^2 &= \frac{dx^2}{1 - Cx^2} + x^2 t(1 + t^2)^{1/2} \\ &\quad \times \left[\left(\frac{1 + t^2}{t^2}\right)^{n/2} dy^2 + \left(\frac{1 + t^2}{t^2}\right)^{-n/2} dz^2\right] - \frac{x^2}{1 + t^2}\, dt^2, \\ u_m &= \left(0, 0, 0, x(1 + t^2)^{-1/2}\right), \\ \varkappa_0\mu - 6C &= \varkappa_0 p + 6C = \frac{1 - n^2}{4x^2t^2(1 + t^2)}, \end{aligned} \tag{20.44}$$

C, n real constants.

Ref.: Letelier and Tabensky (1975a, b), Bray (1971).

Chapter 21. Groups on null orbits. Plane waves

21.1. Introduction

In classifying space-times according to the group orbits in Chapters 9.—20., we postponed the case of null orbits. Groups of motions with null surfaces of transitivity will be the subject of this chapter.

A null surface N_m is geometrically characterized by the existence of a unique null direction $\boldsymbol{k}$ tangent to N_m at any point of N_m. The null congruence $\boldsymbol{k}$ is restricted by the existence of a group of motions acting transitively on N_m, and all space-times considered in this chapter are subject to the condition $R_{ab}k^a k^b = 0$.

The groups G_r, $r \geqq 4$, on N_3 have at least one subgroup G_3 (Theorems 8.5, 8.6 and Petrov (1966), p. 179), which may act on N_3, N_2, S_2. (A G_4 on N_3 cannot contain G_3 on T_2 since the N_3 contains no T_2.) For G_3 on S_2, one obtains special cases of the metric (13.3) admitting either a group G_3 on N_3 or a null Killing vector (see Barnes (1973b)). For G_3 on N_2, the metric and the corresponding Killing vectors can be written in the form (Petrov (1966), p. 154, Barnes (1979))

$$\begin{aligned} ds^2 &= A(x^3, x^4)\left(-2dx^1\, dx^3 + (dx^2)^2\right) + B(x^3, x^4)\, (dx^3)^2 \pm (dx^4)^2, \\ \xi_1 &= \partial_1, \qquad \xi_2 = \partial_2, \qquad \xi_3 = x^2\partial_1 + x^3\partial_2. \end{aligned} \tag{21.1}$$

The group is of Bianchi type II; the Killing vector $\xi_1 = \partial_1$ is null.

Thus we need only consider here the groups G_3 on N_3 (§ 21.2.), G_2 on N_2 (§ 21.3.), and G_1 on N_1 (§ 21.4.). As we study the case of null Killing vectors (G_1 on N_1) separately, we can also restrict ourselves to groups G_3 on N_3 and G_2 on N_2 generated by *non-null* Killing vectors. It will be shown that in these cases, independent of the group structure, there is always a non-expanding, non-twisting, and shearfree null congruence $\boldsymbol{k}$.

For the types of energy-momentum tensor considered in this book (see § 5.2.), all space-times admitting a G_3 on N_3 or a G_2 on N_2 (generated by non-null Killing vectors) are algebraically special and belong to Kundt's class (Chapter 27.).

21.2. Groups G_3 on N_3

In this section we study space-times V_4 for which the orbits of a group G_3 are null hypersurfaces N_3. They can be parametrized by u (u being constant in each N_3). The null vector $k_a = -u_{,a}$ is orthogonal to all vectors tangent to N_3, and therefore it is orthogonal to the three independent Killing vectors ξ_A ($A = 1, \ldots, 3$);

$$\xi^a_A k_a = 0, \qquad \xi_{A(a;b)} = 0, \qquad k_a = -u_{,a}, \qquad k^a k_a = 0, \qquad k^a = c^A(x)\,\xi^a_A. \tag{21.2}$$

From the relations (21.2) it follows immediately that the null congruence $\boldsymbol{k}$ and tensors obtained from it by covariant differentiation have zero Lie derivatives with respect to ξ_A, e.g.

$$k^a_{;b}\xi^b_A - k^b\xi^a_{A;b} = 0, \qquad k^c_{;ca}k^a = 0. \tag{21.3}$$

Now we use the equation (6.32),

$$R_{ab}k^a k^b = k^b{}_{;ab}k^a = -k_{a;b}k^{a;b} = -2(\sigma\bar{\sigma} + \Theta^2) \leqq 0, \tag{21.4}$$

where σ and Θ denote the shear and the expansion, respectively. By continuity, the energy conditions (5.18)

$$T_{ab}u^a u^b \geqq 0, \qquad T_{ab}T^a{}_c u^b u^c \leqq 0, \qquad u^a u_a < 0 \tag{21.5}$$

must still be true if we replace the timelike vector $\boldsymbol{u}$ by the null vector $\boldsymbol{k}$:

$$R_{ab}k^a k^b \geqq 0, \qquad T_{ab}T^a{}_c k^b k^c \leqq 0, \qquad k^a k_a = 0. \tag{21.6}$$

The comparison of (21.4) with (21.6) leads to

Theorem 21.1. *For space-times satisfying* (21.2) *and* (21.6), *the non-twisting* (*and geodesic*) *null congruence* $\boldsymbol{k}$ *is non-expanding and shearfree,*

$$k_{a;b} = 2k_{(a}p_{b)}, \qquad p_a k^a = 0. \tag{21.7}$$

The conditions (21.6) restrict the Ricci tensor so that $\boldsymbol{k}$ is a Ricci eigenvector,

$$R_{ab}k^a k^b = 0 = R_{ab}m^a k^b \Leftrightarrow k_{[c}R_{a]b}k^b = 0. \tag{21.8}$$

Of the types of energy-momentum tensor considered in this book, only vacuum fields, and Einstein-Maxwell and pure radiation fields for which $\boldsymbol{k}$ is an eigendirection of R_{ab}, are compatible with a group of motions acting on N_3. According to Theorem 7.1, such space-times are algebraically special, $\boldsymbol{k}$ being the repeated principal null direction of the Weyl tensor. Perfect fluids (with $\mu + p \neq 0$) are excluded by the condition $R_{ab}k^a k^b = 0$.

Using the Newman-Penrose formalism (Chapter 7.), we now inspect the Einstein-Maxwell equations. For the spin coefficients and tetrad components we have

$$\begin{aligned} &\varrho = \sigma = \varkappa = 0, \qquad \varepsilon + \bar{\varepsilon} = 0, \qquad \tau = \bar{\alpha} + \beta; \\ &\Phi_{00} = \Phi_{01} = \Phi_{02} = 0, \qquad \Psi_0 = \Psi_1 = 0, \qquad R = 0 \end{aligned} \tag{21.9}$$

($\boldsymbol{k}$ is a gradient!). We choose the null tetrad so that its Lie derivatives with respect to ξ_A vanish (invariant basis in N_3). The tetrad vectors $\boldsymbol{k}$, $\boldsymbol{m}$, $\overline{\boldsymbol{m}}$ are linear combinations (with non-constant coefficients) of the Killing vectors ξ_A. Therefore, the intrinsic derivatives D, δ, $\bar{\delta}$ applied to spin coefficients and tetrad components of the field tensors give zero. With these simplifications and (21.9), the Newman-Penrose equations (7.30), (7.43), and the Maxwell equation (7.48) read

$$\tau\varepsilon = 0, \qquad \tau\beta = 0, \qquad \tau\Phi_1 = 0. \tag{21.10}$$

If we assume $\tau \neq 0$, it follows that $\varepsilon = \beta = \Phi_1 = 0$, and the equations (7.39), (7.44),

$$\begin{aligned} \Psi_2 &= \alpha\bar{\alpha} + \beta\bar{\beta} - 2\alpha\beta + \Phi_{11} = \tau\bar{\tau}, \\ \Psi_2 &= \tau(\bar{\beta} - \alpha - \bar{\tau}) = -2\tau\bar{\tau}, \end{aligned} \tag{21.11}$$

give contradictory results. Hence $\tau = 0$ must hold;

$$\tau \equiv -k_{a;b}m^a l^b = p_a m^a = 0, \qquad k_{a;b} = A(u)\, k_a k_b \tag{21.12}$$

(note $\pounds_\xi k_{a;b} = 0$), i.e. there exists a covariantly constant null vector parallel to $\boldsymbol{k}$ (§ 6.1.). For pure radiation fields ($\Phi_{11} = 0$), the same conclusion can be drawn.

Theorem 21.2. *Vacuum fields, Einstein-Maxwell and pure radiation fields admitting a group of motions acting transitively on N_3 are plane waves (see § 21.5.).*

Ref.: Lauten and Ray (1975, 1977), □ Kramer (1980).

21.3. Groups G_2 on N_2

Space-times admitting a null Killing vector will be treated separately in the next section, so we suppose that the group G_2 acting on 2-dimensional null surfaces N_2 does not contain a null Killing vector.

By supposition, the two spacelike Killing vectors $\boldsymbol{\xi}$ and $\boldsymbol{\eta}$ span a null surface N_2, i.e.

$$\xi_{[a}\eta_{b]}\xi^a\eta^b = 0. \tag{21.13}$$

At any point of N_2 we have a unique null direction tangent to N_2:

$$\boldsymbol{k} = \boldsymbol{\xi} - \Omega\boldsymbol{\eta}, \qquad \Omega \equiv (\eta^c\eta_c)^{-1}\,\xi^b\eta_b \tag{21.14}$$

(compare the similar expression (17.18) in the case of stationary axisymmetric fields). The null vector field $\boldsymbol{k}$ is orthogonal to the two Killing vectors,

$$k^a\xi_a = 0 = k^a\eta_a, \tag{21.15}$$

and has the properties

$$\begin{aligned} &k^a{}_{;ab}k^b = 0, \qquad k_{a;b}k^b = 0, \qquad \omega = \mathrm{i}k_{[a;b]}\overline{m}^a m^b = 0, \\ &\sqrt{2}\, m^a = (\eta^c\eta_c)^{-1/2}\,\eta^a + \mathrm{i}q^a, \quad q^a\eta_a = 0, \quad q^a k_a = 0, \quad q^a q_a = 1, \end{aligned} \tag{21.16}$$

which can be verified with the aid of the Killing equations, the commutator relation for $\boldsymbol{\xi}$ and $\boldsymbol{\eta}$, and the formula (21.13). The equations (21.16) hold for both the group structures G_2I and G_2II (§ 8.2.).

We insert the information (21.16) on $\boldsymbol{k}$ into the equation (6.32) and conclude, again from (21.4) together with (21.6), that the null vector field (21.14) is non-expanding and shearfree, $\Theta = \sigma = 0$. (The function Ω in (21.14) obeys the relations

$\Omega_{,a}k^a = 0 = \Omega_{,a}m^a$.) From the conditions (21.6) it follows that $\boldsymbol{k}$ is a Ricci eigenvector. So we can conclude again from Theorem 7.1 that if a vacuum, Einstein-Maxwell, or pure radiation field admits a group of motions G_2 on N_2, then it is algebraically special.

Theorem 21.3. *In space-times admitting a group G_2 on N_2 there is a non-expanding, non-twisting, and shearfree null congruence and the metric can be transformed into* (*cf.* § 27.2.)

$$\begin{gathered} ds^2 = 2P^{-2}\,d\zeta\,d\bar\zeta - 2\,du(dv + W\,d\zeta + \overline{W}\,d\bar\zeta + H\,du), \\ P, H \text{ real}, \qquad W \text{ complex}, \qquad P_{,v} = 0, \qquad W_{,vv} = 0. \end{gathered} \tag{21.17}$$

The gauge transformations

$$u' = h(u), \qquad v' = v/h_{,u} + g(\zeta, \bar\zeta, u), \qquad \zeta' = \zeta'(\zeta, u), \tag{21.18}$$

preserving the form of the line element (21.17) can be used to bring the Killing vectors $\boldsymbol{\xi}$ and $\boldsymbol{\eta}$ satisfying (21.13) into the simple form

$$\boldsymbol{\eta} = \partial_x, \qquad \boldsymbol{\xi} = u\,\partial_x + S(u, y)\,\partial_v, \qquad \sqrt{2}\,\zeta = x + iy. \tag{21.19}$$

These Killing vectors necessarily commute. The metric (21.18) is independent of x, and the rest of the Killing equations imply $W_{,v} \neq 0$.

Vacuum fields admitting a G_2 on N_2 are given by □ Bampi and Cianci (1979).

21.4. Null Killing vectors (G_1 on N_1)

The important relation (6.32),

$$\Theta_{,a}k^a - \omega^2 + \Theta^2 + \sigma\bar\sigma = -\frac{1}{2}\,R_{ab}k^ak^b, \tag{21.20}$$

applied to a null Killing vector $\boldsymbol{k}$ yields

$$R_{ab}k^ak^b = 2\omega^2. \tag{21.21}$$

Obviously, for vacuum solutions, and for Einstein-Maxwell and pure radiation fields for which $\boldsymbol{k}$ is an eigenvector of R_{ab}, the null Killing vector is always twistfree ($\omega = 0$). We know of no Einstein-Maxwell fields which admit a twisting null Killing vector. Perfect fluid solutions cannot admit a non-twisting null Killing vector, except if $\mu + p = 0$. The algebraically special perfect fluid solutions with (twisting) null Killing vectors are treated in Wainwright (1970); they admit an Abelian group G_2. The Gödel universe (10.25) admits the two twisting null Killing vectors $\partial_y \pm \partial_t$.

It is the aim of this section to construct the space-times admitting a *non-twisting* null Killing vector,

$$k_{(a;b)} = 0, \qquad k_ak^a = 0, \qquad k_{[a}k_{b;c]} = 0. \tag{21.22}$$

From these conditions we obtain the relations

$$k_a = -wu_{,a}, \qquad k_{a;b} = w_{,[a}u_{,b]}, \qquad u_{,a}w^{,a} = 0. \tag{21.23}$$

We introduce coordinates $x^i = (x, y, v, u)$, similar to those of (21.17), such that the components of the null Killing vectors are given by

$$k^i = \delta^i_3, \qquad k_i = -w\delta_i{}^4, \tag{21.24}$$

and that the coordinates x and y label the points of the spacelike 2-surfaces V_2 orthogonal to $\boldsymbol{k}$ (Kundt (1961)). Because of it is independent of v, the metric can be transformed into

$$\mathrm{d}s^2 = P^{-2}(\mathrm{d}x^2 + \mathrm{d}y^2) - 2\,\mathrm{d}u(w\,\mathrm{d}v - m\,\mathrm{d}x + H\,\mathrm{d}u), \quad g_{ij,v} = 0 \tag{21.25}$$

(see § 27.2.).

One can show that the conditions (21.22) are incompatible with the field equations for Einstein-Maxwell non-null fields. Hence we can restrict ourselves to *electromagnetic null fields*

$$F_{ab} = 2A_{[b,a]} = 2r_{[a}k_{b]}, \qquad r_a k^a = 0, \tag{21.26}$$

and to vacuum fields. The null Killing vector $\boldsymbol{k}$ is a Ricci eigenvector with zero eigenvalue,

$$R^a{}_b k^b = k^{b;a}{}_{;b} = (w^{[b}u^{,a]})_{;b} = 0. \tag{21.27}$$

Calculation of the divergence in (21.27) with respect to the metric (21.25) shows that the function w satisfies the potential equation in V_2,

$$w_{,xx} + w_{,yy} = 0. \tag{21.28}$$

From the relation (21.26) it follows that the vector potential A_a can always be transformed to the form

$$A_a = \Psi u_{,a}, \qquad u_{,a}\Psi^{,a} = 0, \tag{21.29}$$

by means of a gauge transformation. The Lie derivative of the field tensor $F_{ab} = 2\Psi_{,[a}u_{,b]}$ with respect to $\boldsymbol{k}$ vanishes. In virtue of the Maxwell equations, the real scalar potential defined in (21.29) satisfies the potential equation in V_2,

$$\Psi_{,xx} + \Psi_{,yy} = 0. \tag{21.30}$$

For the complex self-dual field tensor F^*_{ab} we obtain

$$F^*_{ab} = 2F_{,[a}u_{,b]}, \qquad F = F(\zeta, u), \tag{21.31}$$

The potential equation (21.28) leads to two different cases: *I.* $w = 1$ and *II.* $w = x$.

Case I: $(w = 1)$

In this case, $\boldsymbol{k}$ is a constant vector field, $k_{a;b} = 0$, which is an invariant characterization of the *pp* waves (§ 21.5.). In the final form of the metric

$$\begin{aligned} ds^2 &= 2\,d\zeta\,d\bar{\zeta} - 2\,du\,dv - 2H\,du^2, \\ H &= \varkappa_0 F\bar{F} + f + \bar{f}, \qquad f = f(\zeta, u), \qquad F = F(\zeta, u), \end{aligned} \tag{21.32}$$

the functions f and F are analytic in ζ and depend arbitrarily on u.

Case II: $(w = x)$

The Einstein equations and the coordinate transformations preserving the form of the metric (21.25) lead to

$$P^2 = x^{1/2}, \qquad m = 0. \tag{21.33}$$

With $M = x^{-1}H$ the metric then reads

$$ds^2 = \frac{dx^2 + dy^2}{\sqrt{x}} - 2x\,du(dv + M\,du), \tag{21.34}$$

and the remaining Einstein-Maxwell equations are

$$\Psi_{,xx} + \Psi_{,yy} = 0, \qquad (xM_{,x})_{,x} + xM_{,yy} = \varkappa_0(\Psi_{,x}{}^2 + \Psi_{,y}{}^2). \tag{21.35}$$

For a given potential function Ψ, one has to solve an inhomogeneous linear differential equation which has the form of Poisson's equation in cylindrical polar coordinates for axisymmetric solutions. These Einstein-Maxwell null fields are of Petrov type *II* or *D*. The stationary vacuum solutions ($\Psi = 0$) of case *II* with $M_{,u} = 0$ are the van Stockum solutions (18.23). A simple solution of this class is the static type *D* vacuum metric

$$ds^2 = x^{-1/2}(dx^2 + dy^2) - 2x\,du\,dv, \tag{21.36}$$

which admits two null Killing vectors.

For *pure radiation fields* not necessarily obeying the Maxwell equations, H (respectively, M) is an arbitrary function satisfying $(xM_{,A})_{,A} > 0$.

For the energy-momentum tensors we are interested in, the two cases *I* and *II* include all space-times admitting a non-twisting null Killing vector.

Ref.: Kundt and Trümper (1962), Dautcourt (1964), Banerjee (1970a, b), De and Banerjee (1972), Debney (1971, 1972a, b, 1974).

21.5. The plane-fronted gravitational waves with parallel rays (*pp* waves)

In this section we want to investigate *space-times admitting a (covariantly) constant null vector field* $\boldsymbol{k}$,

$$k_{a;b} = 0. \tag{21.37}$$

Such space-times are called plane-fronted gravitational waves with parallel rays (*pp* waves). They were discovered by Brinkmann (1923) and subsequently rediscovered by several authors. Reviews of gravitational waves were given by Ehlers and Kundt (1962), Jordan et al. (1960), Takeno (1961), Zakharov (1972), and Schimming (1974).

The *pp* waves belong to the wider class of solutions admitting a non-expanding, shear- and twistfree null congruence (Kundt's class, see Chapter 27.). Obviously, the *pp* waves always admit a null Killing vector and thus form a subclass of the fields treated in the previous section.

The condition (21.37) implies (see § 21.4. and (31.2)) that electromagnetic non-null fields, perfect fluids, and Λ-term solutions cannot occur and that the metric of vacuum, Einstein-Maxwell null and pure radiation fields can be written in the form

$$\mathrm{d}s^2 = 2\,\mathrm{d}\zeta\,\mathrm{d}\bar{\zeta} - 2\,\mathrm{d}u\,\mathrm{d}v - 2H\,\mathrm{d}u^2, \qquad H = H(\zeta, \bar{\zeta}, u), \tag{21.38}$$

which is preserved under the coordinate transformations

$$\begin{aligned} &\zeta' = e^{i\alpha}\big(\zeta + h(u)\big), \quad v' = a\big(v + \dot{h}(u)\,\bar{\zeta} + \dot{\bar{h}}(u)\,\zeta + g(u)\big), \quad u' = a^{-1}(u + u_0), \\ &H' = a^2\big(H - \ddot{h}(u)\,\bar{\zeta} - \ddot{\bar{h}}(u)\,\zeta + \dot{h}(u)\,\dot{\bar{h}}(u) - \dot{g}(u)\big) \end{aligned} \tag{21.39}$$

(α, a, u_0 real constants; $g(u)$ real, $h(u)$ complex). The metric (21.38) is of Kerr-Schild type (see § 28.1.). We use the complex null tetrad

$$\boldsymbol{m} = \partial_\zeta, \qquad \overline{\boldsymbol{m}} = \partial_{\bar{\zeta}}, \qquad \boldsymbol{l} = \partial_u - H\,\partial_v, \qquad \boldsymbol{k} = \partial_v, \tag{21.40}$$

and compute the Ricci and Weyl tensors:

$$R_{ab} = 2H_{,\zeta\bar{\zeta}}k_ak_b, \quad \frac{1}{2}\,C^*_{abcd} = \Psi_4 V_{ab} V_{cd}, \qquad V_{ab} = 2k_{[a}m_{b]}, \qquad \Psi_4 = H_{,\bar{\zeta}\bar{\zeta}}. \tag{21.41}$$

Space-times satisfying (21.37) are either Petrov type N (with the multiple principal null direction $\boldsymbol{k}$) or conformally flat (if $H_{,\bar{\zeta}\bar{\zeta}} = 0$). The null bivector $V_{ab} = 2k_{[a}m_{b]}$ is constant, $V_{ab;c} = 0$, so that *pp* waves are complex recurrent, $C^*_{abcd;e} = C^*_{abcd}(\ln \Psi_4)_{,e}$ (see § 31.2.). V_{ab} is determined by the metric up to a complex constant coefficient.

If the metric (21.38) is a solution of the Einstein-Maxwell equations, the function H and the electromagnetic (null) field are given by (see (21.31), (21.32))

$$H = f(\zeta, u) + \bar{f}(\bar{\zeta}, u) + \varkappa_0 F(\zeta, u)\,\bar{F}(\bar{\zeta}, u), \qquad F_{ab} = 2k_{[a}F_{,b]}. \tag{21.42}$$

The functions f and F are analytic in ζ and depend arbitrarily on the retarded time coordinate u. For pure radiation fields, H is restricted only by $H_{,\zeta\bar{\zeta}} > 0$.

In general, the group G_1 on N_1 generated by the Killing vector $\boldsymbol{k} = \partial_v$ is the maximal group of motions, but larger groups exist for various special choices of H. For non-flat space-times (21.38), the Killing equations imply

$$\begin{aligned} \xi &= \big(ib\zeta + \beta(u)\big)\,\partial_\zeta + \big(-ib\bar{\zeta} + \bar{\beta}(u)\big)\,\partial_{\bar{\zeta}} + (cu + d)\,\partial_u \\ &\quad + \big(-cv + \dot{\bar{\beta}}(u)\,\zeta + \dot{\beta}(u)\,\bar{\zeta} + a(u)\big)\,\partial_v, \\ &(\xi H) + 2cH + \ddot{\bar{\beta}}(u)\,\zeta + \ddot{\beta}(u)\,\bar{\zeta} + \dot{a}(u) = 0 \end{aligned} \tag{21.43}$$

(b, c, d real constants; $a(u)$ real, $\beta(u)$ complex).

Ehlers and Kundt (1962) investigated the *vacuum pp waves* and found all possible

Table 21.1 Symmetry classes of vacuum *pp* waves $ds^2 = 2\,d\zeta\,d\bar{\zeta} - 2\,du\,dv - 2[f(\zeta, u) + \bar{f}(\bar{\zeta}, u)]\,du^2$ ($\varkappa$, α, a real constants; $A(u)$ and $A(\zeta u^{i\varkappa})$ complex)

$f(\zeta, u)$	Group	Orbits	Killing vectors ξ_A
$f(\zeta, u)$	G_1	N_1	∂_v
$u^{-2}A(\zeta u^{i\varkappa})$	G_2	T_2	$\partial_v,\ u\,\partial_u - v\,\partial_v - i\varkappa(\zeta\,\partial_\zeta - \bar{\zeta}\,\partial_{\bar{\zeta}})$
$f(\zeta\,e^{i\varkappa u})$	G_2	T_2	$\partial_v,\ \partial_u - i\varkappa(\zeta\,\partial_\zeta - \bar{\zeta}\,\partial_{\bar{\zeta}})$
$A(u)\ln\zeta$	G_2	N_2	$\partial_v,\ i(\zeta\,\partial_\zeta - \bar{\zeta}\,\partial_{\bar{\zeta}}) + 2\int \mathrm{Im}\,A(u)\,du\,\partial_v$
$au^{-2}\ln\zeta$	G_3	T_3	$\partial_v,\ i(\zeta\,\partial_\zeta - \bar{\zeta}\,\partial_{\bar{\zeta}}),\ u\,\partial_u - v\,\partial_v$
$\ln\zeta$	G_3	T_3	$\partial_v,\ i(\zeta\,\partial_\zeta - \bar{\zeta}\,\partial_{\bar{\zeta}}),\ \partial_u$
$e^{2\varkappa\zeta}$	G_3	T_3	$\partial_v,\ \partial_u,\ \partial_\zeta + \partial_{\bar{\zeta}} - \varkappa(u\,\partial_u - v\,\partial_v)$
$e^{i\alpha}\zeta^{2i\varkappa}$	G_3	T_3	$\partial_v,\ \partial_u,\ i(\zeta\,\partial_\zeta - \bar{\zeta}\,\partial_{\bar{\zeta}}) + \varkappa(u\,\partial_u - v\,\partial_v)$
$A(u)\,\zeta^2$	G_5	N_3	$\partial_v,\ \beta\,\partial_\zeta + \bar{\beta}\,\partial_{\bar{\zeta}} + \zeta\dot{\bar{\beta}}\,\partial_v + \bar{\zeta}\dot{\beta}\,\partial_v$ $\ddot{\beta} + 2\bar{A}(u)\,\bar{\beta} = 0$
$au^{2i\varkappa-2}\zeta^2$	G_6	V_4	add $u\,\partial_u - v\,\partial_v - i\varkappa(\zeta\,\partial_\zeta - \bar{\zeta}\,\partial_{\bar{\zeta}})$
$e^{2i\varkappa u}\zeta^2$	G_6	V_4	add $\partial_u - i\varkappa(\zeta\,\partial_\zeta - \bar{\zeta}\,\partial_{\bar{\zeta}})$

forms of $H = f(\zeta, u) + \bar{f}(\bar{\zeta}, u)$ and ξ which are compatible with (21.43). Table 21.1 summarizes the results. Note that, in each case, H and ξ_A are determined up to the transformations (21.39).

The *pp* waves with $f = f(\zeta)$ and $F = F(\zeta)$ independent of u admit an Abelian group G_2 (Killing vectors ∂_v, ∂_u) on T_2 (see Table 21.1). These solutions belong to the excluded case $W = 1$ in the metric (17.15) of stationary axisymmetric fields. (The Einstein-Maxwell equations (17.26)—(17.30) are solved by $W = 1$, $\mathscr{E} = \mathscr{E}(\zeta)$, $\Phi = \Phi(\zeta)$.) Hoffman (1969b) considered this case for vacuum fields.

Comparing the formulae $C^*_{abcd} = 2\Psi_4 V_{ab} V_{cd}$ and $F^*_{ab} = 2\Phi_2 V_{ab}$ we expect that the Weyl tensor component Ψ_4 of a vacuum *pp* wave and the field tensor component Φ_2 of an electromagnetic wave permit analogous physical interpretations. Writing $\Psi_4 = Ae^{i\Theta}$, $A > 0$, one calls A the amplitude and associates Θ with the plane of polarization at each space-time point (Ehlers and Kundt (1962)). Vacuum *pp* waves for which Θ is constant are called *linearly polarized*.

In Einstein-Maxwell theory, *plane waves*, first considered by Baldwin and Jeffery (1926), are defined to be *pp* waves in which $\Psi_{4,\bar{\zeta}} = 0 = \Phi_{2,\bar{\zeta}}$. The metric is then given by

$$ds^2 = 2\,d\zeta\,d\bar{\zeta} - 2\,du\,dv - 2\big(A(u)\,\zeta^2 + \bar{A}(u)\,\bar{\zeta}^2 + B(u)\,\zeta\bar{\zeta}\big)\,du^2. \qquad (21.44)$$

($A(u)$ complex, $B(u)$ real); a linear function of ζ and $\bar{\zeta}$ in H can be removed by (21.39). Plane waves admit a group G_5 with an Abelian subgroup G_3 on null hypersurfaces N_3. The electromagnetic term $B(u)\,\zeta\bar{\zeta}$ in (21.44) does not alter the form of the Killing vectors as given in Table 21.1 for $f(\zeta, u) = A(u)\,\zeta^2$, but the equation for $\beta(u)$ is now $\ddot{\beta} + 2\bar{A}(u)\,\bar{\beta} + B(u)\,\beta = 0$. The four integration constants in the solution of this differential equation give rise to four independent Killing vectors. Plane waves admit a G_6 on V_4 if either $A(u) = A_0\,e^{2i\varkappa u}$, $B(u) = B_0$ or $A(u) = A_0 u^{2i\varkappa-2}$, $B(u) = B_0 u^{-2}$ (A_0, B_0 real constants), see also § 10.2.

The metric (21.38) can be cast in the form

$$\mathrm{d}s^2 = g_{MN}(u)\,\mathrm{d}x^M\,\mathrm{d}x^N - 2\,\mathrm{d}u\,\mathrm{d}v', \qquad M, N = 1, 2, \tag{21.45}$$

with the aid of the coordinate transformation

$$\begin{gathered}\zeta = \alpha_M x^M, \qquad v = v' + \frac{1}{4}\,\dot{g}_{MN}x^M x^N, \qquad u' = u, \\ g_{MN}(u) = 2\bar{\alpha}_{(M}\alpha_{N)}, \quad \mathrm{Re}\,[\bar{\alpha}_{(M}\ddot{\alpha}_{N)} + 2A(u)\,\alpha_M\alpha_N + B(u)\,\bar{\alpha}_M\alpha_N] = 0\end{gathered} \tag{21.46}$$

($\alpha_M = \alpha_M(u)$ complex). The calculation of the Ricci tensor in the coordinate system (21.45) yields

$$R_{ab} = -\left(\frac{1}{2}\,g^{MN}\ddot{g}_{MN} + \frac{1}{4}\,\dot{g}^{MN}\dot{g}_{MN}\right)k_a k_b. \tag{21.47}$$

For linearly polarized plane gravitational waves we have $\bar{A}(u) = \mathrm{const}\cdot A(u)$ in the metric (21.44), and $g_{12}(u) = 0$ in (21.45). The plane wave solution (Brdička (1951)),

$$\mathrm{d}s^2 = (1 - \sin\omega u)\,\mathrm{d}x^2 + (1 + \sin\omega u)\,\mathrm{d}y^2 - 2\,\mathrm{d}u\,\mathrm{d}v', \tag{21.48}$$

ω being a real constant, is a conformally flat Einstein-Maxwell field with constant electromagnetic null field.

Plane waves can be interpreted as the gravitational fields at great distances from finite radiating bodies. Peres (1960b), and Bonnor (1969b) considered parallel light beams as the sources of plane waves. Bondi et al. (1959) investigated plane waves from a group-theoretical point of view.

PART III. ALGEBRAICALLY SPECIAL SOLUTIONS

Chapter 22. The various classes of algebraically special solutions. Newman-Tamburino solutions

22.1. Solutions of Petrov type *II*, *D*, *III*, or *N*

Many of the known exact solutions of Einstein's equations are algebraically special, i.e. of Petrov type *II*, *D*, *III*, *N*, or *O*. One common technique for finding such metrics (except those of type *O*) explicitly is to start by considering a null congruence defined by a repeated principal null direction (cp. §§ 4.3., 7.5.). The present section, in particular Table 22.1, gives a brief survey of the different subcases which naturally arise in this approach. As well as being algebraically special,

$$\Psi_0 = \Psi_1 = 0, \tag{22.1}$$

most of the known solutions of types *II*, *D*, *III* and *N* have in common that the multiple null eigenvector $\boldsymbol{k}$ of the Weyl tensor is geodesic and shearfree,

$$\varkappa = \sigma = 0, \tag{22.2}$$

and the Ricci tensor satisfies

$$R_{ab}k^a k^b = R_{ab}k^a m^b = R_{ab}m^a m^b = 0 \tag{22.3}$$

(for the definition of the complex null vector $\boldsymbol{m}$ see § 3.2.). Note that because of the Goldberg-Sachs Theorem and its generalizations (§ 7.5.), and Theorem 7.4 and its corollary, these three sets of assumptions are not independent of each other. In particular, *all* algebraically special vacuum solutions and *all* Einstein-Maxwell null fields obey (22.1)—(22.3).

Table 22.1 The different subcases of the algebraically special (but not conformally flat) solutions which obey (22.1)—(22.3), and corresponding chapter numbers

	$\omega = 0$		$\omega \neq 0$	Kerr-Schild (both $\omega = 0$ and $\omega \neq 0$)
	$\Theta \neq 0$ (Robinson-Trautman)	$\Theta = 0$ (Kundt)		
Vacuum	23, 24	27	23, 25	28
Einstein-Maxwell and pure radiation	23, 24	27	23, 26	28

The bulk of the material presented in the chapters on algebraically special solutions is concerned with solutions which satisfy the assumptions (22.1)—(22.3). The further subdivision of these solutions depends on the complex divergence ϱ of the vector field $\boldsymbol{k}$,

$$\varrho = -(\Theta + \mathrm{i}\omega), \tag{22.4}$$

see Table 22.1.

Let us now have a look at those solutions which are *not* covered by the assumptions (22.2)—(22.3), i.e. at solutions which are of Petrov type *II*, *D*, *III* or *N* but violate (22.2) or (22.3) or both.

For *Einstein-Maxwell-fields* with a *non-null* electromagnetic field, we have to distinguish between two cases: either the two null eigenvectors of the Maxwell field are both distinct from the multiple null eigenvector(s) of the Weyl tensor (*non-aligned case*), or at least one of them is parallel to a repeated Weyl null eigenvector (*aligned case*).

In the non-aligned case, equations (22.3) are not satisfied. Only a few solutions of this kind are known. Type *III* solutions with a twistfree multiple null eigenvector $\boldsymbol{k}$ were considered by Cahen and Spelkens (1967), and the general type N solution (which proves to satisfy $\varkappa = \sigma = \omega = 0$) has been found by Szekeres (1966b), and Cahen and Leroy (1965, 1966). As given by Szekeres, it reads (with $k_a = u_{,a}$)

$$\begin{aligned} \mathrm{d}s^2 = {} & \frac{1}{2}\cos^2 ar(\mathrm{d}x^2 + \mathrm{d}y^2) - 2b(2r + a^{-1}\sin 2ar)\,\mathrm{d}u\,\mathrm{d}x \\ & - 4\,\mathrm{d}u\,\mathrm{d}r + 4a^{-2}(2b^2\sin^2 ar - 2\,\mathrm{e}^{2u} - raa_{,u})\,\mathrm{d}u^2, \end{aligned} \tag{22.5}$$

$$a(x, u) = \mathrm{e}^u \coth\,[\mathrm{e}^u x + f(u)], \qquad b(x, u) = g(u)\,\mathrm{e}^u \sinh\,[\mathrm{e}^u x + f(u)].$$

In the aligned case, (22.3) is satisfied by definition. As shown in § 7.5., the Bianchi identities yield

$$(2\varkappa_0\Phi_1\overline{\Phi}_1 + 3\Psi_2)\,\sigma = 0 = (-2\varkappa_0\Phi_1\overline{\Phi}_1 + 3\Psi_2)\,\varkappa, \tag{22.6}$$

so equation (22.2) can be violated only for Petrov type *II* (or *D*) solutions with a special constant ratio between Ψ_2 and $\Phi\overline{\Phi}_1$ and $\varkappa$ *or* σ must be non-zero (note that because of (6.32) $\varkappa = \varrho = 0$ would induce $\sigma = 0$). The case $\varkappa = 0$, $\sigma \neq 0$, has been excluded by Kozarzewski (1965), so only $\varkappa \neq 0$, $\sigma = 0$, remains to be studied. If the two null eigenvectors of a type D solution are aligned with the eigenvectors of the Maxwell tensor, then they must both be geodesic and shearfree (Plebański and Hacyan (1979)).

For *pure radiation fields*, one again has to distinguish between the aligned and the non-aligned cases. So far, only the aligned case has been treated (the general type N metric is necessarily aligned: Plebański (private communication) and Urbantke (1975)). Due to the theorems given in § 7.5., all aligned type *II*, *D* or *III* solutions have a geodesic and shearfree $\boldsymbol{k}$. The only algebraically special, aligned, pure radiation fields *not* covered by (22.1)—(22.2) are therefore of type N. Among the Weyl and Ricci tensor components only Ψ_4 and $R_{33} = \varkappa_0\Phi^2 = 2\Phi_{22}$, respectively, are different from zero for these type N fields. The Bianchi identities (7.66), (7.69) and (7.70) then give $\varkappa = 0$ and

$$\varrho\Phi_{22} = \sigma\Psi_4, \tag{22.7}$$

and equations (3.23) and the field equations yield

$$\begin{aligned} &\mathrm{d}\boldsymbol{\Gamma}_{41} + \boldsymbol{\Gamma}_{41} \wedge (\boldsymbol{\Gamma}_{21} + \boldsymbol{\Gamma}_{43}) = 0, \\ &\mathrm{d}(\boldsymbol{\Gamma}_{21} + \boldsymbol{\Gamma}_{43}) + 2(\boldsymbol{\Gamma}_{32} \wedge \boldsymbol{\Gamma}_{41}) = 0, \\ &\mathrm{d}\boldsymbol{\Gamma}_{32} - \boldsymbol{\Gamma}_{32} \wedge (\boldsymbol{\Gamma}_{21} + \boldsymbol{\Gamma}_{43}) = \frac{1}{2}\, \omega^3 \wedge (\Phi_{22}\omega^1 + \Psi_4\omega^2). \end{aligned} \tag{22.8}$$

(From (22.7) we see that solutions with $\sigma = 0$ belong to Kundt's class.)

The type N field equations (22.8) have been investigated by Plebański (private communication). In detail, for $\sigma \neq 0$, his results are as follows (for $\sigma = 0$ see Chapter 27.). After a suitable choice of the null tetrad $(\boldsymbol{m}, \overline{\boldsymbol{m}}, \boldsymbol{l}, \boldsymbol{k})$, $\boldsymbol{k}$ being unchanged, (22.8) gives

$$\boldsymbol{\Gamma}_{21} = 0, \qquad \boldsymbol{\Gamma}_{34} = 0, \qquad \mathrm{d}\boldsymbol{\Gamma}_{41} = 0, \qquad \boldsymbol{\Gamma}_{32} \wedge \mathrm{d}\boldsymbol{\Gamma}_{32} = 0 \tag{22.9}$$

(this is true also for $\sigma = 0$). The further integration depends on whether $\boldsymbol{\Gamma}_{41}$ is zero, real, or complex. *For* $\boldsymbol{\Gamma}_{41} = 0$ ($\Rightarrow \varrho = \sigma = \tau = 0$), we regain the *pp* waves (§ 21.5.). *For* $\boldsymbol{\Gamma}_{41}$ *real*, the solutions with non-zero shear σ are (in real coordinates y, u, r, v)

$$\begin{aligned} &\mathrm{d}s^2 = 2\omega^1\omega^2 - 2\omega^3\omega^4, \\ &\omega^1 = \mathrm{d}F + (v + \bar{f})\,\mathrm{d}r = \overline{\omega}^2, \\ &\omega^3 = \mathrm{d}u + (F + \bar{F})\,\mathrm{d}r, \qquad \omega^4 = \mathrm{d}v + (g + Ff_{,u} + \bar{F}\bar{f}_{,u})\,\mathrm{d}r, \\ &F \equiv (-g_{,u} + \mathrm{i}y)/2f_{,uu}, \qquad f_{,uu} \neq 0. \end{aligned} \tag{22.10a}$$

They contain two disposable functions $f(u, r)$ (complex) and $g(u, r)$ (real). The corresponding radiation field is characterized by

$$\begin{aligned} &\boldsymbol{\Gamma}_{41} = -\mathrm{d}r, \quad \sigma = \bar{\varrho} = -Af_{,uu}, \qquad \Phi_{22} = -2Af_{,uu}\bar{f}_{,uu} \\ &A^{-1} = 2\,\mathrm{Re}\,\{f_{,uu}[-(F + \bar{F})\,F_{,u} + F_{,r} + v + \bar{f}]\}. \end{aligned} \tag{22.10b}$$

For $\boldsymbol{\Gamma}_{41}$ *complex and*

$$|\varrho| = |\sigma|, \qquad |\Phi_{22}| = |\Psi_4|, \tag{22.11}$$

the solutions are (in coordinates ζ, $\bar{\zeta}$, r, v)

$$\begin{aligned} &\mathrm{d}s^2 = 2\omega^1\omega^2 - 2\omega^3\omega^4, \\ &\omega^1 = \mathrm{d}F + v\,\mathrm{d}\zeta + \bar{h}\,\mathrm{d}\bar{\zeta} = \overline{\omega}^2, \\ &\omega^3 = \bar{F}\,\mathrm{d}\zeta + F\,\mathrm{d}\bar{\zeta}, \\ &\omega^4 = \mathrm{d}v - (Ff + h_{,\bar{\zeta}})\,\mathrm{d}\zeta - (\bar{F}\bar{f} + \bar{h}_{,\zeta})\,\mathrm{d}\bar{\zeta}, \\ &F = -f_{,\bar{\zeta}}^{-1}(r + h_{,\bar{\zeta}\bar{\zeta}} - f\bar{h}), \qquad f_{,\bar{\zeta}} \neq 0. \end{aligned} \tag{22.12a}$$

They contain two arbitrary complex functions $f(\zeta, \bar{\zeta})$ and $h(\zeta, \bar{\zeta})$. The corresponding radiation field is characterized by (22.11) and

$$\begin{aligned} &\Gamma_{41} = -\mathrm{d}\bar{\zeta}, \quad \varrho = -A\bar{F}\bar{f}_{,\zeta}, \quad \sigma = A\bar{F}f_{,\bar{\zeta}}, \quad \Phi_{22} = -2Af_{,\bar{\zeta}}\bar{f}_{,\zeta}, \\ &A^{-1} = 2\,\mathrm{Re}\,[F(f_{,\bar{\zeta}}v + F_{,\zeta}) - \bar{f}_{,\zeta}(h - \bar{F}_{,\zeta})]. \end{aligned} \tag{22.12b}$$

For complex Γ_{41} not satisfying (22.11), Plebański succeeded in reducing the field equations to the differential equation

$$\mathrm{Im}\,[(h_{,u\bar{\zeta}})^2\,(h_{,uu})^{-1} - h_{,\bar{\zeta}\bar{\zeta}} - \bar{h}h_{,u}] = 0, \qquad h_{,uu} \neq 0. \tag{22.13a}$$

If a solution $h(\zeta, \bar{\zeta}, u)$ of (22.13a) is known, then the metric can be obtained from

$$\begin{aligned} &\mathrm{d}s^2 = 2\omega^1\omega^2 - 2\omega^3\omega^4, \\ &\omega^1 = \mathrm{d}\bar{L} + r\,\mathrm{d}\zeta + \bar{h}\,\mathrm{d}\bar{\zeta} = \bar{\omega}^2, \\ &\omega^3 = \mathrm{d}u + L\,\mathrm{d}\zeta + \bar{L}\,\mathrm{d}\bar{\zeta}, \qquad \omega^4 = \mathrm{d}r + W\,\mathrm{d}\zeta + \bar{W}\,\mathrm{d}\bar{\zeta}, \\ &W = -\bar{\partial}h = -h_{,\bar{\zeta}} + \bar{L}h_{,u}, \qquad \bar{L} = h_{,u\bar{\zeta}}/h_{,uu} \Leftrightarrow \bar{\partial}h_{,u} = 0, \end{aligned} \tag{22.13b}$$

with

$$\begin{aligned} &\varrho = -A(r + \partial\bar{L}), \qquad \sigma = A(h + \partial L), \qquad \Phi_{22} = -2\bar{\sigma}h_{,uu}, \\ &A^{-1} = |r + \partial\bar{L}|^2 - |h + \partial L|^2, \qquad \partial = \partial_\zeta - L\partial_u \end{aligned} \tag{22.13c}$$

($\Phi_{22} > 0$ should be guaranteed). The twisting vacuum type N solutions are included (for $\sigma = 0$) in (22.13).

The general solution of (22.13a) in the case of real $h_{,u}$ (which implies real $\partial\bar{L}$ and hence real ϱ) has been given by Plebański as

$$\begin{aligned} &h(u, \zeta, \bar{\zeta}) = (F - sF_{,s}) + (G - sG_{,s}) - P_{,\zeta\zeta}, \\ &F = F(x, s), \qquad G = G(y, s), \qquad P = P(x, y), \qquad \sqrt{2}\,\zeta = x + \mathrm{i}y, \\ &u = F_{,s} + G_{,s} + P \Rightarrow s = s(u, x, y), \qquad u_{,s} \neq 0, \end{aligned} \tag{22.14}$$

with real functions F, G, and P. This solution includes the special solution found by Köhler and Walker (1975), which, in our notation, would be given by $h = u^2 \times (1 + \zeta\bar{\zeta})^{-1}$.

For real ϱ, a different form of the remaining field equation was given by Köhler and Walker (1975).

Perfect fluid (or dust) solutions necessarily violate $R_{ab}k^ak^b = 0$ (if $\mu + p \neq 0$). If the rest of the assumptions (22.1)—(22.3) are satisfied, then some of the techniques developed for treating (22.1)—(22.3) can be applied; see §§ 29.1., 29.2.

22.2. Conformally flat solutions

By definition, the metric of a conformally flat space-time can be written as

$$\mathrm{d}s^2 = \mathrm{e}^{2U(x,y,z,t)}\,(\mathrm{d}x^2 + \mathrm{d}y^2 + \mathrm{d}z^2 - \mathrm{d}t^2). \tag{22.15}$$

An equivalent coordinate-independent definition makes use of the fact that the Weyl tensor C_{abcd} (3.50) vanishes if and only if space-time is conformally flat: a metric is conformally flat exactly if

$$R_{abcd} = \frac{1}{2}\,(g_{ac}R_{bd} + g_{bd}R_{ac} - g_{ad}R_{bc} - g_{bc}R_{ad}) - \frac{R}{6}\,(g_{ac}g_{bd} - g_{ad}g_{bc}). \tag{22.16}$$

Being zero, the Weyl tensor does not define any null vector which can be used in constructing metrics and solutions, so other techniques have to be applied.

Equation (22.16) shows that conformally flat vacuum solutions ($R_{ab} = 0 = R$) are flat. All conformally flat solutions with a perfect fluid, an electromagnetic field, or a pure radiation field are known. As most of them have been found by applying the techniques of embedding, we will give a more detailed treatment of this subject in Chapter 32. Here we will summarize only the main results.

Conformally flat perfect fluid solutions are either generalized interior Schwarzschild solutions (32.40) or generalized Friedmann solutions (32.46), the only dust solutions being the Friedmann models and the only stationary solution the static interior Schwarzschild solution (14.14). Conformally flat Einstein-Maxwell fields are either the Bertotti-Robinson metric (32.95)—(32.96) (with a non-null electromagnetic field), or they are special plane waves (32.103)—(32.104) (with a null electromagnetic field). Conformally flat pure radiation fields are contained in (32.103); they can always be interpreted in terms of an electromagnetic null field.

For groups of motions in conformally flat space-times, see Levine (1936, 1939).

22.3. Newman-Tamburino solutions

The multiple principal null congruence of the Weyl tensor of an algebraically special vacuum solution is geodesic and shearfree (cp. § 22.1.). Although *vacuum solutions* with a *geodesic but shearing* ($\varkappa = 0$, $\sigma \neq 0$) principal null congruence are algebraically general (non-degenerate Petrov type I), we will list them here, because the method of constructing the (non-twisting) solutions with this property is very similar to that for the Robinson-Trautman class outlined in Chapter 23. (but, in virtue of $\sigma \neq 0$, the calculations are more lengthy; for details see also Carmeli (1977), p. 244). All vacuum metrics with $\varkappa = 0$, $\omega = 0$, $\sigma \neq 0$, $\Theta \neq 0$, were obtained by Newman and Tamburino (1962). Metrics with $\sigma \neq 0$, $\varrho = 0$, are forbidden by the vacuum field equations, and solutions with $\sigma \neq 0$, $\omega \neq 0$, are possible only if $\varrho\bar{\varrho} = \sigma\bar{\sigma}$ (Unti and Torrence (1966)). The twisting case has not been completely solved.

There are two classes of Newman-Tamburino solutions:
$\varrho^2 \neq \sigma\bar{\sigma} \neq 0$ (*spherical class*)

$$\begin{aligned}
g^{11} &= \frac{2(2\zeta\bar{\zeta})^{3/2}}{(r+a)^2}, \quad g^{22} = \frac{2(2\zeta\bar{\zeta})^{3/2}}{(r-a)^2}, \quad g^{34} = -1, \quad g^{12} = g^{14} = g^{24} = g^{44} = 0, \\
g^{13} &= 4A^2(2\zeta\bar{\zeta})^{3/2}\, x \left(\frac{r-a}{R^4} + \frac{r-2a}{2a^2R^2} - \frac{L}{2a^3}\right), \\
g^{23} &= 4A^2(2\zeta\bar{\zeta})^{3/2}\, y \left(\frac{r+a}{R^4} + \frac{r+2a}{2a^2R^2} - \frac{L}{2a^3}\right), \\
g^{33} &= \frac{4A^2r^2(2\zeta\bar{\zeta})^{3/2}}{R^4} - \frac{4Ar^3(\zeta^2+\bar{\zeta}^2)}{R^4} + \frac{2r^2(2\zeta\bar{\zeta})^{1/2}}{R^2} - \frac{2rL}{A}, \\
L &= \frac{1}{2}\ln\left(\frac{r+a}{r-a}\right), \quad a = A(2\zeta\bar{\zeta})^{1/2}, \quad R^2 = r^2 - a^2, \quad A = bu + c,
\end{aligned} \tag{22.17}$$

$\left(x^1 + \mathrm{i}x^2 = x + \mathrm{i}y = \sqrt{2}\,\zeta, \quad x^3 = r, \quad x^4 = u\right)$.

$\varrho^2 = \sigma\bar{\sigma} \neq 0$ (*cylindrical class*)

$$\begin{aligned}
\mathrm{d}s^2 &= 2\omega^1\omega^2 - 2\omega^3\omega^4, \quad \omega^3 = \mathrm{d}u, \quad \omega^4 = \mathrm{d}r + H\,\mathrm{d}u, \\
\omega^1 &= \bar{\omega}^2 = \frac{1}{2} r\,\mathrm{d}x + 4Y\,\mathrm{d}u \\
&\quad + \mathrm{i}2^{-1/2}\,\mathrm{cn}\,(bx)\,(\mathrm{d}y + 8buY\,\mathrm{d}x + 2b\ln r\,\mathrm{d}u), \\
-2H &= 4b^2\,\mathrm{cn}^2\,(bx) + \frac{c + b^2\ln(r^2\,\mathrm{cn}^4\,bx)}{\mathrm{cn}^2\,bx},
\end{aligned} \tag{22.18}$$

where b and c are real constants, $\mathrm{cn}\,(bx)$ is an elliptic function of modulus $k = 2^{-1/2}$ and Y is given by

$$Y = \pm\frac{b(2 - \mathrm{cn}^4\,bx)^{1/2}}{2\sqrt{2}\,\mathrm{cn}\,(bx)}. \tag{22.19}$$

The metric

$$\mathrm{d}s^2 = r^2\,\mathrm{d}x^2 + x^2\,\mathrm{d}y^2 + \frac{4r}{x}\,\mathrm{d}u\,\mathrm{d}x - 2\,\mathrm{d}u\,\mathrm{d}r + [c + \ln(r^2x^4)]\,\mathrm{d}u^2 \tag{22.20}$$

can be obtained from (22.18) by a limiting procedure.

The Newman-Tamburino solutions do not contain arbitrary functions of time; the Robinson-Trautman solutions are not included. The metric (22.17) admits at most one Killing vector ($\xi = \partial_u$ if $b = 0$); there is only one Killing vector ($\xi = \partial_y$) for the metric (22.18), whereas (22.20) admits an Abelian G_2, the ignorable coordinates being y and u (Collinson and French (1967)).

Chapter. 23. The line element for metrics with $\varkappa = \sigma = 0 = R_{11} = R_{14} = R_{44}, \quad \Theta + \mathrm{i}\omega \neq 0$

23.1. The line element in the case with twisting rays ($\omega \neq 0$)

23.1.1. The choice of the null tetrad

In this chapter we will deal with space-times which admit a geodesic non-shearing but diverging null congruence $\boldsymbol{k}$,

$$\varkappa = \sigma = 0, \qquad \varrho = -(\Theta + \mathrm{i}\omega) \neq 0. \tag{23.1}$$

In addition, the Ricci tensor components picked out by the null congruence $\boldsymbol{k}$ and the (complex) tetrad vector $\boldsymbol{m}$ are assumed to satisfy

$$R_{11} = R_{14} = R_{44} = 0. \tag{23.2}$$

Throughout this and the following sections, the null tetrad ($\boldsymbol{m}$, $\overline{\boldsymbol{m}}$, $\boldsymbol{l}$, $\boldsymbol{k}$) and the Newman-Penrose formalism will be used without further warning (see Chapters 3. and 7.). All numerical indices are tetrad indices, e.g. $R_{14} = R_{ab}m^a k^b$; derivatives with respect to coordinates will be abbreviated by explicit use of the coordinate, e.g. $H_{,r} \equiv \partial H/\partial r \equiv \partial_r H$.

Due to the Goldberg-Sachs theorem (Theorem 7.1), all solutions satisfying (23.1) and (23.2) are algebraically special. Note that, because of (6.33), $\Theta = 0 = \sigma$ would imply $\omega = 0$, so that all metrics being discussed here necessarily have an expanding ($\Theta \neq 0$) null congruence $\boldsymbol{k}$.

Following Debney et al. (1969), we will now evaluate the equations (23.1)—(23.2) in order to find a preferred tetrad as well as an appropriate coordinate frame. We start with a tetrad ($\boldsymbol{m}'$, $\overline{\boldsymbol{m}}'$, $\boldsymbol{l}'$, $\boldsymbol{k}'$) in which only the direction of $\boldsymbol{k}'$ is fixed. The conditions (23.1) and (23.2) remain unchanged under the set of tetrad transformations (3.15)—(3.17), i.e. under

$$\boldsymbol{k}' = \boldsymbol{k}, \qquad \boldsymbol{m}' = \boldsymbol{m} + B\boldsymbol{k}, \qquad \boldsymbol{l}' = \boldsymbol{l} + B\overline{\boldsymbol{m}} + \bar{B}\boldsymbol{m} + B\bar{B}\boldsymbol{k}, \tag{23.3}$$

$$\boldsymbol{k}' = \boldsymbol{k}, \qquad \boldsymbol{m}' = \mathrm{e}^{\mathrm{i}C}\boldsymbol{m}, \qquad \boldsymbol{l}' = \boldsymbol{l}, \tag{23.4}$$

$$\boldsymbol{k}' = A\boldsymbol{k}, \qquad \boldsymbol{m}' = \boldsymbol{m}, \qquad \boldsymbol{l}' = A^{-1}\boldsymbol{l}, \tag{23.5}$$

so we may use these transformations to simplify the connection forms.

Under (23.3), the spin coefficient $\tau = -\Gamma_{143} = -k_{a;b}m^a l^b$ transforms to $\tau' = \tau + B\varrho$. Because $\varrho \neq 0$, τ can always be made zero, and from now on we have

$$\varkappa = \tau = \sigma = 0 = \Gamma_{144} = \Gamma_{143} = \Gamma_{141}, \tag{23.6}$$

i.e. the connection form $\boldsymbol{\Gamma}_{14}$ is simply

$$\boldsymbol{\Gamma}_{14} = -\varrho\boldsymbol{\omega}^2 = -\boldsymbol{\Gamma}_{41}, \qquad \Gamma_{142} = -\varrho. \tag{23.7}$$

The tetrad components of the Ricci tensor $R_{ab} = R_{cadb}(m^c \overline{m}^d + \overline{m}^c m^d - k^c l^d - k^d l^c)$ which are zero because of (23.2), can be written in terms of the curvature tensor as

$$R_{44} = 2R_{1424} = 0, \qquad R_{41} = R_{1421} + R_{1434} = 0, \qquad R_{11} = 2R_{1431} = 0. \tag{23.8}$$

As already stated, the space-times under consideration are algebraically special, i.e. Ψ_0 and Ψ_1 vanish:

$$\Psi_0 = R_{1414} = 0, \quad 2\Psi_1 = 2R_{1434} - R_{14} = R_{1434} - R_{1421} = 0. \tag{23.9}$$

Equations (23.8) and (23.9) show that of the tetrad components R_{14cd} of the curvature tensor only R_{1423} survives. Thus the relation (3.23a) between connection and curvature here reads

$$\mathrm{d}\Gamma_{41} + \Gamma_{41} \wedge (\Gamma_{21} + \Gamma_{43}) = R_{4123}\omega^2 \wedge \omega^3. \tag{23.10}$$

From (23.7) and (23.10) we get

$$\Gamma_{41} \wedge \mathrm{d}\Gamma_{41} = 0, \tag{23.11}$$

which is the integrability condition for the existence of a complex function $\bar{\zeta}$ such that $P\Gamma_{41} = -\mathrm{d}\bar{\zeta}$, see (2.43). A rotation (23.4) with $\varrho = \varrho'$, $\Gamma'_{41} = \mathrm{e}^{\mathrm{i}C}\Gamma'_{41}$, can be used to make the function P real, and by a suitable transformation (23.5), $\Gamma'_{41} = A\Gamma_{41}$, we will get $P_{,i}k^i = P_{|4} = 0$. (By means of (23.5) we could arrive at $P = 1$, but at the moment we will not use this special gauge.) Equation (23.7) then reads

$$\omega^2 = \overline{\omega}^1 = -\mathrm{d}\bar{\zeta}/P\varrho = \Gamma_{41}/\varrho, \qquad P = \overline{P}, \qquad P_{|4} = 0. \tag{23.12}$$

Using (2.72) and (23.12), we can compute $\mathrm{d}\omega^2$ as

$$\mathrm{d}\omega^2 = \Gamma^2_{bc}\omega^b \wedge \omega^c = -(\ln P\varrho)_{|b}\, \omega^b \wedge \omega^2 \tag{23.13}$$

and evaluate (23.10), (23.12) and (23.13) and their complex conjugates. The results are

$$\Gamma_{131} = \Gamma_{134} = \Gamma_{124} = \Gamma_{434} = 0 = \lambda = \pi = \varepsilon, \tag{23.14}$$

$$\Gamma_{431} = (\ln \varrho)_{|1}, \quad \Gamma_{121} = (\ln P\varrho)_{|1}, \quad \Gamma_{123} - \Gamma_{132} = (\ln P\varrho)_{|3}, \tag{23.15}$$

$$\varrho_{|4} = \varrho^2, \tag{23.16}$$

$$\varrho(\ln P)_{|3} + \varrho(\Gamma_{213} + \Gamma_{433}) = R_{4123}. \tag{23.17}$$

Equations (23.6) and (23.14) give $\Gamma_{ab4} = 0$, i.e. we have chosen the tetrad ($\boldsymbol{m}, \overline{\boldsymbol{m}}, \boldsymbol{l}, \boldsymbol{k}$) to be parallelly propagated along the rays.

23.1.2. The coordinate frame

We will now choose the coordinate frame as follows. Because of (23.12), the function ζ and its complex conjugate $\bar{\zeta}$ will serve as two (spacelike) coordinates. The third, real, coordinate is the affine parameter r along the rays,

$$k^i = \partial x^i/\partial r = (0, 0, 1, 0). \tag{23.18}$$

Because of (2.17) and (3.11) this implies

$$\mathrm{d}r = WP\bar{\varrho}\omega^1 + \overline{W}P\varrho\omega^2 - H\omega^3 + \omega^4, \tag{23.19}$$

where W is complex and H real. (The coefficients of ω^1 and ω^2 are chosen so that (23.22) will take a simple form.) The fourth, real, coordinate u is introduced by

$$u_{|3} = u_{,n}l^n = 1, \qquad u_{|4} = u_{,n}k^n = 0. \tag{23.20}$$

(23.6) and (23.14) imply that the system (23.20) is integrable: from (7.55) we get $(\Delta D - D\Delta)\,u = (\gamma + \bar{\gamma})\,Du = 0$. Equation (23.20) gives

$$\mathrm{d}u = \bar{\varrho}PL\omega^1 + \varrho P\bar{L}\omega^2 + \omega^3. \tag{23.21}$$

We have thus introduced all four coordinates, the result being

$$\begin{aligned} \omega^1 &= -\mathrm{d}\zeta/P\bar{\varrho} = \overline{m}_n\,\mathrm{d}x^n, \qquad \omega^2 = -\mathrm{d}\bar{\zeta}/P\varrho = m_n\,\mathrm{d}x^n,\\ \omega^3 &= \mathrm{d}u + L\,\mathrm{d}\zeta + \bar{L}\,\mathrm{d}\bar{\zeta} = -k_n\,\mathrm{d}x^n,\\ \omega^4 &= \mathrm{d}r + W\,\mathrm{d}\zeta + \overline{W}\,\mathrm{d}\bar{\zeta} + H\omega^3 = -l_n\,\mathrm{d}x^n. \end{aligned} \tag{23.22}$$

The complex functions ϱ, L, W, and the real functions P, H, are not independent of each other and cannot be chosen arbitrarily, since the metric has to satisfy the field equations (23.10) and the conditions (23.6). We get the restrictions in question by comparing the previous results (23.6) and (23.14)—(23.16) for the Γ_{4bc} (obtained from the field equations (23.10) and from (23.6)) with those coming from the calculation of $\mathrm{d}\omega^3 = -\Gamma_{4bc}\omega^b \wedge \omega^c$. The result is

$$\varrho_{|4} = \varrho^2, \qquad P_{|4} = 0, \qquad L_{|4} = 0, \tag{23.23}$$

$$\varrho P\bar{L}_{|1} - \bar{\varrho}PL_{|2} = \varrho - \bar{\varrho}, \qquad \bar{\varrho}PL_{|3} = (\ln\varrho)_{|1}. \tag{23.24}$$

The evaluation of $\mathrm{d}\omega^4 = -\Gamma_{3bc}\omega^b \wedge \omega^c$ gives an expression for $\Gamma_{314} - \Gamma_{341}$, which, because of (23.14), (23.15) and (23.23), yields the condition

$$\bar{\varrho}PW_{|4} = -(\ln\varrho)_{|1}. \tag{23.25}$$

(Evaluation of the remaining terms, which we will postpone, would give those Γ_{3bc} not yet determined.)

Equations (23.23)—(23.25) are easy to integrate. The result is

$$\begin{aligned} \varrho^{-1} &= -(r + r^0 + \mathrm{i}\Sigma), \qquad 2\mathrm{i}\Sigma = P^2(\bar{\partial}L - \partial\bar{L}),\\ W &= \varrho^{-1}L_{,u} + \partial(r^0 + \mathrm{i}\Sigma), \qquad \partial \equiv \partial_\zeta - L\,\partial_u. \end{aligned} \tag{23.26}$$

(23.26) shows that all metric functions can be given in terms of the coordinate r, the real functions $r^0(\zeta, \bar{\zeta}, u)$, $P(\zeta, \bar{\zeta}, u)$ and $H(\zeta, \bar{\zeta}, r, u)$, and the complex function $L(\zeta, \bar{\zeta}, u)$. We can now summarize the main result in the following theorem:

Theorem 23.1. *A space-time admits a geodesic, shearfree and diverging* ($\varrho \neq 0$) *null congruence* $\boldsymbol{k}$ *and satisfies* $R_{11} = R_{14} = R_{44} = 0$ *if and only if the metric can be written* *in the form*

$$\begin{aligned} ds^2 &= 2\omega^1\omega^2 - 2\omega^3\omega^4 \\ &= \frac{2\,d\zeta\,d\bar\zeta}{P^2\varrho\bar\varrho} - 2[du + L\,d\zeta + \bar L\,d\bar\zeta] \\ &\quad \times \left[dr + W\,d\zeta + \overline{W}\,d\bar\zeta + H\{du + L\,d\zeta + \bar L\,d\bar\zeta\}\right], \end{aligned} \tag{23.27}$$

$$m^i = (-P\bar\varrho, 0, PW\bar\varrho, P\bar\varrho L), \quad \overline{m}^i = (0, -P\varrho, P\varrho\overline{W}, P\varrho\bar L),$$

$$l^i = (0, 0, -H, 1), \quad k^i = (0, 0, 1, 0),$$

where the complex functions ϱ, W, L, *and the real function* P *are subject to* (23.26). (Robinson and Trautman (1962), Debney et al. (1969), Talbot (1969), Robinson et al. (1969a), Lind (1974)).

In (23.27), r is the affine parameter along the rays, u is a retarded time, and ζ and $\bar\zeta$ are (spacelike) coordinates on the 2-surfaces, r, u = const.

23.1.3. Admissible tetrad and coordinate transformations

The vectors $\boldsymbol{l}$ and $\boldsymbol{k}$ are fixed up to a Lorentz transformation (23.5), which will preserve (23.18) and (23.20) only in combination with a coordinate transformation. The metric (23.27) is, therefore, invariant under

$$\begin{aligned} &\boldsymbol{k}' = \boldsymbol{k}F_{,u}, \quad \boldsymbol{l} = \boldsymbol{l}'F_{,u}, \quad F_{,u} > 0, \\ &u' = F(u, \zeta, \bar\zeta), \quad r = r'F_{,u}. \end{aligned} \tag{23.28}$$

This transformation will induce a transformation of the metric functions, e.g.

$$\varrho' = \varrho F_{,u}, \quad P = P'F_{,u}. \tag{23.29}$$

The vectors $\boldsymbol{m}$ (and $\overline{\boldsymbol{m}}$) are fixed up to a rotation (23.4), which keeps the metric invariant and P real if combined with a coordinate transformation $\zeta' = \zeta'(\zeta)$ (ζ' analytic in ζ):

$$\boldsymbol{m}' = e^{iC}\boldsymbol{m}, \quad \zeta' = \zeta'(\zeta), \quad e^{iC} = \left(\frac{d\bar\zeta'/d\bar\zeta}{d\zeta'/d\zeta}\right)^{1/2}, \quad P' = P\left|\frac{d\zeta'}{d\zeta}\right|. \tag{23.30}$$

The possible transformations involving only the coordinates and not the tetrad are (i) changes of the origin of the affine parameter r

$$r' = r + f(u, \zeta, \bar\zeta), \tag{23.31}$$

f being constant along the rays, and (ii) the transformations

$$u' = u + g(\zeta, \bar\zeta). \tag{23.32}$$

The two types of transformations are just the degrees of freedom inherent in the respective definitions (23.18) and (23.20) of the coordinates r and u.

23.2. The line element in the case with non-twisting rays ($\omega = 0$)

If ω vanishes,

$$\varrho = \bar{\varrho} = -\Theta \Leftrightarrow \omega = 0, \tag{23.33}$$

then the line element (23.27) can be further simplified. For $\omega = 0$, the vector field $\boldsymbol{k}$ is normal, i.e. proportional to a gradient, and thus a transformation $\boldsymbol{k}' = A\boldsymbol{k}$ will lead to

$$\boldsymbol{\omega}^3 = -k_i \, \mathrm{d}x^i = \mathrm{d}u. \tag{23.34}$$

(The same result, $L = 0$, could be achieved by starting from (23.22). Because of (23.26) and (23.33), $\boldsymbol{\omega}^2 \wedge \mathrm{d}\boldsymbol{\omega}^3$ is zero, and a transformation (23.28) gives $L = 0$.)

With $L = 0 = \Sigma$, (23.26) gives $\varrho^{-1} = r + r^0(u, \zeta, \bar{\zeta})$; but by a change (23.31) of the affine parameter origin we can always make r^0 vanish. (23.26) then reads

$$r = -\varrho^{-1} = \Theta^{-1}, \qquad L = W = \Sigma = r^0 = 0, \tag{23.35}$$

and instead of (23.22) we get

$$\boldsymbol{\omega}^1 = rP^{-1} \, \mathrm{d}\zeta, \quad \boldsymbol{\omega}^2 = rP^{-1} \, \mathrm{d}\bar{\zeta}, \quad \boldsymbol{\omega}^3 = \mathrm{d}u, \quad \boldsymbol{\omega}^4 = \mathrm{d}r + H \, \mathrm{d}u. \tag{23.36}$$

Theorem 23.2. *A space-time admits a geodesic, shearfree, twistfree and diverging ($\varrho = \bar{\varrho} = -r^{-1}$) null congruence $\boldsymbol{k}$, and satisfies $R_{11} = R_{14} = R_{44} = 0$, exactly if the metric can be written in the form*

$$\begin{aligned}
\mathrm{d}s^2 &= 2\boldsymbol{\omega}^1\boldsymbol{\omega}^2 - 2\boldsymbol{\omega}^3\boldsymbol{\omega}^4 \\
&= 2r^2P^{-2} \, \mathrm{d}\zeta \, \mathrm{d}\bar{\zeta} - 2 \, \mathrm{d}u \, \mathrm{d}r - 2H \, \mathrm{d}u^2, \qquad P_{,r} = 0, \\
m^i &= (P/r, 0, 0, 0), \quad \overline{m}^i = (0, P/r, 0, 0), \\
l^i &= (0, 0, -H, 1), \quad k^i = (0, 0, 1, 0).
\end{aligned} \tag{23.37}$$

This line element is preserved under the transformations

$$\begin{aligned}
&u' = F(u), \quad r = r'F_{,u}, \quad \boldsymbol{k}' = \boldsymbol{k}F_{,u}, \quad \boldsymbol{l} = \boldsymbol{l}'F_{,u}, \\
&\zeta' = \zeta'(\zeta), \quad P' = P\left|\frac{\mathrm{d}\zeta'}{\mathrm{d}\zeta}\right|, \quad \mathrm{e}^{\mathrm{i}C} = \left|\frac{\mathrm{d}\bar{\zeta}'/\mathrm{d}\bar{\zeta}}{\mathrm{d}\zeta'/\mathrm{d}\zeta}\right|^{1/2}, \quad \boldsymbol{m}' = \mathrm{e}^{\mathrm{i}C}\boldsymbol{m}.
\end{aligned} \tag{23.38}$$

Solutions which satisfy the conditions of the above theorem will be called Robinson-Trautman solutions (Robinson and Trautman (1962)).

Chapter 24. Robinson-Trautman solutions

24.1. Robinson-Trautman vacuum solutions

24.1.1. The field equations and their solution

By definition, Robinson-Trautman solutions are solutions meeting the requirements of Theorem 23.2, i.e. they are algebraically special solutions satisfying $\varkappa = \sigma = \omega = 0$, $\Theta \neq 0$, $R_{44} = R_{41} = R_{11} = 0$. To get all vacuum solutions of this class, we start from the metric (23.37) and solve, at least partially, the remaining field equations $R_{12} = R_{13} = R_{33} = R_{34} = 0$.

The connection forms of the 1-forms (23.36) prove to be

$$\begin{aligned} \Gamma_{41} &= -\omega^2/r, \quad \Gamma_{32} = \omega^1[(\ln P)_{,u} + H/r] - \omega^3 P H_{,\bar{\zeta}}/r, \\ \Gamma_{21} + \Gamma_{43} &= -\omega^1 P_{,\zeta}/r + \omega^2 P_{,\bar{\zeta}}/r + \omega^3 H_{,r}, \end{aligned} \tag{24.1}$$

and the surviving tetrad components of the Ricci tensor are

$$R_{12} = R_{1212} - 2R_{1423} = \frac{2P^2}{r^2}(\ln P)_{,\zeta\bar{\zeta}} - \frac{4}{r}\frac{P_{,u}}{P} - \frac{2}{r^2}(Hr)_{,r}, \tag{24.2}$$

$$R_{13} = R_{2113} + R_{4313} = \frac{P}{r}[H_{,r\bar{\zeta}} + (\ln P)_{,u\zeta}], \tag{24.3}$$

$$R_{33} = 2R_{3132} = \frac{2}{r}H_{,u} + \frac{2P^2}{r^2}H_{,\zeta\bar{\zeta}} + \frac{2}{P}H_{,r}P_{,u} + 2P\left(\frac{P_{,u}}{P^2}\right)_{,u} - \frac{4H}{rP}P_{,u}, \tag{24.4}$$

$$R_{34} = R_{3434} + 2R_{1423} = \frac{1}{r^2}(r^2 H_{,r})_{,r} + \frac{2}{r}(\ln P)_{,u}. \tag{24.5}$$

We can now evaluate the vacuum field equations. $R_{12} = 0$ immediately gives

$$2H = \Delta \ln P - 2r(\ln P)_{,u} - 2m/r, \qquad \Delta \equiv 2P^2 \partial_\zeta \partial_{\bar{\zeta}}, \tag{24.6}$$

where m is an arbitrary function of integration (independent of r). $R_{34} = 0$ is then satisfied identically. $R_{13} = 0 = R_{23}$ tells us that m is a function of u alone; by means of a coordinate transformation (23.38) m can be chosen to have the value $m = 0$ or $m = \pm 1$. The last equation $R_{33} = 0$ leads to

$$\Delta\Delta \ln P + 12m(\ln P)_{,u} - 4m_{,u} = 0. \tag{24.7}$$

Theorem 24.1. *The general vacuum solution admitting a geodesic, shearfree, twistfree but diverging null congruence is the Robinson-Trautman metric* (Robinson and

Trautman (1962))

$$ds^2 = \frac{2r^2}{P^2(u, \zeta, \bar{\zeta})} d\zeta\, d\bar{\zeta} - 2\, du\, dr - \left[\Delta \ln P - 2r\, (\ln P)_{,u} - \frac{2m(u)}{r}\right] du^2, \tag{24.8}$$

$$\Delta\Delta(\ln P) + 12m(\ln P)_{,u} - 4m_{,u} = 0, \qquad \Delta \equiv 2P^2\, \partial_\zeta\, \partial_{\bar{\zeta}}.$$

In this metric, r is the affine parameter along the rays of the repeated null eigenvector (r is not necessarily space-like!), and u is a retarded time. The surfaces r, $u = \text{const}$ may be thought of as distorted spheres (if they are closed); the solutions (24.8) are therefore often referred to as describing spherical gravitational radiation. Of course, no exactly spherical gravitational wave exists, since spherical symmetry would imply $\Delta \ln P = K(u)$, and, in the gauge $m = 1$, (24.8) shows that the metric is then static (for $m = 0$, see below under type N). In some special cases, the parameter m has the physical meaning of the system's mass.

For the Robinson-Trautman metric (24.8), the surviving components of the Weyl (curvature) tensor are given by

$$\Psi_2 = -mr^{-3}, \qquad 2\Psi_3 = -r^{-2}P(\Delta \ln P)_{,\bar{\zeta}},$$
$$\Psi_4 = +r^{-2}\left[P^2\left\{\frac{1}{2}\,\Delta \ln P - r(\ln P)_{,u}\right\}_{,\bar{\zeta}}\right]_{,\bar{\zeta}} \tag{24.9}$$

Table 24.1 Petrov types of Robinson-Trautman vacuum solutions ($2H = \Delta \ln P - 2r(\ln P)_{,u} - 2m/r$)

		$6m(P^2H_{,\zeta})_{,\zeta} + rP^2\, (\Delta \ln P)_{,\zeta}$	
		$\neq 0$	$= 0$
m	$\neq 0$	II	D
	$= 0$	III	N, O

24.1.2. Special cases and explicit solutions

Type N solutions, $\Psi_2 = \Psi_3 = 0$

Type N solutions are characterized by $m = 0$ and

$$\Delta \ln P = K(u). \tag{24.10}$$

By a transformation $u' = F(u)$, the Gaussian curvature K of the two-surface $2\, d\zeta\, d\bar{\zeta}/P^2$ could be normalized to $K = 0, \pm 1$. A special solution of (24.10) is

$$P = \alpha(u)\, \zeta\bar{\zeta} + \beta(u)\, \bar{\zeta} + \bar{\beta}(u)\, \zeta + \delta(u), \qquad K = 2(\alpha\delta - \beta\bar{\beta}), \tag{24.11}$$

which gives flat space-time, as Ψ_4 vanishes.

For constant u, (24.11) is the general solution of (24.10). For varying u, the respective $\zeta - \bar{\zeta}$-surfaces of constant curvature may be thought of as mapped onto each other, the mapping having an arbitrary u-dependence. Thus the general solution can be generated from (24.11), by the substitution $\zeta \to \zeta'(u, \zeta)$, $d\zeta \to d\zeta'$, $P \to P\,|d\zeta'/d\zeta|$, in (24.11) and the line element (24.8), where ζ' is an arbitrary function (analytic in ζ). In general, this substitution does not correspond to a coordinate transformation (23.38). For constant u, it represents a mapping of the $\zeta - \bar{\zeta}$-plane into itself. The only one-to-one mapping of this kind is the linear substitution $\zeta' = (a\zeta + b)/(c\zeta + d)$, which leaves (24.11) form-invariant and thus again gives a flat four-dimensional space-time. To get a non-flat type N solution, one has to take a more general (than linear) substitution, but this necessarily gives rise to at least one singular point ζ and thus to a singular line in three-dimensional $(r, \zeta, \bar{\zeta})$-space.

The appearance of singular lines (pipes) is a common feature of many of the Robinson-Trautman solutions, which makes them inadequate for realistically describing the radiation field outside an isolated source.

Type III solutions, $\Psi_2 = 0$

Type *III* solutions are characterized by $m = 0$ and

$$\Delta\Delta \ln P = 0, \qquad (\Delta \ln P)_{,\zeta} \neq 0. \tag{24.12}$$

These conditions immediately give

$$\Delta \ln P = K = -3\left[f(\zeta, u) + \bar{f}(\bar{\zeta}, u)\right], \qquad f_{,\zeta} \neq 0. \tag{24.13}$$

If f is independent of u, then by a coordinate transformation (23.38), $f(\zeta) \to \zeta$, which leaves K invariant, we arrive at

$$\Delta \ln P = K = -3(\zeta + \bar{\zeta}). \tag{24.14}$$

The only known solution of this differential equation is

$$P = (\zeta + \bar{\zeta})^{3/2} \tag{24.15}$$

(Robinson and Trautman (1962)), cp. (3.25). This solution is a Bianchi-type *VI* metric, see § 11.3. and (11.45).

If f depends on ζ *and* u, then, for the purpose of solving the differential equation, we may take u as a parameter in (24.13) and again make the substitution $f(\zeta, u) \to \zeta$ which leads to (24.14). As in the type N case, this substitution is *not* an allowed coordinate transformation.

Theorem 24.2. *To get the most general type III diverging non-twisting vacuum solution one has to* (*a*) *solve* (24.14) *for* P *and* (*b*) *then make the substitution* $\zeta \to f(\zeta, u)$, $P \to P\,|df/d\zeta|$. (Foster and Newman (1967), Robinson (1975)).

A simple example of a metric obtained by this method is

$$P = (\zeta + \bar{\zeta} + u)^{3/2}. \tag{24.16}$$

For all type *III* solutions, because of (24.12), the Gaussian curvature K of the $\zeta - \bar{\zeta}$-surfaces must have singular points.

Type D solutions, $3\Psi_2\Psi_4 = 2\Psi_3{}^2$, $\Psi_2 \neq 0$

The vector $\boldsymbol{l}$ used so far is not necessarily an eigenvector of the Weyl tensor. Thus a type D solution need not have $\Psi_3 = 0 = \Psi_4$; but a null rotation (23.3) which makes $\Psi_3{}'$ and $\Psi_4{}' = 0$ must always exist. This is possible exactly if $3\Psi_2\Psi_4 = 2\Psi_3{}^2$, see (4.30), and the terms with different powers in r entering this condition eventually split it into the two equations

$$P^2[(\Delta \ln P)_{,\zeta}]^2 = 6[P^2(\ln P)_{,u\zeta}]_{,\zeta}, \tag{24.17}$$

$$[P^2(\Delta \ln P)_{,\zeta}]_{,\zeta} = 0, \tag{24.18}$$

which have to be satisfied in addition to the field equation (24.8). Equations (24.18) and (24.17) give

$$P^2(\Delta \ln P)_{,\zeta} = f(u, \bar{\zeta}), \qquad 3\, \partial_u P^{-2} = \partial_{\bar{\zeta}}(fP^{-2}), \tag{24.19}$$

f being an analytic function.

If f is zero, then $K = \Delta \ln P$ is a constant and P is independent of u. The corresponding metrics are

$$ds^2 = \frac{2r^2\, d\zeta\, d\bar{\zeta}}{\left(1 + \frac{K}{2}\,\zeta\bar{\zeta}\right)^2} - 2\, du\, dr - \left(K - \frac{2m}{r}\right) du^2, \quad K = 0, \pm 1. \tag{24.20}$$

For $K = 1$, this is the Schwarzschild metric (13.23). The case $K = 0$ is either ($\varrho > 0$) a special Kasner metric (11.50) or ($\varrho < 0$) a static metric of plane symmetry (13.30) (Bonnor (1970)).

If f is a (nonzero) constant, then by a rotation and a scale transformation in the $\zeta - \bar{\zeta}$-plane it can be chosen to have the value $f = 3\sqrt{2}$. Because of (24.19), P is a function of $u + (\zeta + \bar{\zeta})/\sqrt{2} = u + x$ only and has to satisfy

$$P^2\, \partial_x(P^2\, \partial_x\, \partial_x \ln P) = 6, \quad P = P(x + u), \quad \zeta + \bar{\zeta} = x\sqrt{2}. \tag{24.21}$$

This differential equation can be solved by introducing a new variable η via

$$d\eta/ds = P^{-2}, \qquad s \equiv x + u. \tag{24.22}$$

The solution and the corresponding line element, best written in the coordinates η, y, r, u, are

$$\begin{aligned} ds^2 &= r^2[P^2(\eta)\, d\eta^2 + P^{-2}(\eta)\, dy^2 - 2\, d\eta\, du + P^{-2}(\eta - r^{-1})\, du^2] - 2\, du\, dr, \\ P^{-2}(\eta) &= -2\eta^3 + b\eta + c. \end{aligned} \tag{24.23}$$

This is the static C-metric (given in Table 16.2 in different coordinates).

As can be inferred from the complete list of all type D vacuum solutions (twisting or not) given by Kinnersley (1969b), no solutions with varying f exist.

Type II solutions

In the general case, the field equations (24.8) show that, for a fixed $u = u_0$, the real function $P(\zeta, \bar{\zeta}, u_0)$ can be prescribed arbitrarily.

An explicit type *II* solution can be obtained from (24.15) by allowing $m = \text{const} \neq 0$. It reads (Robinson and Trautman (1962))

$$\mathrm{d}s^2 = 2r^2(\zeta + \bar{\zeta})^{-3}\,\mathrm{d}\zeta\,\mathrm{d}\bar{\zeta} - 2\,\mathrm{d}u\,\mathrm{d}r + [3(\zeta + \bar{\zeta}) + 2m/r]\,\mathrm{d}u^2. \tag{24.24}$$

The solution

$$P = (\zeta + \bar{\zeta} + u)^{3/4}, \qquad m = \frac{1}{4} \tag{24.25}$$

is due to Collinson and French (1967).

Ref.: Kinnersley (1975), Newman (1974), Frolov (1977).

24.2. Robinson-Trautman Einstein-Maxwell fields

24.2.1. Line element and field equations

In this section, we will consider Einstein-Maxwell fields which are algebraically special, the repeated null eigenvector of the Weyl tensor being normal, geodesic and shearfree, and an eigenvector of the Maxwell tensor (aligned case). The conditions of Theorem 23.2 are satisfied, and we can start with the Robinson-Trautman line element (23.37), i.e. with

$$\mathrm{d}s^2 = 2r^2P^{-2}\,\mathrm{d}\zeta\,\mathrm{d}\bar{\zeta} - 2\,\mathrm{d}u\,\mathrm{d}r - 2H\,\mathrm{d}u^2, \qquad P_{,r} = 0. \tag{24.26}$$

By assumption, $\boldsymbol{k}$ is an eigenvector of the Maxwell tensor, and thus the components (7.50)—(7.52) vanish except for

$$\Phi_1 = \frac{1}{2} F_{ab}(k^a l^b + \overline{m}^a m^b), \qquad \Phi_2 = F_{ab}\overline{m}^a l^b. \tag{24.27}$$

The Maxwell equations (7.46)—(7.49) here read

$$r\,\partial_r\Phi_1 = -2\Phi_1, \qquad \partial_\zeta\Phi_1 = 0, \tag{24.28}$$

$$r\,\partial_r\Phi_2 = -\Phi_2 + P\,\partial_{\bar{\zeta}}\Phi_1, \tag{24.29}$$

$$P\,\partial_\zeta\Phi_2 + r(H\,\partial_r - \partial_u)\,\Phi_1 = -2(H + r\,\partial_u \ln P)\,\Phi_1 + P_{,\zeta}\Phi_2. \tag{24.30}$$

Equations (24.28) and (24.29) — in that order — are integrated by

$$\Phi_1 = \frac{\overline{Q}(u, \bar{\zeta})}{2r^2}, \qquad \Phi_2 = -\frac{P\overline{Q}_{,\bar{\zeta}}}{2r^2} + \frac{P}{r}\,h(u, \zeta, \bar{\zeta}). \tag{24.31}$$

$Q(\zeta, u)$ and $h(\zeta, \bar{\zeta}, u)$ are complex functions of their respective arguments and because of (24.30) are restricted by

$$h_{,\zeta} = \partial_u(\bar{Q}/2P^2). \tag{24.32}$$

We can now attack the remaining field equations

$$R_{12} = R_{34} = 2\varkappa_0\Phi_1\bar{\Phi}_1, \tag{24.33}$$

$$R_{13} = 2\varkappa_0\Phi_1\bar{\Phi}_2, \qquad R_{33} = 2\varkappa_0\Phi_2\bar{\Phi}_2, \tag{24.34}$$

cp. (7.16)—(7.21) and (7.53). The expressions for R_{12}, R_{13}, R_{33} and R_{34} in terms of the metric functions can be taken from (24.2)—(24.5).

Equation (24.33) gives the r-dependence of the function H as

$$2H = \Delta \ln P - 2r(\ln P)_{,u} - \frac{2}{r}\, m(\zeta, \bar{\zeta}, u) + \frac{\varkappa_0}{2r^2}\, Q\bar{Q}, \tag{24.35}$$

and equation (24.34) yields the differential equations given below in (24.36d)—(24.36e).

Theorem 24.3. *If an Einstein-Maxwell field admits a geodesic, shearfree, diverging but normal null vector field* $\boldsymbol{k}$*, which is an eigenvector of the Maxwell and Weyl tensors, then the metric is algebraically special and can be written in the form*

$$\mathrm{d}s^2 = 2r^2P^{-2}(\zeta, \bar{\zeta}, u)\, \mathrm{d}\zeta\, \mathrm{d}\bar{\zeta} - 2\, \mathrm{d}u\, \mathrm{d}r$$
$$- \left[\Delta \ln P - 2r(\ln P)_{,u} - \frac{2}{r}\, m(\zeta, \bar{\zeta}, u) + \frac{\varkappa_0}{2r^2}\, Q(\zeta, u)\, \bar{Q}(\bar{\zeta}, u)\right] \mathrm{d}u^2, \tag{24.36a}$$

and the electromagnetic field is given by

$$\Phi_1 = \frac{\bar{Q}}{2r^2}, \quad \Phi_2 = -\frac{P}{2r^2}\, \bar{Q}_{,\bar{\zeta}} + \frac{P}{r}\, h\,(\zeta, \bar{\zeta}, u). \tag{24.36b}$$

P and m are real functions, h is complex and Q analytic in ζ. The four functions have to obey

$$\Delta\Delta \ln P + 12m(\ln P)_{,u} - 4m_{,u} = 4\varkappa_0 P^2 h\bar{h}, \qquad \Delta \equiv 2P^2\, \partial_\zeta\, \partial_{\bar{\zeta}}, \tag{24.36c}$$

$$Q\bar{Q}_{,u} - \bar{Q}Q_{,u} = 2P^2(\bar{h}\bar{Q}_{,\bar{\zeta}} - hQ_{,\zeta}), \tag{24.36d}$$

$$h_{,\zeta} = \left(\bar{Q}/2P^2\right)_{,u}, \qquad m_{,\zeta} = \varkappa_0 \bar{h}\bar{Q}. \tag{24.36e}$$

Like all metrics of the Robinson-Trautman class (23.37), the line element is preserved under the transformations (23.38), i.e. under $\zeta \to \zeta'(\zeta)$ and $u \to F(u)$ and the appropriate change in the metric functions P, Q and m.

To have a rough idea of the physical meaning of the quantities appearing in (24.36), one may think of m and Q as representing, respectively, mass and (electric plus magnetic) charge and of h as an electromagnetic pure radiation field. For $Q = 0$, the Maxwell field is a null field.

The surviving components of the Weyl tensor of the Robinson-Trautman solutions (24.36) prove to be

$$\begin{aligned}\Psi_2 &= -\frac{m}{r^3} + \frac{\varkappa_0 Q\bar{Q}}{2r^4},\\ \Psi_3 &= -\frac{P}{2r^2}(\Delta \ln P)_{,\bar{\xi}} + \frac{3P}{2r^3}m_{,\bar{\xi}} - \frac{\varkappa_0 P}{2r^4}Q\bar{Q}_{,\bar{\xi}},\\ \Psi_4 &= \frac{1}{r^2}\left[P^2\left(\frac{1}{2}\Delta \ln P - r\frac{P_{,u}}{P} - \frac{m}{r} + \frac{\varkappa_0 Q\bar{Q}}{4r^2}\right)_{,\bar{\xi}}\right]_{,\bar{\xi}}.\end{aligned} \tag{24.37}$$

24.2.2. Solutions of type *III*, *N* and *O*

$\Psi_2 = 0$ implies $m = Q = 0$; the Maxwell field is necessarily a null field. For type *III*, because of (24.36e), h is analytic in $\bar{\xi}$ and therefore obeys

$$\Delta \ln (h\bar{h}) = 0, \tag{24.38}$$

which restricts the solutions of the only remaining field equation

$$\partial_\zeta \, \partial_{\bar{\xi}} \, \Delta \ln P = 2\varkappa_0 h\bar{h}, \qquad h\bar{h} \neq 0. \tag{24.39}$$

A special solution of the system (24.38)–(24.39) is (Bartrum (1967))

$$\begin{aligned}P &= l(u)\, k(\zeta)\, \bar{k}(\bar{\zeta})\left(1 + \frac{1}{2}\zeta\bar{\zeta}\right)\\ \sqrt{\varkappa_0}\, h &= \sqrt{2}\, l(u)\, \bar{k}(\bar{\zeta})\, \bar{k}'(\bar{\zeta}).\end{aligned} \tag{24.40}$$

To get type *N* or *O* solutions, we have to require, in addition to $\Psi_2 = 0$, the property $\Psi_3 = 0 = (\Delta \ln P)_{,\zeta}$. Together with the field equations (24.39) this yields $h = 0$.

Theorem 24.4. *There are no non-vacuum Einstein-Maxwell fields of the Robinson-Trautman class which belong to Petrov type N or O.*

24.2.3. Solutions of type *D*

All Robinson-Trautman Einstein-Maxwell fields of type D are known (Debever (1971), Cahen and Sengier (1967), Leroy (1976)). Following Leroy, we will give an exhaustive list of them, without presenting the proofs. The starting point is the field equations (24.36) together with the type D condition $3\Psi_2\Psi_4 = 2\Psi_3^2$. This last is split into five equations by equating to zero the coefficients of the various powers of r. Several cases have to be distinguished.

If $Q = 0$, then Φ_1 vanishes and the electromagnetic field is a null field. The field equation (24.36e) gives $m = m(u)$, and it turns out that the metric of the type D spaces with $Q = 0$ is (Robinson and Trautman (1962))

$$\begin{aligned}ds^2 &= r^2(dx^2 + dy^2) - 2\, du\, dr + 2m(u)\, du^2/r,\\ m(u) &= -\varkappa_0 \int h(u)\, \bar{h}(u)\, du, \qquad \Phi_1 = 0, \qquad \Phi_2 = h(u)/r.\end{aligned} \tag{24.41}$$

If $Q \neq 0$, $Q_{,\zeta} = 0$, then because of (24.36d) Q can be transformed to a complex constant. For $\dot{h} = 0$, the corresponding type D-solutions are

$$\mathrm{d}s^2 = \frac{2r^2\,\mathrm{d}\zeta\,\mathrm{d}\bar{\zeta}}{(1 + K\zeta\bar{\zeta}/2)^2} - 2\,\mathrm{d}u\,\mathrm{d}r - \left(K - \frac{2m}{r} + \frac{\varkappa_0 Q\bar{Q}}{2r^2}\right)\mathrm{d}u^2$$
$$\Phi_1 = \bar{Q}/2r^2, \qquad \Phi_2 = 0, \qquad K = 0, \pm 1, \tag{24.42}$$

Q (complex) and m (real) are arbitrary constants. For $K = 1$, these are the Reissner-Weyl-solutions (13.21). Obviously, (24.42) is the charged counterpart of the vacuum metric (24.20).

For nonzero h, we can set $\varkappa_0 Q\bar{Q} = 1$ by a suitable transformation of u. Then m becomes a function of $s = x + u$ alone. Taking m as a new variable, the corresponding type D solution can be written as

$$\mathrm{d}s^2 = r^2[P^2(m)\,\mathrm{d}m^2 + P^{-2}(m)\,\mathrm{d}y^2 - 2\,\mathrm{d}m\,\mathrm{d}u + P^{-2}(m - r^{-1})\,\mathrm{d}u^2] - 2\,\mathrm{d}r\,\mathrm{d}u,$$
$$P^{-2}(m) = -m^4/2 + am^2 + bm + c,$$
$$\sqrt{\varkappa_0}\,\Phi_1 = \mathrm{e}^{\mathrm{i}q}/2r^2, \qquad \sqrt{\varkappa_0}\,\Phi_2 = \mathrm{e}^{\mathrm{i}q}/rP(m). \tag{24.43}$$

This is the charged C-metric, cp. (24.23) and (19.17); here a, b, c and q are real constants.

It can be shown (Leroy (1976)) that the solutions (24.42) and (24.43) are exactly those for which *both* double eigenvectors of the Weyl tensor are also eigenvectors of the Maxwell field.

If $Q_{,\zeta} \neq 0$, then the type D solutions turn out to be

$$\mathrm{d}s^2 = r^2\,\mathrm{e}^{-x}(\mathrm{d}x^2 + \mathrm{d}y^2) - 2\,\mathrm{d}u\,\mathrm{d}r - \varkappa_0[\mathrm{e}^{3x}(au + b)^2/2r^2 + \mathrm{e}^{2x}a(au + b)/r]\,\mathrm{d}u^2,$$
$$\Phi_1 = \mathrm{e}^{3(x - \mathrm{i}y)/2}(au + b)/2r^2,$$
$$\Phi_2 = -3\mathrm{e}^{2x - 3\mathrm{i}y/2}(au + b)/r^2 2\sqrt{2} - a\,\mathrm{e}^{x - 3\mathrm{i}y/2}/r\sqrt{2}, \tag{24.44}$$

a and b being real constants.

24.2.4. Type *II* solutions

Only some special cases of Petrov type *II* non-twisting Einstein-Maxwell fields have been treated in detail. They refer to special properties of the Maxwell field or to simple functional structures of the metric functions. In particular, the cases $h = 0$, $Q = 0$ and $\Delta \ln P = \text{const}$ have been studied (but in none of these cases is the complete solution known).

If h vanishes, then the field equations (24.36) give (with real q and P_0)

$$m = m(u), \qquad Q = q^2(u)\, f(\zeta), \tag{24.45}$$

$$\Delta\Delta \ln P_0 + 12m(\ln q)_{,u} - 4m_{,u} = 0, \qquad P = q(u)\, P_0(\zeta, \bar\zeta). \tag{24.46}$$

Comparing these equations with the **vacuum** Robinson-Trautman field equations (24.8), we see that the following theorem holds:

Theorem 24.5. *If*

$$\mathrm{d}s_1^2 = 2r^2P^{-2}\,\mathrm{d}\zeta\,\mathrm{d}\bar\zeta - 2\,\mathrm{d}u\,\mathrm{d}r - \left[\Delta \ln P - 2r(\ln P)_{,u} - \frac{2m(u)}{r}\right]\mathrm{d}u^2 \tag{24.47}$$

is a vacuum solution (flat or nonflat) such that P satisfies (24.46), *then*

$$\begin{aligned} \mathrm{d}s_2^2 &= \mathrm{d}s_1^2 - \varkappa_0 q^4(u)\, f(\zeta)\, \bar f(\bar\zeta)\, \mathrm{d}u^2/2r^2, \\ \Phi_1 &= q^2(u)\, \bar f(\bar\zeta)/2r^2, \qquad \Phi_2 = q^3(u) \bar f_{,\bar\zeta}/2r^2 \end{aligned} \tag{24.48}$$

is an Einstein-Maxwell field. $q(u)$ is real, $f(\zeta)$ analytic, and by coordinate transformations $q = 1$ can be achieved. With this choice of coordinates, the field equations (24.46) *give*

$$\Delta\Delta \ln P_0 = k = \mathrm{const}, \qquad 4m = ku. \tag{24.49}$$

Examples of vacuum solutions satisfying (24.46) are (24.15), (24.20) and (24.25).

If the Maxwell field is null ($Q = 0$), the field equations (24.36) give $m = m(u)$, $h = h(\bar\zeta, u)$ and

$$\Delta\Delta \ln P + 12m(\ln P)_{,u} - 4m_{,u} = 4\varkappa_0 P^2 h\bar h. \tag{24.50}$$

Besides the type *III* solution (24.40), and the type *D* solution (24.42), a known solution of this equation is (Bartrum (1967))

$$\begin{aligned} P &= f(\zeta)\, f(\zeta)\, (1 + \zeta\bar\zeta/2), \qquad m = \mathrm{const}, \\ h &= \sqrt{2/\varkappa_0}\, \bar f(\bar\zeta)\, \bar f'(\bar\zeta)\, \mathrm{e}^{\mathrm{i}\varphi(u)}, \qquad \varphi \text{ real.} \end{aligned} \tag{24.51}$$

All type II solutions which (after a suitable choice of the coordinate ζ) *obey*

$$\Delta \ln P = 0, \qquad Q = q(u)\, \mathrm{e}^{\zeta/\sqrt{2}}, \qquad q \text{ real}, \tag{24.52}$$

are explicitly known (Leroy (1976), except (24.53)). Because of the field equations (24.36) and the assumption (24.52), m is a function of u and x alone. It eventually turns out that two cases can occur.

If $3mq_{,u} - 2qm_{,u}$ is zero, then the solution (after a transformation $\zeta \to 4\zeta$) reads

$$\begin{aligned} \mathrm{d}s^2 &= Au\,\mathrm{e}^{-x}r^2(\mathrm{d}x^2 + \mathrm{d}y^2) - 2\,\mathrm{d}u\,\mathrm{d}r - \left[\frac{r}{u} + \frac{2\varkappa_0 A}{3r}\mathrm{e}^{3x} + \frac{\varkappa_0}{2r^2}\mathrm{e}^{4x}\right]\mathrm{d}u^2 \\ \Phi_1 &= \frac{\mathrm{e}^{2(x-\mathrm{i}y)}}{2r^2}, \quad \Phi_2 = -\frac{\sqrt{2Au}}{r^2}\mathrm{e}^{(5x-4\mathrm{i}y)/2} - \frac{\sqrt{2Au}}{2r}\mathrm{e}^{(3x-4\mathrm{i}y)/2} \end{aligned} \tag{24.53}$$

If $3mq_{,u} - 2qm_{,u}$ is nonzero, then the metric and Maxwell field have the form

$$\begin{aligned} ds^2 &= \frac{r^2}{P^2}(dx^2 + dy^2) - 2\,du\,dr - \left(\frac{\varkappa_0 q^2}{2r^2}\,e^x - \frac{2m}{r}\right) du^2, \\ \Phi_1 &= \frac{q}{2r}\,e^{(x-iy)/2}, \quad \Phi_2 = -\frac{Pq}{2r^2\sqrt{2}}\,e^{(x-iy)/2} + \frac{Pm_{,x}}{\varkappa_0 rq\sqrt{2}}\,e^{-(x+iy)/2}, \end{aligned} \tag{24.54a}$$

where P, m and q are given by

$$P = 1, \quad m = m_0 + 2\varepsilon\varkappa_0 u^{-1} q_0^2\,e^x, \quad q = q_0, \quad \varepsilon = 0, 1 \tag{24.54b}$$

or

$$P = e^{x/2}, \quad m = m_0 - \varkappa_0 ax/2 - \varkappa_0 a \ln|q|/8, \qquad q = au + b \tag{24.54c}$$

or

$$P = e^{-x/2}, \quad m = m_0 - a^2(u + ax)/2\varkappa_0 q_0^2, \qquad q = q_0 \tag{24.54d}$$

or

$$\begin{aligned} &P = e^{-x/2}, \quad m = m_0 + \left(\frac{3}{2}\frac{a^2}{\varkappa_0 q_0^4} - \frac{a}{q_0}\,e^x + \frac{\varkappa_0}{6}\,q_0^2\,e^{2x}\right) u^{-1/3}, \\ &q = q_0 u^{1/3} \end{aligned} \tag{24.54e}$$

or

$$\begin{aligned} &P = e^{fx}, \qquad m = m_0 + \frac{2\varkappa_0 a}{(1-2f)(1-10f)}(au+b)^{\frac{1-2f}{10f-1}}\,e^{x(1-2f)} \\ &q = (au+b)^{4f/(10f-1)}, \qquad f(2f-1)(2f+1)(10f-1) \neq 0. \end{aligned} \tag{24.54f}$$

q_0, m_0, a, b, f are real constants. In all these solutions, $P = 1$ or $q = 1$ could be achieved by coordinate transformations (23.38). They all admit at least one Killing vector $\xi = \partial_y$. For special values of the constants of integration, the solutions may be more degenerate than type *II*, e.g. (24.54f) with $m_0 = 0$, $f = 1/6$ gives the type *D* metric (24.44) in somewhat changed coordinates.

All metrics of the form $\Delta \ln P = \varepsilon = \pm 1$, $P_{,u} = 0$, *i.e.*

$$ds^2 = \frac{2r^2\,d\zeta\,d\bar\zeta}{\left(1 + \frac{\varepsilon}{2}\zeta\bar\zeta\right)^2} - 2\,du\,dr + \left(\varepsilon - \frac{2m}{r} + \frac{\varkappa_0 Q\bar Q}{2r^2}\right) du^2, \ \varepsilon = \pm 1 \tag{24.55}$$

have been studied by Kowalczyński (1978). Some of them are covered by Theorem 24.3. The remaining solutions are

$$\begin{aligned} &Q = Q_0\zeta\sqrt{u}, \quad m = m_0 + \varkappa_0 Q_0^2 \ln(1 + \varepsilon\zeta\bar\zeta/2) - \varkappa_0 Q_0^2 \ln u/4, \\ &h = \bar h = \varepsilon Q/(2 + \varepsilon\zeta\bar\zeta)\sqrt{u}, \qquad \varepsilon = \pm 1 \end{aligned} \tag{24.56}$$

and

$$\begin{aligned} &Q = Q_0\sqrt{u}, \quad m = m_0 + \frac{\varkappa_0}{2}\,Q_0\bar Q_0 \ln\left[\frac{1 - \zeta\bar\zeta/2}{\left(1 + \zeta/\sqrt{2}\right)\left(1 + \bar\zeta/\sqrt{2}\right)}\right] - \frac{\varkappa_0}{8}\,Q_0\bar Q_0 \ln u, \\ &h = \frac{Q_0}{2\sqrt{2u}}\,\frac{1 + \zeta/2}{(1 - \zeta\bar\zeta/2)\left(1 + \bar\zeta/\sqrt{2}\right)}, \qquad \varepsilon = -1. \end{aligned} \tag{24.57}$$

Like most of the solutions with a $\zeta - \bar\zeta$-space of constant negative curvature, the solution (24.57) can be interpreted in terms of a tachyon's world line.

24.3. Robinson-Trautman pure radiation solutions

Pure radiation fields

$$T_{mn} = \Phi^2 k_n k_m \tag{24.58}$$

are very similar to electromagnetic null fields. In both cases, the condition $T^{mn}{}_{;n} = 0$ together with $k_{m;n}k^n = 0$ and $k^n{}_{;n} = 2/r$ gives the structure

$$\Phi^2 = n^2(\zeta, \bar{\zeta}, u)/r^2. \tag{24.59}$$

The difference is that in the Maxwell case $n^2 = 2h\bar{h}P^2$ is subject to the further restriction $h_{,\zeta} = 0$.

Starting from (23.37), the calculations run in close analogy with those for the vacuum and Einstein-Maxwell fields and lead to

Theorem 24.6. *All algebraically special solutions with pure radiation fields, the common null eigenvector* $\boldsymbol{k}$ *of the radiation field and the Weyl tensor being geodesic, shearfree, normal but diverging, are given by* (24.58)—(24.59) *and*

$$\mathrm{d}s^2 = \frac{r^2\,\mathrm{d}\zeta\,\mathrm{d}\bar{\zeta}}{P^2(u, \zeta, \bar{\zeta})} - 2\,\mathrm{d}u\,\mathrm{d}r - \left[\Delta \ln P - 2r(\ln P)_{,u} - \frac{2m(u)}{r}\right]\mathrm{d}u^2 \tag{24.60a}$$

$$k^i = (0, 0, 1, 0), \qquad \Delta \equiv 2P^2\,\partial_\zeta\,\partial_{\bar{\zeta}}$$

$$\Delta\Delta \ln P + 12m(\ln P)_{,u} - 4m_{,u} = 4\varkappa_0 n^2(\zeta, \bar{\zeta}, u). \tag{24.60b}$$

In the general case, (24.60b) shows that two functions, say P and m, can be prescribed almost arbitrarily, the limitation being that the left hand side must be positive. One may equally well prescribe n and $m = 0, \pm 1$ (special gauge) for all u and ζ, and $P(\zeta, \bar{\zeta}, u_0)$ for a fixed $u = u_0$.

The components of the Weyl tensor are the same functions of P and m as given in the vacuum case by (24.9). One sees that there are no type N or O ($\Psi_2 = \Psi_3 = 0$) non-vacuum solutions. Type *III* solutions are characterized by $m = 0$, $(\Delta \ln P)_{,\zeta} \neq 0$, $\Delta\Delta \ln P > 0$. Examples of type *III* solutions are

$$P = x^a = \left[(\zeta + \bar{\zeta})\sqrt{2}\right]^a, \qquad 1 < a < 1{,}5. \tag{24.61}$$

All type D solutions are known (Frolov and Khlebnikov (1975)). They are generalizations either of the vacuum solutions (24.20) or of the static C-metric (24.23). In the first case, $K = \Delta \ln P$ is a function only of u, and thus P is given by

$$P = \alpha(u)\,\zeta\bar{\zeta} + \beta(u)\,\zeta + \bar{\beta}(u)\,\bar{\zeta} + \delta(u), \quad K = 2(\alpha\delta - \beta\bar{\beta}). \tag{24.62}$$

$m(u)$ is an arbitrary function, and $n(\zeta, \bar{\zeta}, u)$ can be calculated from (24.60b). For positive K, (24.62) gives the Kinnersley rocket (28.23). If α, β and δ are constant and K is positive, we get the Vaidya solution (13.20). In general, the functions α, β, δ may be interpreted in terms of the acceleration of a "particle" moving along a spacelike, timelike or null world line (Newman and Unti (1963), Frolov and Khlebnikov

(1975), Taub (1976)). In the second case, the metric is given by

$$\mathrm{d}s^2 = r^2 P^2(\eta)\,\mathrm{d}\eta^2 + \frac{r^2\,\mathrm{d}y^2}{P^2(\eta)} - \frac{2r^2}{m(u)}\,\mathrm{d}u\,\mathrm{d}\eta - 2\,\mathrm{d}u\,\mathrm{d}r + \frac{r^2\,\mathrm{d}u^2}{m^2(u)\,P^2(\eta - m/r)}, \tag{24.63}$$

$$P^{-2}(\eta) = -2\eta^3 + b\eta + d$$

(in the coefficient of $\mathrm{d}u^2$, the *argument* of P is $\eta - m/r$). Here b and d are constants, and $m(u)$ is an arbitrary function, from which the radiation field $n(u)$ can be calculated by means of (24.60b), i.e. by

$$-m_{,u} = \varkappa_0 n^2. \tag{24.64}$$

24.4. Robinson-Trautman solutions with a cosmological constant Λ

If we assume an energy-momentum tensor of the form

$$\varkappa_0 T_{mn} = -\Lambda g_{nm}, \qquad \Lambda = \text{const}, \tag{24.65}$$

or add this term to the energy-momentum tensor of the Maxwell field or to that of pure radiation, then the field equations can be solved in a way similar to the case $\Lambda = 0$. The final result of the calculations can be stated as

Theorem 24.7. *If*

$$\mathrm{d}s^2 = 2r^2P^{-2}\,\mathrm{d}\zeta\,\mathrm{d}\bar{\zeta} - 2\,\mathrm{d}u\,\mathrm{d}r - 2H\,\mathrm{d}u^2 \tag{24.66}$$

is a vacuum, Einstein-Maxwell, or pure radiation solution without *the cosmological constant, then*

$$\mathrm{d}s^2 = 2r^2P^{-2}\,\mathrm{d}\zeta\,\mathrm{d}\bar{\zeta} - 2\,\mathrm{d}u\,\mathrm{d}r - (2H - \Lambda r^2/3)\,\mathrm{d}u^2 \tag{24.67}$$

is the corresponding solution including *the cosmological constant Λ.*

Chapter 25. Twisting vacuum solutions

In the preceding chapter, we have treated the *non-twisting* solutions (solutions of the Robinson-Trautman class) in some detail, giving or indicating nearly all proofs, and showing how the variety of known special solutions fits into the framework of the canonical form of metric and field equations.

It is impossible to present the solution with *twisting* degenerate eigenrays in this detailed manner, because it would nearly fill an extra volume. We share this problem with most of the authors writing on the subject. The necessity of presenting the complicated calculations in a compressed form makes some of the papers almost unreadable, and it is sometimes a formidable task merely to check the calculations. What we will try to do here and in the following chapters is to show why, how and how far the integration procedure for the field equations works, and what classes of solutions are known.

25.1. Twisting vacuum solutions — the field equations

25.1.1. The structure of the field equations

To get a better understanding of the structure of the vacuum field equations, we follow Sachs (1962) in dividing them into three sets, namely into

$$\text{Main equations:} \qquad R_{11} = R_{12} = R_{14} = R_{44} = 0 \tag{25.1}$$

$$\text{Trivial equation:} \qquad R_{34} = 0 \tag{25.2}$$

$$\text{Supplementary conditions:} \qquad R_{31} = R_{33} = 0 \tag{25.3}$$

(the indices refers to the tetrad (3.8)). The reason for this splitting is the following property (which can be proved by application of the Bianchi identities): if $\boldsymbol{k}$ is a geodesic and diverging ($k^a{}_{;a} \neq 0$) null congruence, then (a) the trivial equation is satisfied identically and (b) the supplementary conditions hold for all values of the affine parameter r of the null congruence if they hold for a fixed r.

In the context of algebraically special solutions, the most remarkable property of the main equations is that they can be integrated completely, giving the r-dependence of all metric functions in terms of simple rational functions of r. The supplementary conditions then turn out to be differential equations for those constitutive parts of the metric which depend on the three remaining coordinates ζ, $\bar{\zeta}$ and u. In general, these differential equations cannot be solved; they play the role of (and are often called) *the* field equations for algebraically special vacuum metrics.

25.1.2. The integration of the main equations

We have shown in Theorem 23.1. that a space-time admits a geodesic, shearfree and diverging null congruence $\boldsymbol{k}$ (twisting or not) and satisfies $R_{44} = R_{41} = R_{11} = 0$ exactly if the metric can be written in the form

$$\begin{aligned} ds^2 &= 2\omega^1\omega^2 - 2\omega^3\omega^4, \\ \omega^1 &= \overline{m}_n\, dx^n = -d\zeta/P\bar{\varrho}, \qquad \omega^2 = m_n\, dx^n = -d\bar{\zeta}/P\varrho, \\ \omega^3 &= -k_n\, dx^n = du + L\, d\zeta + \bar{L}\, d\bar{\zeta}, \\ \omega^4 &= -l_n\, dx^n = dr + W\, d\zeta + \overline{W}\, d\bar{\zeta} + H\omega^3, \end{aligned} \tag{25.4}$$

with

$$\begin{aligned} m^i &= e_1^i = (-P\bar{\varrho}, 0, PW\bar{\varrho}, P\bar{\varrho}L), \qquad \overline{m}^i = e_2^i, \\ l^i &= e_3^i = (0, 0, -H, 1), \qquad k^i = e_4^i = (0, 0, 1, 0), \end{aligned} \tag{25.5}$$

and

$$\begin{aligned} \varrho^{-1} &= -(r + r^0 + i\Sigma), \qquad 2i\Sigma = P^2(\bar{\partial}L - \partial\bar{L}), \\ W &= \varrho^{-1}L_{,u} + \partial(r^0 + i\Sigma), \qquad \partial \equiv \partial_\zeta - L\,\partial_u, \qquad \bar{\partial} \equiv \partial_{\bar{\zeta}} - \bar{L}\,\partial_u. \end{aligned} \tag{25.6}$$

P, r^0 (both real) and L (complex) are arbitrary functions of ζ, $\bar{\zeta}$ and u, and H (real) is a function of all four coordinates.

Of the main equations (25.1) only $R_{12} = 0$ remains to be integrated. We will do that now, thus obtaining the r-dependence of the function H.

To get the components R_{12} and R_{34} of the Ricci tensor, we start from the connection forms $\boldsymbol{\Gamma}_{ab} = -\boldsymbol{\Gamma}_{ba} = \Gamma_{abc}\omega^c$ of the metric (25.4) and evaluate (3.23c), using the expression (23.17) for R_{4123}. We obtain

$$R_{34} = \varrho^2[\varrho^{-2}(\Gamma_{213} + \Gamma_{433})]_{|4} - 2\varrho(\ln P)_{|3}, \tag{25.7}$$

$$\begin{aligned} R_{12} = {}& (\varrho + \bar{\varrho})\,(\Gamma_{213} + \Gamma_{433}) + 2\varrho\Gamma_{321} + 2\varrho(\ln P)_{|3} + (\ln P)_{|12} \\ & -(\ln P)_{|1}\,(\ln P\bar{\varrho})_{|2} + (\ln P\bar{\varrho}^2)_{|21} - (\ln P\bar{\varrho}^2)_{|2}\,(\ln P\varrho)_{|1}, \end{aligned} \tag{25.8}$$

with

$$\Gamma_{213} = i\,\mathrm{Im}\,[P\varrho\overline{W}_{|1} + H\varrho + (\ln\bar{\varrho})_{|3}], \qquad \Gamma_{433} = H_{|4}, \tag{25.9}$$

$$\Gamma_{321} = -i\,\mathrm{Im}\,[P\varrho\overline{W}_{|1} + H\varrho] + \mathrm{Re}\,[(\ln P\bar{\varrho})_{|3}]. \tag{25.10}$$

To extract the information contained in $R_{12} = 0$ in more readily digestible portions, we start with $R_{34} = 0$ (which is in fact a consequence of the main equations, including $R_{12} = 0$) and integrate it by

$$\Gamma_{213} + \Gamma_{433} = -(\ln P)_{,u} + (m + iM)\,\varrho^2, \qquad (m + iM)_{|4} = 0, \tag{25.11}$$

m and M being real functions of integration. The real part of (25.11) yields

$$H = -(r + r^0)(\ln P)_{,u} + \mathrm{Re}\,[(m + \mathrm{i}M)\,\varrho] + r^0{}_{,u} + K/2, \quad K_{|4} = 0. \quad (25.12)$$

Inserting both results into $R_{12} = 0$ and making repeated use of (23.24) and (25.6), we obtain K in terms of P and L, and the evaluation of the imaginary part of (25.11) eventually gives M in terms of P and L. The result can be summarized as follows:

Theorem 25.1. *A space-time admits a geodesic, shearfree and diverging null congruence* $\boldsymbol{k}$ *and satisfies* $R_{44} = R_{14} = R_{11} = R_{12} = R_{34} = 0$ *exactly if the metric can be written as*

$$\begin{aligned} \mathrm{d}s^2 &= 2\omega^1\omega^2 - 2\omega^3\omega^4, & \omega^1 &= -\mathrm{d}\zeta/P\bar{\varrho} = \overline{\omega}^2, \\ \omega^3 &= \mathrm{d}u + L\,\mathrm{d}\zeta + \bar{L}\,\mathrm{d}\bar{\zeta}, & \omega^4 &= \mathrm{d}r + W\,\mathrm{d}\zeta + \overline{W}\,\mathrm{d}\bar{\zeta} + H\omega^3, \end{aligned} \quad (25.13\mathrm{a})$$

the metric functions satisfying

$$\varrho^{-1} = -(r + r^0 + \mathrm{i}\Sigma), \qquad 2\mathrm{i}\Sigma = P^2(\bar{\partial}L - \partial\bar{L}), \quad (25.13\mathrm{b})$$

$$W = L_{,u}/\varrho + \partial(r^0 + i\Sigma), \qquad \partial \equiv \partial_\zeta - L\,\partial_u, \quad (25.13\mathrm{c})$$

$$H = -(r + r^0)(\ln P)_{,u} - \frac{m(r + r^0) + M\Sigma}{(r + r^0)^2 + \Sigma^2} + \frac{K}{2} + r^0{}_{,u}, \quad (25.13\mathrm{d})$$

$$K = 2P^2\,\mathrm{Re}\,[\partial(\bar{\partial}\ln P - \bar{L}_{,u})], \quad (25.13\mathrm{e})$$

$$M = \Sigma K + P^2\,\mathrm{Re}\,[\partial\,\bar{\partial}\Sigma - 2\bar{L}_{,u}\,\partial\Sigma - \Sigma\,\partial_u\,\partial\bar{L}] \quad (25.13\mathrm{f})$$

(Kerr (1963a), Debney et al. (1969), Robinson et al. (1969a), Trim and Wainwright (1974)).

The line element (25.13) shows a remarkably simple r-dependence. Furthermore, all remaining functions of ζ, $\bar{\zeta}$ and u are given in terms of the complex function L and the real functions r^0, P, and m. As r^0 and P can be removed by coordinate transformations (see § 25.1.4.), L (complex) and m (real) can be considered to represent the metric.

25.1.3. The remaining field equations

We will now see what conditions the remaining field equations, i.e. the supplementary conditions $R_{13} = R_{33} = 0$, impose on the functions L, m, P, and r^0. The metric being given by (25.13a), it is a matter of straightforward but lengthy calculation to compute the curvature tensor and to formulate the field equations $R_{13} = R_{33} = 0$ in terms of the above mentioned functions.

For R_{13} we obtain the expression

$$R_{13} = \varrho^{-1}[\varrho(\Gamma_{213} + \Gamma_{433})]_{|1} + (\ln P)_{|13} - (\ln P)_{|1}\,(\ln P\bar{\varrho})_{|3}, \quad (25.14)$$

and because of (25.11) and (23.24) the field equation $R_{13} = 0$ immediately yields

$$\partial(m + \mathrm{i}M) = 3(m + \mathrm{i}M)\,L_{,u}. \quad (25.15)$$

The calculation of $R_{33} = 0$ is rather lengthy, and we therefore will not give any details. The final consequence proves to be (Kerr (1963a), Debney et al. (1969), Robinson et al. (1969), Trim and Wainwright (1974))

$$[P^{-3}(m + iM)]_{,u} = P[\partial + 2(\partial \ln P - L_{,u})]\, \partial[\bar\partial(\bar\partial \ln P - \bar L_{,u}) + (\bar\partial \ln P - \bar L_{,u})^2]. \tag{25.16}$$

As M is already given in terms of L and P by (25.13f), the four real equations (25.15)–(25.16) form a system of partial differential equations for P, m and L. Note that, because of the definition $\partial = \partial_\zeta - L\,\partial_u$, the function L appears in the differential operator too!

By use of the commutator relations

$$\partial\bar\partial - \bar\partial\partial = (\bar\partial L - \partial \bar L)\,\partial_u, \qquad \partial\partial_u - \partial_u\,\partial = L_{,u}\,\partial_u, \tag{25.17}$$

the field equations, as well as the definitions (25.13) of the metric functions M, W and H, can be put into rather different forms. A formal simplification can be achieved by the introduction of a real "potential" V for the function P by means of

$$V_{,u} = P \tag{25.18}$$

(Robinson and Robinson (1969)). From (25.18) it follows that

$$I \equiv \bar\partial(\bar\partial \ln P - \bar L_{,u}) + (\bar\partial \ln P - \bar L_{,u})^2 = P^{-1}(\bar\partial\bar\partial V)_{,u}, \tag{25.19}$$

and the system of equations (25.13f), (25.15) and (25.16) reads

$$[P^{-3}(m + iM) - \partial\partial\bar\partial\bar\partial V]_{,u} = -P^{-1}(\partial\partial V)_{,u}\,(\bar\partial\bar\partial V)_{,u}, \tag{25.20a}$$

$$\partial(m + iM) = 3(m + iM)\,L_{,u}, \tag{25.20b}$$

$$P^{-3}M = \operatorname{Im} \partial\partial\bar\partial\bar\partial V. \tag{25.20c}$$

In the special gauge $P = 1$, $V = u \Rightarrow \partial V = -L$, these equations take the form given by Kerr (1963), i.e.

$$[(m + iM) + \partial\partial\bar\partial\bar L]_{,u} = -(\partial L)_{,u}\,(\bar\partial\bar L)_{,u}, \tag{25.21a}$$

$$\partial(m + iM) = 3(m + iM)\,L_{,u}, \tag{25.21b}$$

$$M = \operatorname{Im} \bar\partial\bar\partial\partial L. \tag{25.21c}$$

We have given the field equations in three different forms, as (25.21), as (25.20), and as (25.13f), (25.15) and (25.16), which we will use as alternatives. As usual, we have listed the definition (25.13f) or (25.20c) or (25.21c) of M in terms of L and P among the field equations. It makes the structure of the equations more transparent, and in the process of integration one sometimes tries to solve the other two equations first and impose the definition of M as an additional constraint. If a solution m, L, P of the field equations is known, then the metric can be determined from (25.13).

If $L = 0$, M and Σ and W vanish also (if we transform r^0 to zero); because of

(25.20b) and its complex conjugate, m is a function only of u, and (25.16) turns into the field equation (24.7) of the Robinson-Trautman vacuum metrics.

For later use, we list the surviving components of the Weyl tensor. They have the form (Trim and Wainwright (1974), Weir and Kerr (1977))

$$\begin{aligned} \Psi_2 &= (m + \mathrm{i}M)\,\varrho^3, \qquad \Psi_3 = -P^3\varrho^2\,\partial I + O(\varrho^3), \\ \Psi_4 &= P^2\varrho\,\partial_u I + O(\varrho^2), \\ I &\equiv \bar\partial(\bar\partial \ln P - \bar L_{,u}) + (\bar\partial \ln P - \bar L_{,u})^2 = P^{-1}(\bar\partial\bar\partial V)_{,u}. \end{aligned} \tag{25.22}$$

The terms of higher order in ϱ occurring in Ψ_3 or Ψ_4 vanish identically if $\Psi_2 = 0$ or $\Psi_2 = \Psi_3 = 0$, respectively.

It can be inferred from (25.22) that a solution is flat exactly if $m + \mathrm{i}M = 0 = \partial I = \partial_u I$.

25.1.4. Coordinate freedom and transformation properties

We will now have a look at the possible coordinate transformations and the transformation properties of the metric functions and the field equations. As shown in § 23.1.3., there are essentially three types of coordinate transformations.

The first coordinate transformation we will discuss is the freedom (23.31) in the choice of the affine parameter origin $r^0(\zeta, \bar\zeta, u)$. As r^0 enters only into W, ϱ and H, cp. (25.13), but not into those functions L, P, M and m which appear in the field equations, the choice of r^0 is more a matter of taste than of practical value for the integration of the field equations. A possible choice (Debney et al. (1969)) is $r^0 + \mathrm{i}\Sigma = P^2\,\partial L$, but most authors take

$$r^0(\zeta, \bar\zeta, u) = 0, \tag{25.23}$$

and we will stick to that gauge from now on.

The remaining degrees of freedom

$$\zeta' = f(\zeta), \tag{25.24}$$

$$u' = F(\zeta, \bar\zeta, u), \qquad r' = rF_{,u}^{-1}, \tag{25.25}$$

are more important; they change L, $m + \mathrm{i}M$ and P and will often be used in the integration procedure for the field equations. In particular, the transformations (25.24), (25.25), give, together with the appropriate changes (23.28), (23.30) of the tetrad, the transformation laws

$$\omega^{1'} = (f'\bar f'^{-1})^{1/2}\,\omega^1, \qquad \omega^{3'} = F_{,u}\omega^3, \qquad \omega^{4'} = F_{,u}^{-1}\omega^4, \tag{25.26}$$

which imply

$$\begin{aligned} &\varrho' = F_{,u}\varrho, \qquad \Sigma = F_{,u}\Sigma', \qquad \partial' = f'^{-1}\,\partial, \qquad P' = F_{,u}^{-1}\,|f'|\,P, \\ &L' = f'^{-1}(LF_{,u} - F_{,\zeta}), \qquad (m + \mathrm{i}M)' = F_{,u}^{-3}(m + \mathrm{i}M), \\ &(\partial \ln P - L_{,u})' = f'^{-1}\left(\partial \ln P - L_{,u} + \frac{1}{2}\,f''f'^{-1}\right), \\ &I' = \bar f'^{-2}(I + \bar f'''/2\bar f' - 3\bar f''^2/4\bar f'^2). \end{aligned} \tag{25.27}$$

As is to be expected, the field equations are either explicitly invariant or can easily be written in an invariant way. A list of (other) invariants can be found in the paper of Robinson et al. (1969a); among them are $r^{-1}\Sigma$, $\varrho^2 K$, $(m + \mathrm{i}M)\,\varrho^3$.

If we examine the definition $V_{,u} = P$ of the function V in the light of the transformation laws, we can recognize the meaning of this function: V is exactly that transformation F $(f' = 1)$ which transforms an arbitrary P into $P' = 1$. So the timplification of the field equations by introducing V is on an equal footing with she simplification by $P = 1$ and the Kerr form (25.21).

25.2. Some general classes of solutions

25.2.1. Characterization of the solutions

The field equations of the algebraically special vacuum solutions become less complicated if a special dependence of the metric functions on the coordinates is assumed. So far, this has been done in two different ways (leading to all known solutions except Hauser's type N solution (25.71)).

One possible assumption is that all metric functions are independent of ζ and $\bar{\zeta}$; we will discuss this class in § 25.2.6.

The second way is to assume, on the contrary, a rather special dependence of the metric functions on the variable u. To make this dependence explicit, we have to restrict the possible transformations (25.25) of u, which we will do by

$$P_{,u} = 0 \Leftrightarrow u' = uk(\zeta, \bar{\zeta}) + h(\zeta, \bar{\zeta}). \tag{25.28}$$

The main assumptions are now (Robinson and Robinson (1969)) that $L_{,u} - \partial \ln P$ and $m + \mathrm{i}M$ are independent of u. In view of the field equation (25.16) and the gauge $P_{,u} = 0$, these assumptions are equivalent to

$$L_{,u} - (\ln P)_{,\zeta} = G(\zeta, \bar{\zeta}), \tag{25.29}$$

$$(\partial_\zeta - 2G)\,\partial_\zeta I = 0, \qquad I \equiv -\partial_{\bar{\zeta}}\bar{G} + \bar{G}^2. \tag{25.30}$$

The main idea then runs as follows. Assume we are able to solve the differential equation (25.30), which is in fact one of the field equations, and have found a solution G. It then turns out that the second field equation (25.20b) can be solved in the sense that $m + \mathrm{i}M$ is given in terms of G, P and arbitrary functions or constants of integration. The third (and last) field equation (25.20c) can now be reduced to a linear homogeneous differential equation for a real function, or can be solved explicitly. G is sometimes called the background, because $m + \mathrm{i}M = 0$ and $L = [G + (\ln P)_{,\zeta}]\,u$ also give algebraically special solutions. Explicit solutions have been found for several cases, see Table 25.1 and the following sections.

It is possible to reduce and simplify the field equations in a similar way even if $m + \mathrm{i}M$ depends on u (so that (25.30) is no longer true). But no solution is so far known which satisfies (25.29) but not (25.30).

Table 25.1 Twisting algebraically special vacuum solutions with P, $m + \mathrm{i}M$ and $L_{,u} - \partial_\zeta \ln P = G$ independent of u ($I \equiv -\partial_{\bar\zeta}\bar G + \bar G^2$). In all cases the coordinate frame of equation (25.13) is used

	$\partial_\zeta I \neq 0$	$\partial_\zeta I = 0 = I$
$L_{,u} \neq 0$	Equations only: (25.31)–(25.34)	Flat background. All solutions are known and given by (25.50), (25.51), (25.55).
$L_{,u} = 0$	Equations: (25.38)–(25.41) Background: non-twisting type III. Solutions (not exhaustive): (25.42), (25.43)–(25.47) and (33.6)	All solutions are known and given by (25.58)–(25.59).

25.2.2. The case $\partial_\zeta I = \partial_\zeta(\bar G^2 - \partial_{\bar\zeta}\bar G) \neq 0$

The condition (25.30) is a differential equation for the complex function $G(\zeta, \bar\zeta)$, which in full reads

$$(\partial_\zeta - 2G)\, \partial_\zeta(\bar G^2 - \partial_{\bar\zeta}\bar G) = 0. \tag{25.31}$$

If it is satisfied, then, due to the field equations (25.16), the "mass aspect" $m + \mathrm{i}M$ is independent of u; we will write it as $m + \mathrm{i}M = 2P^3 A(\zeta, \bar\zeta)\, [\partial_\zeta I]^{3/2}$. Substituting this expression into (25.20b) and eliminating $L_{,u}$ by means of (25.29), one sees that A is a function only of $\bar\zeta$:

$$m + \mathrm{i}M = 2P^3 A(\bar\zeta)\, [\partial_\zeta(\bar G^2 - \partial_{\bar\zeta}\bar G)]^{3/2}. \tag{25.32}$$

Now only the third and last field equation (25.20c) remains to be solved. If we make the ansatz

$$\begin{aligned} L = {} & (G + \partial_\zeta \ln P)\, u - A(\bar\zeta)\, P^{-1}\, [\partial_\zeta(\bar G^2 - \partial_{\bar\zeta}\bar G)]^{1/2} \\ & + P^{-1}(G + \partial_\zeta)\, [\Phi + \mathrm{i}\Psi] \end{aligned} \tag{25.33}$$

for L, then this field equation shows that $\Phi(\zeta, \bar\zeta)$ is an arbitrary (real) function and that the real function $\Psi(\zeta, \bar\zeta)$ has to obey

$$\begin{aligned} & \Psi_{,\zeta\zeta\bar\zeta\bar\zeta} - [(\bar G^2 - \bar G_{,\bar\zeta})\, \Psi]_{,\zeta\zeta} - [(G^2 - G_{,\zeta})\, \Psi]_{,\bar\zeta\bar\zeta} \\ & + (G^2 - G_{,\zeta})\, (\bar G^2 - \bar G_{,\bar\zeta})\, \Psi = 0. \end{aligned} \tag{25.34}$$

The four equations (25.31)–(25.34) show how to construct an explicit solution: one has to solve (25.31) for $G(\zeta, \bar\zeta)$ and then (25.34) for $\Psi(\zeta, \bar\zeta)$. The functions $P(\zeta, \bar\zeta)$, $\Phi(\zeta, \bar\zeta)$ (both real) and $A(\bar\zeta)$ are disposable; these functions being chosen, L and $m + \mathrm{i}M$ can be determined from (25.33) and (25.32), respectively. The full metric is then obtainable from (25.13); in general, it admits no Killing vector.

Unfortunately, no explicit solution is known in the general case. All known solutions belong to subcases, with either $L_{,u} = 0$ or $\partial_\zeta(\bar G^2 - \bar G_{,\bar\zeta}) = 0$.

25.2.3. The case $\partial_\zeta I = \partial_\zeta(\overline{G}^2 - \partial_{\bar{\zeta}}\overline{G}) \neq 0$, $L_{,u} = 0$

If the function $G(\zeta, \bar{\zeta})$ in (25.29) can be written as the derivative with respect to ζ of a real function, then we can choose the coordinate u (i.e. the function P) so that $L_{,u} = 0$ holds. We then have

$$G(\zeta, \bar{\zeta}) = -(\ln P)_{,\zeta}. \tag{25.35}$$

The coordinate u is now fixed up to a change of its origin ($k = 1$ in (25.28)). All metric functions are independent of u, $\xi = \partial_u$ is a Killing vector.

The "background equation" (25.31) for G now reads

$$\Delta\Delta \ln P \equiv 4P^2\, \partial_\zeta\, \partial_{\bar{\zeta}} P^2\, \partial_\zeta\, \partial_{\bar{\zeta}} \ln P = 0, \tag{25.36}$$

and the restriction $\partial_\zeta I \neq 0$ can be written as

$$2P^2\, \partial_\zeta I = (\Delta \ln P)_{,\zeta} \neq 0. \tag{25.37}$$

Surprisingly, equation (25.36) is exactly the field equation for non-twisting diverging vacuum solutions of type *III* and *N*, and (25.37) ensures that the solutionl is of type *III*, cp. (24.12). So the background for all solutions with $\partial_\zeta I \neq 0$, $L_{,u} = 0 = P_{,u}$, is an arbitrary type *III* twistfree vacuum solution in which the real function P obeys

$$\Delta \ln P(\zeta, \bar{\zeta}) = -3(\zeta + \bar{\zeta}), \tag{25.38}$$

cp. (24.14). The corresponding twisting solutions can be generated as follows (Robinson (1975)).

We start with an arbitrary solution P of (25.38). Because of (25.32)—(25.33) and (25.37)—(25.38), $m + \mathrm{i}M$ does not depend on ζ,

$$m + \mathrm{i}M = 3\mathrm{i}A(\bar{\zeta})\sqrt{3/2}, \tag{25.39}$$

and L can be written as

$$L = (m + \mathrm{i}M)/3P^2 + \partial_\zeta[(\Psi + \mathrm{i}\Phi)/P], \tag{25.40}$$

where Ψ obeys

$$\Psi_{,\zeta\zeta\bar{\zeta}\bar{\zeta}} - (\Psi P_{,\bar{\zeta}\bar{\zeta}}/P)_{,\zeta\zeta} - (\Psi P_{,\zeta\zeta}/P)_{,\bar{\zeta}\bar{\zeta}} + \Psi P_{,\zeta\zeta} P_{,\bar{\zeta}\bar{\zeta}}/P^2 = 0. \tag{25.41}$$

Equation (25.41) has to be solved, and then P, $m + \mathrm{i}M$ and L represent the metric. $\Phi(\zeta, \bar{\zeta})$ and $A(\bar{\zeta})$ are disposable functions, but Φ can be eliminated by a coordinate transformation $u' = u + h(\zeta, \bar{\zeta})$.

A simple solution of this kind is

$$L = (m + \mathrm{i}M)/3P^2 + a/P^2, \qquad a = \text{const (real)}. \tag{25.42}$$

A large class of solutions can be constructed from the only known solution

$$P = (\zeta + \bar{\zeta})^{3/2} = \left(x\sqrt{2}\right)^{3/2} \tag{25.43}$$

of the background equation (25.38). If we choose in (25.40) the function Φ such that $(\Phi/P)_{,y} = (\Psi/P)_{,x}$ is satisfied, then L takes the form

$$L = (m + \mathrm{i}M)/3P^2 + l(x, y), \tag{25.44}$$

in which the real function l is defined by

$$l_{,y} \equiv w = 2^{-3/2}(\Psi/P)_{,\zeta\bar{\zeta}}. \tag{25.45}$$

Instead of (25.41), we then have to solve

$$w_{,xx} + w_{,yy} + 6x^{-1}w_{,x} + 3x^{-2}w = 0. \tag{25.46}$$

The solution can be found by standard separation: the most general solution is a linear superposition of

$$\begin{aligned} w_0 &= x^{(\sqrt{13}-5)/2}(a_0 + b_0 y), \\ w_s &= x^{-5/2} J_{\sqrt{13/4}}(sx)\,[a_s\,\mathrm{e}^{sy} + b_s\,\mathrm{e}^{-sy}], \qquad s^2 \neq 0, \end{aligned} \tag{25.47}$$

$J_n(sx)$ being a Bessel function. Equations (25.13), (25.39), (25.43)–(25.45) and (25.47) exhibit the class of solutions. Special cases have been published earlier (Robinson and Robinson (1969), Held (1974a)); cp. also (33.6).

25.2.4. The case $I = 0$

In the previous two sections we dealt with the case $\partial_\zeta I \neq 0$. For $\partial_\zeta I = 0$, i.e. for $I = I(\bar{\zeta})$, the transformation law (25.27) implies that the coordinate ζ can be chosen so that I vanishes. We will now take that gauge and discuss solutions which obey

$$P_{,u} = 0, \qquad L_{,u} - \partial_\zeta \ln P = G(\zeta, \bar{\zeta}), \qquad I - \bar{G}^2 - \partial_{\bar{\zeta}}\bar{G} = 0. \tag{25.48}$$

The coordinate transformations which preserve (25.48) are

$$\zeta' = (p_1\zeta + q_1)/(p_2\zeta + q_2), \qquad u' = uk(\zeta, \bar{\zeta}) + h(\zeta, \bar{\zeta}). \tag{25.49}$$

Because of the form (25.22) of the Weyl tensor components, $m + \mathrm{i}M = 0$, together with $I = 0$, would give flat space-time. So we have to assume $m + \mathrm{i}M \neq 0$; that excludes the Petrov types *III*, *N* and *O*.

The integration procedure of the field equation runs as follows. As the function $G = \partial_\zeta \ln P - L_{,u}$ undergoes an inhomogeneous transformation law because of (25.27), G can always be made non-zero. The differential equations (25.48) are then integrated by

$$G(\zeta, \bar{\zeta}) = -[\zeta + g(\bar{\zeta})]^{-1}, \qquad L(\zeta, \bar{\zeta}, u) = (G + \partial_\zeta \ln P)\,u + P^{-1}Gl(\zeta, \bar{\zeta}). \tag{25.50}$$

Having thus evaluated the conditions (25.48), we can now attack the three field equations.

The first field equation (25.16) shows that $m + \mathrm{i}M$ does not depend on u, and can therefore be written as $m + \mathrm{i}M = 2P^3G^3A(\zeta, \bar{\zeta})$. The second field equation

(25.20b) then yields $A = A(\bar\zeta)$, i.e.

$$m + \mathrm{i}M = 2P^3G^3A(\bar\zeta). \tag{25.51}$$

With these results, the last field equation (25.20c) can be written in the form

$$\mathrm{Im}\,[\partial_\zeta\,\partial_\zeta(AG - \bar\partial\bar\partial V)] = 0 = (AG - \bar\partial\bar\partial V)_{,\zeta\zeta} - (\bar A\bar G - \partial\partial V)_{,\bar\zeta\bar\zeta}. \tag{25.52}$$

This is exactly the condition for the existence of a real function $B(\zeta, \bar\zeta)$ such that

$$AG - \bar\partial\bar\partial V = B_{,\bar\zeta\bar\zeta}. \tag{25.53}$$

Because of $V_{,u} = P$ and $P_{,u} = 0$, V has the structure $V = Pu + v(\zeta, \bar\zeta)$, with an arbitrary real function v. Inserting this result into equation (25.53), and using (25.51), we obtain

$$AG + \bar G l_{,\bar\zeta} = B_{,\bar\zeta\bar\zeta} + v_{,\bar\zeta\bar\zeta}. \tag{25.54}$$

If we choose $v = -B$, then (25.54) is solved by

$$l(\zeta, \bar\zeta) = -\int \bar A(\zeta)\,\bar G(\bar\zeta, \zeta)\,G^{-1}(\zeta, \bar\zeta)\,\mathrm{d}\zeta + l_1(\bar\zeta). \tag{25.55}$$

To get the explicit form of the metric, one has to prescribe the functions $P(\zeta, \bar\zeta)$, $A(\bar\zeta)$, $g(\bar\zeta)$ and $l_1(\bar\zeta)$. The formulae (25.50), (25.51) and (25.55) will give the functions L and $m + \mathrm{i}M$, which together with P give all the information needed to construct the full metric according to (25.13). As P can be made equal to one by a coordinate transformation (25.49), the above vacuum solutions contain three disposable analytic functions. In general they will not admit any Killing vector.

Physically the class of algebraically special diverging vacuum solutions which satisfy the conditions (25.48) can be characterized (Trim and Wainwright (1974)) as the only solutions which are non-radiative in the sense that the Weyl tensor asymptotically (for large r) behaves as

$$C_{abcd} = II_{abcd}/r^3 + O(r^4). \tag{25.56}$$

25.2.5. The case $I = 0 = L_{,u}$

If we add $L_{,u} = 0$ to the conditions (25.48), then the metric becomes independent of u; $\boldsymbol{\xi} = \partial_u$ is a Killing vector. It can be shown (Trim and Wainwright (1974)) that instead of $L_{,u} = 0$ the equivalent condition $\Sigma_{,u} = 0$ may be imposed. Like $I = 0$, the condition $L_{,u} = 0$ is not invariant (strictly speaking, the condition is that $L_{,u}$ can be made zero). So the coordinate transformations are now restricted to

$$\zeta' = f(\zeta) = (p_1\zeta + q_1)/(p_2\zeta + q_2), \qquad u' = h_0 u + h(\zeta, \bar\zeta). \tag{25.57}$$

Because of (25.48), $L_{,u} = 0$ yields $G = -(\ln P)_{,\zeta}$, which is compatible with $\bar I = G^2 - G_{,\zeta} = 0$ only if $P_{,\zeta\zeta} = 0$, i.e. with real P only if

$$P = \alpha\zeta\bar\zeta + \beta\zeta + \bar\beta\bar\zeta + \delta. \tag{25.58}$$

Equation (25.58) shows that the 2-space $2\,\mathrm{d}\zeta\,\mathrm{d}\bar\zeta/P^2$ has constant curvature $K = 2(\alpha\delta - \beta\bar\beta)$.

Instead of (25.50), (25.51) and (25.55), the constitutive parts of the metric are now given by (25.58) and

$$m + \mathrm{i}M = Z(\bar{\zeta}), \qquad L = -\frac{1}{2P^2}\int \frac{\bar{Z}(\zeta)\,\mathrm{d}\zeta}{(\alpha\zeta + \bar{\beta})^2} + \frac{l_1(\bar{\zeta})}{P^2} \tag{25.59}$$

(note that we have chosen a gauge with $G = -(\alpha\bar{\zeta} + \beta)/P \neq 0$, so that in the case $K = 0$ we have to stick to a P with α or $\beta \neq 0$). The solutions contain the disposable functions $Z(\bar{\zeta})$ and $l_1(\bar{\zeta})$. The value of K can be transformed to $0, \pm 1$, and L may be simplified by a transformation $u' = u + h(\zeta, \bar{\zeta})$, $L' = L - h_{,\zeta}$, e.g. so that L does not contain terms in m.

A subcase of these solutions is the class $m + \mathrm{i}M = \mathrm{const}$. It contains some of the well-known type D solutions such as Kerr and NUT (see below in § 25.5.) as well as the Kerr-Debney (1970)/Demiański (1972) four-parameter (m, M, a, c) solution, which corresponds to

$$L = -\frac{2\mathrm{i}M}{P^2\zeta} - \frac{\mathrm{i}\bar{\zeta}(M + a)}{P^2} - \frac{\mathrm{i}c\bar{\zeta}}{4P^2}\ln\frac{\bar{\zeta}}{\sqrt{2}}. \tag{25.60}$$

For $m = \mathrm{const}$, $M = 0$ the solutions (25.58)—(25.59) specialize to the Kerr-Schild class of vacuum solutions (see § 28.2.), which can be characterized by $\partial\partial V = 0$, or, in the gauge $P = 1$, by $\partial L = 0$.

25.2.6. Solutions independent of ζ and $\bar{\zeta}$

Following Weir and Kerr (1977), we choose coordinates such that $P = 1$ and assume that in this system the metric functions do not depend on ζ and $\bar{\zeta}$. The field equation (25.21 b) then reads

$$-L\,\partial_u(m + \mathrm{i}M) = 3(m + \mathrm{i}M)\,L_{,u} \tag{25.61}$$

(because $\partial = -L\,\partial_u$) and is integrated by

$$m + \mathrm{i}M = \mu_0 L^{-3}(u), \tag{25.62}$$

μ_0 being a complex constant. The second field equation (25.16) can be written as

$$\partial_u(L^2\,\partial I) = 3\mu_0 L^{-3}L_{,u}, \qquad \partial I \equiv -L\,\partial_u\,\partial_u\bar{L}L_{,u}, \tag{25.63}$$

and yields

$$-2\,\partial I = 2L\,\partial_u\,\partial_u\bar{L}L_{,u} = 3\mu_0 L^{-4} + \nu_0 L^{-2}, \tag{25.64}$$

ν_0 being another complex constant.

This last equation and the third field equation $M = \mathrm{Im}\,\bar{\partial}\bar{\partial}\partial L$ give a system of three real differential equations for the complex function $L(u)$. This system can be simplified if we introduce a new real variable w by

$$L = g^{-1/2}(u)\,\mathrm{e}^{-\mathrm{i}\varphi(u)}, \qquad \mathrm{d}w = g^{-3/2}\,\mathrm{d}u, \qquad \partial_u = g^{3/2}\,\partial_w. \tag{25.65}$$

In terms of g and φ, the remaining field equations then read

$$\varphi' g'' - \varphi'' g' + (2\varphi'^3 - \varphi''')\, g = -\mathrm{Im}\, \mu_0\, \mathrm{e}^{3\mathrm{i}\varphi}, \tag{25.66a}$$

$$(2\varphi'^3 + \varphi''')\, g = -\mathrm{Im}\, \nu_0\, \mathrm{e}^{\mathrm{i}\varphi}, \tag{25.66b}$$

and

$$g''' + 6g'\varphi'^2 + 12g\varphi'\varphi'' = -\mathrm{Re}\, (3\mu_0\, \mathrm{e}^{3\mathrm{i}\varphi} + \nu_0 g^{-1}\, \mathrm{e}^{\mathrm{i}\varphi}), \tag{25.67}$$

where g', φ', etc., denote the derivatives with respect to w.

The solutions $g = \text{const}$, $\varphi = \text{const}$ and $\mu_0 = \nu_0 = 0$ give $m + \mathrm{i}M = 0 = \partial I = \partial_u I$ and are therefore flat space-times, cp. (25.22).

Solutions with $\varphi' = 0$ ($\Rightarrow \varphi = 0$ by a $\zeta - \bar{\zeta}$-rotation) have zero twist ($\Sigma = 0$), belong to the Robinson-Trautman class and can easily be transformed to a coordinate system with $L = 0 = W$, $P = P(\zeta, \bar{\zeta}, u)$. Examples of non-twisting solutions of this kind are the C-metric (24.23) ($\Leftrightarrow \nu_0 = 0$), and the type III metric (24.16) ($\Leftrightarrow \mu_0 = 0$).

If $\varphi' \neq 0$, the third differential equation (25.67) is a consequence of the first two (25.66) and need not be considered. The only known solution of this kind (twisting and independent of ζ and $\bar{\zeta}$) is the general solution in the case $\nu_0 = 0$. With $\nu_0 = 0$, the differential equation (25.66b) is integrated by $\mathrm{e}^{-2\mathrm{i}\varphi}(\varphi'^2 - \mathrm{i}\varphi'') = 1/2$, and the complete solution of the system (25.66) is

$$\begin{aligned} &\mathrm{e}^{\mathrm{i}\varphi} = \mathrm{dn}\,(w) + \mathrm{i}\,\mathrm{sn}\,(w)/\sqrt{2}, \qquad \sqrt{2}\,\varphi' = \mathrm{cn}\,(w) = \sqrt{\cos 2\varphi}, \\ &g(w) = \mathrm{Re}\, \alpha_0\, \mathrm{e}^{2\mathrm{i}\varphi} - 4\, \mathrm{Im}\, \mu_0 \varphi'\, \mathrm{e}^{-\mathrm{i}\varphi}. \end{aligned} \tag{25.68}$$

The functions dn, sn and cn are the Jacobian elliptic functions of modulus $1/\sqrt{2}$, and α_0 and μ_0 are complex constants. This solution is the twisting generalization of the C-metric (24.23).

25.3. Solutions of type N ($\Psi_2 = \Psi_3 = 0$)

Due to the structure (25.22) of the components of the Weyl tensor, solutions of Petrov type N are characterized by

$$\begin{aligned} &m + \mathrm{i}M = 0, \qquad \partial_u I \neq 0, \\ &\partial I = \partial[\bar{\partial}(\bar{\partial} \ln P - \bar{L}_{,u}) + (\bar{\partial} \ln P - \bar{L}_{,u})^2] = \partial[P^{-1}(\bar{\partial}\bar{\partial} V)_{,u}] = 0 \end{aligned} \tag{25.69}$$

If these conditions are satisfied, then the field equations reduce to the single equation

$$\mathrm{Im}\, \partial\partial\bar{\partial}\bar{\partial} V = 0. \tag{25.70}$$

Note that no solution of the classes studied in the preceding sections can be of type N, because all such solutions violate $\partial_u I \neq 0$ if satisfying the rest of (25.69).

So far only one (one-parameter) class of solutions of type N has been found (Hauser (1974, 1978)). The calculations which lead to the construction of this solution were *not* carried out in the tetrad and coordinate system used so far, but in a system adapted to the preferred null congruence and the type N conditions in a different way (the Newman-Penrose coefficient τ being nonzero). In detail, the solution is

$$
\begin{aligned}
&\mathrm{d}s^2 = 2\omega^1\omega^2 - 2\omega^3\omega^4,\\
&\omega^1 = -\mathrm{d}\zeta/\bar{\varrho} - A\omega^3 = \overline{\omega}^2,\\
&\omega^3 = -\Sigma\,\mathrm{d}u + \mathrm{i}\sqrt{2}\,\Sigma x(\mathrm{d}\zeta - \mathrm{d}\bar{\zeta}),\\
&\omega^4 = \mathrm{d}r + 3\mathrm{i}\Sigma(A\,\mathrm{d}\zeta - \bar{A}\,\mathrm{d}\bar{\zeta}),\\
&\zeta \equiv (x + \mathrm{i}y)/\sqrt{2}, \qquad \varrho^{-1} = -(r + \mathrm{i}\Sigma),
\end{aligned}
\tag{25.71 a}
$$

(compare also Sommers and Walker (1976)) where Σ and A are given by

$$\Sigma = -(-x)^{3/2} f, \qquad A = \frac{\sqrt{2}}{x}\left[\left(\mathrm{i} - \frac{u}{x^2}\right)\frac{f'}{f} + \frac{3}{4}\right], \tag{25.71 b}$$

in terms of a function $f = f(u/x^2)$ which obeys the differential equation

$$16(1 + u^2/x^4)\, f'' + 3f = 0, \tag{25.71 c}$$

cp. Novoseller (1975). In the notation of (25.13), the Hauser solution is given by $L = 2(u + \mathrm{i})/(\zeta + \bar{\zeta})$, $P = (\zeta + \bar{\zeta})^{7/2} f(u)$ (□ Stephani (1980)).

The type N solution (25.71) admits a Killing vector $\boldsymbol{\xi} = \partial_y$. It contains one essential parameter, which can be chosen to be the ratio of the coefficients of two appropriately chosen solutions of the hypergeometric differential equation (25.71c). The solutions are not asymptotically flat and do not describe the radiation field of an isolated source. The mathematical structure of the general type N field equations has been studied by Sommers (1977) and Ernst and Hauser (1978).

25.4. Solutions of type III ($\Psi_2 = 0$, $\Psi_3 \neq 0$)

Solutions of type III are characterized by

$$m + \mathrm{i}M = 0, \qquad \partial I \neq 0, \tag{25.72}$$

cp. (25.22). All twisting solutions known so far are subcases of the general class treated in § 25.2.3., i.e. in addition to (25.72) they all satisfy $P_{,u} = 0 = \Delta\Delta \ln P$, $L_{,u} = 0$, and can therefore be generated from non-twisting type III solutions. These known solutions (Robinson and Robinson (1969), Held (1974a), Robinson (1975)) are given by (25.43)—(25.47), with, of course, $m + \mathrm{i}M = 0$; they are twisting exactly if $\bar{L}_{,\zeta} \neq L_{,\bar{\zeta}}$.

25.5. Solutions of type D ($3\Psi_2\Psi_4 = 2\Psi_3^2$, $\Psi_2 \neq 0$)

All vacuum solutions of Petrov type D are known (Kinnersley (1969b)). They were found by applying the Newman-Penrose formalism, the main difference from the method outlined in this book being that *both* vectors $\boldsymbol{k}$ and $\boldsymbol{l}$ were chosen to be eigenvectors of the Weyl tensor. A systematic approach on the basis of the canonical tetrad and coordinate system used in this chapter can be found in Weir and Kerr (1977).

From our point of view, the diverging type D solutions divide into two classes, the first covering all solutions of Kinnersley's classes I and II and the second Kinnersley's class III (twisting C-metric).

In detail, the *first class* is contained in the subcase $m + \mathrm{i}M = \text{const}$ of the solutions with $I = 0 = L_{,u}$ discussed in § 25.2.5. These type D solutions are given by (25.58)–(25.59) with $l_1''' = 0$, i.e. they satisfy

$$\begin{aligned} &P = \alpha\zeta\bar{\zeta} + \beta\zeta + \bar{\beta}\bar{\zeta} + \delta, \qquad K = 2(\alpha\delta - \bar{\beta}\beta) = 0, \pm 1, \\ &m + \mathrm{i}M = \text{const}, \qquad\qquad (25.73)\\ &L = -\frac{m - \mathrm{i}M}{2P^2} \int \frac{\mathrm{d}\zeta}{(\alpha\zeta + \bar{\beta})^2} + \frac{1}{P^2}(\lambda_0 + \lambda_1\bar{\zeta} + \lambda_2\bar{\zeta}^2). \end{aligned}$$

Among the constants α, δ, m, M (all real) and β, λ_0, λ_1, λ_2 (all complex) only three real numbers (including m and M) are essential, i.e. cannot be removed by a coordinate transformation.

For nonzero curvature, $K = \pm 1$, the metric can be written as

$$\begin{aligned} &P = 1 + \frac{K}{2}\zeta\bar{\zeta}, \quad L = -\frac{2\mathrm{i}M}{P^2\zeta} - \frac{\mathrm{i}\bar{\zeta}(M + a)}{P^2}, \quad W = \frac{Ka\zeta}{P^2}, \\ &\Sigma = KM - a\frac{1 - K\zeta\bar{\zeta}/2}{1 + K\zeta\bar{\zeta}/2}, \quad H = \frac{K}{2} - \frac{mr + M\Sigma}{r^2 + \Sigma^2}, \quad K = \pm 1, \qquad (25.74) \\ &\mathrm{d}s^2 = 2(r^2 + \Sigma^2)P^{-2}\,\mathrm{d}\zeta\,\mathrm{d}\bar{\zeta} - 2(\mathrm{d}u + L\,\mathrm{d}\zeta + \bar{L}\,\mathrm{d}\bar{\zeta}) \\ &\qquad \times [\mathrm{d}r + W\,\mathrm{d}\zeta + \bar{W}\,\mathrm{d}\bar{\zeta} + H(\mathrm{d}u + L\,\mathrm{d}\zeta + \bar{L}\,\mathrm{d}\bar{\zeta})] \end{aligned}$$

($\lambda_1 \equiv \mathrm{i}(a + M)$; in L the m-term has been transformed to zero.) The Kerr solution (Kerr (1963a)) with $M = 0$, $K = 1$, the Schwarzschild solution $M = a = 0$ and the NUT-solutions (Newman et al. (1963)) $a = 0$ are special cases. The parameters m, M and a are called mass, NUT-parameter and Kerr parameter, respectively.

The *second class* of type D diverging vacuum solutions are those for which the metric does not depend on ζ and $\bar{\zeta}$, cp. § 25.2.6. It turns out (Weir and Kerr (1977)) that the type D-condition exactly reduces to $\nu_0 = 0$. The metric is therefore (25.13) with $P = 1$ and $m + \mathrm{i}M$ and L given by (25.62), (25.65) and (25.68).

Obviously, the coordinate frame best adapted to type D metrics is one which gives a single expression for the metric of *all* diverging vacuum type D metrics. In this frame, the type D vacuum solutions are

$$\begin{aligned} \mathrm{d}s^2 = (1 - pq)^{-2}\Bigg[&\frac{p^2 + q^2}{X}\,\mathrm{d}p^2 + \frac{p^2 + q^2}{Y}\,\mathrm{d}q^2 \\ &+ \frac{X}{p^2 + q^2}(\mathrm{d}\tau + q^2\,\mathrm{d}\sigma)^2 - \frac{Y}{p^2 + q^2}(\mathrm{d}\tau - p^2\,\mathrm{d}\sigma)^2\Bigg], \qquad (25.75) \\ X = X(p) &= \gamma(1 - p^4) + 2lp - \varepsilon p^2 + 2mp^3, \\ Y = Y(q) &= \gamma(1 - q^4) - 2mq + \varepsilon q^2 - 2lq^3 \end{aligned}$$

(Carter (1968b), Debever (1971), Plebański and Demiański (1976), Weir and Kerr (1977)). In (25.75) m, l, ε, and γ are arbitrary real constants. Note that the m of (25.75) is related to, but not always equal to, the mass parameter m used in this chapter; in the case of the twisting C-metric, the m of (25.75) corresponds to Re μ_0 in (25.62). A detailed discussion of the metric (25.75), its various subcases, and its generalization to Einstein-Maxwell-fields is given in § 19.1.

The comparatively simple form (25.75) of the general diverging type D metric indicates that the canonical frame (25.13), which in principle covers all diverging algebraically special vacuum metrics, is not the best frame for type D. It is an open question whether different coordinates would also facilitate the integration of the field equations in the other Petrov types.

25.6. Solutions of type II

The type II solutions explicitly known all belong to the classes treated in §§ 25.2.3.—25.2.5. and its subclasses (e.g. Kerr-Schild). They contain at most three arbitrary analytic functions.

To get an impression of the degrees of freedom inherent in the general solution of the field equations, we write them down in the gauge (25.21), i.e. (with $P = 1$) as

$$[(m + \mathrm{i}M) + \partial\partial\bar{\partial}\bar{L}]_{,u} = -(\partial L)_{,u}\,(\bar{\partial}\bar{L})_{,u}, \tag{25.76}$$

$$\partial(m + \mathrm{i}M) = 3(m + \mathrm{i}M)\,L_{,u}, \tag{25.77}$$

$$M = \mathrm{Im}\,\bar{\partial}\bar{\partial}\partial L, \tag{25.78}$$

with $\partial \equiv \partial_\zeta - L\partial_u$. These equations show that on some surface $u = \text{const}$ the following 5 (complex) functions of ζ and $\bar{\zeta}$ can be prescribed (subject only to the constraint (25.78)): $m + \mathrm{i}M$, L, $L_{,u}$, $L_{,uu}$, $L_{,uuu}$. Equation (25.76) guarantees that the constraint is satisfied for all u if it is for a fixed u. All higher derivatives (with respect to u) of $m + \mathrm{i}M$ and the complex function L can be computed from the field equations. Due to the constraint (one real function) and the freedom $u' = u + h(\zeta, \bar{\zeta})$ in the choice of u (another real function), 4 complex functions of 2 real variables remain as essential degrees of freedom. For type N, the result would be 2 complex functions, and for type III 3 complex functions of two real variables. In all cases, it is the same freedom as in the linearized theory (Sommers (1976)).

Chapter 26. Twisting Einstein-Maxwell and pure radiation field solutions

26.1. The structure of the Einstein-Maxwell field equations

In the following five sections, we are looking for exact solutions of the coupled system,

$$2R_{ab} = \varkappa_0 F^{*c}_{a} \overline{F^{*}_{bc}}, \tag{26.1}$$

$$F^{*ab}{}_{;b} = (F^{ab} + \mathrm{i}\tilde{F}^{ab})_{;b} = 0, \tag{26.2}$$

of Einstein-Maxwell equations in the case when the Weyl tensor possesses a (multiple) null eigenvector $\boldsymbol{k}$ which is twisting, geodesic and shearfree, and is also an eigenvector of the Maxwell tensor F_{ab} (aligned case). The latter condition implies (a) that of the tetrad components (7.50)—(7.52) of the Maxwell tensor only

$$\Phi_1 = \frac{1}{2} F_{ab}(k^a l^b + \overline{m}^a m^b), \qquad \Phi_2 = F_{ab}\overline{m}^a l^b \tag{26.3}$$

can be nonzero (Φ_0 must vanish), and (b) that (therefore)

$$R_{11} = R_{14} = R_{44} = 0 \tag{26.4}$$

holds.

The conditions of Theorem 23.1 being satisfied, we can integrate the equations (26.4) by (23.26)—(23.27), i.e. by

$$\begin{aligned} \mathrm{d}s^2 &= 2\omega^1\omega^2 - 2\omega^3\omega^4, \\ \omega^1 &= -\mathrm{d}\zeta/P\bar{\varrho} = \overline{\omega}^2, \qquad \omega^3 = \mathrm{d}u + L\,\mathrm{d}\zeta + \bar{L}\,\mathrm{d}\bar{\zeta}, \\ \omega^4 &= \mathrm{d}r + W\,\mathrm{d}\zeta + \overline{W}\,\mathrm{d}\bar{\zeta} + H\omega^3, \end{aligned} \tag{26.5}$$

with (in the gauge $r^0 = 0$)

$$\begin{aligned} \varrho^{-1} &= -(r + \mathrm{i}\Sigma), \qquad 2\mathrm{i}\Sigma = P^2(\bar{\partial}L - \partial\bar{L}), \\ W &= \varrho^{-1}L_{,u} + \mathrm{i}\,\partial\Sigma, \qquad \partial \equiv \partial_\zeta - L\,\partial_u. \end{aligned} \tag{26.6}$$

All metric functions are given in terms of r, $P(\bar{\zeta}, \zeta, u)$, $L(\zeta, \bar{\zeta}, u)$ and $H(\bar{\zeta}, \zeta, r, u)$ (for more details see Chapter 23.).

The metric (26.5)—(26.6) is of course subject to the rest of Einstein's equations (26.1),

$$R_{12} = R_{34} = 2\varkappa_0\Phi_1\overline{\Phi}_1, \tag{26.7}$$

$$R_{13} = 2\varkappa_0\Phi_1\overline{\Phi}_2, \qquad R_{33} = 2\varkappa_0\Phi_2\overline{\Phi}_2, \tag{26.8}$$

and Maxwell's equations (26.2) have to be solved too. In close analogy with the vacuum case (§ 25.1.) the calculations naturally divide in two steps. In the first step, the radial dependence of the metric (i.e. in addition to (26.6) that of the function H) and of the Maxwell field will be completely determined. In the second step, the remaining field equations (both Einstein and Maxwell) will be reduced to a system of partial differential equations for the as yet undetermined functions of ζ, $\bar{\zeta}$, u which enter into the metric and the electromagnetic field.

26.2. Determination of the radial dependence of the metric and the Maxwell field

In the metric (26.5), the spin coefficients τ, λ, π, ε, $\varkappa$, and σ vanish (cp. § 23.1.1.), and Φ_0 is zero by assumption. So the first part (7.46)—(7.47) of the Maxwell equations reads

$$\partial_r \Phi_1 = 2\varrho\Phi_1, \qquad \partial_r \Phi_2 = \varrho\Phi_2 - P\varrho\,\bar{\partial}\Phi_1 + P\varrho\overline{W}\,\partial_r\Phi_1. \tag{26.9}$$

Using (26.6), these equations can be integrated and yield

$$\Phi_1 = \varrho^2\Phi_1^0(\zeta, \bar{\zeta}, u), \tag{26.10 a}$$

$$\Phi_2 = \varrho\Phi_2^0(\zeta, \bar{\zeta}, u) + \varrho^2 P(2\bar{L}_{,u}\Phi_1^0 - \bar{\partial}\Phi_1^0) + 2\mathrm{i}\varrho^3 P(\Sigma\bar{L}_{,u} - \bar{\partial}\Sigma)\,\Phi_1^0. \tag{26.10 b}$$

As in the non-twisting case, one may think of Φ_1^0 and Φ_2^0 as representing the charge and a pure radiation field, respectively.

We can now evaluate the field equations (26.7). Substituting the expressions (25.7) and (26.10a) for R_{34} and Φ_1, we obtain

$$\varrho^2[\varrho^{-2}(\Gamma_{213} + \Gamma_{433})]_{|4} = 2\varrho(\ln P)_{,u} + 2\varkappa_0\varrho^2\bar{\varrho}^2\Phi_1^0\overline{\Phi}_1^0. \tag{26.11}$$

This equation yields

$$\Gamma_{213} + \Gamma_{433} = -(\ln P)_{,u} + 2\varkappa_0\bar{\varrho}\varrho^2\Phi_1^0\overline{\Phi}_1^0 + (m + \mathrm{i}M)\,\varrho^2, \tag{26.12}$$

(because $\varrho_{|4} = \varrho^2$), and, as Γ_{213} is purely imaginary and Γ_{433} equals $H_{|4}$, we can determine H,

$$H = -r(\ln P)_{,u} - (mr + M\Sigma - \varkappa_0\Phi_1^0\overline{\Phi}_1^0)/(r^2 + \Sigma^2) + K/2, \tag{26.13}$$

the real functions m, M and K being independent of r. After some calculations it then turns out that the term proportional to $\Phi_1^0\overline{\Phi}_1^0$ does not enter into those equations which as a consequence of the field equations (26.7) give M and K in terms of P and L. So we can take M and K from (25.13); they are the same functionals as in the vacuum case.

Theorem 26.1. *An (algebraically special) Einstein-Maxwell field admits a diverging, geodesic and shearfree null congruence* $\boldsymbol{k}$ *and satisfies* $R_{11} = R_{14} = R_{44} = 0$ *(aligned case),* $R_{12} = R_{34} = 2\varkappa_0\Phi_1\overline{\Phi}_1$ *and the radial part* (7.46)—(7.47) *of Maxwell's equations* *exactly if the metric and the electromagnetic field can be given by*

$$\begin{aligned} ds^2 = \frac{2\,d\zeta\,d\bar{\zeta}}{P^2\varrho\bar{\varrho}} - 2(du + L\,d\zeta + \bar{L}\,d\bar{\zeta})\,[dr + W\,d\zeta + \bar{W}\,d\bar{\zeta} \\ + H(du + L\,d\zeta + \bar{L}\,d\bar{\zeta})], \end{aligned}$$

$$\begin{aligned} &H = -r(\ln P)_{,u} - (mr + M\Sigma - \varkappa_0\Phi_1^0\bar{\Phi}_1^0)\,\varrho\bar{\varrho} + K/2, \\ &\varrho^{-1} = -(r + i\Sigma), \qquad 2i\Sigma = P^2(\bar{\partial}L - \partial\bar{L}), \\ &W = \varrho^{-1}L_{,u} + i\,\partial\Sigma, \qquad K = 2P^2\,\mathrm{Re}\,[\partial(\bar{\partial}\ln P - \bar{L}_{,u})], \\ &M = \Sigma K + P^2\,\mathrm{Re}\,[\partial\bar{\partial}\Sigma - 2\bar{L}_{,u}\,\partial\Sigma - \Sigma\partial_u\,\partial\bar{L}], \end{aligned} \tag{26.14}$$

and

$$\Phi_1 = \varrho^2\Phi_1^0, \quad \Phi_2 = \varrho\Phi_2^0 + \varrho^2P(2\bar{L}_u - \bar{\partial})\,\Phi_1^0 + 2i\varrho^3P(\Sigma\bar{L}_{,u} - \bar{\partial}\Sigma)\,\Phi_1^0 \tag{26.15}$$

(Robinson et al. (1969b); Trim and Wainwright (1974)).

Equations (26.14)—(26.15) show that all field functions can be constructed from the real functions P and m and the complex functions L, Φ_1^0, Φ_2^0, which are all independent of r. These (so far arbitrary) functions of ζ, $\bar{\zeta}$ and u are of course subject to the remaining field equations, both Einstein and Maxwell, which we will now formulate. Concerning the coordinate freedom and the transformation properties of the various functions we refer to § 25.1.4. The transformation properties given there should be completed by

$$\Phi_1^{0\prime} = F_{,u}^{-2}\Phi_1^0, \qquad \Phi_2^{0\prime} = (\bar{f}'^{-1}f')^{1/2}\,F_{,u}^{-2}\Phi_2^0. \tag{26.16}$$

26.3. The remaining field equations

The second part (7.48)—(7.49) of Maxwell's equations, which has to be satisfied in addition to (26.9), reads

$$\delta\Phi_1 = 0, \qquad \delta\Phi_2 - \Delta\Phi_1 - 2\mu\Phi_1 + 2\beta\Phi_2 = 0. \tag{26.17}$$

Taking $\mu = -\Gamma_{321}$ from (25.10), and substituting $2\beta = -(\ln P)_{|1}$ and the expressions (26.14)—(26.15) for the metric and the Maxwell field, a straightforward calculation yields the simple equations

$$(\partial - 2L_{,u})\,\Phi_1^0 = 0, \tag{26.18a}$$

$$(\partial - L_{,u})\,(P^{-1}\Phi_2^0) + (P^{-2}\Phi_1^0)_{,u} = 0. \tag{26.18b}$$

The two Einstein equations (26.8) not yet taken into account can be simplified in a way analogous to that in the vacuum case. The final result (Robinson et al. (1969b), Lind (1974), Trim and Wainwright (1974)) reads

$$P(3L_{,u} - \partial)\,(m + iM) = 2\varkappa_0\Phi_1^0\bar{\Phi}_2^0, \tag{26.19a}$$

$$\begin{aligned} P^4(\partial - 2L_{,u} + 2\,\partial\ln P)\,\partial[\bar{\partial}(\bar{\partial}\ln P - \bar{L}_{,u}) + (\bar{\partial}\ln P - \bar{L}_{,u})^2] \\ - P^3[P^{-3}(m + iM)]_{,u} = \varkappa_0\Phi_2^0\bar{\Phi}_2^0, \end{aligned} \tag{26.19b}$$

$$P^{-3}M = \mathrm{Im}\,(\partial\partial\bar{\partial}\bar{\partial}V), \qquad V_{,u} \equiv P. \tag{26.19c}$$

The five equations (26.18)—(26.19) form a system of partial differential equations for the functions P, m (real) and L, Φ_1^0, Φ_2^0 (complex). If a solution has been found, then the full metric and the Maxwell field can be obtained from (26.3), (26.14) and (26.15). Different forms of the field equations may easily be derived from (25.20) and (25.21).

The equations (26.18)—(26.19) generalize the vacuum equations (25.15)—(25.16) as well as the Einstein-Maxwell equations (24.36) of the non-twisting case. To achieve conformity with the notation in the non-twisting case, we have to put

$$\Phi_1^0 = \bar{Q}/2, \qquad \Phi_2^0 = -Ph. \tag{26.20}$$

The detailed expressions for the nonzero components of the Weyl tensor can be found in the paper of Trim and Wainwright (1974). Here we mention only that they have the structure

$$\begin{aligned} \Psi_2 &= (m + \mathrm{i}M)\,\varrho^3 + \varkappa_0 Q\bar{Q}\bar{\varrho}\varrho^3/2, \\ \Psi_3 &= -P^3\varrho^2\,\partial I + O(\varrho^3), \qquad \Psi_4 = P^2\varrho\,\partial_u I + O(\varrho^2), \\ I &\equiv \bar{\partial}(\bar{\partial}\ln P - \bar{L}_{,u}) + (\bar{\partial}\ln P - \bar{L}_{,u})^2 = P^{-1}(\bar{\partial}\bar{\partial}V)_{,u}, \end{aligned} \tag{26.21}$$

where the terms of higher order in ϱ occurring in Ψ_3 or Ψ_4 vanish identically if Ψ_2 or Ψ_2 and Ψ_3, respectively, vanish.

26.4. Charged vacuum metrics

By inspecting the field equations (26.19) one immediately sees that they reduce to the vacuum case if no free radiation field is present, i.e. if

$$\Phi_2^0 = 0, \qquad \Phi_1^0 = \bar{Q}(\zeta, \bar{\zeta}, u)/2 \tag{26.22}$$

holds. In that case, Maxwell's equations (26.18) yield

$$Q = q(\zeta, \bar{\zeta})\,P^2 \tag{26.23}$$

and (in the gauge $P_{,u} = 0$)

$$P^2[\partial_\zeta - 2(L_{,u} - \partial_\zeta \ln P)]\,q = 0. \tag{26.24}$$

As q and P do not depend on u, the same is true for $L_{,u} - \partial_\zeta \ln P$, i.e. as a consequence of Maxwell's equations we obtain

$$L_{,u} - \partial \ln P = G(\zeta, \bar{\zeta}), \qquad P_{,u} = 0. \tag{26.25}$$

This condition coincides with the assumption (25.29) we made in the vacuum case to get special classes of solutions, cp. § 25.2.1. We thus have proved the following generalization of Theorem 24.5 (on Robinson-Trautman Einstein-Maxwell fields):

Theorem 26.2. *All Einstein-Maxwell fields (aligned case) admitting a diverging, geodesic and shearfree null congruence with a non-radiative ($\Phi_2^0 = 0$) Maxwell field are* *given by*

$$\mathrm{d}s^2 = \mathrm{d}s_0^2 - \frac{\varkappa_0}{2}\,\bar{Q}Q\varrho\bar{\varrho}(\mathrm{d}u + L\,\mathrm{d}\zeta + \bar{L}\,\mathrm{d}\bar{\zeta})^2,$$

$$\bar{Q} = \alpha(\bar{\zeta})\,P^2 \exp\left(2\int G(\zeta,\bar{\zeta})\,\mathrm{d}\zeta\right), \tag{26.26}$$

$$\Phi_1 = \bar{Q}\varrho^2/2, \qquad \Phi_2 = \frac{1}{2}\,P\varrho^2(2\bar{L}_{,u} - \partial_{\bar{\zeta}})\,\bar{Q} + \mathrm{i}P\varrho^3\bar{Q}(\Sigma\bar{L}_{,u} - \bar{\partial}\Sigma),$$

where $\mathrm{d}s_0^2$ *is an algebraically special* vacuum *metric subject to* (26.25), *and* $\alpha(\bar{\zeta})$ *is a disposable function* (Robinson et al. (1969b), Trim and Wainwright (1974)).

In § 25.2. we gave a survey of all explicitly known vacuum solutions which satisfy the above conditions (and, moreover, $(m + \mathrm{i}M)_{,u} = 0$). For the subcases treated there, the complex (electric plus magnetic) charge Q is given by

$$\bar{Q}(\bar{\zeta}) = \alpha(\bar{\zeta}) \qquad \text{for } L_{,u} = 0, \tag{26.27}$$

$$\bar{Q}(\zeta,\bar{\zeta}) = \alpha(\bar{\zeta})\,P^2G^2 \qquad \text{for } I = 0. \tag{26.28}$$

Special cases included here are the Reissner-Weyl solution (13.21), the charged NUT and Kerr-Schild solutions etc. Note that the charged C-metric (19.17) and its twisting generalization are *not* covered by Theorem 26.2 as in both cases the Maxwell field contains a pure radiation field ($\Phi_2^0 \neq 0$).

The Einstein-Maxwell fields covered by (26.28) together with (26.26) are exactly those fields which are non-radiative in the sense that the Weyl tensor and the Maxwell tensor do not contain terms with r^{-n}, where $n < 3$ and $n < 2$, respectively (Trim and Wainwright (1974)). Solutions which furthermore are regular and stationary must be type D (Lind (1975a, b), Held (1976a)).

26.5. Remarks concerning solutions of the different Petrov types

Several *type D* Einstein-Maxwell fields are known, all depending on (at most six) arbitrary parameters. Most of these solutions are contained in the Plebański-Demiański solution (which also includes some Petrov type O metrics); see § 19.1.2. for a detailed discussion. In addition to the charged Kerr-NUT metrics covered by Theorem 26.2, the charged and twisting C-metric is included. All solutions of the Plebański-Demiański class have the property that *both* null eigenvectors of the Maxwell field are (multiple) eigenvectors of the Weyl tensor (Debever (1976), Leroy (1978)). Besides these solutions, one more type D solution can be generated by means of Theorem 26.2, namely (Leroy (1978))

$$\begin{aligned} \mathrm{d}s^2 &= 2\,\mathrm{d}\zeta\,\mathrm{d}\bar{\zeta}/\varrho\bar{\varrho} - 2(\mathrm{d}u + L\,\mathrm{d}\zeta + \bar{L}\,\mathrm{d}\bar{\zeta}) \\ &\quad \times \left[\mathrm{d}r + \frac{\varkappa_0}{4}\,Q\bar{Q}\varrho\bar{\varrho}(\mathrm{d}u + L\,\mathrm{d}\zeta + \bar{L}\,\mathrm{d}\bar{\zeta})\right], \\ L &= \mathrm{i}\Sigma\zeta, \qquad \bar{Q} = \bar{\zeta}^{-3}\,\mathrm{e}^{\mathrm{i}b}, \qquad b, \Sigma = \text{const}. \end{aligned} \tag{26.29}$$

This metric is the twisting generalization of the Einstein-Maxwell field (24.44) in the subcase $m = 0$; the corresponding vacuum solution is flat. For (26.29), only one eigenvector of the Maxwell field coincides with an eigenvector of the Weyl tensor.

Comparing the list of known twisting (and diverging) type D Einstein-Maxwell fields with the exhaustive lists of type D twisting vacuum or non-twisting Einstein-Maxwell solutions, one sees that for each non-twisting or vacuum solution a charged and twisting counterpart exists. Any as yet unknown Type D Einstein-Maxwell fields cannot be generated from vacuum solutions by simply adding a charge in the sense of Theorem 26.2; i.e. they must be radiative.

Concerning *type* N (*or* O) Einstein-Maxwell fields, an inspection of the structure (26.21) of the Weyl tensor shows that $\Psi_2 = \Psi_3 = 0$ implies $\Phi_1^0 = 0 = m + iM$ and $\partial I = 0$. But under these assumptions, the field equation (26.19b) yields $\Phi_2^0 = 0$; there is no Maxwell field at all. As all aligned type N (or O) Einstein-Maxwell fields must have $\varkappa = \sigma = 0$, cp. § 7.5., we thus obtain

Theorem 26.3. *There are no non-vacuum diverging Einstein-Maxwell fields (aligned case) of Petrov type N or O.*

As a consequence of this theorem, type N Einstein-Maxwell fields are either non-aligned (i.e. the repeated null eigenvector of the Weyl tensor is not an eigenvector of the Maxwell field), cf. § 22.1. and (22.5), or they are aligned and non-diverging, cf. Chapter 27.

We close this section with two remarks. It can be shown that no *regular* diverging (aligned) type III Einstein-Maxwell field exists, the only regular diverging, geodesic and shearfree non-radiating fields being the Kerr-Newman-solutions (Lind (1975a)). All known algebraically special, diverging and twisting Einstein-Maxwell solutions (aligned case) are given by the type D metrics listed above and the charged metrics covered by Theorem 26.2. To the authors' knowledge, no solution at all has been published which corresponds to a purely radiative Maxwell field ($\Phi_0 = 0 = \Phi_1$).

26.6. Pure radiation fields

26.6.1. The field equations

Pure radiation fields

$$T_{mn} = \Phi^2 k_n k_m, \tag{26.30}$$

k being a geodesic and shearfree multiple eigenvector of the Weyl tensor, are similar to electromagnetic null fields ($\Phi_0 = \Phi_1 = 0$) in so far as (because of $T^{mn}{}_{;n} = 0$, $k_{m;n}k^n = 0$, $k^n{}_{;n} = -(\varrho + \bar{\varrho})$) the factor Φ^2 has the same ϱ-dependence

$$\Phi^2 = n^2(\zeta, \bar{\zeta}, u)\, \varrho\bar{\varrho} \tag{26.31}$$

as the corresponding expression $2\Phi_2\bar{\Phi}_2$ in the electromagnetic case. The difference is that in the electromagnetic case $n^2 = 2P^2 h\bar{h}$ is subject to the additional restriction $(\partial - L_{,u})\, h = 0$. Note that in the Maxwell case a null field is *necessarily* geodesic and shearfree and the metric must be algebraically special (§ 7.5.), whereas here we have to assume these properties.

As a consequence of this similarity, the metric of a pure radiation field has exactly the form (26.14) of an Einstein-Maxwell field, with, of course, $\Phi_1^0 = 0$ (so that in fact it looks like a vacuum solution), and the field equations read

$$(3L_{,u} - \partial)(m + \mathrm{i}M) = 0, \tag{26.32a}$$

$$M = P^3 \operatorname{Im} \partial\partial\bar{\partial}\bar{\partial}V = \Sigma K + P^2 \operatorname{Re}[\partial\bar{\partial}\Sigma - 2\bar{L}_{,u}\,\partial\Sigma - \Sigma\,\partial_u\,\partial\bar{L}], \tag{26.32b}$$

and

$$\begin{aligned} &P^4(\partial - 2L_{,u} + 2\,\partial \ln P)\,\partial I - P^3[P^{-3}(m + \mathrm{i}M)]_{,u} = \varkappa_0 n^2(\zeta, \bar{\zeta}, u)/2, \\ &I \equiv \bar{\partial}(\bar{\partial} \ln P - \bar{L}_{,u}) + (\bar{\partial} \ln P - \bar{L}_{,u})^2. \end{aligned} \tag{26.33}$$

As $n^2(\zeta, \bar{\zeta}, u)$ is an arbitrary (positive) function, equation (26.33) is in fact only the definition of n^2, and not an equation which needs to be integrated. The actual field equations are (26.32), which are the same as in the vacuum case. Accordingly, one can try to generate radiation field solutions from vacuum metrics (possibly flat) by simply changing one or more of the metric functions in such a way that (26.32) remains satisfied, but (26.33) gives a nonzero n^2. Before we discuss this idea in detail, we note that all known twisting (and shearfree) pure radiation field solutions have either been found by application of this generation procedure (see below) or by the Kerr-Schild ansatz (see § 28.4. for details and the list of solutions). In the non-twisting case, the equations (26.32) are satisfied identically, and (26.33) simplifies to (24.60b), see § 24.3.

26.6.2. Generating pure radiation field solutions from vacuum by changing P

In dealing with the coupled system (26.32) of field equations it simplifies the task if we confine ourselves to metrics which satisfy (26.32a) identically by $L_{,u} = 0 = M$, $m = \text{const}$. If we start from a vacuum (or pure radiation) metric (L^0, P^0, m^0, $M^0 = 0$) of this type, leave L^0, m^0 and M^0 unchanged, but make the ansatz $P = P^0 A(\zeta, \bar{\zeta}, u)$ for P, then the remaining field equation (26.32b), i.e. $M = 0$, reduces to a single linear partial differential equation for the function A. By straightforward calculation we arrive at the following result (Stephani (1979)):

Theorem 26.4. *If (L^0, P^0, m^0, M^0) is an algebraically special diverging vacuum solution satisfying*

$$L^0{}_{,u} = 0, \qquad M^0 = 0, \qquad m^0 = \text{const}, \tag{26.34}$$

then

$$L = L^0, \qquad M = 0, \qquad m = \text{const}, \qquad P = P^0 A(\zeta, \bar{\zeta}, u], \tag{26.35}$$

represents a pure radiation field exactly if the real function A obeys

$$\begin{aligned} &\partial(\Sigma^0\,\bar{\partial}A) + \bar{\partial}(\Sigma^0\,\partial A) = 0, \\ &2\mathrm{i}\Sigma^0 \equiv P^{02}(\bar{\partial}L^0 - \partial\bar{L}^0), \qquad \partial \equiv \partial_\zeta - L^0\,\partial_u. \end{aligned} \tag{26.36}$$

This new solution is twisting if the original one is, i.e. if $\Sigma^0 \neq 0$, and it is non-vacuum if

$$(\partial + 2\partial \ln P)\, \bar{\partial}[\bar{\partial}\partial \ln P + (\bar{\partial} \ln P)^2] + 3mP_{,u}P^{-4} > 0. \tag{26.37}$$

For $P_{,u} \neq 0$, m can always be chosen so that the inequality (26.37) holds for at least some region of space-time.

If we look for vacuum solutions satisfying the condition (26.34), we fall back on the two classes discussed in §§ 25.2.3. and 25.2.5. In both cases not only L^0 but also P^0 and Σ^0 are independent of u.

In the first case ($L_{,u} = 0$, $\partial_\zeta I \neq 0$), the vacuum solutions which satisfy the assumptions of Theorem 26.4 are included in (25.43)—(25.47). So far no non-vacuum solutions with positive n^2 have been found.

In the second case ($L_{,u} = 0 = I$), the starting point is the Kerr-Schild class

$$L^0 = l_1(\bar{\zeta})\, P^{0-2}, \quad P^0 = \alpha\zeta\bar{\zeta} + \beta\zeta + \bar{\beta}\bar{\zeta} + \delta, \quad m^0 = \text{const}, \quad M = 0, \tag{26.38}$$

of vacuum solutions. Explicit solutions of the resulting differential equation for A have been found in several subcases.

If A is a function only of u, then we find

$$\begin{aligned} &A(u) = a_1(u + a_2), \qquad L^0 = -\mathrm{i}a^0\bar{\zeta}/(P^0)^2, \quad P^0 = 1 + K\zeta\bar{\zeta}/2, \\ &K = 0, \pm 1, \qquad a^0 \text{ real}, \end{aligned} \tag{26.39}$$

and

$$A(u) = a_1(u + a_2), \quad L^0 = (b_1\bar{\zeta}^2 + b_2\bar{\zeta} + b_3)^{1/2}, \quad P^0 = 1, \quad b_1 \text{ real}, \tag{26.40}$$

as radiating solutions. In both cases, the intensity of the radiation field is given by

$$\varkappa_0 n^2 = 6m(u + a_2)^{-1}. \tag{26.41}$$

In particular, (26.39) with $K = +1$ is a radiating Kerr metric (asymptotically flat) first given by Kramer (1972); it belongs to the class of Kerr-Schild metrics. If for this solution we make a coordinate transformation $u \to F(u) = a_1(u + a_2)^2/2$, which transforms P back into P^0, then (26.39) transforms into

$$\begin{aligned} &L = -\mathrm{i}a^0(2a_1u)^{1/2}\, \bar{\zeta}/(P^0)^2, \qquad m = m^0(2a_1u)^{-3/2}, \\ &P^0 = 1 + \zeta\bar{\zeta}/2, \qquad K = 1, \end{aligned} \tag{26.42}$$

which differs from the Kerr metric exactly by a (special) time-(u-) dependence of the parameters m and $a = a^0(2a_1u)^{1/2}$.

If we assume axial symmetry (Σ^0 and A dependent only on $\zeta\bar{\zeta}$ and u), then we have to solve

$$\begin{aligned} &(\Sigma^0\zeta\bar{\zeta}A')' + \Sigma^0 a^2\zeta\bar{\zeta}(P^0)^{-4} \quad \ddot{A} = 0, \qquad A = A(\zeta\bar{\zeta}, u). \\ &L^0 = -\mathrm{i}a\bar{\zeta}/(P^0)^2, \qquad P^0 = 1 + K\zeta\bar{\zeta}/2. \end{aligned} \tag{26.43}$$

For $K = 0$, the general solution is a superposition (with different α) of

$$\begin{aligned} &A(\zeta\bar{\zeta}, u) = (a_1\, \mathrm{e}^{\alpha u} + a_2\, \mathrm{e}^{-\alpha u})\, J_0(a\alpha\zeta\bar{\zeta}) \qquad \alpha \neq 0, \\ &A(\zeta\bar{\zeta}, u) = (a_1u + a_2)\, (a_3 \ln \zeta\bar{\zeta} + a_4) \qquad \alpha = 0, \end{aligned} \tag{26.44}$$

J_0 being Bessel functions. For $K = \pm 1$, equation (26.43) can be separated, but except for (26.39) only the solution

$$A(\zeta\bar{\zeta}) = a_1[\ln \zeta\bar{\zeta} - 2 \ln (1 - K\zeta\bar{\zeta}/2)] + a_2 \tag{26.45}$$

has been found.

26.6.3. Generating pure radiation field solutions from vacuum by changing m

If we let P, L and M stay fixed and change only m,

$$m = m^0 + B(\zeta, \bar{\zeta}, u), \qquad B \text{ real}, \tag{26.46}$$

then (26.32b) remains valid and the field equations reduce to

$$(3L_{,u} - \partial)\, B = 0 = (3\bar{L}_{,u} - \bar{\partial})\, B \tag{26.47}$$

(Hughston (1971)). The new solution would be non-vacuum exactly if $(P^{-3}B)_{,u}$ were nonzero.

For twisting solutions ($\bar{\partial}L - \partial\bar{L} \neq 0$), the integrability condition of equations (26.47) imposes severe restrictions on the vacuum metric. From (26.47) and the commutator relations (25.17) one gets

$$(\partial\bar{\partial} - \bar{\partial}\partial)\, B = 3B\, \partial_u(\partial\bar{L} - \bar{\partial}L) = (\bar{\partial}L - \partial\bar{L})\, \partial_u B, \tag{26.48}$$

i.e. the real function B has the form

$$B(\zeta, \bar{\zeta}, u) = b(\zeta, \bar{\zeta})\, (\bar{\partial}L - \partial\bar{L})^{-3}, \tag{26.49}$$

and because of (26.47) the function $b(\zeta, \bar{\zeta})$ has to satisfy

$$(\ln b)_{,\zeta} = 3L_{,u} + 3\, \partial \ln (\bar{\partial}L - \partial\bar{L}). \tag{26.50}$$

The main implication of equation (26.50) is that the right-hand side must be independent of u.

We will now inspect the known classes of vacuum solutions to see in which cases equation (26.50) can be satisfied. For $L_{,u} = 0$, we know two classes of vacuum solutions, (§§ 25.2.3. and 25.2.5.). In both cases P is independent of u, and because of (26.47) B must be a constant, so $(P^{-3}B)_{,u}$ is zero; the change (26.46) is only a (trivial) change of the constant mass parameter. Thus only vacuum solutions with $L_{,u} \neq 0$ can meet all the above conditions. Inspecting the solutions of § 25.2.4. we see that P is independent of u, and L and Σ are linear in u. If we choose the origin of u so that Σ is proportional to u, then (26.50) implies that L is also proportional to u. Because of the structure (25.50)—(25.55) of the class $L_{,u} \neq 0$, $I = 0$, of vacuum solutions, the only radiation solution we can generate is given by

$$\begin{aligned} &L = \{\partial_\zeta \ln P - [\zeta + g(\bar{\zeta})]^{-1}\}\, u, \qquad P_{,u} = 0, \qquad m^0 = M = I = 0 \\ &m = B = bu^{-3}, \qquad \varkappa_0 n^2 = -6bu^{-4}, \end{aligned} \tag{26.51}$$

where $g(\bar{\zeta})$ is a disposable function and b is a negative constant. The corresponding vacuum solution ($b = 0$) is flat. For a pure radiation field generated from Hauser's type N solution (25.71), see □ Stephani (1980).

Chapter 27. Non-diverging solutions (Kundt's class)

27.1. Introduction

In the preceding Chapters 23.—26. we dealt with those algebraically special solutions for which $\boldsymbol{k}$, the multiple principal null direction of the Weyl tensor, is diverging ($\varrho \neq 0$) and shearfree. Here we treat the non-diverging case, i.e. we assume $\varrho = -(\Theta + \mathrm{i}\omega) = 0$ throughout this chapter.

Since physically reasonable energy-momentum tensors have to satisfy the energy condition $T_{ab}k^ak^b \geqq 0$ (§ 5.3.), one sees from the equation (6.32), i.e. from

$$\Theta_{,a}k^a - \omega^2 + \Theta^2 + \sigma\bar{\sigma} = -\frac{1}{2} R_{ab}k^ak^b \leqq 0, \tag{27.1}$$

that $\Theta + \mathrm{i}\omega = 0$ implies

$$\sigma = 0 = R_{ab}k^ak^b. \tag{27.2}$$

Thus the non-twisting (and therefore geodesic) and non-expanding null congruence must be shearfree, and $R_{ab}k^ak^b = 0$ implies that the relations

$$R_{ab}k^ak^b = R_{ab}k^am^b = R_{ab}m^am^b = 0 \tag{27.3}$$

are satisfied for vacuum, Einstein-Maxwell, and pure radiation fields. Hence, by Theorem 7.1, these space-times are algebraically special. Einstein-Maxwell and pure radiation fields are aligned, i.e. they have a common eigendirection $\boldsymbol{k}$ of the Weyl and Ricci tensor. Perfect fluid solutions violate $R_{ab}k^ak^b = 0$ unless $p + \mu = 0$.

Kundt (1961) was the first to study the class of solutions with $\varrho = 0$ (for vacuum and an energy-momentum tensor of the form $T_{ab} = \Phi^2 k_a k_b$). The class of solutions admitting a non-expanding and non-twisting null congruence is therefore called Kundt's class.

Ref.: Kundt and Trümper (1962), Kundt (1962).

27.2. The line element for metrics with $\Theta + \mathrm{i}\omega = 0$

The non-twisting null vector field may be chosen to be a gradient field, and coordinates u and v are then naturally introduced by

$$\boldsymbol{e}_4 = k^i\partial_i = \partial_v, \qquad \boldsymbol{\omega}^3 = -k_i\,\mathrm{d}x^i = \mathrm{d}u. \tag{27.4}$$

As a result of (27.4) the line element has the form

$$ds^2 = g_{AB}\,dx^A\,dx^B - 2\,du(dv + m_A\,dx^A + H\,du), \qquad A, B = 1, 2 \tag{27.5}$$

(Kundt (1961)). As coordinates in the null hypersurfaces $u = \text{const}$ we use the affine parameter v and two spacelike coordinates x^1, x^2, and the spin coefficients ϱ and σ are then

$$\begin{aligned} \varrho &= -k_{a;b}m^a\overline{m}^b = -\frac{1}{2}\,g_{ij,v}m^i\overline{m}^j = -\frac{1}{2}\,g_{AB,v}m^A\overline{m}^B, \\ \sigma &= -k_{a;b}m^a m^b = -\frac{1}{2}\,g_{ij,v}m^i m^j = -\frac{1}{2}\,g_{AB,v}m^A m^B. \end{aligned} \tag{27.6}$$

Hence $\varrho = \sigma = 0$ leads to $g_{AB,v} = 0$. Performing a coordinate transformation $x^{A'} = x^{A'}(x^B, u)$, and using the complex conjugate coordinates $\zeta = (x^1 + ix^2)/\sqrt{2}$ and $\bar{\zeta} = (x^1 - ix^2)/\sqrt{2}$, one can write the line element in the form

$$ds^2 = 2P^{-2}\,d\zeta\,d\bar{\zeta} - 2\,du(dv + W\,d\zeta + \overline{W}\,d\bar{\zeta} + H\,du), \quad P_{,v} = 0 \tag{27.7}$$

(P and H being real and W complex).

In terms of 1-forms, this line element can be written as

$$\begin{aligned} &ds^2 = 2\omega^1\omega^2 - 2\omega^3\omega^4, \\ &\omega^1 = d\zeta/P, \quad \omega^2 = d\bar{\zeta}/P, \quad \omega^3 = du, \quad \omega^4 = dv + W\,d\zeta + \overline{W}\,d\bar{\zeta} + H\,du. \end{aligned} \tag{27.8}$$

This choice of 1-forms is very similar to the choice (23.22) ($L = 0$) in the case of diverging algebraically special fields. In order to calculate the Riemann tensor we found it helpful to first perform a null rotation (23.3) and to use the basis of 1-forms

$$\omega^1 = \overline{\omega}^2 = d\zeta/P - P\overline{W}\,du, \quad \omega^3 = du, \quad \omega^4 = dv + (H + P^2W\overline{W})\,du, \tag{27.9}$$

and corresponding tetrad vectors

$$\boldsymbol{e}_1 = \bar{\boldsymbol{e}}_2 = P\,\partial_\zeta, \quad \boldsymbol{e}_3 = \partial_u + P^2(\overline{W}\,\partial_\zeta + W\,\partial_{\bar{\zeta}}) - (H + P^2W\overline{W})\,\partial_v, \quad \boldsymbol{e}_4 = \partial_v. \tag{27.10}$$

The coordinate transformations preserving the form (27.7) of the metric, and the associated transformations of the metric functions P, H and W are

(i)
$$\begin{aligned} &\zeta' = f(\zeta, u), \\ &P'^2 = P^2 f_{,\zeta}\bar{f}_{,\bar{\zeta}}, \qquad W' = W/f_{,\zeta} + \bar{f}_{,u}/(P^2 f_{,\zeta}\bar{f}_{,\bar{\zeta}}), \\ &H' = H - (f_{,u}\bar{f}_{,u}/P^2 + Wf_{,u}\bar{f}_{,\bar{\zeta}} + \overline{W}\bar{f}_{,u}f_{,\zeta})/(f_{,\zeta}\bar{f}_{,\bar{\zeta}}), \end{aligned} \tag{27.11a}$$

(ii)
$$\begin{aligned} &v' = v + g(\zeta, \bar{\zeta}, u), \\ &P' = P, \qquad W' = W - g_{,\zeta}, \qquad H' = H - g_{,u}, \end{aligned} \tag{27.11b}$$

(iii)
$$\begin{aligned} &u' = h(u), \qquad v' = v/h_{,u}, \\ &P' = P, \qquad W' = W/h_{,u}, \qquad H' = (H + h_{,uu}/h_{,u})/(h_{,u})^2. \end{aligned} \tag{27.11c}$$

In order to maintain the form (27.10) of the null tetrad, the effect of (27.11) must be compensated by the following tetrad rotations ($\boldsymbol{e}_2 = \bar{\boldsymbol{e}}_1$)

$$\text{(i)} \qquad \boldsymbol{e}_1' = (\bar{f}_{,\bar{\zeta}}/f_{,\zeta})^{1/2}\, \boldsymbol{e}_1; \qquad \boldsymbol{e}_3' = \boldsymbol{e}_3, \qquad \boldsymbol{e}_4' = \boldsymbol{e}_4, \tag{27.12a}$$

$$\text{(ii)} \qquad \boldsymbol{e}_1' = \boldsymbol{e}_1 - Pg_{,\zeta}\boldsymbol{e}_4; \quad \boldsymbol{e}_3' = \boldsymbol{e}_3 - Pg_{,\bar{\zeta}}\boldsymbol{e}_1 - Pg_{,\zeta}\boldsymbol{e}_2 + P^2 g_{,\zeta} g_{,\bar{\zeta}}\boldsymbol{e}_4, \quad \boldsymbol{e}_4' = \boldsymbol{e}_4, \tag{27.12b}$$

$$\text{(iii)} \qquad \boldsymbol{e}_1' = \boldsymbol{e}_1; \qquad \boldsymbol{e}_3' = \boldsymbol{e}_3/h_{,u}, \qquad \boldsymbol{e}_4' = h_{,u}\boldsymbol{e}_4. \tag{27.12c}$$

If $(\ln P)_{,\zeta\bar{\zeta}} = 0$, then one can always transform P to $P = 1$ by means of (27.11a). The condition $W_{,v} = 0$ is invariant under the transformations (27.11), and so characterizes a special subclass of solutions.

The 2-surfaces $u, v = \text{const}$ with metric $2\,\mathrm{d}\zeta\,\mathrm{d}\bar{\zeta}/P^2$ are called *wave surfaces.* The vector fields $\boldsymbol{e}_1 = P\,\partial_\zeta$ and $\boldsymbol{e}_2 = P\,\partial_{\bar{\zeta}}$ in (27.10) are surface-forming, i.e. their commutator is a linear combination of themselves (see (6.11)). They are tangent to the wave surfaces, whereas the vector fields $\boldsymbol{e}_1 = P(\partial_\zeta - W\,\partial_v)$ and $\boldsymbol{e}_2 = P \times (\partial_{\bar{\zeta}} - \overline{W}\,\partial_v)$ associated with the basis (27.8) are not.

The existence of (spacelike) 2-surfaces orthogonal to $\boldsymbol{k}$ implies $\omega = 0$ (Kundt (1961)). The proof is short:

$$0 = k_a(m^b\overline{m}^a{}_{;b} - \overline{m}^b m^a{}_{;b}) = 2k_{a;b}m^{[a}\overline{m}^{b]} = 2\mathrm{i}\omega. \tag{27.13}$$

Conversely, we have seen that $\omega = 0$ implies that 2-surfaces (wave surfaces) orthogonal to $\boldsymbol{k}$ exist.

The space-time geometry uniquely determines the null congruence $\boldsymbol{k}$ and the wave surfaces. Therefore the Gaussian curvature

$$K = 2P^2(\ln P)_{,\zeta\bar{\zeta}} = \Delta(\ln P) \tag{27.14}$$

of the wave surfaces is a space-time invariant.

Ref.: Hall (1974, 1978).

27.3. The Ricci tensor components

In this section we list the tetrad components of the Ricci tensor with respect to the basis (27.9). As in Chapters 23.—26., we shall use the conventions that numerical indices are tetrad indices, and partial derivatives are denoted by a comma.

In the basis (27.9), the independent connection forms $\boldsymbol{\Gamma}_{ab}$ are

$$\Gamma_{41} = \frac{P}{2}\, W_{,v}\omega^3,$$

$$\Gamma_{21} + \Gamma_{43} = -\left(P_{,\zeta} - \frac{P}{2}\, W_{,v}\right)\omega^1 + \left(P_{,\bar{\zeta}} + \frac{P}{2}\,\overline{W}_{,v}\right)\omega^2 + \left[(H + P^2 W\overline{W})_{,v} - \frac{1}{2}\,(P^2\overline{W})_{,\zeta} + \frac{1}{2}\,(P^2 W)_{,\bar{\zeta}}\right]\omega^3, \tag{27.15}$$

$$\Gamma_{31} = \left[-\frac{P^2}{2}(\overline{W}_{,\zeta} + W_{,\bar\zeta}) + (\ln P)_{,u}\right]\omega^1 - (P^2\overline{W})_{,\bar\zeta}\,\omega^2$$
$$-P(H + P^2W\overline{W})_{,\bar\zeta}\,\omega^3 - \frac{P}{2}\overline{W}_{,v}\omega^4. \tag{27.15}$$

Inserting (27.15) into the equations (3.23) we obtain the following expressions for the tetrad components of the Ricci tensor:

$$R_{44} = 0, \tag{27.16a}$$

$$R_{41} = -\frac{P}{2}W_{,vv}, \tag{27.16b}$$

$$R_{11} = (P^2W_{,v})_{,\zeta} - \frac{P^2}{2}(W_{,v})^2, \tag{27.16c}$$

$$R_{12} = \Delta \ln P + \frac{P^2}{2}(\overline{W}_{,v\zeta} + W_{,v\bar\zeta} - W_{,v}\overline{W}_{,v}), \tag{27.16d}$$

$$R_{34} = H_{,vv} - \frac{P^2}{2}(\overline{W}_{,v\zeta} + W_{,v\bar\zeta} - 2W_{,v}\overline{W}_{,v}) + P^2(W_{,vv}\overline{W} + W\overline{W}_{,vv}), \tag{27.16e}$$

$$R_{31} = P(P^2W)_{,\zeta\bar\zeta} - 2P_{,\bar\zeta}(P^2W)_{,\zeta} + P\left[-\frac{1}{2}P^2\overline{W}_{,\zeta} - \frac{1}{2}P^2W_{,\bar\zeta} + (\ln P)_{,u}\right]_{,\zeta}$$
$$+ PH_{,v\zeta} + \frac{P}{2}[(P^2W_{,v})_{,\zeta}\overline{W} + (P^2\overline{W}_{,v})_{,\zeta}W] + \frac{P^3}{2}[(W\overline{W}_{,v})_{,\zeta} - (WW_{,v})_{,\bar\zeta}]$$
$$+ P_{,u}W_{,v} - \frac{1}{2}PW_{,vu} + \frac{1}{2}P(H + P^2W\overline{W})W_{,vv}, \tag{27.16f}$$

$$R_{33} = 2P^2S_{,\zeta\bar\zeta} + P^2\overline{W}_{,v}S_{,\zeta} + P^2W_{,v}S_{,\bar\zeta} - 2(P^2\overline{W})_{,\bar\zeta}(P^2W)_{,\zeta}$$
$$-2[\mu_{,\zeta}P^2\overline{W} + \mu_{,\bar\zeta}P^2W + \mu_{,u} - \mu_{,v}S + \mu S_{,v} + \mu^2)], \tag{27.16g}$$
$$\mu \equiv \frac{P^2}{2}(\overline{W}_{,\zeta} + W_{,\bar\zeta}) - (\ln P)_{,u}, \qquad S \equiv H + P^2W\overline{W}.$$

27.4. The structure of the vacuum and Einstein-Maxwell equations

In this section we shall consider the Einstein-Maxwell equations including, of course, the vacuum field equations as a special case.

As mentioned in § 27.1., the existence of a non-expanding and non-twisting null congruence $\boldsymbol{k}$ implies that $\boldsymbol{k}$ is also an eigendirection of the Maxwell tensor, i.e. that the tetrad component $\Phi_0 = F_{41}$ must be zero. With $\Phi_0 = 0$, the Einstein equations are

$$R_{11} = R_{14} = R_{44} = 0, \tag{27.17a}$$

$$R_{12} = R_{34} = 2\varkappa_0\Phi_1\overline{\Phi}_1, \tag{27.17b}$$

$$R_{31} = 2\varkappa_0\Phi_1\overline{\Phi}_2, \qquad R_{33} = 2\varkappa_0\Phi_2\overline{\Phi}_2, \tag{27.17c}$$

and the (sourcefree) Maxwell equations (7.46)—(7.49) read

$$\Phi_{1,v} = 0, \qquad \Phi_{1,\zeta} = W_{,v}\Phi_1, \tag{27.18a}$$

$$\Phi_{2,v} = P(\Phi_{1,\bar{\zeta}} + \overline{W}_{,v}\Phi_1), \tag{27.18b}$$

$$P\Phi_{2,\zeta} = \Phi_{1,u} + P^2[(\overline{W}\Phi_1)_{,\zeta} + (W\Phi_1)_{,\bar{\zeta}}] - 2(\ln P)_{,u}\,\Phi_1 + P_{,\zeta}\Phi_2. \tag{27.18c}$$

Using $R_{11} = 0$, the set of Maxwell equations (27.18a) can be integrated, giving

$$\begin{aligned} \Phi_1 &= \Phi_1{}^0 = P^4(W_{,v})^2\,\overline{F}(\bar{\zeta}, u) \qquad \text{for} \qquad W_{,v} \neq 0, \\ \Phi_1 &= \Phi_1{}^0 = \overline{F}(\bar{\zeta}, u) \qquad \text{for} \qquad W_{,v} = 0. \end{aligned} \tag{27.19}$$

For a *null field* ($\Phi_0 = \Phi_1 = 0$) the general solution of the Maxwell equations (27.18) is

$$\Phi_2 = \Phi_2{}^0 = P\bar{g}(\bar{\zeta}, u). \tag{27.20}$$

We now study the structure of the Einstein-Maxwell equations (27.17), (27.18). From $R_{14} = 0$ one infers that W is a linear function of v,

$$W = W_{,v}(\zeta, \bar{\zeta}, u)\,v + W^0(\zeta, \bar{\zeta}, u). \tag{27.21}$$

The equations

$$\begin{aligned} R_{11} &= (P^2 W_{,v})_{,\zeta} - \frac{P^2}{2}(W_{,v})^2 = 0, \\ R_{12} &= \Delta \ln P + \frac{P^2}{2}(\overline{W}_{,v\zeta} + W_{,v\bar{\zeta}} - W_{,v}\overline{W}_{,v}) = 2\varkappa_0\Phi_1\overline{\Phi}_1, \\ \Phi_{1,\zeta} &= W_{,v}\Phi_1 \end{aligned} \tag{27.22}$$

form a system of simultaneous differential equations for the v-independent functions P, $W_{,v}$ and Φ_1. No general solution is known, even for the vacuum case.

Because P, $W_{,v}$ and Φ_1 do not depend on v, the equation $R_{34} = 2\varkappa_0\Phi_1\overline{\Phi}_1$ tells us that H is a quadratic function of v,

$$\begin{aligned} H &= \frac{1}{2} H_{,vv}(\zeta, \bar{\zeta}, u)\,v^2 + G^0(\zeta, \bar{\zeta}, u)\,v + H^0(\zeta, \bar{\zeta}, u), \\ H_{,vv} &= \frac{P^2}{2}(\overline{W}_{,v\zeta} + W_{,v\bar{\zeta}} - 2W_{,v}\overline{W}_{,v}) + 2\varkappa_0\Phi_1\overline{\Phi}_1, \end{aligned} \tag{27.23}$$

and from the Maxwell equation (27.18b) it follows that Φ_2 is linear in v,

$$\Phi_2 = P(\Phi_{1,\bar{\zeta}} + \overline{W}_{,v}\Phi_1)\,v + \Phi_2{}^0(\zeta, \bar{\zeta}, u). \tag{27.24}$$

The (complex) field equation $R_{31} = 2\varkappa_0\Phi_1\overline{\Phi}_2$ and the Maxwell equation (27.18c) contain parts linear in v and v-independent parts. If the Einstein equations (27.17a, b) and the Maxwell equations (27.18a, b) are satisfied, then it follows from the Bianchi identity (7.70) that the v-dependent parts of the equation $R_{31} = 2\varkappa_0\Phi_1\overline{\Phi}_2$ and the Maxwell equation (27.18c) are identically satisfied. The v-independent parts of these equations, namely

$$R^0_{31} = P(P^2W^0)_{,\zeta\bar{\zeta}} - 2P_{,\bar{\zeta}}(P^2W^0)_{,\zeta} + P\left[-\frac{1}{2}P^2\overline{W}^0{}_{,\zeta} - \frac{1}{2}P^2W^0_{,\bar{\zeta}} + (\ln P)_{,u}\right]_{,\zeta}$$

$$+ PG^0_{,\zeta} + \frac{P}{2}[(P^2W_{,v})_{,\zeta}\overline{W}^0 + (P^2\overline{W}_{,v})_{,\zeta}W^0] + \frac{P^3}{2}[(W^0\overline{W}_{,v})_{,\zeta} - (W^0W_{,v})_{,\bar{\zeta}}]$$

$$+ P_{,u}W_{,v} - \frac{1}{2}PW_{,vu} = 2\varkappa_0\Phi_1\overline{\Phi}_2{}^0, \tag{27.25a}$$

$$P\Phi^0_{2,\zeta} = \Phi_{1,u} + P^2[(\overline{W}^0\Phi_1)_{,\zeta} + (W^0\Phi_1)_{,\bar{\zeta}}] - 2(\ln P)_{,u}\Phi_1 + P_{,\zeta}\Phi_2{}^0, \tag{27.25b}$$

determine the functions W^0, G^0 and $\Phi_2{}^0$, once P, $W_{,v}$ and Φ_1 are known. The equations (27.25) are linear in W^0, G^0, $\Phi_2{}^0$, and their derivatives.

The last Einstein equation, $R_{33} = 2\varkappa_0\Phi_2\overline{\Phi}_2$, contains terms up to second order in v. Provided that the other Einstein-Maxwell equations are already satisfied, the Bianchi identity (7.71) gives us $(R_{33} - 2\varkappa_0\Phi_2\overline{\Phi}_2)_{,v} = 0$, that is, the v-dependent part of the last field equation is identically satisfied. Finally, the v-independent part,

$$\begin{aligned} R^0_{33} &= 2P^2H^0{}_{,\zeta\bar{\zeta}} + P^2(\overline{W}_{,v}H^0)_{,\zeta} + P^2(W_{,v}H^0)_{,\bar{\zeta}} + \text{(known function)} \\ &= 2\varkappa_0\Phi^0_2\overline{\Phi}^0_2 \end{aligned} \tag{27.26}$$

(the "known function" being obtainable from the expression (27.16g)), is a linear partial differential equation of the second order for the remaining function H^0.

The Einstein-Maxwell equations split into the set (27.22) determining P, $W_{,v}$ and Φ_1, the set (27.25) determining W^0, G^0 and Φ^0_2 and the equation (27.26) determining H^0. It should be emphasized that the functions W^0, G^0, $\Phi_2{}^0$ and H^0 do not occur in the field equations (27.22) for P, $W_{,v}$ and Φ_1. This remarkable fact gives us the possibility of constructing new solutions:

Theorem 27.1. *From a known solution ("background" metric) one can generate other solutions with the same* P, $W_{,v}$ *and* Φ_1, *by choosing new functions* W^0, G^0, $\Phi_2{}^0$ *satisfying the linear equations* (27.25) *and by then choosing a new function* H^0 *satisfying the linear equation* (27.26).

To conclude this section we give the tetrad components Ψ_2 and Ψ_3 of the Weyl tensor for Einstein-Maxwell (and pure radiation) fields:

$$\Psi_2 = \frac{P^2}{2}\left(\overline{W}_{,v\zeta} - \frac{1}{2}W_{,v}\overline{W}_{,v}\right),$$

$$\Psi_3 = -P(P^2\overline{W})_{,\zeta\bar{\zeta}} + 2P_{,\zeta}(P^2\overline{W})_{,\bar{\zeta}} + P\left[\frac{P^2}{2}(\overline{W}_{,\zeta} + W_{,\bar{\zeta}}) - (\ln P)_{,u}\right]_{,\bar{\zeta}} \tag{27.27}$$

$$+ \frac{P}{2}\left[\frac{P^2}{2}(\overline{W}_{,\zeta} + W_{,\bar{\zeta}}) - (\ln P)_{,u}\right]\overline{W}_{,v} - \frac{P}{2}W_{,v}(P^2\overline{W})_{,\bar{\zeta}} + \frac{1}{2}R_{32}.$$

For $W = 0$, the last expression reduces to

$$\Psi_3 = \frac{P}{2}[H_{,v} - (\ln P)_{,u}]_{,\bar{\zeta}}. \tag{27.28}$$

27.5. Vacuum solutions

27.5.1. Solutions of types *III* and *N*

The vacuum solutions of types *III* and *N* in Kundt's class are completely known (Kundt (1961)). The field equation $R_{12} = 0$ and the type *III* condition $\Psi_2 = 0$ give $(\ln P)_{,\zeta\bar\zeta} = 0$. With the aid of a coordinate transformation (27.11a) one can make $P = 1$. The field equation $R_{34} = 0$ determines $H_{,vv}$ as

$$H_{,vv} = -\frac{1}{2} W_{,v}\overline{W}_{,v}. \tag{27.29}$$

The field equation $R_{11} = 0$ and the condition $\Psi_2 = 0$ lead to

$$W_{,v} = -2n_{,\zeta}, \qquad \bar n = n, \qquad (\mathrm{e}^n)_{,\zeta\zeta} = 0 = (\mathrm{e}^n)_{,\zeta\bar\zeta}. \tag{27.30}$$

Either $W_{,v} = 0$ or one can, remembering the transformations (27.11a, c), put $W_{,v} = -2/(\zeta + \bar\zeta)$ without loss of generality.

First we treat the *case* $W_{,v} = 0$. With $P = 1$, $W_{,v} = 0$, the equation $R_{31} = 0$ gives

$$R_{31} = \left(H_{,v} + \frac{1}{2} W_{,\bar\zeta} - \frac{1}{2} \overline{W}_{,\zeta}\right)_{,\zeta} = 0. \tag{27.31}$$

This equation implies $H_{,v} + \frac{1}{2}(W_{,\bar\zeta} - \overline{W}_{,\zeta}) = \bar f(\bar\zeta, u)$. Because $H_{,v}$ is real, it follows that

$$W_{,\bar\zeta} - \overline{W}_{,\zeta} = \bar f(\bar\zeta, u) - f(\zeta, u), \tag{27.32}$$

which implies

$$W = \int \bar f(\bar\zeta, u)\,\mathrm{d}\bar\zeta - g_{,\zeta}, \qquad g = g(\zeta, \bar\zeta, u) \text{ real}, \tag{27.33}$$

and so, finally, in virtue of the admissible coordinate transformation (27.11b), one obtains $W = W(\bar\zeta, u)$. The class of solutions is determined by

$$\begin{gathered} \mathrm{d}s^2 = 2\,\mathrm{d}\zeta\,\mathrm{d}\bar\zeta - 2\,\mathrm{d}u(\mathrm{d}v + W\,\mathrm{d}\zeta + \overline{W}\,\mathrm{d}\bar\zeta + H\,\mathrm{d}u), \\ W = W^0(\bar\zeta, u), \qquad H = \frac{1}{2}(W_{,\bar\zeta} + \overline{W}_{,\zeta})\,v + H^0, \\ H^0_{,\zeta\bar\zeta} - \mathrm{Re}\,(W_{,\bar\zeta}{}^2 + W W_{,\bar\zeta\bar\zeta} + W_{,\bar\zeta u}) = 0, \end{gathered} \tag{27.34}$$

where $\overline{W}(\zeta, u)$ is a disposable function. In general, these solutions are of Petrov type *III*. They are type *N* if $\Psi_3 = \frac{1}{2} W_{,\bar\zeta\bar\zeta} = 0$ (see (27.27)), but if W is a linear function of $\bar\zeta$, then we can use the remaining freedom in the coordinate transformation (27.11) to make W zero. The solutions (27.34) with $W = 0$ are the plane-fronted waves studied in detail in § 21.5.

The *case* $W_{,v} = -2/(\zeta + \bar\zeta)$ can be solved in a similar way. With $P = 1$ and $W_{,v} = -2/(\zeta + \bar\zeta)$ the equation $R^0_{31} = 0$ reads

$$\left[\frac{1}{2}(W^0{}_{,\bar\zeta} - \overline{W}^0{}_{,\zeta}) + G^0 - \frac{W^0 + \overline{W}^0}{\zeta + \zeta}\right]_{,\zeta} = -\frac{1}{\zeta + \bar\zeta}(W^0_{,\bar\zeta} + \overline{W}^0_{,\zeta}). \tag{27.35}$$

By means of a coordinate transformation (27.11 b) which induces the change

$$W^{0\prime} = W^0 - g_{,\zeta} + \frac{2g}{\zeta + \bar{\zeta}} \tag{27.36}$$

of W^0, one can arrange that the right-hand side of (27.35) vanishes. Then $(W^0_{,\bar{\zeta}} - \overline{W}^0_{,\zeta})$ is the imaginary part of an analytic function. One obtains

$$W^0 = \int \bar{f}(\bar{\zeta}, u)\, \mathrm{d}\bar{\zeta} - \bar{\zeta} f(\zeta, u) + h(\zeta, u), \tag{27.37}$$

and the remaining freedom in the transformation (27.11 b) can be used to put $W^0 = W^0(\zeta, u)$. The class of solutions is now determined by

$$\begin{aligned} \mathrm{d}s^2 &= 2\, \mathrm{d}\zeta\, \mathrm{d}\bar{\zeta} - 2\, \mathrm{d}u(\mathrm{d}v + W\, \mathrm{d}\zeta + \overline{W}\, \mathrm{d}\bar{\zeta} + H\, \mathrm{d}u), \\ W &= -\frac{2v}{\zeta + \bar{\zeta}} + W^0(\zeta, u), \qquad H = -\frac{v^2}{(\zeta + \bar{\zeta})^2} + \frac{W^0 + \overline{W}^0}{\zeta + \bar{\zeta}}\, v + H^0, \\ &(\zeta + \bar{\zeta}) \left(\frac{H^0 + W^0 \overline{W}^0}{\zeta + \bar{\zeta}} \right)_{,\zeta\bar{\zeta}} = \overline{W}^0{}_{,\bar{\zeta}} W^0{}_{,\zeta} \end{aligned} \tag{27.38}$$

(Cahen and Spelkens (1967)), where the function $W^0(\zeta, u)$ is disposable. In general, these solutions are Petrov type *III*; they are of type *N* if $\Psi_3 = \overline{W}^0{}_{,\bar{\zeta}}/(\zeta + \bar{\zeta}) = 0$. In this case, W^0 can be transformed to zero.

Theorem 27.2. *The classes* (27.34) *and* (27.38) *exhaust all type III and N vacuum solutions with* $\varrho = 0$. *They are characterized by* $\Delta \ln P = 0$, *that is, by plane wave surfaces.*

These two classes were originally given by Kundt (1961), who chose W^0 to be real. For the class (27.38), W^0 is then given by

$$W^0 = \overline{W}^0 = \varphi(\zeta + \bar{\zeta})^{-1}, \qquad \varphi_{,\zeta\bar{\zeta}} = 0. \tag{27.39}$$

The type *III* solution,

$$\mathrm{d}s^2 = x(v - \mathrm{e}^x)\, \mathrm{d}u^2 - 2\, \mathrm{d}u\, \mathrm{d}v + \mathrm{e}^x(\mathrm{d}x^2 + \mathrm{e}^{-2u}\, \mathrm{d}z^2), \tag{27.40}$$

given by Petrov (1962) and its generalizations (Harris and Zund (1975), Kaigorodov (1967)) are particular members of the class (27.34), written in other coordinate systems. The metric (27.40) admits a non-Abelian group G_2.

Ref.: Kerr and Goldberg (1961), Debever (1960), Pandya and Vaidya (1961), Wyman and Trollope (1965).

27.5.2. Solutions of types *D* and *II*

The *type D* vacuum solutions are completely known (Kinnersley (1969 b), Carter (1968 b)). The non-diverging ($\varrho = 0$) solutions admit a group of motions G_4 on T_3 and are given in § 11.3.1. In our present notation this class of solutions can be written as

$$ds^2 = 2\,d\zeta\,d\bar{\zeta}/P^2 - 2\,du(dv + W\,d\zeta + \overline{W}\,d\bar{\zeta} + H\,du),$$

$$\sqrt{2}\,\zeta = x + iy, \qquad P^2\,dx = dz, \qquad P^2 = \frac{z^2 + l^2}{k(z^2 - l^2) + 2mz}, \tag{27.41}$$

$$W = -\frac{\sqrt{2}\,v}{P^2(z - il)}, \qquad H = -\left[\frac{k}{2(z^2 + l^2)} + \frac{2l^2}{P^2(z^2 + l^2)^2}\right] v^2,$$

with constant m, l and $k = 0, \pm 1$. Note that the solutions can only be given implicitly ($P^2\,dx = dz$) if one sticks to an isotropic form of the wave surface metric, and that only the terms with the highest possible powers of v occur in the metric functions W and H.

The *type II* solutions are only partially known. If a vacuum field admits a *null Killing vector* $\boldsymbol{\xi} = e^{-\sigma}\boldsymbol{k}$, then the space-time clearly belongs to Kundt's class (see Chapter 21.). In virtue of the Killing equations for the metric (27.7) the real function σ has to obey the relations

$$\sigma_{,\zeta} = W_{,v}, \qquad \sigma_{,u} = H_{,v}, \qquad \sigma_{,v} = 0. \tag{27.42}$$

The corresponding integrability conditions restrict the metric functions in such a way that one can construct the general solution of Einstein's field equations. We have treated these space-times in § 21.4.

If an algebraically special vacuum field admits a hypersurface-normal *spacelike Killing vector* orthogonal to $\boldsymbol{k}$, then this solution belongs to Kundt's class. All these (in general, type *II*) solutions are known (Kramer and Neugebauer (1968a)).

Using Theorem 27.1, one can generate type *II* solutions from the type *D* solution (27.41). If one changes only the metric function H, to $H + H^0$ say, then H^0 has to satisfy the equation

$$2H^0{}_{,\zeta\bar{\zeta}} + (\overline{W}_{,v}H^0)_{,\zeta} + (W_{,v}H^0)_{,\bar{\zeta}} = 0. \tag{27.43}$$

($W_{,v}$ as in (27.41)). The van Stockum class (18.23) can be generated in this way from the type D solution (27.41) with the parameters $k = l = 0$, $m = 1$. It is likely that Theorem 27.1 also leads to other solutions not contained in previously known classes.

Ref.: Debney (1971).

27.6. Einstein-Maxwell null fields and pure radiation fields

For Einstein-Maxwell null fields (aligned case, $\Phi_0 = \Phi_1 = 0$) and for pure radiation fields the field equations have the same form as in the vacuum case, except that $R^0_{33} = 0$ has to be replaced by

$$R^0_{33} = \varkappa_0 \Phi^2(\zeta, \bar{\zeta}, u). \tag{27.44}$$

Φ^2 is positive for pure radiation, and for Einstein-Maxwell fields Φ^2 has the structure

$$\Phi^2 = 2\Phi_2\overline{\Phi}_2 = 2P^2 g(\zeta, u)\,\bar{g}(\bar{\zeta}, u), \tag{27.45}$$

see (27.20). (With the aid of a coordinate transformation (27.11a) one can set $\Phi^2 = P^2$.) The function H^0, which occurs in (27.44) but in no other field equation, is essentially disposable for pure radiation solutions so long as $\Phi^2 > 0$. For Einstein-Maxwell fields, H^0 must be chosen so that (27.45) holds. (In the literature such functions H^0 are not usually explicitly determined.) So we have demonstrated that there are pure radiation and Einstein-Maxwell null fields which differ from vacuum solutions only in the function H^0. All pure radiation fields of Petrov type *III* and more special types can be obtained from the vacuum solutions (27.34) and (27.38) by omitting the differential equations for H^0. The *pp* waves (§ 21.5.) are a well-known class of Einstein-Maxwell null fields.

The Einstein-Maxwell fields of Kundt's class which are also Kerr-Schild metrics (§ 28.3.) are of Petrov type N, and the electromagnetic field is null (Debney (1973, 1974)).

Ref.: Wyman and Trollope (1965).

27.7. Einstein-Maxwell non-null fields

The non-diverging ($\varrho = 0$) *type N* Einstein-Maxwell non-null fields (aligned case) are completely known (Cahen and Leroy (1965, 1966), Szekeres (1966b)). This class of solutions is given by setting

$$\begin{aligned} &2P^2(\ln P)_{,\zeta\bar{\zeta}} = K, \qquad H = \frac{K}{2}v^2 + G^0(u)\,v + H^0, \\ &W = -F_{,\zeta}, \qquad F_{,\zeta\bar{\zeta}} + \frac{K}{P^2}F = 0, \qquad F + \bar{F} = \frac{2}{K}(\ln P)_{,u}, \\ &F = F(\zeta, \bar{\zeta}, u), \\ &\Phi_0 = \Phi_2 = 0, \qquad \sqrt{2\varkappa_0}\,\Phi_1 = \sqrt{K}, \qquad H^0_{,\zeta\bar{\zeta}} + K^2P^{-2}F\bar{F} = KW\overline{W}, \end{aligned} \tag{27.46}$$

in the metric (27.7), the Gaussian curvature K of the wave surfaces being a (non-zero) constant. The (conformally flat) Bertotti-Robinson solution (10.16),

$$ds^2 = 2\,d\zeta\,d\bar{\zeta}(1 + K\zeta\bar{\zeta}/2)^{-2} - 2\,du\,dv - Kv^2\,du^2, \tag{27.47}$$

is contained in (27.46) as a special case ($F = G^0 = H^0 = 0$).

For *type III solutions*, $W_{,v}$ is zero. This we now prove. (The same holds for the general type N solution (27.46).)

Theorem 27.3. *In the non-null case ($\Phi_1 \neq 0$), there are no type III Einstein-Maxwell fields with $W_{,v} \neq 0$.*

Proof: Substituting P^2 from the field equation $R_{11} = 0$ into the type *III* condition $\Psi_2 = 0$, one obtains

$$[\ln(W_{,v}P^2/\overline{W}_{,v})]_{,\zeta} = 0 \tag{27.48}$$

if $W_{,v} \neq 0$. This equation implies

$$P^2 W_{,v} = \overline{W}_{,v} \bar{f}(\bar{\zeta}, u) = W_{,v} \bar{f}(\bar{\zeta}, u)\, f(\zeta, u)/P^2 \Rightarrow (\ln P)_{,\zeta\bar{\zeta}} = 0. \tag{27.49}$$

Then the field equation $R_{12} = 2\varkappa_0 \Phi_1 \overline{\Phi}_1$ would lead to $\Phi_1 = 0$.

For $W_{,v} = 0$, one gets (from $R_{12} = 2\varkappa_0 \Phi_1 \overline{\Phi}_1$ and $\Phi_1 = \overline{F}(\bar{\zeta}, u)$) the differential equation

$$P^2 (\ln P)_{,\zeta\bar{\zeta}} = \varkappa_0\, |F(\zeta, u)|^2 \tag{27.50}$$

(the Liouville equation) which has the general solution

$$P^2 = \varkappa_0\, |F|^2 \frac{(1 + f\bar{f})^2}{f_{,\zeta} \bar{f}_{,\bar{\zeta}}}, \qquad f = f(\zeta, u), \tag{27.51}$$

containing an arbitrary function $f(\zeta, u)$. Hacyan and Plebański (1975) investigated the subcase $W = 0$. The functions Φ_2^0, G^0, H^0, in

$$\Phi_1 = \overline{F}(\bar{\zeta}, u), \quad \Phi_2 = P\overline{F}_{,\bar{\zeta}} v + \Phi_2^0, \quad H = \varkappa_0 F\overline{F} v^2 + G^0 v + H^0, \tag{27.52}$$

with P given by (27.51), are then subject to the rest of the Einstein-Maxwell equations, i.e. to

$$\begin{aligned} &(\Phi_2^0/P)_{,\zeta} = (\overline{F}/P^2)_{,u}, \qquad [G^0 + (\ln P)_{,u}]_{,\zeta} = 2\varkappa_0 \overline{F} \overline{\Phi}_2^0/P, \\ &P^2 H^0{}_{,\zeta\bar{\zeta}} + G^0 (\ln P)_{,u} + (\ln P)_{,uu} - (\ln P)_{,u}{}^2 = \varkappa_0 \Phi_2^0 \overline{\Phi}_2^0. \end{aligned} \tag{27.53}$$

An obvious particular solution is given by

$$\begin{aligned} &\mathrm{d}s^2 = 2\, \mathrm{d}\zeta\, \mathrm{d}\bar{\zeta}/(1 + f\bar{f})^2 - 2\, \mathrm{d}u[\mathrm{d}v + (f'\bar{f}' v^2 + H^0)\, \mathrm{d}u], \\ &\sqrt{\varkappa_0}\, \Phi_1 = \bar{f}', \qquad \sqrt{\varkappa_0}\, \Phi_2 = (1 + f\bar{f})\, \bar{f}'' v, \qquad H^0{}_{,\zeta\bar{\zeta}} = 0, \\ &f = f(\zeta), \qquad f' = f_{,\zeta} = \sqrt{\varkappa_0}\, F. \end{aligned} \tag{27.54}$$

From (27.28), this metric is Petrov type *III* if $f'' \neq 0$. The electromagnetic field is determined up to a constant duality rotation. Other particular solutions of (27.53) are given in Hacyan and Plebański (1975). It is a general feature of the type *III* solutions (aligned case) that the second eigendirection of the electromagnetic non-null field is not parallel to the single principal null direction of the Weyl tensor.

A class of *type D* Einstein-Maxwell fields can be obtained from the general type *D* vacuum class (27.41) by simply modifying the function P^2 to

$$P^2 = \frac{z^2 + l^2}{k(z^2 - l^2) + 2mz - e^2}, \tag{27.55}$$

the rest of the metric remaining unchanged. The tetrad components of the electromagnetic field tensor are

$$\Phi_0 = 0, \qquad \sqrt{2\varkappa_0}\, \Phi_1 = \frac{e}{(z - \mathrm{i}l)^2}, \qquad \sqrt{\varkappa_0}\, \Phi_2 = \frac{-2}{P} \frac{ezv}{(z - \mathrm{i}l)^2 (z^2 + l^2)}. \tag{27.56}$$

The two null eigendirections of the Maxwell field coincide with the eigendirections of the type D Weyl tensor. Like their uncharged counterparts (27.41), these solutions admit a group of motions G_4 on T_3 (compare (11.42)). Unlike the vacuum case this class does not exhaust all type D solutions: it does not, for instance, contain an Einstein-Maxwell field given by Kowalczyński and Plebański (1977) in the form

$$\mathrm{d}s^2 = \frac{2}{x^2}\left[\left(\frac{\mathrm{d}x}{A}\right)^2 + A^2\,\mathrm{d}y^2 + \left(\frac{\mathrm{d}z}{B}\right)^2 - B^2\,\mathrm{d}t^2\right] \tag{27.57}$$

$$A^2 = -2e^2x^4 + cx^3 - ax^2, \qquad B^2 = az^2 + b, \qquad a, b, c, e \text{ constants}.$$

The same procedure concerning the generation of new solutions from known ones works as for vacuum fields (see Theorem 27.1).

Chapter 28. Kerr-Schild metrics

28.1. General properties of Kerr-Schild metrics

28.1.1. The origin of the Kerr-Schild-Trautman ansatz

In General Relativity, the field equations are often simplified by considering null vector fields. One important case is the Kerr-Schild metrical ansatz, which is given by

$$g_{ab} = \eta_{ab} - 2Vk_a k_b, \tag{28.1}$$

where η_{ab} is the Minkowski metric, V is a scalar function and k_a is a null vector with respect to both the metrics g_{ab} and η_{ab},

$$g_{ab}k^a k^b = \eta_{ab}k^a k^b = 0. \tag{28.2}$$

The ansatz (28.1)—(28.2) was first studied by Trautman (1962). His idea was that a solution representing wave motion has to have the possibility of propagation of information. This can be achieved if such a solution depends on an arbitrary function. Trautman applied this idea to the case when both the covariant and the contravariant components of the metric tensor depend linearly on the same function V of the coordinates

$$g_{ab} = \eta_{ab} - 2Vh_{ab}; \qquad g^{ab} = \eta^{ab} + 2VL^{ab}. \tag{28.3}$$

From the identity $g^{ad}g_{dc} = g^a_c$ we get the relations

$$L^{af} = \eta^{ad}\eta^{cf}h_{cd}, \qquad L^{ad}h_{dc} = 0. \tag{28.4}$$

These relations imply the following structure for the symmetric matrix h_{ab}

$$h_{ab} = V_a\lambda_b = V_b\lambda_a, \qquad \eta^{ab}V_aV_b = \eta^{ab}\lambda_a\lambda_b = \eta^{ab}V_a\lambda_b = 0. \tag{28.5}$$

The two null vectors V_a and λ_a have to be parallel and we are led to the ansatz (28.1)—(28.2).

28.1.2. The Ricci tensor, Riemann tensor and Petrov type

Kerr-Schild metrics have been studied by several authors, using either the Newman-Penrose formalism or a more conventional procedure. We start with a covariant form of the second method. The calculations are greatly simplified by the fact that the 4-vector $\boldsymbol{k}$ is null.

First we note that the Christoffel symbols and the determinant of the metric tensor (28.1)—(28.2) satisfy the conditions

$$\begin{Bmatrix} c \\ ef \end{Bmatrix} k^e k^f = 0, \qquad \begin{Bmatrix} c \\ ef \end{Bmatrix} k_c k^f = 0, \qquad (-g)^{1/2} = 1. \tag{28.6}$$

These relations imply that $\boldsymbol{k}$ has the following properties:

$$k_{a;c}k^c = k_{a,c}k^c, \qquad k^a{}_{;c}k^c = k^a{}_{,c}k^c, \tag{28.7}$$

$$\begin{Bmatrix} c \\ ab \end{Bmatrix} k_c = (Vk_ak_b)_{;c}k^c; \qquad \begin{Bmatrix} c \\ ab \end{Bmatrix} k^a = -(Vk^ck_b)_{;a}k^a. \tag{28.8}$$

(A semicolon denotes the covariant derivative with respect to g_{ab}.) $\boldsymbol{k}$ is geodesic with respect to g_{ab} if and only if it is geodesic with respect to the flat metric η_{ab}, and it has the same expansion with respect to both metrics,

$$2\Theta = k^c{}_{;c} = k^c{}_{,c}. \tag{28.9}$$

The definition of the Ricci tensor together with the relations (28.6)—(28.8) lead to the equation

$$R_{bd}k^bk^d = 2Vg^{bd}(k_{d;a}k^a)\,(k_{b;c}k^c) = \varkappa_0 T_{bd}k^bk^d. \tag{28.10}$$

This gives rise to

Theorem 28.1. *The null vector of a Kerr-Schild metric is geodesic if and only if the energy-momentum tensor obeys the condition*

$$T_{bd}k^bk^d = 0. \tag{28.11}$$

We proceed further with energy-momentum tensors of the type (28.11). The geodesic null vector $\boldsymbol{k}$, $k^a{}_{;b}k^b = 0$, has the same twist and shear with respect to both the metrics,

$$2\omega^2 = k_{[a;b]}k^{a;b} = k_{[a,b]}k^{a,b}; \qquad 2\Theta^2 + 2\sigma\bar{\sigma} = k_{(a;b)}k^{a;b} = k_{(a,b)}k^{a,b}. \tag{28.12}$$

The Ricci tensor has the simple structure ($D \equiv k^i\,\partial_i$)

$$R_{bd} = (Vk_bk_d)^{;a}{}_{;a} - (Vk^ak_d)_{;ab} - (Vk^ak_b)_{;ad} + 2V(D^2V)\,k_bk_d. \tag{28.13}$$

It obeys the eigenvalue equation

$$R_{cd}k^d = -[4\omega^2V + (k^aV_{,b}k^b)_{;a}]\,k_c. \tag{28.14}$$

By means of the field equations we are led to

Theorem 28.2. *The geodesic null vector* $\boldsymbol{k}$ *of a Kerr-Schild metric is an eigenvector of the energy-momentum tensor.*

Further we get, from the definition of the Riemann tensor,

$$k^ck^aR_{abcd} = (D^2V)\,k_bk_d. \tag{28.15}$$

A straightforward calculation leads to the main result of this section:

Theorem 28.3. *The geodesic null vector* $\boldsymbol{k}$ *of a Kerr-Schild space-time obeying* (28.11) *is a multiple principal null direction of the Weyl tensor; thus the space-time is algebraically special,*

$$T_{ab}k^ak^b = 0 \leftrightarrow k^a{}_{;b}k^b = 0 \rightarrow k^ck^aC_{abcd} = Hk_bk_d \tag{28.16}$$

(Gürses and Gürsey (1974)). The scalar H is given by $H = D^2V - \frac{1}{6}R$.

28.1.3. Field equations and the energy-momentum tensor

The field equations of a Kerr-Schild space-time possessing a geodesic null vector $\boldsymbol{k}$ take the "linear" form (Gürses and Gürsey (1974))

$$\varkappa_0 T^d{}_b = \frac{1}{2}\,\eta_{cb}(\eta^{ae}g^{cd} - \eta^{ca}g^{de} - \eta^{da}g^{ce} + \eta^{cd}g^{ae})_{,a,e}. \tag{28.17}$$

This result follows from the relations (28.7), (28.13). Every Kerr-Schild solution is a solution of the linear field equations (28.17). The reverse is not true, because the Kerr-Schild conditions (28.1)—(28.2) have to be fulfilled.

In terms of a complex null tetrad $\{\boldsymbol{e}_a\} = (\boldsymbol{m}, \overline{\boldsymbol{m}}, \boldsymbol{l}, \boldsymbol{k})$ the tetrad components S_{44} and S_{41} of the traceless part of the Ricci tensor vanish because of Theorem 28.2. Usually one considers Kerr-Schild space-times satisfying the additional restriction

$$S_{ab}m^a m^b = S_{11} = 0. \tag{28.18}$$

With this assumption, among the energy-momentum tensor types considered in this book only

$$\varkappa_0 T_{ab} = -\frac{1}{4}\,Rg_{ab} + \lambda_1 k_a k_b + \lambda_2(m_a\overline{m}_b + \overline{m}_a m_b + k_a l_b + k_b l_a) \tag{28.19}$$

survives. The form (28.19) includes (i) electromagnetic non-null fields ($\lambda_1 = R = 0$), (ii) pure radiation fields ($\lambda_2 = R = 0$). Perfect fluid distributions cannot occur.

28.1.4. A geometrical interpretation of the Kerr-Schild ansatz

Newman and Unti (1963) introduced a coordinate system $\{x^a\}$ attached to an arbitrary particle world line $y^a(u)$ in flat space-time. Kinnersley and Walker (1970), and Bonnor and Vaidya (1972) have used this coordinate system to find accelerated particle solutions in General Relativity. Here we confine the investigation to Kerr-Schild metrics with a geodesic, shearfree and twistfree null congruence, following closely the paper of Bonnor and Vaidya (1972).

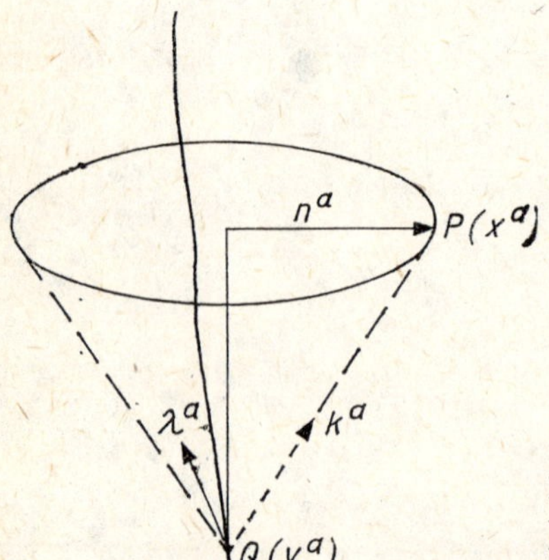

Fig. 28.1 A geodesic and shearfree null vector $\boldsymbol{k}$ and an arbitrary particle world line in flat space-time

Let $y^a(u)$ be a particle world line in Minkowski space and $\lambda^a \equiv \mathrm{d}y^a/\mathrm{d}u$ be its unit tangent vector; u is the proper time of the particle. If $P(x^a)$ is any event, the backward null cone from P will cut $y^a(u)$ in a single event $Q(y^a)$, see Fig. 28.1. We can extend the definition of u off the world line by taking $u(P) = u(Q)$. Similarly, we can extend the definition of λ^a by $\lambda^a(P) = \lambda^a(Q)$. The null cone at P and the world line intersect where

$$\eta_{ab}(x^a - y^a)(x^b - y^b) = 0, \tag{28.20}$$

which determines $u(Q)$. We identify the null vector

$$k^a = r^{-1}(x^a - y^a) \tag{28.21}$$

with the null vector in the Kerr-Schild ansatz (28.1)—(28.2) and get

$$g^{ab} = \eta^{ab} + 2Vr^{-2}(x^a - y^a)(x^b - y^b). \tag{28.22}$$

Here we have introduced the retarded distance $r \equiv \lambda_a(x^a - y^a)$. The null congruence given by (28.21) is shearfree, geodesic and twistfree with respect to both the Minkowski metric η_{ab} and the Kerr-Schild metric g_{ab}. In order to find the energy-momentum tensor of the Kerr-Schild metric (28.22), we need the identities $k_{a,b} = r^{-1}[\eta_{ab} - k_a\lambda_b - k_b\lambda_a - k_ak_b(1 + rk_c\dot{\lambda}^c)]$ and $r_{,b} = k_b(1 + r\dot{\lambda}_ck^c) + \lambda_b$. (The dot denotes $\mathrm{d}/\mathrm{d}u$ and $k_c\dot{\lambda}^c$ may be interpreted as the acceleration of the particle at Q in the direction n^a, see Fig. 28.1.).

The particular scalar function $V = m(u)\,r^{-1} - e^2(u)\,r^{-2}/2$ leads by means of the relations (28.17)—(28.22) to the energy-momentum tensor

$$\begin{aligned} \varkappa_0 T_b{}^d &= -\frac{e^2}{r^4}\, g_b^d + \frac{2e^2}{r^4}\,(k^d\lambda_b + k_b\lambda^d) + \Phi^2 k^d k_b, \\ \Phi^2 &\equiv \frac{2e^2}{r^4} + \frac{4}{r}\left(\frac{m}{r} - \frac{e^2}{2r^2}\right)\dot{\lambda}_c k^c - \frac{2}{r}\left(\frac{m_{,u}}{r} - \frac{ee_{,u}}{r^2}\right). \end{aligned} \tag{28.23}$$

The Kerr-Schild solutions (28.23) contain (i) the Schwarzschild solution (13.19) $e = 0$, $m = \text{const}$, $\dot{\lambda}^c k_c = 0$, (ii) Vaidya's shining star solution (13.20) $e = 0$, $m = m(u)$, $\dot{\lambda}^c k_c = 0$, (iii) the Reissner-Nordström solution (13.21) $e = \text{const}$, $m = \text{const}$, $\dot{\lambda}^c k_c = 0$, (iv) Kinnersley's photon rocket solution $e = 0$, $m = m(u)$, $\dot{\lambda}^c k_c \neq 0$ (Kinnersley (1969a)), (v) the general case $e = e(u)$, $m = m(u)$, $\dot{\lambda}_c k^c \neq 0$.

28.1.5. The Newman-Penrose formalism for shearfree and geodesic Kerr-Schild metrics

In solving the field equations, the Newman-Penrose formalism is more convenient than the direct procedure outlined in § 28.1.2. Throughout this section the vector $\boldsymbol{k}$ in the null tetrad $(\boldsymbol{m}, \overline{\boldsymbol{m}}, \boldsymbol{l}, \boldsymbol{k})$ is the null vector of the Kerr-Schild metric (28.1)—(28.2); all numerical indices are tetrad indices.

Following Debever (1974), we will deal only with those special Kerr-Schild space-times which admit a geodesic, diverging and shearfree null congruence with tangent vector $\boldsymbol{k}$,

$$\varkappa = \sigma = 0; \qquad \Gamma_{142} = -\varrho = \Theta + \mathrm{i}\omega \neq 0, \tag{28.24}$$

and have a Ricci tensor satisfying the relations

$$R_{44} = R_{41} = R_{11} = R_{22} = 0. \tag{28.25}$$

(Ricci tensors of this type are general enough for most of the applications of the Kerr-Schild ansatz.)

As in § 23.1., we can infer from (28.24)—(28.25) and $\Psi_0 = \Psi_1 = 0$ (Theorem 28.3) that

$$\Gamma_{41} = \varrho\omega^2 + \tau\omega^3 \tag{28.26}$$

can be written as

$$\Gamma_{41} = \mathrm{d}\overline{Y}. \tag{28.27}$$

The difference from § 23.1. is that τ may be non-zero, and P is transformed to unity. Comparing (28.26) and (28.27) we see that

$$D\overline{Y} = 0, \qquad \delta\overline{Y} = 0, \qquad \overline{\delta}\overline{Y} = \varrho \neq 0, \qquad \Delta\overline{Y} = \tau \tag{28.28}$$

holds. Using the commutation relations and the Newman-Penrose equations, we find that ε and β vanish. A null rotation $\boldsymbol{k}' = \boldsymbol{k}$, $\boldsymbol{m}' = \boldsymbol{m} + B\boldsymbol{k}$, $\boldsymbol{l}' = \boldsymbol{l} + B\overline{\boldsymbol{m}} + \overline{B}\boldsymbol{m} + B\overline{B}\boldsymbol{k}$, preserves the relations $\varkappa = \sigma = \varepsilon = \beta = 0$ and can be used to make α vanish, which because of $\alpha' = \alpha + \overline{B}\varrho$ and $\varrho \neq 0$ is always possible. As a consequence of the Newman-Penrose equations, the coefficient π vanishes too. Thus we have

$$\varkappa = \sigma = \varepsilon = \beta = \alpha = \pi = 0. \tag{28.29}$$

At this stage, Debever (1974) proves the

Theorem 28.4. *A space-time is a special Kerr-Schild space-time (defined by the properties* (28.1)—(28.2) *and* (28.24)—(28.25)) *if and only if its Newman-Penrose coefficients can be transformed to*

$$\begin{pmatrix} \varkappa & \tau & \sigma & \varrho \\ \pi & \nu & \mu & \lambda \\ \varepsilon & \gamma & \beta & \alpha \end{pmatrix} = \begin{pmatrix} 0 & \Delta\overline{Y} & 0 & \overline{\delta}\overline{Y} \\ 0 & \nu & \mu & 0 \\ 0 & \gamma & 0 & 0 \end{pmatrix}; \quad V = \frac{\mu}{\varrho} = \frac{\overline{\mu}}{\overline{\varrho}} \tag{28.30}$$

with respect to a suitably chosen null tetrad.

The details of the proof are given by Debever (1974). To show that (28.30) is sufficient, the main steps are: (a) One determines the remaining Newman-Penrose coefficients by means of the Newman-Penrose equations and the commutation relations, cp. §§ 7.1., 7.2. The result reads

$$\gamma = 2^{-1}\big(DV + V(\varrho - \overline{\varrho})\big), \qquad \nu = \overline{\delta}V - V\Delta Y, \qquad \mu = V\varrho. \tag{28.31}$$

(b) Solving the first Cartan equations (2.74)

$$\begin{aligned} d\omega^1 &= Y_{,a}\omega^a \wedge (\omega^4 - V\omega^3) = d\overline{\omega}^2, \\ d\omega^3 &= \overline{Y}_{,a}\omega^a \wedge \omega^1 + Y_{,a}\omega^a \wedge \omega^2, \\ d\omega^4 &= VY_{,a}\omega^a \wedge \omega^2 + V\overline{Y}_{,a}\omega^a \wedge \omega^1 + V_{,a}\omega^a \wedge \omega^3, \end{aligned} \tag{28.32}$$

one obtains the dual basis

$$\begin{aligned} &\omega^1 = d\zeta + Y\,dv, \qquad \omega^2 = d\bar{\zeta} + \overline{Y}\,dv, \\ &\omega^3 = \overline{Y}\,d\zeta + Y\,d\bar{\zeta} + Y\overline{Y}\,dv + du, \qquad \omega^4 = V\omega^3 + dv. \end{aligned} \tag{28.33}$$

The associated metric

$$ds^2 = 2(d\zeta\,d\bar{\zeta} - du\,dv) - 2V(\overline{Y}\,d\zeta + Y\,d\bar{\zeta} + Y\overline{Y}\,dv + du)^2 \tag{28.34}$$

is clearly of the Kerr-Schild type.

If we start with a line element (28.34), then the null congruence

$$\omega^3 = -k_a\,dx^a = \overline{Y}\,d\zeta + Y\,d\bar{\zeta} + Y\overline{Y}\,dv + du \tag{28.35}$$

will be geodesic and shearfree only if Y obeys (28.28). Because of

$$\begin{aligned} D &= -(Y\,\partial_\zeta + \overline{Y}\,\partial_{\bar{\zeta}} - \partial_v - Y\overline{Y}\,\partial_u), \\ \Delta &= V(Y\,\partial_\zeta + \overline{Y}\,\partial_{\bar{\zeta}} - \partial_v - Y\overline{Y}\,\partial_u) + \partial_u, \\ \delta &= \partial_\zeta - \overline{Y}\,\partial_u, \qquad \bar{\delta} = \partial_{\bar{\zeta}} - Y\,\partial_u, \end{aligned} \tag{28.36}$$

these conditions read

$$\overline{Y}_{,\zeta} - \overline{Y}\overline{Y}_{,u} = 0, \qquad Y_{,v} - YY_{,\zeta} = 0. \tag{28.37}$$

They show that Y satisfies an equation

$$F(Y, \bar{\zeta}Y + u, vY + \zeta) = 0, \tag{28.38}$$

where F is an arbitrary function analytic in the three complex variables Y, $\bar{\zeta}Y + u$ and $vY + \zeta$ (Cox and Flaherty (1976)). The geodesic and shearfree null congruence (28.35) of the Kerr-Schild space-time is simultaneously a geodesic and shearfree null congruence in Minskowski space, see above. We have arrived at

Theorem 28.5 (Kerr's Theorem). *Any analytic, geodesic and shearfree null congruence in Minkowski space is given by* $\omega^3 = dv$ *or by*

$$\omega^3 = \overline{Y}\,d\zeta + Y\,d\bar{\zeta} + Y\overline{Y}\,dv + du, \tag{28.39}$$

where $Y(u, v, \zeta, \bar{\zeta})$ *is defined implicitly by the relation* (28.38) *with arbitrary* F (Cox and Flaherty (1976)).

To get explicitly all special Kerr-Schild metrics one has to solve the Newman-Penrose equations. Using the basis (28.33) we find that the following Newman-Penrose equations are already fulfilled:

$$D\varrho = \varrho^2; \quad D\tau = \tau\varrho; \quad \delta\tau = \tau^2; \quad \delta\varrho = (\varrho - \bar{\varrho})\,\tau; \quad \bar{\delta}\tau - \Delta\varrho = \tau\bar{\tau} + V\varrho^2. \tag{28.40}$$

The Newman-Penrose equations one really has to solve read

$$D^2V = 2(\varrho + \bar{\varrho})\, DV - 2V\varrho\bar{\varrho} - \frac{1}{2} R, \qquad (\varrho + \bar{\varrho})\, DV - V(\varrho^2 + \bar{\varrho}^2) = \frac{1}{4} R + 2\Phi_{11}, \tag{28.41}$$

$$\delta\gamma = \tau(\mu + \gamma) + \Phi_{12}, \quad \delta\nu - \Delta\mu - \mu^2 - (\gamma + \bar{\gamma})\,\mu - \tau\nu = \Phi_{22}. \tag{28.42}$$

The remaining Newman-Penrose equations determine the non-vanishing components of the Weyl tensor, namely

$$\Psi_2 = \mu\varrho + \gamma(\varrho - \bar{\varrho}) + \frac{1}{24} R + \Phi_{11}, \quad \Psi_3 = \bar{\delta}\gamma + \varrho\nu - \bar{\tau}\gamma, \quad \Psi_4 = \bar{\delta}\nu - \bar{\tau}\nu. \tag{28.43}$$

The Newman-Penrose equations (28.41) can be solved immediately in the case of vanishing Ricci scalar ($R = 0$). The solution reads

$$V = \frac{1}{2} \tilde{M}(\varrho + \bar{\varrho}) + \tilde{B}\varrho\bar{\varrho}, \qquad \Phi_{11} = \tilde{B}\varrho^2\bar{\varrho}^2, \qquad D\tilde{M} = D\tilde{B} = 0, \tag{28.44}$$

$\tilde{M}$ and $\tilde{B}$ being real functions. So far $R = 0$, only (28.42) remains to be solved.

28.2. Application of the Kerr-Schild ansatz to Einstein's vacuum field equations

28.2.1. The case $\varrho = -(\Theta - i\omega) \neq 0$

The general Kerr-Schild solution of Einstein's vacuum field equations was found by Kerr and Schild (1965, 1967). The remaining field equations (28.42) require an intricate calculation to get the final result. For details see Debney et al. (1969) and § 28.3.1. The solutions are

$$\begin{aligned} ds^2 &= 2(d\zeta\, d\bar{\zeta} - du\, dv) + \tilde{m}P^{-3}(\varrho + \bar{\varrho})\,[du + \bar{Y}\, d\zeta + Y\, d\bar{\zeta} + Y\bar{Y}\, dv]^2, \\ P &= pY\bar{Y} + qY + \bar{q}\bar{Y} + c, \qquad \tilde{M} = \tilde{m}P^{-3}, \end{aligned} \tag{28.45}$$

where $\tilde{m}$, p, c (real) and q (complex) are constants. The function Y is given by the implicit relation

$$\Phi(Y) + (qY + c)(\zeta + Yv) - (pY + \bar{q})(u + Y\bar{\zeta}) = 0, \tag{28.46}$$

where $\Phi(Y)$ is an arbitrary analytic function of the complex variable Y. The complex expansion of the null vector $\boldsymbol{k}$ is given by

$$\varrho = -P[\{\Phi + (qY + c)(\zeta + Yv) - (pY + \bar{q})(u + Y\bar{\zeta})\}_{,Y}]^{-1}. \tag{28.47}$$

The solutions have the following properties.

(i) They are all algebraically special, $\boldsymbol{k}$ being a multiple principal null direction of the Weyl tensor. Thus $\boldsymbol{k}$ is shearfree and geodesic.

(ii) They all admit at least a one-parameter group of motions with Killing vector

$$\xi = c\,\partial_u + q\,\partial_{\bar\zeta} + \bar q\,\partial_\zeta + p\,\partial_v, \tag{28.48}$$

which is simultaneously a Killing vector of flat space-time. The solutions can be simplified by performing a Lorentz transformation. One can thus assume that if $\eta_{ab}\xi^a\xi^b < 0$, $P\sqrt{2} = 1 + Y\bar Y$; if $\eta_{ab}\xi^a\xi^b > 0$, $P\sqrt{2} = 1 - Y\bar Y$; and if $\eta_{ab}\xi^a\xi^b = 0$, $P = 1$.

(iii) The points $(P\varrho^{-1}) = 0$ are singularities of the Riemannian space. In general Y is a multivalued function with these singularities as branch points.

(iv) The particular case $\Phi = -iaY$ leads to the Kerr solution (18.25). Its Kerr-Schild form reads (with r given by $(x^2 + y^2)(r^2 + a^2)^{-1} + z^2r^{-2} = 1$)

$$\begin{aligned} ds^2 &= dx^2 + dy^2 + dz^2 - dt^2 \\ &+ \frac{2mr^3}{r^4 + a^2z^2}\left[dt + \frac{z}{r}\,dz + \frac{r}{r^2 + a^2}(x\,dx + y\,dy)\right. \\ &\left. - \frac{a}{r^2 + a^2}(x\,dy - y\,dx)\right]^2, \end{aligned} \tag{28.49}$$

the double Debever-Penrose null vector being

$$\begin{aligned} \omega^3 = 2^{1/2}\frac{r}{r + z}&\left[dt + \frac{z}{r}\,dz + \frac{r}{r^2 + a^2}(x\,dx + y\,dy)\right. \\ &\left. - \frac{a}{r^2 + a^2}(x\,dy - y\,dx)\right] = -k_i\,dx^i. \end{aligned} \tag{28.50}$$

For a different characterization of the Kerr-Schild vacuum solutions see § 25.2.5.

28.2.2. The case $\varrho = -(\Theta - i\omega) = 0$

The case of a non-expanding and non-twisting congruence $\boldsymbol{k}$ in a Kerr-Schild metric was solved by Trautman (1962) and later treated by Urbantke (1972) and Debney (1973). Their results are summarized in

Theorem 28.6. *The Kerr-Schild vacuum fields with a non-expanding and non-twisting null congruence are the plane-fronted gravitational waves* (§ 21.5.), *and conversely.*

28.3. Application of the Kerr-Schild ansatz to the Einstein-Maxwell equations

28.3.1. The case $\varrho = -(\Theta - i\omega) \neq 0$

Electromagnetic solutions of the Kerr-Schild type were studied by Debney et al. (1969), again assuming the null vector $\boldsymbol{k}$ to be geodesic. By Theorem 28.1, the energy-momentum tensor of the electromagnetic field fulfils the conditions

$$k_{a;b}k^b = 0 \leftrightarrow T_{ab}k^ak^b = 0 \leftrightarrow F_{ab}k^a = \lambda k_b. \tag{28.51}$$

Hence $\boldsymbol{k}$ is a principal null direction of F_{ab} if and only if it is geodesic. The space-time is algebraically special with $\boldsymbol{k}$ as the multiple principal null direction. The energy-momentum tensor obeys (28.25), and the Bianchi identities imply that the null vector $\boldsymbol{k}$ is shearfree. So all results of § 28.1.5. are true for the electromagnetic case too. Compare also Chapter 26!

The result (28.51) implies that of the tetrad components of the Maxwell tensor only Φ_1 and Φ_2 do not vanish,

$$\Phi_1 = 2^{-1}F_{ab}(k^a l^b + \overline{m}^a m^b), \qquad \Phi_2 = F_{ab}\overline{m}^a l^b, \qquad \Phi_0 = 0. \tag{28.52}$$

The field equations therefore read

$$R_{12} = R_{34} = 2\varkappa_0 \Phi_1 \overline{\Phi}_1, \qquad R_{13} = 2\varkappa_0 \Phi_1 \overline{\Phi}_2, \qquad R_{33} = 2\varkappa_0 \Phi_2 \overline{\Phi}_2, \tag{28.53}$$

and Maxwell's equations (7.46)—(7.49) are

$$D\Phi_1 - 2\varrho\Phi_1 = 0, \qquad \delta\Phi_1 - 2\tau\Phi_1 = 0, \tag{28.54}$$

$$-\delta\Phi_2 + \Delta\Phi_1 + 2V\varrho\Phi_1 + \tau\Phi_2 = 0, \qquad \overline{\delta}\Phi_1 - D\Phi_2 + \varrho\Phi_2 = 0. \tag{28.55}$$

Using (28.40) and the commutation relations (7.55)—(7.58), the field equations (28.53)—(28.55) can be partially integrated. The result is

$$\Phi_1 = \varrho^2\Phi_1^0, \quad \Phi_2 = \varrho\Phi_2^0 + \overline{\delta}(\varrho\Phi_1^0), \quad D\Phi_1^0 = D\Phi_2^0 = 0, \tag{28.56}$$

$$V = \frac{1}{2}\tilde{M}(\varrho + \overline{\varrho}) + \varkappa_0\Phi_1^0\overline{\Phi}_1^0\varrho\overline{\varrho}, \qquad D\tilde{M} = 0. \tag{28.57}$$

In terms of the four unknown functions Y, $\tilde{M}$, Φ_1^0 and Φ_2^0, the remaining Maxwell equations and Einstein equations are found to be

$$\begin{aligned} &D\Phi_1^0 = D\Phi_2^0 = 0, \qquad \delta\Phi_1^0 - 2\overline{\varrho}\varrho^{-1}\tau\Phi_1^0 = 0,\\ &\Delta\Phi_1^0 - \tau\varrho^{-1}\overline{\delta}\Phi_1^0 - \overline{\tau}\overline{\varrho}^{-1}\delta\Phi_1^0 = \tau\varrho^{-1}\Phi_2^0 - \overline{\varrho}^{-1}\delta\overline{\Phi}_2^0 \end{aligned} \tag{28.58}$$

and

$$\begin{aligned} &D\tilde{M} = DY = \overline{\delta}Y = 0,\\ &\delta\tilde{M} - \tilde{M}\overline{\varrho}\varrho^{-1}\tau = 2\varkappa_0\overline{\varrho}\Phi_1^0\overline{\Phi}_2^0,\\ &\Delta\tilde{M} - \overline{\tau}\overline{\varrho}^{-1}\delta\tilde{M} - \tau\varrho^{-1}\overline{\delta}\tilde{M} = -\varkappa_0\Phi_2^0\overline{\Phi}_2^0. \end{aligned} \tag{28.59}$$

No general solution of this system is yet known. Debney et al. (1969) have completely solved the case where the electromagnetic field is further restricted by the condition

$$\Phi_2^0 = 0. \tag{28.60}$$

Their solutions are

$$\begin{aligned} \mathrm{d}s^2 = 2(\mathrm{d}\zeta\,\mathrm{d}\overline{\zeta} - \mathrm{d}u\,\mathrm{d}v) + P^{-3}\left[\tilde{m}(\varrho + \overline{\varrho}) - \frac{\varkappa_0}{2}\Psi\overline{\Psi}P^{-1}\varrho\overline{\varrho}\right]&\\ \times[\overline{Y}\,\mathrm{d}\zeta + Y\,\mathrm{d}\overline{\zeta} + Y\overline{Y}\,\mathrm{d}v + \mathrm{d}u]^2.& \end{aligned} \tag{28.61}$$

As in the vacuum case, Y, ϱ and P are given by (28.45)—28.47), and Φ and Ψ are arbitrary (analytic) functions of Y. The solutions admit the Killing vector (28.48). Compare also Theorem 26.2! $\Psi = 0$ gives the vacuum solution (28.45).

The particular choice

$$P = 2^{-1/2}(1 + Y\bar{Y}), \qquad \Phi = -\mathrm{i}aY, \qquad \Psi = e \tag{28.62}$$

leads to the field of a rotating charged body first obtained by Newman et al. (1965) (cp. § 19.1.3.). Using the special null coordinates $\sqrt{2}\,\zeta = x + \mathrm{i}y$, $\sqrt{2}\,\bar{\zeta} = x - \mathrm{i}y$, $\sqrt{2}\,u = t - z$, $\sqrt{2}\,v = t + z$ the solution and the electromagnetic field are found to be

$$\begin{aligned} \mathrm{d}s^2 = \mathrm{d}x^2 + \mathrm{d}y^2 + \mathrm{d}u^2 - \mathrm{d}t^2 + \frac{2\tilde{m}r^3 - e^2r^2}{r^4 + a^2z^2}\Bigg[\mathrm{d}t + \frac{z}{r}\,\mathrm{d}t \\ + \frac{r}{r^2 + a^2}\,(x\,\mathrm{d}x + y\,\mathrm{d}y) - \frac{a}{r^2 + a^2}\,(x\,\mathrm{d}y - y\,\mathrm{d}x)\Bigg]^2, \end{aligned} \tag{28.63}$$

$$(F_{xt} - \mathrm{i}F_{yz}, F_{yt} - \mathrm{i}F_{zx}, F_{zt} - \mathrm{i}F_{xy}) = er^3(r^2 + \mathrm{i}az)^{-3}\,(x, y, z + \mathrm{i}a),$$

where the function r is defined by the implicit relation

$$(x^2 + y^2)\,(r^2 + a^2)^{-1} + z^2r^{-2} = 1. \tag{28.64}$$

Taking $e = 0$ in (28.63), we recover the Kerr solution (28.49).

28.3.2. The case $\varrho = -(\Theta + \mathrm{i}\omega) = 0$

The non-diverging solutions were considered in detail in Chapter 27. (Kundt's class). In § 27.1. it was shown that $\varrho = 0$ implies that $\boldsymbol{k}$ is both a multiple principal null direction of the Weyl tensor and the electromagnetic field tensor and that $\boldsymbol{k}$ is shearfree.

Expansionfree electromagnetic solutions of the Kerr-Schild class were treated first by Białas (1963), and later by Debney (1974), again assuming $\boldsymbol{k}$ to be a geodesic null congruence. All these solutions belong to Kundt's class.

Choosing the basis (28.33), the only non-zero connection coefficients are given by

$$\Gamma_{233} = \bar{\delta}V - V\Delta Y, \qquad \Gamma_{143} = -\Delta\bar{Y}, \qquad \Gamma_{433} = DV, \tag{28.65}$$

and the vector fields $(\boldsymbol{e}_1, \boldsymbol{e}_2, \boldsymbol{e}_3, \boldsymbol{e}_4)$ obey the commutation relations $[\boldsymbol{e}_1, \boldsymbol{e}_2] = [\boldsymbol{e}_1, \boldsymbol{e}_4] = [\boldsymbol{e}_2, \boldsymbol{e}_4] = 0$. The non-zero Newman-Penrose coefficients and the non-vanishing components of the Weyl and Ricci tensors read

$$\begin{aligned} &2\gamma = DV, \qquad \nu = \bar{\delta}V - V\Delta Y, \qquad \tau = \Delta\bar{Y}, \\ &R_{33} = 2\varkappa_0\Phi_2\bar{\Phi}_2, \qquad \Psi_4 = \bar{\delta}\bar{\delta}V - 2\bar{\delta}V\,\Delta Y. \end{aligned} \tag{28.66}$$

The only non-zero component of the electromagnetic field tensor is Φ_2.

Theorem 28.7 (Debney (1974)). *All source-free Einstein-Maxwell fields of the Kerr-Schild class with an expansionfree (geodesic) null congruence* $\boldsymbol{k}$ *are of Petrov type N (or O).* $\boldsymbol{k}$ *is both a shearfree quadruple principal null direction of the Weyl tensor and a principal null direction of the Maxwell field. The electromagnetic field is null.*

These results reduce the Einstein-Maxwell equations for expansionfree Kerr-Schild fields to the equations

$$D^2V = 0, \quad D\bar{\delta}V - \varDelta YDV = 0, \quad \delta\bar{\delta}V - \delta V\varDelta\bar{Y} - \bar{\delta}V\varDelta\bar{Y} = \varkappa_0\Phi_2\bar{\Phi}_2,$$
$$D\Phi_2 = 0, \quad \bar{\delta}\Phi_2 - \varDelta Y\bar{\Phi}_2 = 0, \quad \delta Y = \bar{\delta}Y = DY = 0. \tag{28.67}$$

In solving the field equations new coordinates

$$z = \zeta + Yv, \qquad \bar{z} = \bar{\zeta} + \bar{Y}v, \qquad u' = u + \bar{Y}\zeta + Y\bar{\zeta} + Y\bar{Y}v, \quad v' = v, \tag{28.68}$$

are more convenient than the old set of coordinates $\zeta, \bar{\zeta}, u, v$. In terms of the new coordinates the dual basis (28.33) and tetrad basis read

$$\omega^1 = \bar{\omega}^2 = \mathrm{d}z - r^{-1}v\bar{\tau}\,\mathrm{d}u', \quad \omega^3 = r^{-1}\,\mathrm{d}u', \quad \omega^4 = r^{-1}V\,\mathrm{d}u' + \mathrm{d}v,$$
$$r = [1 - \dot{\bar{Y}}z - \dot{Y}\bar{z}]^{-1}, \quad r\,\mathrm{d}\bar{Y}/\mathrm{d}u' = r\dot{\bar{Y}} = \tau; \tag{28.69}$$

$$\boldsymbol{e}_1 = \partial_z, \quad \boldsymbol{e}_2 = \partial_{\bar{z}}, \quad \boldsymbol{e}_3 = v[\bar{\tau}\partial_z + \tau\,\partial_{\bar{z}}] + r\,\partial_{u'} - V\,\partial_v, \quad \boldsymbol{e}_4 = \partial_v. \tag{28.70}$$

The field equations are easier to solve in terms of the new coordinates. They are, with $Y = Y(u')$,

$$V_{,vv} = 0, \quad \bar{\Phi}_{2,v} = 0, \quad V_{,\bar{z}v} - \bar{\tau}V_{,v} = V_{,zv} - V_{,v}\tau = 0, \tag{28.71a}$$
$$\bar{\Phi}_{2,\bar{z}} - \tau\bar{\Phi}_2 = 0, \quad V_{,z\bar{z}} - V_{,z}\bar{\tau} - V_{,\bar{z}}\tau = \varkappa_0\Phi_2\bar{\Phi}_2. \tag{28.71b}$$

From the equations (28.71a) we get

$$\bar{\Phi}_2 = \bar{\Phi}_2(z, \bar{z}, u'); \qquad V = vra(u') + g(z, \bar{z}, u'), \tag{28.72}$$

where $a(u')$ and $g(z, \bar{z}, u')$ are real functions. By a coordinate transformation one can set $a(u') = 0$ (Debney (1974)). Then the electromagnetic Kerr-Schild solutions admitting a geodesic, shearfree and expansionfree null congruence become

$$\mathrm{d}s^2 = 2(\mathrm{d}z - v\dot{Y}\,\mathrm{d}u')\,(\mathrm{d}\bar{z} - v\dot{\bar{Y}}\,\mathrm{d}u') - 2r^{-1}(gr^{-1}\,\mathrm{d}u' + \mathrm{d}v)\,\mathrm{d}u'. \tag{28.73}$$

The remaining field equations (28.71b) restrict the electromagnetic field and the real function $g(z, \bar{z}, u')$ by the equations

$$(r^{-1}\bar{\Phi}_2)_{,\bar{z}} = 0; \qquad (r^{-1}g)_{,z\bar{z}} = \varkappa_0 r^{-1}\Phi_2\bar{\Phi}_2. \tag{28.74}$$

The vector $\boldsymbol{k}$ is a null Killing vector (see § 21.4.) if and only if the rotation coefficient $\tau = r\dot{\bar{Y}}$ vanishes. In this case the null vector is covariantly constant; all solutions of this type are plane-fronted gravitational waves (see § 21.5.). To get all solutions in the general case $\tau \neq 0$ means solving the 2-dimensional Poisson equation (28.74).

28.4. Application of the Kerr-Schild ansatz to pure radiation fields

Kerr-Schild metrics with the energy-momentum tensor of pure radiation, $T_{mn} = \Phi^2 k_m k_n$, were investigated by Kinnersley (1969a), Bonnor and Vaidya (1972), Vaidya (1973), Urbantke (1975), and Herlt (1979). Unlike the vacuum or Einstein-Maxwell case the geodesic null vector $\boldsymbol{k}$ need not be shearfree.

28.4.1. The case $\varrho \neq 0$, $\sigma = 0$

This case was investigated by Vaidya (1973, 1974). We will give briefly the translation from the dual bases (28.33) and the coordinates $(\zeta, \bar{\zeta}, u, v)$ to the more convenient terms Vaidya works with.

After a null rotation ($\boldsymbol{l}$ fixed) and a special Lorentz transformation in the $(\boldsymbol{k}, \boldsymbol{l})$-plane, (28.33) is transformed into

$$\begin{aligned} \omega^{1'} &= \omega^1 + \bar{\varrho}^{-1} Y_{,u}\omega^3 = \bar{\varrho}^{-1}\,\mathrm{d}Y, \qquad \omega^{2'} = \overline{\omega}^{1'}, \\ \omega^{3'} &= 2^{1/2}(1 + Y\overline{Y})\,\omega^3, \\ \omega^{4'} &= 2^{-1/2}(1 + Y\overline{Y})^{-1}\{\omega^4 + Y_{,u}\bar{\varrho}^{-1}\omega^2 + \overline{Y}_{,u}\varrho^{-1}\omega^1 + (\varrho\bar{\varrho})^{-1}\,Y_{,u}\overline{Y}_{,u}\,\omega^3\}. \end{aligned} \tag{28.75}$$

Here, and throughout, $\varrho = \bar{\delta}\overline{Y}$ holds and $Y(\zeta, \bar{\zeta}, u, v)$ is the complex function defining the null congruence (28.35). The new coordinates $(\alpha, \beta, \hat{u}, v)$ are defined by

$$Y = \tan\frac{\alpha}{2}\,\mathrm{e}^{\mathrm{i}\beta}, \quad \hat{u} = 2^{-1/2}[\zeta \sin\alpha\,\mathrm{e}^{-\mathrm{i}\beta} + \bar{\zeta}\sin\alpha\,\mathrm{e}^{\mathrm{i}\beta} + (u - v)\cos\alpha + u + v],\ v = v. \tag{28.76}$$

$\hat{u}$ can be interpreted as the proper time for messages conveyed along $\boldsymbol{k}$. Finally two new field functions U and W are introduced by

$$U = 2^{-1/2}[\zeta\cos\alpha\,\mathrm{e}^{-\mathrm{i}\beta} + \bar{\zeta}\cos\alpha\,\mathrm{e}^{\mathrm{i}\beta} + (v - u)\sin\alpha], \quad W = \mathrm{i}2^{-1/2}[\zeta\,\mathrm{e}^{-\mathrm{i}\beta} - \bar{\zeta}\,\mathrm{e}^{\mathrm{i}\beta}]. \tag{28.77}$$

By means of the relations (28.28) and (28.36) and the definition (28.76) we can prove that the functions U and W depend on the new coordinates $\hat{u}, \alpha, \beta$ only,

$$U = U(\alpha, \beta, \hat{u}), \qquad W = W(\alpha, \beta, \hat{u}). \tag{28.78}$$

We look upon (28.78) as giving us α and β as functions of the coordinates $(\zeta, \bar{\zeta}, \hat{u}, v)$, once the forms of the functions U and W are known. With this understanding a tedious calculation leads from the shearfree and geodesic conditions (28.37) to the following two partial differential equations for the functions U and W:

$$\begin{aligned} &UU_{,\hat{u}} - WW_{,\hat{u}} + U_{,\alpha} - U\cot\alpha + W_{,\beta}\sin^{-1}\alpha = 0, \\ &-WU_{,\hat{u}} - UW_{,\hat{u}} - W_{,\alpha} + W\cot\alpha + U_{,\beta}\sin^{-1}\alpha = 0. \end{aligned} \tag{28.79}$$

Evaluating the null tetrad (28.75) in terms of the new coordinates and field functions, one gets the line element of Kerr-Schild space-times in the form (Vaidya (1974))

$$\begin{aligned} \mathrm{d}s^2 = {} & (p^2 + q^2)(\mathrm{d}\alpha^2 + \sin^2\alpha\,\mathrm{d}\beta^2) + 2^{3/2}(1 + Y\overline{Y})^{-1}(k_a\,\mathrm{d}x^a)\,\mathrm{d}t \\ & - 2^{3/2}[(qU_{,\hat{u}} + pW_{,\hat{u}})\,\mathrm{d}\alpha + (pU_{,\hat{u}} - qW_{,\hat{u}})\sin\alpha\,\mathrm{d}\beta](1 + Y\overline{Y})^{-1}k_a\,\mathrm{d}x^a \\ & + 2[1 - 2q(p^2 + q^2)^{-1}\hat{M} + U_{,\hat{u}}{}^2 + W_{,\hat{u}}{}^2](1 + Y\overline{Y})^{-2}(k_a\,\mathrm{d}x^a)^2, \end{aligned} \tag{28.80}$$

where the abbreviations q and φ mean

$$\begin{aligned} p &= UW_{,\hat{u}} + W_{,\alpha}, \qquad q = UU_{,\hat{u}} + U_{,\alpha} + \hat{u} - t, \\ u &= u(\hat{u}, v, \alpha, \beta), \qquad t = 2^{-1/2}(u + v), \end{aligned} \tag{28.81}$$

$\hat{M}$ is related to $\tilde{M}$ of (28.44) by $\hat{M} = 2^{-3/2}(1 + Y\bar{Y})^3 \tilde{M}$, and the 1-form $k_a \, dx^a$ turns out to be

$$-2^{1/2}(1 + Y\bar{Y})^{-1} k_a \, dx^a = d\hat{u} - U \, d\alpha + W \sin\alpha \, d\beta. \tag{28.82}$$

After a tedious calculation the remaining field equations become

$$R_{33} = 4(p^2 + q^2)^{-1} (1 + Y\bar{Y}) \hat{M}_{,\hat{u}} = \varkappa_0 \Phi^2, \tag{28.83}$$

$$\hat{M}_{,v} = 0, \qquad \hat{M} = \hat{M}(\alpha, \beta, \hat{u}), \tag{28.84}$$

$$U\hat{M}_{,\hat{u}} + 3\hat{M}U_{,\hat{u}} + \hat{M}_{,\alpha} = 0, \qquad W\hat{M}_{,\hat{u}} + 3\hat{M}W_{,\hat{u}} - \hat{M}_{,\beta} \sin^{-1}\alpha = 0. \tag{28.85}$$

As $\Phi^2(\alpha, \beta, \hat{u})$ is an arbitrary (positive) function, equation (28.83) is only the definition of $R_{33} = \varkappa_0 \Phi^2$, and not an equation which needs to be integrated. The actual equations which have to be integrated to obtain the three real field functions U, W, $\hat{M}$ (depending on three coordinates α, β, $\hat{u}$ only) are the geodesic and shearfree conditions (28.79) and the two equations (28.85). The most general solution of the geodesic and shearfree conditions (28.79) can be found by using Theorem 28.5. It reads (Herlt 1980))

$$F\left[e^{i\beta} \tan\frac{\alpha}{2}, \quad e^{i\beta}\left(U - iW + \hat{u} \tan\frac{\alpha}{2}\right), \quad \hat{u} - \tan\frac{\alpha}{2}(U - iW)\right] = 0, \tag{28.86}$$

where F is an arbitrary analytic function of its three complex arguments. The functions F and $\hat{M}$ have to be chosen so that the field equations (28.85) are satisfied.

Twistfree null congruences $\boldsymbol{k}$ (i.e. $p = 0$, $U = 0$, $W = 0$) leave $\hat{M}$ as an arbitrary function of $\hat{u}$. In the twisting case, the integrability condition of (28.85), $\hat{M}_{,\alpha,\beta} = \hat{M}_{,\beta,\alpha}$ leads to

$$p\hat{M}_{,\hat{u}} + 3\hat{M}p_{,\hat{u}} = 0, \quad \text{where} \quad p = UW_{,\hat{u}} + W_{,\alpha} \tag{28.87}$$

and

$$\hat{M} = f(\alpha, \beta) \, p^{-3}. \tag{28.88}$$

When this is substituted in (28.85) one obtains

$$\begin{aligned} f_{,\beta} f^{-1} &= 3p^{-1}[W_{,\hat{u}} U_{,\beta} - WU_{,\beta\hat{u}} + p_{,\beta} + W^2 U_{,\hat{u}\hat{u}} \sin\alpha], \\ f_{,\alpha} f^{-1} &= 3p^{-1}[p_{,\alpha} + Up_{,\hat{u}} - pU_{,\hat{u}}]. \end{aligned} \tag{28.89}$$

The (so far unknown) function f must be independent of $\hat{u}$. Therefore the following consistency condition has to be fulfilled (Vaidya (1974)):

$$[p^{-1}(W_{,\hat{u}} U_{,\beta} - WU_{,\beta\hat{u}} + p_{,\beta} + W^2 U_{,\hat{u}\hat{u}} \sin^{-1}\alpha)]_{,\hat{u}} = 0. \tag{28.90}$$

Summarizing all the results, we see that one obtains all Kerr-Schild fields possessing a shearfree, geodesic and twisting null congruence and a pure radiation energy-momentum tensor by solving the equation (28.90) and the integrability condition $f_{,\beta,\alpha} = f_{,\alpha,\beta}$ of (28.89), where the functions U and W have to simultaneously fulfil the implicit relation (28.86). If the functions U and W are known, the functions p and q follow by differentiation, cp. (28.81), and the function $\hat{M}$ is obtainable from a line integral according to (28.89) and (28.88).

In the axisymmetric case,

$$\hat{M}_{,\beta} = U_{,\beta} = W_{,\beta} = 0, \tag{28.91}$$

the most general twisting and shearfree Kerr-Schild solution with pure radiation is known. It reads (Herlt □ (1980))

$$\begin{aligned} U &= -\sin\alpha\,[a_0(\cos\alpha + a_1)^2 + a_2]^{-1}\,[a_0(\cos\alpha + a_1)\,\hat{u} + b_1 + b_2\cos\alpha], \\ W^2 &= \sin^2\alpha[a_0(\cos\alpha + a_1)^2 + a_2]^{-2} \\ &\quad \times \{a_0a_2\hat{u}^2 + 2[a_2b_2 + a_0(\cos\alpha + a_1)\,(a_1b_2 - b_1)]\,\hat{u} \\ &\quad - (b_1 + b_2\cos\alpha)^2 - c_0a_0(\cos\alpha + a_1)^2 - c_0a_2\}, \\ \hat{M} &= m_0\{a_0a_2\hat{u}^2 + 2[a_2b_2 + a_0(\cos\alpha + a_1)\,(a_1b_2 - b_1)]\,\hat{u} \\ &\quad - (b_1 + b_2\cos\alpha)^2 - c_0a_0(\cos\alpha + a_1)^2 - c_0a_2\}^{-3/2}; \end{aligned} \tag{28.92}$$

a_1, a_2, a_0, b_1, b_2, and m_0 are arbitrary (real) constants. The particular choice $a_0 = b_1 = 0$, $2b_2a_2^{-1} = b$, $-c_0a_2^{-1} = a^2$, i.e. $U = -\frac{1}{2}b\sin\alpha\cos\alpha$, $W^2 + U^2 = (bu + a^2)\sin^2\alpha$, leads to the radiating Kerr metric given by Vaidya and Patel (1973). For $b = 0$, we recover the Kerr solution (28.49).

As yet the general case $U_{,\beta} \neq 0$ seems to be unsolved.

28.4.2. The case $\sigma \neq 0$

In the aligned case (28.11), space-time is algebraically special because of Theorem 28.3, and the theorems given in § 7.5. show that it must be of type N. So the Kerr-Schild solutions in question are contained in the solutions discussed in § 22.1. To extract them from the classes considered there, we have to incorporate the Kerr-Schild condition (28.1), i.e. we have to demand that the type N solution has non-zero shear and can be written as

$$\begin{aligned} \mathrm{d}s^2 &= 2\omega^1\omega^2 - 2\omega^3\omega^4, \\ \omega^1 &= \mathrm{d}\zeta + Y\,\mathrm{d}v = \overline{\omega}^2, \qquad \omega^4 = V\omega^3 + \mathrm{d}v, \\ \omega^3 &= -k_a\,\mathrm{d}x^a = \overline{Y}\,\mathrm{d}\zeta + Y\,\mathrm{d}\bar{\zeta} + Y\overline{Y}\,\mathrm{d}v + \mathrm{d}u. \end{aligned} \tag{28.93}$$

Starting with (28.93), and using $\Psi_0 = \Psi_1 = \Psi_2 = \Psi_3 = 0$ and the field equations $R_{ab} = \varkappa_0\delta_a^3\delta_b^3\Phi^2$, the Newman-Penrose equations show that the null congruence $\boldsymbol{k}$ must be geodesic and twistfree,

$$\varkappa = \varrho - \bar{\varrho} = 0, \qquad \varrho^2 = \sigma\bar{\sigma}, \tag{28.94}$$

and the spin coefficients ϱ, σ and τ have to satisfy

$$\delta(\varrho\tau\sigma^{-1} + \tau) = \bar{\delta}(\varrho\bar{\tau}\sigma^{-1} + \tau) \tag{28.95}$$

(Urbantke (1975)). All type N pure radiation fields satisfying (28.94) and $\sigma \neq 0$ are known and contained in (22.10) and (22.12), but only special (Kerr-Schild) metrics have been determined which obey in addition (28.95), cp. Urbantke (1975).

28.4.3. The case $\varrho = \sigma = 0$

Kerr-Schild pure radiation solutions possessing a geodesic, expansionfree and shear-free null congruence belong to Petrov type N, the vector $\boldsymbol{k}$ being the quadruple principal null direction. All solutions of this class are given by (28.73) with arbitrary functions $Y(u')$ and (real) $g(z, z, u')$.

Table 28.1 Kerr-Schild space-times

	Vacuum	Einstein-Maxwell	Pure radiation	Perfect fluid
$\varrho \neq 0$	A (28.45)	S (28.61)	S (28.80), (28.96)	∄
$\varrho = 0$	A Th. 28.6 (21.38)	A (28.73)	A (28.73)	∄

28.5. Generalizations of the Kerr-Schild ansatz

Space-times with metrics g_{ab} and $\tilde{g}_{ab}$, both non-flat, which are related by

$$\tilde{g}_{ab} = g_{ab} + 2Vk_ak_b, \qquad \tilde{g}^{ab} = g^{ab} - 2Vk^ak^b, \tag{28.96}$$

have been considered by Thompson (1966) and Dozmorov (1971). The vector $\boldsymbol{k}$ is a null vector with respect to both metrics, and in the vacuum case, say $R_{ab} = 0$, a multiple null eigenvector of both the Weyl tensors C_{abcd} and $\tilde{C}_{abcd}$.

The class (19.11) of solutions of the Einstein-Maxwell equations can be written in the double Kerr-Schild form

$$g_{ab} = \eta_{ab} + Kk_ak_b + Ll_al_b, \tag{28.97}$$

where $\boldsymbol{k}$ and $\boldsymbol{l}$ mean independent complex null vectors satisfying

$$\eta^{ab}k_ak_b = \eta^{ab}l_al_b = \eta^{ab}k_al_b = 0 \tag{28.98}$$

(Plebański and Schild (1976)).

Chapter 29. Algebraically special perfect fluid solutions

29.1. Generalized Robinson-Trautman solutions

Generalized Robinson-Trautman solutions are characterized by the following set of assumptions. (i) The multiple null eigenvector $\boldsymbol{k}$ of the Weyl tensor is geodesic, shearfree, and twistfree but expanding

$$\Psi_0 = \Psi_1 = 0, \qquad \varkappa = \sigma = \omega = 0, \qquad \varrho = \bar{\varrho} \neq 0. \tag{29.1}$$

(ii) The energy-momentum tensor is that of a perfect fluid,

$$R_{ab} = \varkappa_0(\mu + p)\, u_a u_b + \varkappa_0(\mu - p)\, g_{ab}/2. \tag{29.2}$$

(iii) The four-velocity $\boldsymbol{u}$ of the fluid obeys

$$u_{[a;b}u_{c]} = 0, \qquad k_{[c}k_{a];b}u^b = 0. \tag{29.3}$$

Introducing the null tetrad $(\boldsymbol{m}, \overline{\boldsymbol{m}}, \boldsymbol{l}, \boldsymbol{k})$, one sees that (29.1) and (29.3) imply that τ can be made zero by choice of $\boldsymbol{l}$ and that then $\boldsymbol{u}$ lies in the plane spanned by $\boldsymbol{l}$ and $\boldsymbol{k}$. With this choice, (29.2) yields

$$R_{11} = R_{14} = R_{13} = 0. \tag{29.4}$$

From now on, the calculations run in close analogy with those for the Robinson-Trautman solutions, cp. Chapters 23. and 24. We will present here only the main results, all due to Wainwright (1974).

As a first step, one can infer from (29.1) and (29.4) that coordinates $\zeta, \bar{\zeta}, r, u$ can be introduced such that the metric takes the form

$$\begin{aligned} ds^2 &= 2\chi^2(r, u)\, P^{-2}(\zeta, \bar{\zeta}, u)\, d\zeta\, d\bar{\zeta} - 2\, du\, dr - 2H(\zeta, \bar{\zeta}, r, u)\, du^2, \\ \varrho &= -\partial_r \ln \chi; \qquad H = -r\, \partial_u \ln P - H^0(\zeta, \bar{\zeta}, u) - S(r, u), \end{aligned} \tag{29.5}$$

the (irrotational) four-velocity $\boldsymbol{u}$ having the components

$$B\sqrt{2}\, u_i = (0, 0, -1, -H - B^2). \tag{29.6}$$

As a second step, the so far unknown real functions χ, P, H^0, S and B should be determined from the rest of the equations (29.2)—(29.3). Two of the field equations (29.2) can be regarded as giving the energy density μ and the pressure p in terms of the metric functions, so only three of the equations (29.2) need to be considered. These ensure the correct algebraic structure for R_{ab} (i.e. the isotropy of pressure). It eventually turns out that three different cases occur.

The first case is characterized by the Newman-Penrose coefficient ν being zero, i.e. by

$$\partial_u P = 0, \qquad \partial_\zeta H^0 = 0, \qquad K = 2P^2\, \partial_\zeta\, \partial_{\bar\zeta} \ln P = \text{const}. \tag{29.7}$$

Obviously, these solutions admit a G_3 acting on the 2-spaces $r = \text{const}$, $u = \text{const}$, of constant curvature K, and are therefore of Petrov type D or 0. The corresponding metrics (including the spherically symmetric solutions) are discussed in detail in Chapters 13. and 14.

In the second case ($\nu \neq 0$, χ^2 quadratic in r), the metric is given by (29.5)—(29.6) with

$$\begin{aligned} &\chi^2 = \varepsilon(r^2 - a^2), \quad H^0 = -\varepsilon P^2\, \partial_\zeta\, \partial_{\bar\zeta} \ln P, \quad B^2 = -H + c\chi^2, \\ &S = \chi^2\left(b - 3m \int \chi^{-4}\, \mathrm{d}r\right) + \varepsilon a^2 c, \qquad \varepsilon = \pm 1, \end{aligned} \tag{29.8a}$$

where the real constants a, b, c and m and the function P have to satisfy

$$\begin{aligned} &a^2 cm = 0, \qquad a^2 c(c + 2b) = 0, \\ &P^2[\partial_\zeta\, \partial_{\bar\zeta} P^2\, \partial_\zeta\, \partial_{\bar\zeta} \ln P + a^2 \partial_u\, \partial_u (2P^2)^{-1}] \\ &+ 3\varepsilon m\, \partial_u \ln P + ca^2(P^2 \partial_\zeta\, \partial_{\bar\zeta} \ln P - ca^2/2) = 0. \end{aligned} \tag{29.8b}$$

In the third case ($\nu \neq 0$, χ^2 linear in r), the metric is given by (29.5)—(29.6) with

$$\chi^2 = \varepsilon r, \qquad S = \varepsilon ra + 2\varepsilon br \ln \varepsilon r, \qquad B^2 = -H, \tag{29.9a}$$

where P and H^0 have to satisfy

$$2P^2\, \partial_\zeta\, \partial_{\bar\zeta} \ln P - \varepsilon\, \partial_u \ln P = 2b, \qquad 2\partial_\zeta\, \partial_{\bar\zeta} H^0 + \varepsilon\, \partial_u (H^0 P^{-2}) = 0. \tag{29.9b}$$

The (lengthy) expressions for μ, p and the non-vanishing components Ψ_2, Ψ_3 and Ψ_4 of the Weyl tensor in the two cases (29.8) and (29.9) are given in Wainwright's paper. They show that no dust solutions and no solutions of type III or N are possible.

Besides the Robinson-Trautman vacuum solutions (see Chapter 24.), which are contained in (29.8) as the subcase $a = b = 0$ ($\Rightarrow \varepsilon = 1$, $S = m$), the only explicitly known solutions of (29.8) or (29.9) are the subcase $\partial_u P = b = 0$ of (29.9). With the choice $\varepsilon = 1$ ($r > 0$) and $S = 2r$ they are given by

$$\begin{aligned} &\mathrm{d}s^2 = 2r\, \mathrm{d}\zeta\, \mathrm{d}\bar\zeta - 2\, \mathrm{d}u\, \mathrm{d}r + 2[H^0(\zeta, \bar\zeta, u) + 2r]\, \mathrm{d}u^2, \\ &\varkappa_0 \mu = 3r^{-1} + H^0 r^{-2}/2, \qquad \varkappa_0 p = -r^{-1} + H^0 r^{-2}/2, \\ &\sqrt{2} u_a = -(H^0 + 2r)^{-1/2}\, \delta_a^3, \end{aligned} \tag{29.10}$$

$$(2\partial_\zeta\, \partial_{\bar\zeta} + \partial_u)\, H^0 = 0. \tag{29.11}$$

If $\partial_\zeta H^0 = 0$, we return to (29.7). Non-trivial ($\partial_\zeta H^0 \neq 0$) real solutions of (29.11) can easily be constructed.

Algebraically special solutions with a geodesic and twistfree, but expanding and *shearing* multiple null eigenvector have been considered by □ Oleson (1972).

29.2. Solutions with a geodesic, shearfree, non-expanding multiple null eigenvector

In this section, we assume that the multiple null eigenvector $\boldsymbol{k}$ satisfies

$$\Psi_0 = \Psi_1 = 0, \qquad \varkappa = \sigma = \Theta = 0, \qquad \varrho = -\bar{\varrho} \neq 0. \tag{29.12}$$

Note that, because of (6.32), $\varkappa = \sigma = \varrho = 0$ implies $\mu + p = 0$. We will exclude this case here, i.e. in what follows $\boldsymbol{k}$ is necessarily twisting ($\omega \neq 0$). We will again present only the main results, all due to Wainwright (1970).

As a first step in solving the field equations, one uses (29.12) and parts of (29.2) to introduce a suitable tetrad and coordinate system. These are found to be

$$\begin{aligned} ds^2 &= 2P^{-2}\, d\zeta\, d\bar{\zeta} - 2[du + L\, d\zeta + \bar{L}\, d\bar{\zeta}] \\ &\quad \times [dr + W\, d\zeta + \overline{W}\, d\bar{\zeta} + H\{du + L\, d\zeta + \bar{L}\, d\bar{\zeta}\}], \\ m^i &= P(-1, 0, W, L), \qquad \overline{m}^i = P(0, -1, \overline{W}, \bar{L}), \\ & \hspace{8cm} \varrho = \mathrm{i}, \\ l^i &= (0, 0, -H, 1), \qquad k^i = (0, 0, 1, 0), \end{aligned} \tag{29.13}$$

where $L(\zeta, \bar{\zeta})$ is purely imaginary and determined by the real function $P(\zeta, \bar{\zeta})$ as

$$\partial_x L = -\partial_x \bar{L} = \mathrm{i}P^{-2}\sqrt{2}, \qquad x\sqrt{2} = \zeta + \bar{\zeta}. \tag{29.14}$$

The coordinates and the tetrad (29.13) are of the Robinson-Trautman type (23.27) if one inserts into (23.27) the value $\varrho = \mathrm{i}$ and makes P in (23.12) imaginary instead of real. Equation (29.14) corresponds to the first part of (23.26), the second part being invalid here. The four-velocity $\boldsymbol{u}$ turns out to have the form

$$u^a\sqrt{2} = Bk^a + B^{-1}l^a, \qquad B^2 = \varkappa_0(\mu + p)/4. \tag{29.15}$$

As a second step, one has to solve the remaining field equations to obtain explicit expressions for the so far unknown functions $W(\zeta, \bar{\zeta}, u)$ and $H(\zeta, \bar{\zeta}, r, u)$. Three different cases can occur.

The *first case* is characterized by $W = 0$, $\partial_r \partial_r(H + \varkappa_0 p/2) \neq 0$. These solutions are the case *IIIa*, $\alpha + E = \tau = 0$ of the locally rotationally symmetric solutions of Stewart and Ellis (1968).

The *second case* is characterized by

$$k_{a;b} + k_{b;a} = 0. \tag{29.16}$$

It includes all algebraically special (not conformally flat) perfect fluid solutions whose multiple null eigenvector $\boldsymbol{k}$ is a (twisting) Killing vector. $W(\zeta, \bar{\zeta})$, chosen to be purely imaginary, and $H(\zeta, \bar{\zeta})$ are given by

$$\partial_x W = -\partial_x \overline{W} = i\varkappa_0 p P^{-2}\sqrt{2}, \qquad 2H = -\varkappa_0 p, \tag{29.17}$$

and $P(\zeta, \bar{\zeta})$, $p(\zeta, \bar{\zeta})$ and $\mu(\zeta, \bar{\zeta})$ have to obey

$$\begin{aligned} 4P^2\, \partial_\zeta \partial_{\bar{\zeta}} \ln P &= \varkappa_0(\mu - 5p), \\ 8P^2\, \partial_\zeta \partial_{\bar{\zeta}}\, p &= \varkappa_0(p - \mu)(\mu + 3p). \end{aligned} \tag{29.18}$$

To get the explicit metric, one has to solve (29.18) for P, p and μ (one function can be prescribed) and then to determine H, L and W from (29.17) and (29.14). Besides $\boldsymbol{k}$, the metric admits a second Killing vector $\boldsymbol{\xi} = \partial_u$. Unless space-time is conformally flat, the four-velocity $\boldsymbol{u}$ has non-zero rotation.

Particular solutions include the Gödel universe (10.25), the Einstein universe (10.23), and the Petrov type II solution

$$\mu = p = x, \quad 3P^2 = 4\varkappa_0 x^3, \quad 2H = -\varkappa_0 x, \quad W = \mathrm{i}3\sqrt{2}/4x, \quad L = -\mathrm{i}3\sqrt{2}/8\varkappa_0 x^3, \tag{29.19}$$

which admits a G_3 simply transitive on $x = \text{const}$, see § 11.4.

The *third case* is characterized by $\partial_r(H + \varkappa_0 p/2) = 0$, $\partial_r H \neq 0$. The only non-trivial field equation remaining to be solved is

$$\begin{aligned} P^2 \partial_\zeta \partial_{\bar\zeta} K &= \varkappa_0 m(\varkappa_0 m + K), \\ K \equiv 2P^2 \partial_\zeta \partial_{\bar\zeta} \ln P, &\qquad 2m \equiv \mu + 3p = \text{const}. \end{aligned} \tag{29.20}$$

Having found a solution $P(\zeta, \bar\zeta)$ of this equation, one can obtain the other metric functions, and the pressure p, from (29.14) and the following set of equations

$$\partial_x W(\zeta, \bar\zeta) = -\mathrm{i}(\varkappa_0 m + K)/2P^2\sqrt{2}, \qquad W = -\overline{W}, \tag{29.21}$$

$$\partial_\zeta \ln R(\zeta, \bar\zeta) = 2\mathrm{i}W, \tag{29.22}$$

$$\varkappa_0 4p(\zeta, \bar\zeta, r) = 2(R\,\mathrm{e}^{2\mathrm{i}r} + \bar R\,\mathrm{e}^{-2\mathrm{i}r}) + \varkappa_0 m - K, \tag{29.23}$$

$$2H(\zeta, \bar\zeta, r) = \varkappa_0(m - p). \tag{29.24}$$

These metrics admit a Killing vector collinear with the four-velocity

$$u^i = (0, 0, 0, [\varkappa_0(m - p)]^{-1/2}), \tag{29.25}$$

which is, therefore, expansion- and shearfree. Only Petrov types II and D can occur.

A special solution is given by

$$K + \varkappa_0 m = 0 \;\Rightarrow\; W = 0, \qquad R = R(\bar\zeta). \tag{29.26}$$

For $R = \text{const}$ it contains a solution admitting a G_4 on $r = \text{const}$, cp. § 11.4.

29.3. Type D solutions

Most of the algebraically special perfect fluid solutions are of Petrov type D, but, as the following review shows, only a minor subset of all such type D solutions is known.

In the type D solutions, the two multiple null eigenvectors $\boldsymbol{l}$ and $\boldsymbol{k}$ define a preferred 2-space Σ. The only known solution with a four-velocity $\boldsymbol{u}$ *not* lying in Σ is the axially symmetric Wahlquist solution (19.40), see Wainwright (1977a). For all other known solutions

$$u_{[a}k_b l_{c]} = 0 \tag{29.27}$$

holds, and the rotation $\omega^a = \varepsilon^{abcd} u_b u_{c;d}$ and shear σ_{ab} of the velocity field furthermore obey

$$\omega_{[a} k_b l_{c]} = 0, \quad k^d \sigma_{d[a} k_b l_{c]} = 0. \tag{29.28}$$

Following Wainwright (1977a), we classify all solutions satisfying (29.27) and (29.28) according to the acceleration $\dot{\boldsymbol{u}}$ and the Newman-Penrose coefficients $\varkappa$, ν, σ and λ (it can be shown that for the solutions in question $\varkappa = 0$ ($\sigma = 0$) if and only if $\nu = 0$ ($\lambda = 0$)). For none of the subcases is the complete list of solutions known.

29.3.1. Solutions with $\varkappa = \nu = 0$

The subcase $\sigma = \lambda = 0$, $\dot{u}_{[a} k_b l_{c]} = 0$ contains all the locally rotationally symmetric perfect fluid solutions considered by Stewart and Ellis (1968), see Wainwright (1977b). The type D solutions covered by §§ 29.1., 29.2 also belong to this subcase. The spherically symmetric perfect fluid and dust solutions and the Kantowski-Sachs models (see Chapters 13. and 14. and § 12.3.) are important examples.

The subcase $\sigma = \lambda = 0$, $\dot{u}_{[a} k_b l_{c]} \neq 0$ contains a solution found by Barnes (1973a); it can be obtained from the solution (16.66) by allowing A and B to be t-dependent.

The subcase $\sigma \neq 0$, $\lambda \neq 0$, $\dot{u}_{[a} k_b l_{c]} = 0$ contains the Ozsváth solution (10.28), admitting a G_4 on V_4, the Bianchi type I solution (12.21b), and the solution (20.44), admitting an Abelian G_2.

No solutions are known for which σ, λ and $\dot{u}_{[a} k_b l_{c]}$ are all non-zero.

29.3.2. Solutions with $\varkappa \neq 0$, $\nu \neq 0$

No solutions are known for which $\varkappa, \nu$, σ and λ are all non-zero. The Barnes solutions (16.67) and (16.68) are examples with $\sigma = \lambda = 0$, but $\dot{u}_{[a} k_b l_{c]} \neq 0$.

A large variety of explicitly known solutions is contained in the subcase $\sigma = \lambda = \dot{u}_{[a} k_b l_{c]} = 0$. They all have a metric of the form

$$ds^2 = 2e^{2b} \, d\zeta \, d\bar{\zeta} + e^{2a} \, dr^2 - dt^2, \qquad \boldsymbol{u} = \partial_t, \tag{29.29}$$

and have been considered by Szekeres (1975) (for dust), Szafron (1977), Tomimura (1977), Szafron and Wainwright (1977), and Wainwright (1977b). To ensure $\varkappa \neq 0$, $\sigma \neq 0$, $a_{,\zeta}$ must be non-zero. Two cases have to be distinguished, $b_{,r} \neq 0$ and $b_{,r} = 0$. In both cases, the mass density μ can be computed from

$$\varkappa_0(\mu + 3p) = -2(\ddot{a} + \dot{a}^2 + 2\ddot{b} + 2\dot{b}^2). \tag{29.30}$$

If b depends on r, then the metric must have the form

$$ds^2 = \Phi^2(r, t) \, [2P^{-2}(\zeta, \bar{\zeta}, r) \, d\zeta \, d\bar{\zeta} + (\partial_r \ln \{\Phi P^{-1}\})^2 \, dr^2] - dt^2, \tag{29.31}$$

where the $\zeta - \bar{\zeta}$-space is a space of constant curvature $K(r)$,

$$\begin{aligned} P(\zeta, \bar{\zeta}, r) &= \alpha(r) \, \zeta\bar{\zeta} + \beta(r) \, \zeta + \bar{\beta}(r) \, \bar{\zeta} + \delta(r), \\ K(r) &= 2(\alpha\delta - \beta\bar{\beta}), \end{aligned} \tag{29.32}$$

and the function $\Phi(r, t)$ is a solution of the ordinary (Friedmann-type) differential equation

$$2\Phi\ddot{\Phi} + \dot{\Phi}^2 + \varkappa_0 p(t)\,\Phi^2 = 1 - K(r). \tag{29.33}$$

To get an explicit solution, one has to prescribe the pressure $p(t)$ and the functions $\alpha(r)$, $\delta(r)$ (both real) and $\beta(r)$ (complex), and then to solve (29.33).

For dust ($p = 0$), the differential equation (29.33) is integrated by

$$\dot{\Phi}^2 - 2m(r)\,\Phi^{-1} = 1 - K(r), \tag{29.34}$$

which can be completely solved (Szekeres (1975)), the solutions being of the form (13.38). In general, these dust solutions contain five arbitrary functions of r, and admit no Killing vector (Bonnor et al. (1977)). Surprisingly, they can be matched to the spherically symmetric, static, exterior Schwarzschild solution (Bonnor (1976)).

For a perfect fluid (with $p \neq 0$), solutions of (29.33) with

$$\varkappa_0 p(t) = \varphi^2(r)\,\Phi^{-2}(r, t) \tag{29.35}$$

can easily be obtained from the dust solutions with $1 - K(r) = \text{const}\ \varphi^2(r)$, $m = \text{const}\ \varphi^3$. A (different) example of an explicit perfect fluid solution has been given by Szafron (1977) as

$$K = 1, \qquad \Phi = [g(r)\,t^{1-q} + h(r)\,t^q]^{2/3}, \qquad 3\varkappa_0 p = 4q(1-q)\,t^{-2} \tag{29.36}$$

($q = 1/2$ leads to a conformally flat space-time). It admits at least one Killing vector.

If b does not depend on r, then the metric must have the form

$$\begin{aligned}\mathrm{d}s^2 = \Phi^2(t)\,\big[&2P^{-2}(\zeta,\bar{\zeta})\,\mathrm{d}\zeta\,\mathrm{d}\bar{\zeta}\\ &+ \{A(r,t)\,P^{-1}\big(U(r)\,\zeta\bar{\zeta} + V(r)\,\zeta + \bar{V}(r)\,\bar{\zeta} + W(r)\big)\}^2\,\mathrm{d}r^2\big] - \mathrm{d}t^2,\end{aligned} \tag{29.37}$$

with

$$P = 1 + k\zeta\bar{\zeta}/2, \qquad k = 0, \pm 1, \tag{29.38}$$

$$2\Phi\ddot{\Phi} + \dot{\Phi}^2 + \varkappa_0 p(t)\,\Phi^2 = -k, \tag{29.39}$$

$$\ddot{A}\Phi^2 + 3\dot{A}\dot{\Phi}\Phi - Ak = U + kW. \tag{29.40}$$

To get an explicit solution, one has to specify the pressure $p(t)$ and the functions $U(r)$, $W(r)$ (both real) and $V(r)$ (complex), and then to determine $\Phi(t)$ from (29.39) and $A(r, t)$ from (29.40). The energy density μ can be obtained from (29.30).

For dust, the general solution is given in Szekeres (1975), compare also Bonnor and Tomimura (1976), and the perfect fluid case $\varkappa_0 p(t) = \text{const}\ \Phi^{-2}$ has been solved by Tomimura (1977). Some other perfect fluid solutions, corresponding to the choice $k = 0$, $p(t) = \text{const}\ t^{-2}$, have been constructed by Szafron and Wainwright (1977); the solution (32.51)—(32.52) of embedding class one is also included here.

Ref.: □ Collins and Szafron (1979)

29.4. Type *III* and type *N* solutions

Apparently, no type *III* perfect fluid (or dust) solution is known. Concerning type N solutions, the following theorems restrict the properties of possible solutions ($\mu + p = 0$ is always excluded).

Theorem 29.1. *No type N perfect fluid (or dust) solutions with zero acceleration* ($\dot{u}_a = 0$) *exist.* (□ Oleson (1972); cp. also Kundt and Trümper (1962) and Szekeres (1966b)).

Theorem 29.2. *The principal null congruence of a type N perfect fluid solution is geodesic if and only if the fluid is irrotational and pressure and energy density are related by*

$$\mu = p + A(t), \qquad u_a = (-2H)^{-1/2}\, t_{,a} \tag{29.41}$$

(Kundt and Trümper (1962)).

Oleson (1971, □ 1972) determined all metrics covered by Theorem 29.2, i.e. all type N perfect fluid solutions with a geodesic (but twisting) principal null congruence. He gives the following two classes of solutions.

Class I: $\varrho^{-1} = -2t$, $\sigma^{-1} = -4t$

$$\begin{aligned} \mathrm{d}s^2 &= t^{3/2}(\mathrm{d}x - 2t^{-1/2}G_{,x}\,\mathrm{d}u)^2 + t^{1/2}(\mathrm{d}y + 2t^{1/2}G_{,y}\,\mathrm{d}u)^2 \\ &\quad -2G\,\mathrm{d}t\,\mathrm{d}u - 2G^2H\,\mathrm{d}u^2, \\ H(t,u) &= -2[a^2(u)\,t^{1/2} + b^2t^{3/2}], \qquad G(x,y,u) = g(x,u)\,h(y,u), \\ g_{,xx} &+ a^2(u)\,g = 0, \qquad h_{,yy} + b^2h = 0, \qquad b = \text{const}, \\ 4p &= 3t^{-3/2}(a^2 - 7b^2t), \qquad A = 12b^2t^{-1/2}. \end{aligned} \tag{29.42}$$

Class II: $\varrho^{-1} = (1 - t^2)/t$, $\quad \sigma^{-1} = 2(t^2 - 1)$

$$\begin{aligned} \mathrm{d}s^2 &= \varepsilon(1 - t^2)\,R^{-1}(\mathrm{d}x - \varepsilon RG_{,x}\,\mathrm{d}u)^2 + \varepsilon(1 - t^2)\,R(\mathrm{d}y + \varepsilon R^{-1}G_{,y}\,\mathrm{d}u)^2 \\ &\quad -2G\,\mathrm{d}t\,\mathrm{d}u - 2G^2H\,\mathrm{d}u^2, \qquad \varepsilon = \pm 1, \\ H(t,u) &= -\varepsilon[\varepsilon(1 - t^2)]^{1/2}\,[\lambda a^2(u) - b^2(t + 1)], \\ R(r) &= [\varepsilon(1 - t)/(1 + t)]^{1/2}, \qquad G(x,y,u) = g(x,u)\,h(y,u), \\ g_{,xx} &+ \lambda a^2(u)\,g = 0, \qquad h_{,yy} + [\varepsilon(a^2 - 2\lambda b^2)]^{1/2}\,h = 0, \\ (\varepsilon, \lambda) &= (1, 1) \quad \text{or} \quad (-1, \pm 1), \qquad b = \text{const}, \\ 2p &= 3\varepsilon[\varepsilon(1 - t^2)]^{-3/2}\,[b^2(4t^3 - 5t - 1) + \lambda a^2], \\ A &= 12b^2t[\varepsilon(1 - t^2)]^{-1/2}. \end{aligned} \tag{29.43}$$

For both classes, the fluid has non-zero shear, acceleration and expansion provided that $a'(u) \neq 0$; $a = \text{const}$ gives conformally flat solutions. Killing vectors (maximum number two) are possible only in class I for metrics with $b = 0$.

PART IV. SPECIAL METHODS

Chapter 30. Generation techniques

30.1. Introduction

It is extremely difficult to solve Einstein's field equations, so it would be very helpful if one could generate new solutions from known ones on the basis of internal symmetries of the field equations. In the main part of the present chapter we shall deal with a generation method which works only if space-time admits a non-null Killing vector field. The basic idea of this method will be explained in § 30.2. and the most important case, that of Einstein-Maxwell fields, will be treated in § 30.3. An additional Killing vector in the space-time enlarges the internal symmetry group, which leads to new possibilities for generating solutions (§ 30.4.). To illustrate the power of these tools, some new metrics are given (§ 30.5.). Current research attacks the problem of which classes of solutions can in principle be obtained by these generation techniques.

Other generation methods are summarized at the end of this chapter (§ 30.6.).

30.2. The potential space

In the present section we shall consider space-times admitting a *non-null* Killing vector field $\boldsymbol{\xi}$,

$$\xi_{(a;b)} = 0, \qquad F = \xi^a \xi_a \neq 0. \tag{30.1}$$

It is convenient to introduce the 3-metric γ_{ab} defined by

$$\gamma_{ab} = |F| \, (g_{ab} - F^{-1} \xi_a \xi_b), \qquad \gamma = \det (\gamma_{ab}). \tag{30.2}$$

For stationary fields ($F < 0$) (see Chapter 16.), γ_{ab} is a positive definite metric, and for a spacelike Killing vector ($F > 0$) γ_{ab} has the signature $(++-)$. For some physically relevant cases, e.g. Einstein-Maxwell fields (see § 30.3.), it turns out that one can introduce scalar potentials φ^A ($A = 1 \dots N$) such that the Euler-Lagrange equations

$$\delta L / \delta \gamma_{ab} = 0, \qquad \delta L / \delta \varphi^A = 0, \tag{30.3}$$

of a variational principle with a Lagrangian of the form

$$L = \sqrt{\gamma} \, [\hat{R} + G_{AB}(\varphi^C) \, \gamma^{ab} \varphi^A{}_{,a} \varphi^B{}_{,b}] \tag{30.4}$$

are exactly the field equations. ($\hat{R}$ denotes the scalar curvature with respect to γ_{ab}; $\det (G_{AB}) \neq 0$.) The quantities φ^A and γ_{ab} completely determine the space-time metric g_{ab} and the non-metric fields entering the energy-momentum tensor.

Suppose the Lagrangian (30.4) is *invariant* under certain transformations (invariance transformations) of the potentials φ^A,

$$\varphi^{A'} = \varphi^{A'}(\varphi^B), \qquad L' = L(\gamma'_{ab}, \varphi^{A'}, \ldots), \tag{30.5}$$

(the metric γ_{ab} remaining unchanged, $\gamma'_{ab} = \gamma_{ab}$). The invariance of the Lagrangian implies that the field equations (30.3) are also invariant under (30.5). Therefore, if the original set (φ^A, γ_{ab}) satisfies the field equations, the new set $(\varphi^{A'}, \gamma_{ab})$ does so, and in general the set $(\varphi^{A'}, \gamma_{ab})$ gives a really *new* solution.

To find out the invariance transformations for a given L, we assign to the second term of the Lagrangian (30.4) the line element (Neugebauer and Kramer (1969))

$$\mathrm{d}S^2 = G_{AB}(\varphi^C)\,\mathrm{d}\varphi^A\,\mathrm{d}\varphi^B \tag{30.6}$$

of an N-dimensional abstract Riemannian space, the *potential space* V_N, with coordinates φ^A. The potentials φ^A are functions of the space-time coordinates x^i, but this dependence does not concern us here. The transition to the line element (30.6) enables us to investigate the invariance of the Lagrangian by applying familiar methods from Riemannian geometry. We simply have to solve the Killing equations

$$X_{(A;B)} = 0 \tag{30.7}$$

in the metric G_{AB}; every solution of (30.7) determines an infinitesimal transformation of the potentials,

$$\varphi^A \to \varphi^A + \varepsilon X^A(\varphi), \tag{30.8}$$

which leaves $\mathrm{d}S^2$ invariant. The R independent Killing vectors form the isometry group (group of motions) G_R $\left(R \leqq N(N+1)/2\right)$ of the metric (30.6). The corresponding finite transformations are invariance transformations of L and, in general, will generate new exact solutions.

Discrete symmetries of the Lagrangian (30.4) or the metric (30.6) may be discovered by taking appropriate limits of the continuous invariance transformations.

Besides the study of symmetry groups, other geometrical investigations of potential space (geodesics, subspaces) are important in simplifying the field equations and in generating exact solutions. In what follows we shall consider briefly these aspects.

Let us assume that the potentials φ^A depend only on a single potential λ. The ansatz

$$\varphi^A = \varphi^A(\lambda) \tag{30.9}$$

reduces the potential space to a one-dimensional subspace. The field equations $\delta L/\delta\varphi^A = 0$,

$$\begin{gathered} \varphi^A{}_{,a}{}^{;a} + \begin{Bmatrix} A \\ B\,C \end{Bmatrix} \varphi^B{}_{,a}\varphi^{C,a} = 0, \\ \begin{Bmatrix} A \\ B\,C \end{Bmatrix} = \frac{1}{2}\,G^{AD}\left(\frac{\partial G_{BD}}{\partial\varphi^C} + \frac{\partial G_{CD}}{\partial\varphi^B} - \frac{\partial G_{BC}}{\partial\varphi^D}\right), \end{gathered} \tag{30.10}$$

take the simplified form

$$\dot{\varphi}^A \lambda_{,a}{}^{;a} = \left(\ddot{\varphi}^A + \begin{Bmatrix} A \\ B\,C \end{Bmatrix} \dot{\varphi}^B \dot{\varphi}^C \right) \lambda_{,a} \lambda^{,a}, \qquad \dot{\varphi}^A = \mathrm{d}\varphi^A/\mathrm{d}\lambda. \tag{30.11}$$

The potential λ is determined up to a scale transformation $\lambda' = \lambda'(\lambda)$. From the equation (30.11) it follows that $\lambda_{,a}{}^{;a} = f(\lambda)\, \lambda_{,a}\lambda^{,a}$ holds; therefore one can always choose a new λ such that $\lambda_{,a}{}^{;a} = 0$. For $\lambda_{,a}\lambda^{,a} \neq 0$, the field equations (30.3) reduce to

$$\hat{R}_{ab} = \varepsilon \lambda_{,a} \lambda_{,b}, \qquad D^2\varphi^A/\mathrm{d}\lambda^2 = 0, \qquad \varepsilon = \pm 1, 0 \tag{30.12}$$

(Neugebauer and Kramer (1969)), and the equation $\lambda_{,a}{}^{;a} = 0$ follows from the Bianchi identities provided that $\varepsilon \neq 0$. (For $\varepsilon = 0$, the 3-space with the metric γ_{ab} is flat.) Because of (30.12) the potentials φ^A can be determined as solutions of the geodesic equation in potential space. If $\lambda_{,a}\lambda^{,a} = 0$, the potentials φ^A are arbitrary functions of λ.

In many cases we can reduce the problem to a 2-dimensional subspace of V_N $(N > 2)$,

$$\varphi^A = \varphi^A(\lambda_1, \lambda_2). \tag{30.13}$$

If we insert this parametric representation of the 2-space into the Lagrangian (30.4), the field equations (30.3) take the form

$$\delta L/\delta \gamma_{ab} = 0, \qquad \delta L/\delta \lambda_1 = 0, \qquad \delta L/\delta \lambda_2 = 0. \tag{30.14}$$

In § 30.3.4. we shall consider 2-dimensional subspaces of the potential space for stationary Einstein-Maxwell fields.

Ref.: Hoenselaers (1976).

30.3. The invariance transformations for Einstein-Maxwell fields

30.3.1. The potential space and its symmetry group

The methods outlined in the previous section can be applied immediately to Einstein-Maxwell fields admitting a *non-null Killing vector* field. (We assume that the electromagnetic field shares the space-time symmetry.) Vacuum fields are, of course, included as a special case.

For our purposes, the existence of scalar potentials is of crucial importance. In §§ 16.1.—16.3. it is shown in detail that the complete set of the Einstein-Maxwell equations can be written in terms of complex scalar potentials Φ and $\mathscr{E}$ and the 3-metric γ_{ab} defined by

$$\gamma_{ab} = |F|\,(g_{ab} - F^{-1}\xi_a\xi_b), \qquad F = \xi_a\xi^a. \tag{30.15}$$

Theorem 30.1. *In source-free Einstein-Maxwell fields admitting a non-null Killing vector* $\boldsymbol{\xi}$, *there exists a set* $(\Phi, \mathscr{E}, \gamma_{ab})$ *such that the Einstein-Maxwell equations* (16.38)

— (16.40) *follow from a variational principle with the Lagrangian*

$$L = \sqrt{\gamma}\left[\hat{R} + \frac{1}{2}F^{-2}\gamma^{ab}(\mathscr{E}_{,a} + 2\overline{\Phi}\Phi_{,a})(\bar{\mathscr{E}}_{,b} + 2\Phi\overline{\Phi}_{,b}) + 2F^{-1}\gamma^{ab}\Phi_{,a}\overline{\Phi}_{,b}\right]. \tag{30.16}$$

($\hat{R}$ denotes the scalar curvature with respect to γ_{ab}, $\gamma \equiv \det(\gamma_{ab})$.)

The two complex scalar potentials Φ and $\mathscr{E}$ are defined by (16.32) and (16.36):

$$\begin{aligned} &\Phi_{,a} = \sqrt{\varkappa_0/2}\,\xi^c F^*_{ca}, \qquad \mathscr{E}_{,a} = \xi^c K^*_{ca},\\ &K^*_{ab} \equiv -2\xi^*_{a;b} - \sqrt{2\varkappa_0}\,\overline{\Phi}F^*_{ab}. \end{aligned} \tag{30.17}$$

The complex self-dual electromagnetic field tensor F^*_{ab} satisfies

$$F^*_{[ab;c]} = 0 \Leftrightarrow F^{*ab}{}_{;b} = 0, \tag{30.18}$$

$$£_\xi F^*_{ab} = F^*_{ab;c}\xi^c + F^*_{cb}\xi^c{}_{;a} + F^*_{ac}\xi^c{}_{;b} = 0. \tag{30.19}$$

These relations imply that the integrability condition for $\Phi_{,a}$ is satisfied,

$$2\Phi_{,[a;b]} = \sqrt{\varkappa_0/2}\,(\xi^c F^*_{[ca;b]} - £_\xi F^*_{ab}) = 0. \tag{30.20}$$

The complex self-dual bivector K^*_{ab} obeys exactly the same relations as F^*_{ab}. Its Lie derivative vanishes, and $K^{*ab}{}_{;b} = 0$ follows from part of the Einstein field equations,

$$\begin{aligned} \frac{1}{2}K^{*ac}{}_{;c} &= -\xi^{*a;c}{}_{;c} - \sqrt{\varkappa_0/2}\,\overline{\Phi}_{,c}F^{*ac} = R^a_b\xi^b - (\varkappa_0/2)\,F^{*ac}\overline{F}^*_{bc}\xi^b\\ &= (R^a_b - \varkappa_0 T^a_b)\,\xi^b = 0, \end{aligned} \tag{30.21}$$

T_{ab} being the energy-momentum tensor (5.7) of the electromagnetic field. Thus the integrability condition for $\mathscr{E}_{,a}$ is satisfied too, and the existence of the potentials Φ and $\mathscr{E}$ has been verified.

From a given solution (Φ, $\mathscr{E}$, γ_{ab}) of the Einstein-Maxwell equations

$$\delta L/\delta\gamma_{ab} = 0, \qquad \delta L/\delta\Phi = 0, \qquad \delta L/\delta\mathscr{E} = 0, \tag{30.22}$$

F, ξ_a, the space-time metric and the electromagnetic field tensor can be determined (in that order) via (see § 16.2.)

$$\begin{aligned} &-F = \frac{1}{2}(\mathscr{E} + \bar{\mathscr{E}}) + \overline{\Phi}\Phi, \qquad K^*_{ab} = 2F^{-1}(\xi_{[a}\mathscr{E}_{,b]})^*,\\ &g_{ab} = |F|^{-1}\gamma_{ab} + F^{-1}\xi_a\xi_b, \qquad \sqrt{\varkappa_0/2}\,F^*_{ab} = 2F^{-1}(\xi_{[a}\Phi_{,b]})^*. \end{aligned} \tag{30.23}$$

For Einstein-Maxwell fields admitting a non-null Killing vector field, the real and imaginary parts of the complex potentials $\mathscr{E}$ and Φ are the coordinates φ^A of a 4-dimensional potential space. Because of the form (30.16) of the Lagrangian, the potential space metric (30.6) takes the form

$$\mathrm{d}S^2 = \frac{1}{2}F^{-2}\,|\mathrm{d}\mathscr{E} + 2\overline{\Phi}\,\mathrm{d}\Phi|^2 + 2F^{-1}\,\mathrm{d}\Phi\,\mathrm{d}\overline{\Phi}. \tag{30.24}$$

For stationary fields ($F < 0$), this potential space has an indefinite metric G_{AB} with signature $(++--)$. For a space-time with a spacelike Killing vector ξ, the metric (30.24) is positive definite.

To obtain the invariance transformations of the Lagrangian (30.16), we have to solve the Killing equations (30.7) for the metric (30.24). As the potential space V_4 with metric (30.24) is not a space of constant curvature, the order of its group of motions G_R is $R \leqq 8$, according to Theorem 8.17. It turns out that there are 8 independent Killing vectors whose associated *finite invariance transformations* (Neugebauer and Kramer (1969)) are

$$\mathscr{E}' = \alpha\bar{\alpha}\mathscr{E}, \qquad \Phi' = \alpha\Phi, \tag{30.25a}$$

$$\mathscr{E}' = \mathscr{E} + \mathrm{i}b, \qquad \Phi' = \Phi, \tag{30.25b}$$

$$\mathscr{E}' = \mathscr{E}(1 + \mathrm{i}c\mathscr{E})^{-1}, \qquad \Phi' = \Phi(1 + \mathrm{i}c\mathscr{E})^{-1}, \tag{30.25c}$$

$$\mathscr{E}' = \mathscr{E} - 2\bar{\beta}\Phi - \beta\bar{\beta}, \qquad \Phi' = \Phi + \beta, \tag{30.25d}$$

$$\mathscr{E}' = \mathscr{E}(1 - 2\bar{\gamma}\Phi - \gamma\bar{\gamma}\mathscr{E})^{-1}, \qquad \Phi' = (\Phi + \gamma\mathscr{E})(1 - 2\bar{\gamma}\Phi - \gamma\bar{\gamma}\mathscr{E})^{-1}. \tag{30.25e}$$

The complex constants α, β, γ and the real constants b, c are the 8 real parameters of the isometry group G_8. Because the potential space admits a G_8, the Ricci tensor R_{AB} is proportional to the metric G_{AB} (Egorov (1955)).

The Einstein-Maxwell equations (30.22) are invariant under an arbitrary combination of the transformations (30.25). Hence we have proved

Theorem 30.2. *Given a solution* $(\Phi, \mathscr{E}, \gamma_{ab})$ *of the Einstein-Maxwell equations (with a non-null Killing vector field), then any set* $(\Phi', \mathscr{E}', \gamma_{ab})$ *obtained by application of an arbitrary product of the invariance transformations* (30.25) *is again a solution.*

Note that $\gamma'_{ab} = \gamma_{ab}$ holds; only the potentials are transformed.

The transformations (30.25b) and (30.25d) are simply gauge transformations of the potentials; the space-time metric g_{ab} and the electromagnetic field tensor F_{ab} remain unchanged. The transformation (30.25a) describes a *duality rotation*,

$$F^{*}_{ab}{}' = (\alpha/\bar{\alpha})^{1/2} F^{*}_{ab}, \tag{30.26}$$

combined with a rescaling of the coordinate x^n along the Killing vector $\xi = \partial_n$. Only the transformations (30.25c) and (30.25e) affect the space-time and Maxwell field in a non-trivial manner.

The product of the transformations of (30.25b, c) and 30.25a) (in the limit $b \to \infty$, $c = b^{-1}$, $\alpha = \mathrm{i}b^{-1}$) leads to the *inversion*

$$\mathscr{E}' = \mathscr{E}^{-1}, \qquad \Phi' = \mathscr{E}^{-1}\Phi, \tag{30.27}$$

which maps the gauge transformations (30.25b) and (30.25d) into the non-trivial symmetry operations (30.25c) and (30.25e), respectively.

Ref.: Kinnersley (1975), Kramer et al. (1972).

30.3.2. Special cases of the invariance transformations

Some transformation theorems given earlier in the literature are special cases of the general invariance transformations (30.25):

Theorem 30.3 (Ehlers (1957)). *From any static vacuum solution (metric g_{ij}) one can obtain a stationary vacuum solution (g'_{ij}) by the transformation*

$$\begin{gathered} g'_{ij} = \cosh 2U(e^{2U}g_{ij} + \xi_i\xi_j) - (\cosh 2U)^{-1}\, u_i u_j, \\ \varepsilon_{ijkl}U^{,k}\xi^l = u_{[i;j]}, \qquad u_i\xi^i = -1, \qquad e^{2U} = -\xi_i\xi^i. \end{gathered} \tag{30.28}$$

The equation for u_i is integrable because of $U_{,k}\xi^k = 0$ and the equations (16.55) for static vacuum fields. The corresponding invariance transformation is (30.25c).

Theorem 30.4 (Buchdahl (1954)). *From any static vacuum solution (metric g_{ij}) one can obtain another static vacuum solution (g'_{ij}) by the transformation*

$$g'_{ij} = e^{4U}g_{ij} + (e^{2U} - e^{-6U})\,\xi_i\xi_j, \qquad e^{2U} = -\xi_i\xi^i. \tag{30.29}$$

The corresponding invariance transformation is the inversion (30.27), which takes the simple form $U' = -U$ for static vacuum fields.

Theorem 30.5 (Harrison (1968)). *From any vacuum solution admitting a non-null Killing vector one can obtain a solution of the Einstein-Maxwell equations by the transformation*

$$\mathscr{E}' = \mathscr{E}(1 - \gamma\bar{\gamma}\mathscr{E})^{-1}, \qquad \Phi' = \gamma\mathscr{E}(1 - \gamma\bar{\gamma}\mathscr{E})^{-1}, \tag{30.30}$$

γ being a complex constant.

The transformation (30.30) is just (30.25e) with $\Phi = 0$. Harrison (1965, 1968) postulated a functional dependence between the gravitational and electromagnetic potentials, i.e., in terms of Φ and $\mathscr{E}$, a relation $\Phi = \Phi(\mathscr{E})$. He did not systematically investigate all possible invariance transformations. Obviously, for the Einstein-Maxwell fields generated from vacuum fields via transformation (30.30), the potential Φ is a *linear* function of $\mathscr{E}$. Woolley (1973b) assumed that the bivectors F^*_{ab} and K^*_{ab} in (30.17) satisfying the Maxwell equations $F^{*ab}{}_{;b} = K^{*ab}{}_{;b} = 0$ determine the same geometry. According to the Rainich problem (§ 5.4), this assumption means that $F^*_{ab} = e^{i\varphi}K^*_{ab}$, for non-null fields, and consequently also leads to a linear relation between the potentials Φ and $\mathscr{E}$.

Theorem 30.6 (Bonnor (1954)). *There is a one-to-one correspondence between electrostatic and magnetostatic Einstein-Maxwell fields,*

$$\mathscr{E}' = \mathscr{E}, \qquad \Phi' = i\Phi. \tag{30.31}$$

This transformation is a special case of the duality rotation (30.26). (For stationary Maxwell fields in flat space-time, the so-called Fitzgerald transformation $\Phi' = i\Phi$ assigns to an electric field a magnetic field with the same source structure (and vice versa). For instance, the field of an electric dipole can be transformed into the field of a magnetic dipole.)

Asymptotically flat Einstein-Maxwell fields generated from vacuum fields by means of the transformation (30.30) exhibit a typical feature: the sources of the gravitational and electromagnetic fields have a very similar structure. This follows from an analysis of the far-field behaviour (see e.g. Kramer et al. (1972)).

30.3.3. The group $SU(2, 1)$

The invariance transformations (30.25) yield a non-linear representation of the group G_8 in question. A *linear* representation of this group will be used in the proof of the following theorem.

Theorem 30.7 (Kinnersley (1973)). *The group of symmetry transformations of the Einstein-Maxwell equations (with a non-null Killing vector field) is the group* $SU(2, 1)$.

Proof: The complex potentials $\mathscr{E}$ and Φ can be expressed by the components $Y^\Sigma = (u, v, w)$ of a vector $\boldsymbol{Y}$ in a 3-dimensional complex vector space,

$$\mathscr{E} = \frac{u - w}{u + w}, \qquad \Phi = \frac{v}{u + w}, \qquad Y^\Sigma = (u, v, w). \tag{30.32}$$

The redundancy of this description can be removed by adding a suitable differential equation for the complex quantity w. Then the field equations (16.38)–(16.40) take the form

$$\delta L/\delta\varphi^A = 0: \quad \overline{Y}_\Theta Y^\Theta Y^\Sigma{}_{,a}{}^{;a} = 2\overline{Y}_\Theta \gamma^{ab} Y^\Theta{}_{,a} Y^\Sigma{}_{,b}, \tag{30.33}$$

$$\delta L/\delta\gamma_{ab} = 0: \quad \hat{R}_{kl} = (\overline{Y}_\Theta Y^\Theta)^{-2}\, \overline{V}_{\Sigma(k} V^\Sigma_{l)}, \qquad V^\Sigma_k = \varepsilon^{\Sigma\Theta\Gamma} Y_\Theta Y_{\Gamma,k}. \tag{30.34}$$

For convenience, we have introduced a metric $\eta_{\Sigma\Gamma} = \operatorname{diag}(1, 1, -1)$ in the vector space, so that the norm of the vector $\boldsymbol{Y}$ can be written as

$$\overline{Y}_\Sigma Y^\Sigma = \overline{u}u + \overline{v}v - \overline{w}w. \tag{30.35}$$

A linear homogeneous transformation

$$Y^{\Sigma'} = U^\Sigma{}_\Gamma Y^\Gamma \tag{30.36}$$

is said to be pseudo-unitary if it preserves the norm (30.35). In matrix notation, the pseudo-unitary matrices $\boldsymbol{U}$ satisfy the relation

$$\boldsymbol{U}^+\eta\boldsymbol{U} = \eta, \qquad \eta = \operatorname{diag}(1, 1, -1). \tag{30.37}$$

Obviously the equations (30.33), (30.34), are invariant under the transformations (30.36), (30.37), with constant matrices $\boldsymbol{U}$. According to (30.32), the potentials $\mathscr{E}$ and Φ are determined by the ratios of u, v, w. Therefore a common factor in u, v, w is not significant, and we may restrict ourselves to *unimodular* transformations, i.e. to the group $SU(2, 1)$.

The group $SU(2, 1)$ has 8 independent generators $\boldsymbol{X}$ (with $\boldsymbol{U} = 1 - \mathrm{i}\boldsymbol{X}$):

$$\boldsymbol{X} = \begin{pmatrix} a_1 & b & c \\ \bar{b} & a_2 & d \\ -\bar{c} & -\bar{d} & a_3 \end{pmatrix} \qquad \begin{array}{l} a_1, a_2, a_3 \text{ real}, \; a_1 + a_2 + a_3 = 0, \\ b, c, d \text{ complex}. \end{array} \tag{30.38}$$

The subgroup of invariance transformations of *vacuum* solutions ($\Phi = v = 0$) is $SU(1, 1)$ (Geroch (1971)).

Ref.: Hauser and Ernst (1978).

30.3.4. Subgroups of $SU(2, 1)$ and 2-dimensional susbpaces of the potential space

Simplifications arise when we succeed in reducing the potential space to a 2-dimensional subspace, $\varphi^A = \varphi^A(\lambda_1, \lambda_2)$. All known solutions of the Einstein-Maxwell equations depend on at most two (real) potentials only. For stationary Einstein-Maxwell fields, each 2-dimensional subspace (real coordinates λ_1, λ_2) of the potential space (30.24) must be either a null surface or a space of constant curvature (Neugebauer, private communication). The stationary Einstein-Maxwell fields with two independent potentials can be divided into four *inequivalent classes:* $\Phi = 0$ (or $\mathscr{E} = -1$) (vacuum fields), $\mathscr{E} = +1$, the class $\mathscr{E} = \bar{\mathscr{E}}$, $\Phi = \bar{\Phi}$ (electrostatic fields), and $\mathscr{E} = 0$. The space-times with $\mathscr{E} = +1$ become flat when the electromagnetic field is switched off.

Table 30.1 The subspaces of the potential space for stationary Einstein-Maxwell fields, and the corresponding subgroups of $SU(2, 1)$

Subspace	Metric	Curvature and signature	Subgroup of $SU(2, 1)$ and invariant	Potentials $\mathscr{E}$ and Φ
$\Phi = 0$ $(v = 0)$ or:	$\mathrm{d}S^2 = \dfrac{2\,\mathrm{d}\xi\,\mathrm{d}\bar{\xi}}{(1 - \xi\bar{\xi})^2}$	$K = -2$ $(++)$	$SU(1, 1)$ $\bar{u}u - \bar{w}w$	$\mathscr{E} = \dfrac{1 - \xi}{1 + \xi}$
$\mathscr{E} = -1$ $(u = 0)$	$\mathrm{d}S^2 = \dfrac{2\,\mathrm{d}\Phi\,\mathrm{d}\bar{\Phi}}{(1 - \Phi\bar{\Phi})^2}$	$K = -2$ $(++)$	$SU(1, 1)$ $\bar{v}v - \bar{w}w$	
$\mathscr{E} = +1$ $(w = 0)$	$\mathrm{d}S^2 = \dfrac{-2\,\mathrm{d}\Phi\,\mathrm{d}\bar{\Phi}}{(1 + \Phi\bar{\Phi})^2}$	$K = -2$ $(--)$	$SU(2)$ $\bar{u}u + \bar{v}v$	
$\mathscr{E} = \bar{\mathscr{E}}, \Phi = \bar{\Phi}$ $(\bar{u} = u, \bar{v} = v,$ $\bar{w} = w)$	$\mathrm{d}S^2 = \dfrac{8\,\mathrm{d}\zeta\,\mathrm{d}\eta}{(1 - \zeta\eta)^2}$	$K = -\dfrac{1}{2}$ $(+-)$	$O(2, 1)$ $u^2 + v^2 - w^2$	$\mathscr{E} = \dfrac{(1 - \zeta)(1 - \eta)}{(1 + \zeta)(1 + \eta)}$ $\Phi = \dfrac{\zeta - \eta}{(1 + \zeta)(1 + \eta)}$
$\mathscr{E} = 0$ $(u = w)$	$\mathrm{d}S^2 = 0$	—	$\bar{v}v$	

In Table 30.1 the metrics, signatures, and Gaussian curvatures of the corresponding 2-spaces of constant curvature are listed. The exceptional case $\mathscr{E} = 0$ is a null surface in the potential space. All subspaces are planes in the space of the variables u, v, w (see also Tanabe (1977)).

The isometry groups of the 2-spaces of constant curvature are 3-parameter subgroups of $SU(2, 1)$. In fact, the symmetry groups of the vacuum fields, the electrostatic fields and the fields of class $\mathscr{E} = +1$ are $SU(1, 1)$, $O(2, 1)$ and $SU(2)$, respectively.

The subgroups of $SU(2, 1)$ have been classified by Montgomery et al. (1969) up to conjugation, i.e., the generators of the inequivalent subgroups are determined up to $SU(2, 1)$ transformations. Hence, the inequivalent classes given in Table 30.1 can be extended by the application of arbitrary invariance transformations (30.25).

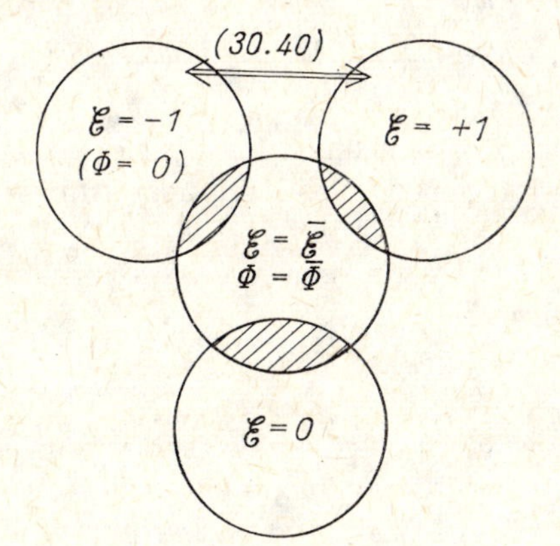

Fig. 30.1 The inequivalent classes of stationary Einstein-Maxwell fields

Fig. 30.1 is to be understood as follows. The interior of a circle represents the class of Einstein-Maxwell fields generated from the fields indicated ($\mathscr{E} = 0$, $\mathscr{E} = +1$, $\mathscr{E} = -1$, and $\mathscr{E} = \bar{\mathscr{E}}$, $\Phi = \bar{\Phi}$) by the symmetry operations of $SU(2, 1)$. For instance, the class $\mathscr{E} = -1$ discovered by Demiański (1976) is equivalent to the class of vacuum fields ($\Phi = 0$); the two cases are linked by elements of $SU(2, 1)$. It is impossible to pass from one class to another inequivalent class by the invariance transformations (30.25) if the subspace of the potential space is really 2-dimensional.

The overlapping areas (hatched regions in Fig. 30.1) represent those Einstein-Maxwell fields for which the potentials are mutually dependent on each other. In this case, the 2-spaces degenerate into one-dimensional subspaces of the potential space (geodesics, see § 30.2.). The invariance transformations map each hatched region onto itself. Electrostatic fields with only one independent potential belong to one of the classes $\mathscr{E} = -1$, $\mathscr{E} = +1$ or $\mathscr{E} = 0$. The Reissner-Nordström solution (13.21) is an example: this electrostatic Einstein-Maxwell field belongs to the class $\mathscr{E} = -1$ for $e^2 < m^2$, to the class $\mathscr{E} = +1$ for $e^2 > m^2$, and to the class $\mathscr{E} = 0$ for $e^2 = m^2$. These three branches of the solution are inequivalent with respect to the $SU(2, 1)$ transformations, which leave the quantity $m^2 - e^2$ invariant.

30.3.5. Complex invariance transformations

Allowing a *complex* space-time metric and complex electromagnetic fields, but maintaining real space-time coordinates, we need four complex potentials $\mathscr{E}_1$, $\mathscr{E}_2$, Φ_1, Φ_2. In the potential space metric (30.24) we have to substitute $\mathscr{E} \to \mathscr{E}_1$, $\bar{\mathscr{E}} \to \mathscr{E}_2$; $\Phi \to \Phi_1$, $\bar{\Phi} \to \Phi_2$. After this complexification the isometry group of the potential space has 8 complex parameters. The potentials, and the group parameters in the invariance transformations (30.25), are now considered to be independent of their complex conjugates. For instance, the complexified version of the invariance transformation (30.25e) reads

$$\begin{aligned} \mathscr{E}_1' &= \frac{\mathscr{E}_1}{1 - 2\gamma_2\Phi_1 - \gamma_1\gamma_2\mathscr{E}_1}, \qquad & \Phi_1' &= \frac{\Phi_1 + \gamma_1\mathscr{E}_1}{1 - 2\gamma_2\Phi_1 - \gamma_1\gamma_2\mathscr{E}_1}, \\ \mathscr{E}_2' &= \frac{\mathscr{E}_2}{1 - 2\gamma_1\Phi_2 - \gamma_1\gamma_2\mathscr{E}_2}, \qquad & \Phi_2' &= \frac{\Phi_2 + \gamma_2\mathscr{E}_2}{1 - 2\gamma_1\Phi_2 - \gamma_1\gamma_2\mathscr{E}_2} \end{aligned} \tag{30.39}$$

($\gamma \to \gamma_1$, $\bar{\gamma} \to \gamma_2$; γ_1 and γ_2 being complex parameters). In some cases, complex symmetry transformations can be used to generate real solutions from complex ones (see Herlt (1978)).

The two classes $\Phi = 0$ and $\mathscr{E} = +1$ (having the same Gaussian curvature) are related by the *complex* invariance transformation

$$\Phi_1' = \mathrm{i}\xi_1, \qquad \Phi_2' = \mathrm{i}\xi_2. \tag{30.40}$$

(In Fig. 30.1 this correspondence is indicated by a double arrow.) Starting with a real vacuum solution ($\xi_2 = \bar{\xi}_1$), we obtain a corresponding complex electromagnetic solution ($\Phi_2 = -\bar{\Phi}_1$) which can be converted into a real solution by complex continuation of the parameters. For instance, the potentials Φ and ξ of the Reissner-Nordström solution (13.21) are $\Phi = e/r$ and $\xi = (1 - \mathscr{E})/(1 + \mathscr{E}) = m/r$. The complex transformation (30.40) and the substitution $\mathrm{i}m \to e$ take the Schwarzschild solution into the Reissner-Nordström solution with vanishing mass parameter. There is a one-to-one relation between stationary vacuum fields with purely mechanical sources (masses, rotation) and stationary Einstein-Maxwell fields with corresponding purely electromagnetic sources (charges, currents). The complex invariance transformations can change the sign of $m^2 - e^2$.

The subclass of $\mathscr{E} = +1$ solutions given by (up to invariance transformations)

$$\begin{aligned} &\mathrm{d}s^2 = \mathrm{e}^{2k}(\mathrm{d}\varrho^2 + \mathrm{d}z^2) + \varrho^2\,\mathrm{d}\varphi^2 - (\mathrm{d}t + A\,\mathrm{d}\varphi)^2, \\ &A_{,\varrho} = \varrho V_{,z}, \qquad A_{,z} = -\varrho V_{,\varrho}, \\ &k_{,\varrho} = -\frac{\varrho}{4}(V_{,\varrho}{}^2 - V_{,z}{}^2), \qquad k_{,z} = -\frac{\varrho}{2} V_{,\varrho} V_{,z}, \\ &V_{,\varrho\varrho} + \varrho^{-1} V_{,\varrho} + V_{,zz} = 0, \quad \mathscr{E} = 1/2, \quad \Phi = 2^{-1/2}\,\mathrm{e}^{\mathrm{i}V(\varrho,z)}, \quad \bar{V} = V, \end{aligned} \tag{30.41}$$

contains the homogeneous solution (10.21) for $V = 2\ln\varrho$, and a solution mentioned by McIntosh (1978) for $V = -2bz$. Tanabe (1978) constructed fields of massless charges; these solutions belong to the class $\mathscr{E} = +1$.

The class $\mathscr{E} = 0$ (§ 16.7.) cannot be transformed into the classes $\mathscr{E} = +1$ and $\mathscr{E} = -1$ even by complex invariance transformations.

Ref.: Bitsadze (1977).

30.4. The generation theorems for Einstein-Maxwell fields admitting an Abelian group G_2

The generation formalism (invariance transformations) outlined in the previous sections requires that space-time admits a non-null Killing vector $\boldsymbol{\xi}$. Now we assume that there exists an additional non-null Killing vector η commuting with $\boldsymbol{\xi}$, and that there are 2-spaces orthogonal to the group orbits. These assumptions give rise to new generation principles which can be combined with the invariance transformations already discussed.

Among the space-times to be considered the most important are the stationary axisymmetric Einstein-Maxwell fields, but we could also have both Killing vectors spacelike. In § 17.4., the former problem was reduced to the complex differential equations (17.26), (17.27), for Φ and $\mathscr{E}$, which can be written as

$$\delta L/\delta\Phi = 0, \qquad \delta L/\delta\mathscr{E} = 0, \tag{30.42}$$

with the Lagrangian (30.16). For γ_{ab} substitute Γ_{ab} which is covariantly defined by

$$\begin{aligned} \Gamma_{ab} &= -\xi_c\xi^c[g_{ab} + W^{-2}(\eta_c\eta^c\xi_a\xi_b + \xi_c\xi^c\eta_a\eta_b - 2\xi_c\eta^c\eta_{(a}\xi_{b)})], \\ \Gamma_{ab}\xi^b &= 0 = \Gamma_{ab}\eta^b, \qquad -W^2 = 2\xi_{[a}\eta_{b]}\xi^a\eta^b. \end{aligned} \tag{30.43}$$

In the coordinate system (17.15), Γ_{ab} coincides with the 2-metric γ_{MN}.)

To generate new solutions from old ones, we have to hand the isometry transformations of the potential space metric (30.16), Γ_{ab} and W being fixed. Moreover, the existence of the second Killing vector gives rise to further generation methods. In what follows we will describe these new possibilities.

(i) Comparing the 2-dimensional potential space metrics of the vacuum fields ($\Phi = 0$) and the electrostatic fields ($\mathscr{E} = \bar{\mathscr{E}}$, $\Phi = \bar{\Phi}$) (listed in Table 30.1) we conclude that the complex transformation $\zeta = \xi$, $\eta = \bar{\xi}$ or, equivalently, $\mathscr{E}' = \mathscr{E}\bar{\mathscr{E}}$, $\Phi' = (\bar{\mathscr{E}} - \mathscr{E})/2$ takes the field equations $\delta L/\delta\varphi^A = 0$ for vacuum fields into the corresponding field equations for electrostatic fields (see also Theorem 17.3).

Theorem 30.8 (Bonnor (1961)). *From any stationary axisymmetric vacuum solution* ($\mathscr{E}$) *one obtains an electrostatic Einstein-Maxwell field* ($\mathscr{E}'$, Φ') (*or vice versa*) *by the substitution*

$$\mathscr{E}' = \mathscr{E}\bar{\mathscr{E}}, \qquad \Phi' = (\bar{\mathscr{E}} - \mathscr{E})/2 = \chi. \tag{30.44}$$

The electrostatic potential χ is imaginary and a *real* Einstein-Maxwell field can be generated from a real vacuum solution only if it is possible to perform a complex continuation of the parameters in the solution. To obtain the new metric, one has to compute the 2-metric Γ_{ab} from the transformed potentials; Γ_{ab} does not remain invariant. Theorem 30.8 can be used in combination with the invariance transformations (see § 30.5.2.).

(ii) The vacuum field equations for the metric (17.15) can be derived from either of the Lagrangians

$$L = W\left(U_{,M}U^{,M} + \frac{1}{4}\,\mathrm{e}^{-4U}\omega_{,M}\omega^{,M}\right), \tag{30.45}$$

$$L = W\left(S_{,M}S^{,M} - \frac{1}{4}\,\mathrm{e}^{-4S}A_{,M}A^{,M}\right), \qquad S = -U + \frac{1}{2}\ln W. \tag{30.46}$$

Up to a sign, these Lagrangians have exactly the same form. Therefore we can formulate (see also Theorem 17.3)

Theorem 30.9 (Kramer and Neugebauer (1968b)). *From a given stationary axisymmetric vacuum solution* (U, ω) *one gets another solution* (S', A') *by the substitution* $S' = U$, $A' = \mathrm{i}\omega$. *One obtains a real solution if it is possible to continue the parameters in the solution analytically so that the imaginary unit can be compensated.*

The transformations preserving (30.45) are the invariance transformations studied in § 30.3. The symmetry of (30.46) manifests the freedom of choice of linear combinations of the two Killing vectors $\xi_1 = \xi$ and $\xi_2 = \eta$:

$$\xi_1' = a\xi_1 + b\xi_2, \qquad \xi_2' = c\xi_2 + d\xi_1, \qquad ac - bd = 1. \tag{30.47}$$

The transitions to the linear combinations (30.47) are equivalent to linear transformations of the coordinates φ and t along the trajectories of the Killing vectors.

Theorem 30.9 exhibits the isomorphism between the two Lie groups preserving the forms of the Lagrangians (30.45) and (30.46), respectively.

(iii) The potentials Φ and $\mathscr{E}$ were defined with respect to the Killing vector ξ. In order to define such complex potentials, one could instead select an arbitrary (non-null) linear combination of the two Killing vectors ξ_1 and ξ_2. In general, different linear combinations, when inserted in an invariance transformation and applied to a given solution, lead to different solutions.

The product of the two Lie groups (invariance transformations and transition to other linear combinations of the Killing vectors) forms an infinite-parameter group $\mathscr{K}$. The infinitesimal action of $\mathscr{K}$ can be described with the aid of an infinite number of potentials (Geroch (1972), Kinnersley (1977), Kinnersley and Chitre (1977—78)). For special subgroups of $\mathscr{K}$ which preserve asymptotic flatness, the exponentiation of the infinitesimal transformations is known (Kinnersley and Chitre (1978), □ Hoenselaers et al. (1979, 1980)) and generating function methods were used to construct new vacuum solutions with any number of parameters. □ Hauser and Ernst (1979a, b) found a formulation in terms of a linear integral equation for the generating function.

(iv) The Bäcklund transformations (which provide us with solutions of some non-linear differential equations, e.g. the sine-Gordon equation) have their counterparts in the Einstein theory of stationary axisymmetric vacuum fields (Harrison (1978), Neugebauer (1979)). The repeated application of these transformations to a given solution leads to a large class of new solutions with any number of parameters. To find a closed formula expressing the new Ernst potential Γ (note that $\mathscr{E} = \Gamma, \Phi = 0$, for vacuum fields) in terms of the given Γ^0 and quantities obtained from Γ^0, it is convenient to introduce two complex functions ψ and χ which satisfy the linear equations (□ Neugebauer (1980b))

$$\begin{aligned} \psi_{,\zeta} &= \frac{\bar{\Gamma}_{,\zeta}}{\Gamma + \bar{\Gamma}} \left(\psi + \sqrt{\gamma}\,\chi\right), \qquad \psi_{,\bar{\zeta}} = \frac{\bar{\Gamma}_{,\bar{\zeta}}}{\Gamma + \bar{\Gamma}} \left(\psi + \frac{1}{\sqrt{\gamma}}\,\chi\right), \\ \chi_{,\zeta} &= \frac{\Gamma_{,\zeta}}{\Gamma + \bar{\Gamma}} \left(\chi + \sqrt{\gamma}\,\psi\right), \qquad \chi_{,\bar{\zeta}} = \frac{\Gamma_{,\bar{\zeta}}}{\Gamma + \bar{\Gamma}} \left(\chi + \frac{1}{\sqrt{\gamma}}\,\psi\right), \end{aligned} \tag{30.48}$$

where γ is given by

$$\gamma = -\frac{\bar{\zeta} + \mathrm{i}c}{\zeta - \mathrm{i}c}, \qquad \zeta = \varrho + \mathrm{i}z, \tag{30.49}$$

and (ϱ, z) are Weyl's canonical coordinates. The integrability conditions of (30.48) imply the Ernst equation (17.37) (note that the arbitrary parameter c does not enter Γ). To obtain Γ from ψ and χ, put $\gamma = 1$ in (30.48).

A given Γ^0 leads to a set of solutions $(\psi^0{}_m, \chi^0{}_m)$ of (30.48) which differ by integration constants (indicated by the index m). The Bäcklund transformations give rise to a recursion formula for ψ and χ, incorporating at each step a new member of the set $(\psi^0{}_m, \chi^0{}_m)$. After N recursion steps the Ernst potential is explicitly given by (□ Neugebauer (1980a))

$$\begin{vmatrix} \Gamma - \Gamma^0 & 1 & 1 & \dots & 1 \\ \Gamma + \bar{\Gamma}^0 & a_1\lambda_1 & a_2\lambda_2 & \dots & a_N\lambda_N \\ \Gamma - \Gamma^0 & \lambda_1^2 & \lambda_2^2 & \dots & \lambda_N^2 \\ \Gamma + \bar{\Gamma}^0 & a_1\lambda_1^3 & a_2\lambda_2^3 & \dots & a_N\lambda_N^3 \\ \dots & \dots & \dots & \dots & \dots \\ \Gamma - \Gamma^0 & \lambda_1^N & \lambda_2^N & \dots & \lambda_N^N \end{vmatrix} = 0, \tag{30.50}$$

$$\lambda_m = (c_m - \mathrm{i}\bar{\zeta})^{1/2} (c_m + \mathrm{i}\zeta)^{-1/2}, \quad a_m = -\psi^0{}_m/\chi^0{}_m, \quad m = 1, \dots, N, \tag{30.51}$$

c_m constants.

It follows from (30.48) that $a \equiv -\psi^0/\chi^0$ is a solution of the total Riccati equation

$$\begin{aligned} a_{,\zeta} &= \frac{1}{\Gamma^0 + \bar{\Gamma}^0} \left[\bar{\Gamma}^0{}_{,\zeta} \left(a - \sqrt{\gamma}\right) + \Gamma^0{}_{,\zeta} \left(a^2 \sqrt{\gamma} - a\right)\right], \\ a_{,\bar{\zeta}} &= \frac{1}{\Gamma^0 + \bar{\Gamma}^0} \left[\bar{\Gamma}^0{}_{,\bar{\zeta}} \left(a - \frac{1}{\sqrt{\gamma}}\right) + \Gamma^0{}_{,\bar{\zeta}} \left(\frac{a^2}{\sqrt{\gamma}} - a\right)\right]. \end{aligned} \tag{30.52}$$

The metric generated from a real metric is again real if we apply an even number of recursion steps and if

$$\bar{a}_m = a_m^{-1}, \qquad \bar{\gamma}_m = \gamma_m^{-1} \tag{30.53}$$

holds. From $\psi^0{}_m$ and $\chi^0{}_m$, the complete metric (17.20) of the new solution can be obtained by purely algebraic operations. Every recursion step corresponds to a Harrison (1978) transformation and is also related to an operation derived by □ Cosgrove (1979a, b).

□ Belinski and Zakharov (1978, 1979) start with linear equations equivalent to (30.48) (their spectral parameter λ is associated with c in (30.49)). The approach of Belinski and Zakharov (expansions in terms of poles in the complex λ-plane, solution of a system of algebraic equations) is equivalent to the method of Bäcklund transformations, the explicit relation has been discussed by □ Cosgrove (1980). Maison (1978) treated another form of linear eigenvalue equations, see also □ Maison (1979).

30.5. Applications

In the present section we give some examples of applications of the generation techniques. All these applications concern stationary axisymmetric vacuum and Einstein-Maxwell fields.

30.5.1. Vacuum fields from vacuum fields

The Kerr solution (18.25) results from flat space-time if one applies Bäcklund transformations ($N = 2$ in (30.50)) or equivalent methods. The Kerr solution has been also generated from the Schwarzschild solution (Kinnersley and Chitre (1978)), and from a special *complex* solution of the van Stockum class (18.23) (Herlt (1978)).

The application of Bäcklund transformations ($N = 2$ in (30.50)) to Weyl's class (§ 18.1) yields a new class of stationary metrics containing an arbitrary potential function (□ Neugebauer (1980b), see also □ Cosgrove (1979b)).

For $N = 4$, one obtains from flat space-time a nonlinear superposition of two Kerr-NUT-solutions (□ Kramer and Neugebauer (1980)). This solution is determined by the potential

$$\xi = \frac{1 - \Gamma}{1 + \Gamma} = \frac{\det \boldsymbol{A}}{\det \boldsymbol{B}},$$

$$\boldsymbol{A} = \begin{pmatrix} s_1 & s_2 & s_3 & s_4 \\ 1 & 1 & 1 & 1 \\ c_1 & c_2 & c_3 & c_4 \\ c_1{}^2 & c_2{}^2 & c_3{}^2 & c_4{}^2 \end{pmatrix}, \qquad \boldsymbol{B} = \begin{pmatrix} s_1 & s_2 & s_3 & s_4 \\ 1 & 1 & 1 & 1 \\ c_1 & c_2 & c_3 & c_4 \\ c_1 s_1 & c_2 s_2 & c_3 s_3 & c_4 s_4 \end{pmatrix}, \tag{30.54}$$

$$s_m = e^{i\omega_m} \sqrt{\varrho^2 + (z - c_m)^2}, \qquad m = 1 \cdots 4,$$

where c_m and ω_m are real parameters. The Tomimatsu-Sato $\delta = 2$ solution (§ 18.5) is contained as the limiting case $c_3 \to c_1$, $c_4 \to c_2$.

Kinnersley and Chitre (1978), and ☐ Hoenselaers et al. (1979) generated asymptotically flat solutions which are generalizations of the Tomimatsu-Sato $\delta = 2$ solution and the extreme Kerr solution, respectively.

The pure invariance transformation (30.25c) takes Weyl's class (§ 18.1) into Papapetrou's class (§ 18.3). Starting with the Kerr solution one gets the Demiański-Newman (1966) solution

$$\mathrm{d}s^2 = \mathrm{e}^{-2U}\left[(r^2 - 2mr + a^2\cos^2\vartheta - l^2)\left(\mathrm{d}\vartheta^2 + \frac{\mathrm{d}r^2}{R^2}\right) + R^2\sin^2\vartheta\,\mathrm{d}\varphi^2\right]$$

$$- \mathrm{e}^{2U}\left(\mathrm{d}t + \frac{2a\sin^2\vartheta(mr + l) + 2l\cos\vartheta R^2}{r^2 - 2mr + a^2\cos^2\vartheta - l^2}\,\mathrm{d}\varphi\right)^2, \qquad (30.55)$$

$$R^2 = r^2 - 2mr + a^2 - l^2, \qquad \mathrm{e}^{2U} = 1 - 2\,\mathrm{Re}\left[\frac{m + \mathrm{i}l}{r + \mathrm{i}(a\cos\vartheta + l)}\right].$$

This metric belongs to the type D vacuum solutions (§ 25.5); it contains both the Kerr ($l = 0$) and the NUT-solution ($a = 0$).

Theorem 30.9 maps solutions of Papapetrou's class into solutions of the class treated in § 18.4.

Ref.: Sneddon (1975), ☐ Tanabe (1979).

30.5.2. Einstein-Maxwell fields from vacuum fields

Here we have to hand the general invariance transformations (30.23), in particular the Harrison transformation (30.30) (Theorem 30.5), the Bonnor transformation (30.44) (Theorem 30.8), and the operation (30.47).

The transformation (30.30) leads from the Schwarzschild solution ($\xi = \partial_t$) to the Brill solution (§ 11.3.1.), and from flat space-time ($\xi = \partial_\varphi$) to the Melvin solution (20.10) or its electric analogue. Taking $\xi = \partial_\varphi$ as the preferred Killing vector field in the Kerr metric (18.25), the complex scalar potentials are given by

$$\Phi = 0, \quad \mathscr{E} = -(r^2 + a^2)\sin^2\vartheta + 2ma\mathrm{i}(3 - \cos^2\vartheta) - \cos\vartheta\,\frac{2ma^2\sin^4\vartheta}{r + \mathrm{i}a\cos\vartheta}. \qquad (30.56)$$

The transformation (30.30) ($\gamma = B_0/2$) generates the "magnetized" Kerr solution which was determined explicitly by Ernst and Wild (1976). For large values of the radial coordinate r, this solution approaches the Melvin magnetic universe (magnetic field B_0). There is a non-singular event horizon. Thus this solution describes a Kerr black hole in a magnetic field.

The application of invariance transformations ($\xi = \partial_t$), with suitably chosen parameters, to the Kerr solution gives the Kerr-Newman solution (19.19); the NUT parameter can be included too.

Using Theorem 30.8, Bonnor (1966) generated from the Kerr solution the field of a massive magnetic dipole,

$$
\begin{aligned}
&\mathrm{d}s^2 = \mathrm{e}^{-2U}(\gamma_{AB}\,\mathrm{d}x^A\,\mathrm{d}x^B + W^2\,\mathrm{d}\varphi^2) - \mathrm{e}^{2U}\,\mathrm{d}t^2,\\
&\gamma_{MN}\,\mathrm{d}x^M\,\mathrm{d}x^N = \frac{(r^2 - a^2\cos^2\vartheta - 2mr)^4}{[(r-m)^2 - (a^2+m^2)\cos^2\vartheta]^3}\left(\mathrm{d}\vartheta^2 + \frac{\mathrm{d}r^2}{r^2 - a^2 - 2mr}\right),\\
&W^2 = (r^2 - a^2 - 2mr)\sin^2\vartheta,\\
&\mathrm{e}^{2U} = \left(1 - \frac{2mr}{r^2 - a^2\cos^2\vartheta}\right)^2, \quad \Phi = \frac{2\mathrm{i}am\cos\vartheta}{r^2 - a^2\cos^2\vartheta}.
\end{aligned}
\tag{30.57}
$$

Starting with this solution, Kramer and Neugebauer (1969), and Esposito and Witten (1973) generated an Einstein-Maxwell field by means of invariance transformations (30.25),

$$
\Phi = \sqrt{\frac{\bar{\gamma}}{\gamma}}\,\frac{(1+\gamma\bar{\gamma})\,\Phi_0 + \gamma(\mathscr{E}_0 - 1)}{1 - 2\bar{\gamma}\Phi_0 - \gamma\bar{\gamma}\mathscr{E}_0}, \qquad \mathscr{E} = \frac{\mathscr{E}_0 + 2\bar{\gamma}\Phi_0 - \gamma\bar{\gamma}}{1 - 2\bar{\gamma}\Phi_0 - \gamma\bar{\gamma}\mathscr{E}_0}, \tag{30.58}
$$

$\mathscr{E}_0$, Φ_0 being the potentials of the Bonnor solution (30.57); γ_{AB} and W are unchanged and the function A in the metric (17.15) is given by

$$
A = \frac{4\mathrm{i}am(\bar{\gamma} - \gamma)\sin^2\vartheta}{r^2 - 2mr - a^2\cos^2\vartheta}\cdot\frac{(1 - \gamma\bar{\gamma})\,r + 2m\gamma\bar{\gamma}}{(1 - \gamma\bar{\gamma})^2}. \tag{30.59}
$$

This solution describes the asymptotically flat external gravitational field of a rotating charged source, but it is different from the Kerr-Newman solution. The static limit ($a = 0$) is the Darmois solution (18.9). The magnetic potential does not contain a monopole term.

Theorem 30.8 takes Weyl's vacuum class (§ 18.1.) into Weyl's class of electrostatic fields (§ 19.1.1.). The combination of Theorems 30.8 and 30.9 gives rise to a class of electrostatic Einstein-Maxwell fields which was constructed from Weyl's vacuum class by Gautreau and Hoffman (1970)).

Herlt's class (19.4) of electrostatic Einstein-Maxwell fields was generated, by repeated application of the inversion $\mathscr{E} \to \mathscr{E}^{-1}$ and the substitution $A \to \mathrm{i}\omega$, from real solutions of the van Stockum class (18.23).

Charged versions of the Tomimatsu-Sato solutions (§ 18.5.) have been given by several authors (e.g. Ernst (1973), Wang (1974)). The successive application of Theorem 30.8 and of the $SU(2, 1)$ invariance transformation to the TS-solutions yields a five parameter stationary solution of the Einstein-Maxwell equations (Önengüt and Serdaroğlu (1975)).

Hauser and Ernst (1979) systematically studied the generation of electrovac solutions from Minskowski space-time by means of the $SU(2, 1)$ invariance transformations.

Finally, we note that for asymptotically flat stationary Einstein-Maxwell fields generated from the corresponding vacuum fields by means of invariance transformations, the gyromagnetic factor is equal to that of an electron, $g = 2$, $\mu = eJ/m$ (μ = dipole moment, J = angular momentum) (Reina and Treves (1975)).

Ref.: Voorhees (1972), Das and Banerji (1978), Ray and Wei (1977).

30.5.3. Generalized invariance transformations

Up to now we have applied the potential space formalism of § 30.2. only to Einstein-Maxwell fields outside the sources.

Now we generalize the method of invariance transformations to the case of *charged perfect fluids* under the condition that the 4-velocity is proportional to the timelike Killing vector, $\xi_{[a}u_{b]} = 0$. In this case, the Lagrangian (30.16) is to be complemented by the additive term $2\varkappa_0 \sqrt{\gamma}\, F^{-1}p$ (Kramer et al. (1972)). Under the invariance transformations of (30.16), this additional term also remains invariant provided that we put

$$p/F = p_0/F_0 \tag{30.60}$$

for the pressure p, where p_0, F_0 are the quantities of the original metric from which we start. Obviously, the transformation (30.60) maps a surface $p_0 = 0$ of the original solution into a surface $p = 0$ of the new solution generated by invariance transformations. For electrostatic Einstein-Maxwell fields containing a charged perfect fluid, the energy density μ and the charge density σ can be calculated from the pressure $p = p(F, \chi)$ by differentiating,

$$p + \mu = 2F\, \partial p/\partial F, \qquad \sigma = \sqrt{-\varkappa_0 F/2}\, \partial p/\partial \chi. \tag{30.61}$$

(These relations follow from $T^{ab}{}_{;b} = 0$.) This method has been applied to the interior Schwarzschild metric (14.14). The result is a regular interior Reissner-Nordström solution

$$\begin{aligned}
\mathrm{d}s^2 &= \frac{4C^2R^2}{(C^2+r^2)^2}\left(\frac{1+b^2F_0}{1-b^2}\right)^2 (\mathrm{d}x^2+\mathrm{d}y^2+\mathrm{d}z^2) + \left(\frac{1-b^2}{1+b^2F_0}\right)^2 F_0\,\mathrm{d}t^2,\\
-F_0 &= \left(B - \frac{1}{2}\frac{C^2-r^2}{C^2+r^2}\right)^2, \qquad \chi = \frac{b(1+F_0)}{1+b^2F_0},\\
R^2 &= 3/(\varkappa_0\mu_0), \quad C^2 = 2a^3/m, \quad 2B = 3(C^2-a^2)\,(C^2+a^2)^{-1},\\
&(1+m/2a)^3 = 2RC^{-1}, \qquad \mu_0,\ a,\ b \text{ constants}
\end{aligned} \tag{30.62}$$

(Kramer and Neugebauer (1971)). At the boundary $r = a$ ($p = 0$), this solution can be matched to the exterior Reissner-Nordström solution. Inserting the pressure p_0 of the interior Schwarzschild solution ($b = 0$) into (30.60), we get p and then μ and σ from (30.61).

The potential space formalism can be extended so that a (real) *scalar field* Ψ is included. In this case, the line-element (30.24) has an additional term $a \cdot \mathrm{d}\Psi^2$ (a = const); the potential space is 5-dimensional. In general, only the trivial change $\Psi' = \Psi + \text{const}$ is a possible invariance transformation. However, if the potentials $\varphi^A = (\Phi, \mathscr{E}, \Psi)$ depend mutually on each other, $\varphi^A = \varphi^A(\lambda)$, a non-trivial isometry transformation of the potential space metric exists. Another procedure for generating whole classes of solutions with scalar fields from a given stationary axisymmetric Einstein-Maxwell field uses the fact that Ψ satisfies the potential equation $\Delta\Psi = 0$ which is decoupled from the other field equations (see also Theorem 15.1).

Theorem 30.4 was generalized to include *perfect fluids* (Buchdahl (1956)): from a static solution with $-F = \mathrm{e}^{2U}$, pressure p and energy density μ, one obtains the *reciprocal* static solution with

$$\check{U} = -U, \qquad \check{p} = \mathrm{e}^{-4U}p, \qquad \check{\mu} = -\mathrm{e}^{-4U}(\mu + 6p). \tag{30.63}$$

The above considerations on generating new solutions rest on the basic assumption that a *non-null* Killing vector exists. The case of a *null Killing vector* has been excluded so far. However, the field equations for plane-fronted Einstein-Maxwell wave solutions (§ 21.5.) can also be derived from a variational principle (Kramer (1977)). The Lagrangian is given by

$$L = \Theta_{,\zeta}\overline{\Theta}_{,\bar{\zeta}} + \Theta_{,\bar{\zeta}}\overline{\Theta}_{,\zeta} \qquad \Theta = H - (\varkappa_0/2)\,\Psi^2 + \mathrm{i}\Psi, \tag{30.64}$$

where Ψ denotes the electromagnetic potential, $F_{ab} = 2\Psi_{,[a}u_{,b]}$, and H is the function in the metric (21.38). The invariance transformation $\Theta' = \mathrm{e}^{\mathrm{i}\gamma}\Theta$, $\gamma = \gamma(u)$, generates wave solutions of the Einstein-Maxwell equations from corresponding vacuum solutions.

Ref.: Ehlers (1962).

30.5.4. Concluding remarks

The invariance transformations together with linear combinations of the Killing vectors and Theorem 30.8 provide us with powerful tools for generating solutions. The main area of applications is the class of stationary axisymmetric Einstein-Maxwell fields. Many solutions or classes of solutions were rediscovered by the generation methods (e.g., the Kerr-Newman solution, Melvin's universe, the Brill solution, Weyl's electrovac class) but, in some cases, really new solutions were found, for instance, the Kerr black hole in a magnetic field, an exterior field of a rotating charged source, an interior Reissner-Nordström solution, and a new class of electrostatic Einstein-Maxwell fields.

The possibilities of generating new solutions are not at all exhausted. The repeated application of the various procedures is limited in practice to a few steps because the (straightforward) calculations are very lengthy.

The generation methods, in general, change the invariant characteristics (Petrov type, group of motions) so that one can generate non-trivial solutions even from flat space. However, if space-time admits two commuting Killing vectors and 2-surfaces orthogonal to the group orbits, then these properties are preserved under the generation methods mentioned above.

30.6. Other generation methods

30.6.1. Limiting procedures for space-times

It is possible that one can find new solutions of Einstein's equations as limits of known solutions. Limits of space-times were investigated by Geroch (1969). The following remarks are based on this paper.

If in the Schwarzschild metric

$$\mathrm{d}s^2 = (1 - \lambda^{-3}r^{-1})^{-1}\,\mathrm{d}r^2 + r^2(\mathrm{d}\vartheta^2 + \sin^2\vartheta\,\mathrm{d}\varphi^2) - (1 - \lambda^{-3}r^{-1})\,\mathrm{d}t^2 \quad (30.65)$$

we perform the transformation

$$\tau = \lambda r, \qquad R = \lambda^{-1}t, \qquad \varrho = \lambda^{-1}\vartheta, \quad (30.66)$$

containing the parameter λ explicitly, we obtain the line element

$$\mathrm{d}s^2 = (\lambda^2 - \tau^{-1})^{-1}\,\mathrm{d}\tau^2 + \tau^2(\mathrm{d}\varrho^2 + \lambda^{-2}\sin^2\lambda\varrho\,\mathrm{d}\varphi^2) - (\lambda^2 - \tau^{-1})\,\mathrm{d}R^2, \quad (30.67)$$

and in the limit $\lambda \to 0$ the Kasner metric (11.50) ($p_1 = p_2 = 2/3$, $p_3 = -1/3$),

$$\mathrm{d}s^2 = \tau^{-1}\,\mathrm{d}R^2 + \tau^2(\mathrm{d}\varrho^2 + \varrho^2\,\mathrm{d}\varphi^2) - \tau\,\mathrm{d}\tau^2. \quad (30.68)$$

On the other hand, if in place of (30.66) we use the transformation $x = r + \lambda^{-4}$, $\varrho = \lambda^{-4}\vartheta$, we obtain flat space-time in the limit $\lambda \to 0$. The limits are not uniquely determined.

Other examples of obtaining new solutions by taking appropriate limits are given in §§ 18.6. and 19.1.2. One has to choose the transformation so that the limit $\lambda \to 0$ exists. There is no general recipe for this procedure. The examples given in the literature have been found intuitively.

Of course, there are "trivial" limits obtained simply by putting a parameter in the metric equal to zero. In the more interesting cases, the limit $\lambda \to 0$ is connected with a transformation which contains the parameter λ.

30.6.2. The "complex trick"

Some new metrics were discovered by a formal procedure which can be roughly described as follows. A given metric is first complexified and then a complex coordinate transformation is performed in such a way that the result is a new real metric.

Such a "complex trick" was used by Newman et al. (1965) to "derive" a new solution of the Einstein-Maxwell equations from the Reissner-Nordström metric

$$\mathrm{d}s^2 = r^2(\mathrm{d}\vartheta^2 + \sin^2\vartheta\,\mathrm{d}\varphi^2) - 2\,\mathrm{d}u\,\mathrm{d}r - (1 - 2m/r + e^2/r^2)\,\mathrm{d}u^2. \quad (30.69)$$

The radial coordinate $x^1 = r$ and the retarded time $x^4 = u$ were allowed to take complex values and the null tetrad was formally replaced by the expressions

$$\boldsymbol{k} = \partial_r, \quad \boldsymbol{l} = \partial_u - \frac{1}{2}M\,\partial_r, \quad \boldsymbol{m} = 2^{-1/2}\bar{r}^{-1}(\partial_\vartheta + \mathrm{i}(\sin\vartheta)^{-1}\,\partial_\varphi),$$
$$M = M(r, \bar{r}) = 1 - m/r - m/\bar{r} + e^2/(r\bar{r}). \quad (30.70)$$

For real values of the coordinate r, (30.70) is a null tetrad in the metric (30.69). Originally, no convincing reason for choosing the tetrad in just this way was known. The "trick" was subsequently justified within the formalism for algebraically special fields (Talbot (1969)). After the *complex* coordinate transformation

$$r' = r + \mathrm{i}a \cos \vartheta, \qquad u' = u - \mathrm{i}a \cos \vartheta, \tag{30.71}$$

one obtains from (30.70) the new tetrad components

$$\begin{aligned} \boldsymbol{k}' &= \partial_r, \quad \boldsymbol{l}' = \partial_u - \frac{1}{2}\left(1 - \frac{2mr' - e^2}{r'^2 + a^2 \cos^2 \vartheta}\right) \partial_r, \\ \boldsymbol{m}' &= 2^{-1/2}(r' + \mathrm{i}a \cos \vartheta)^{-1} [\partial_\vartheta + \mathrm{i}(\sin \vartheta)^{-1} \partial_\varphi + \mathrm{i}a \sin \vartheta(\partial_u - \partial_r)]. \end{aligned} \tag{30.72}$$

For real values of the coordinates r', u', the associated metric is the Kerr-Newman metric (19.19).

Demiański (1972) found the most general vacuum solution which results from the complexified null tetrad (30.70), $M = \overline{M} = M(r, r')$ still being unspecified, when the complex coordinate transformation $r' = r + \mathrm{i}F(\vartheta, \varphi)$, $u' = u + \mathrm{i}G(\vartheta, \varphi)$, $\vartheta' = \vartheta$, $\varphi' = \varphi$ (F and G being real functions of their real arguments ϑ and φ) is performed and the new coordinates r', u', are restricted to be real. The resulting solution is given by (25.60); in general it is of Petrov type II. For a non-zero cosmological constant, $\Lambda \neq 0$, Demiański (1972) obtains the metric

$$\begin{aligned} \mathrm{d}s^2 = &-[1 - (2mr + 2b^2)(r^2 + b^2)^{-1} - \Lambda(r^2 + 5b^2)/3] (\mathrm{d}u + 2b \cos \vartheta \, \mathrm{d}\varphi)^2 \\ &- 2(\mathrm{d}u + 2b \cos \vartheta \, \mathrm{d}\varphi) \, \mathrm{d}r + (r^2 + b^2)(\mathrm{d}\vartheta^2 + \sin^2 \vartheta \, \mathrm{d}\varphi^2). \end{aligned} \tag{30.73}$$

Twisting type D vacuum metrics (§ 25.5.) may be obtained from corresponding non-twisting metrics by means of the "complex trick" (Basey (1975)).

Putting $e = 0$ in (30.69), (30.70), one generates the Kerr solution from the Schwarzschild solution. In quite another sense, the Kerr solution can be considered as a complexified Schwarzschild solution. Some vacuum Kerr-Schild metrics (§ 28.2.) can be computed from a generating (complex) potential γ which simultaneously satisfies the two equations $\Delta\gamma = 0$, $(\nabla\gamma)^2 = \gamma^4$ in flat 3-space. The solution $\gamma = r^{-1} = (x^2 + y^2 + z^2)^{-1/2}$ of these equations yields the Schwarzschild solution; the Kerr solution results from $\gamma = r^{-1}$ by an imaginary translation of the origin, $z \to z - \mathrm{i}a$ (Schiffer et al. (1973)). For a corresponding treatment of Einstein-Maxwell fields, see Finkelstein (1975).

Ref.: Demiański (1973).

Chapter 31. Special vector and tensor fields

31.1. Space-times that admit constant vector and tensor fields

31.1.1. Constant vector fields

Because of the definition of the curvature tensor, the very existence of a constant vector field $\boldsymbol{a}$,

$$a_{b;c} = 0, \tag{31.1}$$

imposes severe conditions on the curvature tensor and the metric: $\boldsymbol{a}$ is (proportional to) a constant vector field if and only if it satisfies

$$a_b R^b{}_{cde} = 0 \tag{31.2}$$

and all equations obtained by repeated differentation of (31.2).

Equation (31.2) shows that a four-dimensional space admitting *four constant vectors* is necessarily flat. The constancy of the metric g_{ab} and the existence of *three constant vectors* imply the existence of a fourth constant vector (which completes the system of the three) via (3.2), and therefore the space is again flat.

If *two constant vectors* $\boldsymbol{a}$ and $\boldsymbol{b}$ exist (and are linearly independent), it follows from (31.2) by considering a tetrad representation of the curvature tensor that this tensor can be given in terms of a simple bivector A_{cd}:

$$R_{cdef} = A_{cd} A_{ef}, \qquad A_{cd} a^c = 0 = A_{cd} b^c. \tag{31.3}$$

Two different cases occur, depending on whether $\boldsymbol{a}$ and $\boldsymbol{b}$ can both be chosen to be non-null, in which case

$$\begin{gathered} A_{cd} = p_c q_d - q_c p_d, \qquad a_c b^c = 0, \qquad p_c q^c = 0, \\ (p_a p^a)(q_b q^b) \neq 0, \qquad (a^c a_c)(b_d b^d) \neq 0, \end{gathered} \tag{31.4}$$

or not, in which case one of them is necessarily a null vector, say $\boldsymbol{b} = \boldsymbol{k}$, and their product vanishes,

$$A_{ab} = p_a k_b - k_a p_b, \quad p_a p^a \neq 0, \quad a_c a^c \neq 0, \quad k_a k^a = a_a k^a = k_a p^a = 0. \tag{31.5}$$

In the first case (31.4), the metric can be transformed into

$$\begin{gathered} ds^2 = g_{AB}(x^1, x^2)\, dx^A\, dx^B + \varepsilon_1 (dx^3)^2 + \varepsilon_2 (dx^4)^2, \\ a^c = \delta^c_4, \qquad b^c = \delta^c_3, \qquad A, B = 1, 2, \qquad \varepsilon_1, \varepsilon_2 = \pm 1. \end{gathered} \tag{31.6}$$

Because of (31.3), the energy-momentum tensor T_{cd} is proportional to $(a_c a_d/a^2 + b_c b_d/b^2)$. In the second case (31.5), we get

$$\begin{aligned} ds^2 &= g_{AB}(x^1, x^2)\, dx^A\, dx^B + 2\, dx^1\, dx^3 + (dx^4)^2, \\ a^c &= \delta^c_4, \qquad k^c = \delta^c_3, \qquad A, B = 1, 2, \end{aligned} \tag{31.7}$$

the energy-momentum tensor $\varkappa_0 T_{ab} p^2 k_a k_b$ being that of a pure radiation field. (For both cases, cf. Eisenhart (1949), p. 262.)

Finally, if the space-time admits *one constant vector* $a^c = \delta^c_4$, this vector is either non-null, the metric being

$$ds^2 = g_{\alpha\beta}(x^\nu)\, dx^\alpha\, dx^\beta + \varepsilon(dx^4)^2, \quad \varepsilon = \pm 1, \quad \alpha, \beta, \nu = 1, 2, 3, \tag{31.8}$$

or the vector $a^c = \delta^c_4$ is null,

$$ds^2 = g_{\alpha\beta}(x^\nu)\, dx^\alpha\, dx^\beta - 2\, dx^3\, dx^4, \qquad \alpha, \beta, \nu = 1, 2, 3. \tag{31.9}$$

The existence of a non-null constant vector $a^c = \delta^c_4$ implies $R_{4knm} = 0$ and $\overset{4}{R}_{\alpha\beta\gamma\delta} = \overset{3}{R}_{\alpha\beta\gamma\delta}$. For solutions of the vacuum equations $R_{ab} = 0$ this yields $\overset{3}{R}_{\alpha\beta} = 0$, which is equivalent to $\overset{3}{R}_{\alpha\gamma\beta\delta} = 0$. Thus we can state the theorem that if a solution of the vacuum equations admits a constant vector field, this vector is a null vector or the space-time is flat.

31.1.2. Constant tensor fields

Decomposable space-times. A four-dimensional space-time is *decomposable* if it is the product of a three- and a one-dimensional space

$$ds^2 = g_{\alpha\beta}(x^\nu)\, dx^\alpha\, dx^\beta + \varepsilon(dx^4)^2, \qquad \varepsilon = \pm 1, \tag{31.10}$$

or a product of two two-dimensional spaces

$$ds^2 = g_{AB}(x^C)\, dx^A\, dx^B + g_{MN}(x^P)\, dx^M\, dx^N. \tag{31.11}$$

Decomposable space-times can be characterized (Petrov (1966), p. 398) by the existence of a symmetric tensor h_{mn} which is idempotent and constant,

$$h_{ab} = h_{ba}, \qquad h_{ab} h^b{}_c = h_{ac}, \qquad h_{ab;e} = 0. \tag{31.12}$$

This tensor can be used to split the metric tensor into two parts,

$$g_{ab} = \overset{(1)}{g}_{ab} + \overset{(2)}{g}_{ab} = h_{ab} + (g_{ab} - h_{ab}), \tag{31.13}$$

both satisfying (31.12). The rank of the matrix h_{ab} is three (one) or two, respectively.

The metric (31.10) is that of (31.8), admitting a constant non-null vector. In the metric (31.11), the only surviving components of the curvature tensor are R_{1212} and R_{3434}, and the Ricci tensor obeys

$$R^1_1 = R^2_2, \qquad R^3_3 = R^4_4, \qquad R^b_a = 0 \quad \text{otherwise.} \tag{31.14}$$

Correspondingly, the energy-momentum tensor should be that of a Maxwell field, which is possible only if $R = 0$, i.e. for the conformally flat space-times (31.42).

Constant non-null bivectors. A constant non-null bivector F_{ab} (which need not be the actual electromagnetic field, but trivially satisfies the Maxwell equations) implies the existence of a constant self-dual bivector F^*_{ab}, which because of (5.11) has the form

$$F^*_{ab} = AW_{ab}, \qquad A = \text{const}. \tag{31.15}$$

W_{ab} is constant, and so is the tensor

$$2h_{ab} = g_{ab} - W_{ac}\overline{W}^c{}_b = 2(g_{ab} + k_a l_b + k_b l_a), \tag{31.16}$$

which obeys (31.12). Consequently, if a V_4 admits a constant non-null bivector, it is decomposable and its metric can be written in the form (31.11) (Debever and Cahen (1960)).

Conversely, if a V_4 is decomposable, the constant constituent parts of the metric being

$$\overset{1}{g}_{ab} = 2m_{(a}\overline{m}_{b)}, \qquad \overset{2}{g}_{ab} = -2k_{(b}l_{a)}, \tag{31.17}$$

then $(m_a\overline{m}_b)_{;c} = 0 = (k_a l_b)_{;c}$ holds and

$$W_{ab} = 2m_{[a}\overline{m}_{b]} + 2l_{[a}k_{b]} \tag{31.18}$$

is constant too.

Constant null bivectors. A null bivector F_{ab} can be written in the form

$$F_{ab} = p_a k_b - p_b k_a, \qquad p^n p_n = 1, \qquad k^a k_a = 0, \qquad p^n k_n = 0. \tag{31.19}$$

If F_{ab} is constant, so is $F_{ab}F_c{}^b = k_a k_c$, which implies $k_{a;b} = 0$. It can be shown (Ehlers and Kundt (1962)) that a V_4 admits a constant null bivector if and only if its fundamental form can be written in the form

$$\mathrm{d}s^2 = \mathrm{d}x^2 + \mathrm{d}y^2 - 2\,\mathrm{d}u\,\mathrm{d}v - 2H(x, y, u)\,\mathrm{d}u^2, \qquad k_a = -u_{,a}. \tag{31.20}$$

This metric is a *pp* wave, cf. § 21.5.

31.2. Complex recurrent, conformally recurrent, recurrent and symmetric space-times

31.2.1. The definitions

A *complex recurrent* space-time V_4 is a space for which the self-dual Weyl tensor (3.53) satisfies the condition

$$C^*_{abcd;e} = C^*_{abcd}K_e. \tag{31.21}$$

Generally, the recurrence vector K_e is complex; if it is real, then the space is a *conformally recurrent* space

$$C_{abcd;e} = C_{abcd}K_e, \tag{31.22}$$

and if K_e is zero, one gets a *conformally symmetric* space

$$C_{abcd;e} = 0. \tag{31.23}$$

A *recurrent* space is a space in which the Riemann tensor satisfies

$$R_{abcd;e} = R_{abcd}K_e. \tag{31.24}$$

For a recurrent space, the identities

$$R_{abcd;[mn]} + R_{cdmn;[ab]} + R_{mnab;[cd]} = 0 \tag{31.25}$$

yield

$$K_e = K_{,e}; \tag{31.26}$$

the recurrence vector is a gradient. If instead of (31.24) only

$$R_{ac;e} = R_{ac}K_e \tag{31.27}$$

holds, the space is Ricci recurrent (for Ricci recurrent space-times see Hall (1976a)).
A recurrent space is said to be *symmetric* if K_e vanishes,

$$R_{abcd;e} = 0. \tag{31.28}$$

Obviously, each recurrent (symmetric) space-time is conformally recurrent (conformally symmetric) and hence a complex recurrent space too.

Using the canonical forms (§ 4.2.) of the self-dual Weyl tensor C^*_{abcd}, and evaluating (31.21), it can easily be shown that there are no complex recurrent space-times of Petrov types I, II and III. Therefore, we have only to deal with types D, N and O. We will do this in the following sections, but without giving all details of the proofs, which can be found in McLenaghan and Leroy (1972), Kaigorodov (1972), Sciama (1961), Cahen and McLenaghan (1968), and the references given there.

31.2.2. Space-times of Petrov type D

The canonical form (cp. Table 4.2)

$$C^*_{abcd} = 2\Psi_2(V_{ab}U_{cd} + U_{ab}V_{cd} + W_{ab}W_{cd}) \tag{31.29}$$

of a type D Weyl tensor is compatible with (31.21) only if

$$\Psi_{2,e} = \Psi_2 K_e, \qquad U_{ab;e}W^{ab} = V_{ab;e}W^{ab} = 0 \tag{31.30}$$

holds. Equation (31.30) implies $W_{ab} = \text{const}$, which means that a *type D complex recurrent* space-time is necessarily decomposable, and a product (31.11) of two two-dimensional spaces.

Because of (31.21) and (31.30), $2C^*_{abcd;[ef]} = 0$ holds. Inserting (31.29), expressing the derivatives in terms of the curvature tensor and using the decomposition (3.44) of the curvature tensor, we obtain

$$\begin{gathered} \Psi_2 = -R/12, \qquad 4E_{abcd} = E\overline{W}_{ab}W_{cd}, \\ 4R_{ab} = -(E+R)\,(l_ak_b + k_al_b) + (R-E)\,(m_a\overline{m}_b + \overline{m}_am_b). \end{gathered} \tag{31.31}$$

Thus, $K_e = \Psi_{2,e}/\Psi_2$ being real, all complex recurrent space-times of type D are *conformally recurrent*. Furthermore, (31.31) tells us that the scalar curvatures $\overset{1}{K}$ and $\overset{2}{K}$ of the $(\boldsymbol{m}, \overline{\boldsymbol{m}})$- and the $(\boldsymbol{k}, \boldsymbol{l})$-spaces are given by

$$4\overset{1}{K}(x^1, x^2) = R - E, \qquad 4\overset{2}{K}(x^3, x^4) = R + E. \tag{31.32}$$

If the space is *recurrent*, with $K_e = (\ln R)_{,e}$ non-zero, the Bianchi identities together with (31.27) and (31.31) imply that either $R = E$ or $R = -E$ holds, i.e. the space-time is a product of a flat and a curved two-dimensional space.

For a *conformally symmetric* space-time, R is constant (K_n zero), and the full set of Bianchi identities yields $E = \text{const}$: the space is symmetric and has the line element (10.8),

$$ds^2 = \frac{2\,d\zeta\,d\bar{\zeta}}{[1 + (R - E)\,\zeta\bar{\zeta}/8]^2} - \frac{2\,du\,dv}{[1 - (R + E)\,uv/8]^2}. \tag{31.33}$$

31.2.3. Space-times of type N

Inserting the canonical form

$$C^*_{abcd} = -4V_{ab}V_{cd} \tag{31.34}$$

of a Weyl tensor of type N (cp. Table 4.2) into (31.21), one obtains

$$k_{a;b} = k_a p_b, \tag{31.35}$$

the (eigen-) null vector $\boldsymbol{k}$ is recurrent. From (31.35) $m_{a;b} = k_a r_b + m_a s_b$ follows, and therefore (31.35) and (31.34) yield (31.21): if a complex recurrent space-time is of type N, it contains a recurrent null vector, and, conversely, if a space-time of type N contains a recurrent null vector, it is complex recurrent, the recurrence vector being

$$K_b = p_b + m^a \overline{m}_{a;b}. \tag{31.36}$$

The metric of these complex recurrent spaces of type N can be proved to be

$$\begin{aligned} ds^2 &= 2k^{-2}(1 + \varepsilon\zeta\bar{\zeta})^{-2}\,(d\zeta + b\,du)\,(d\bar{\zeta} + \bar{b}\,du) \\ &\quad - 2\,du[dv + du(\varepsilon k^2 v^2 + lv + H)], \\ l &= \frac{1}{2}\,(b_{,\zeta} + \bar{b}_{,\bar{\zeta}}) - \varepsilon(1 + \varepsilon\zeta\bar{\zeta})^{-1}\,(\bar{\zeta}b + \zeta\bar{b}), \\ H &= H(u, \zeta, \bar{\zeta}) = \overline{H}, \qquad b = b(u, \zeta), \\ \varepsilon &= 0, \pm 1, \qquad k^2 = 1 + \varepsilon^2 K^2(u). \end{aligned} \tag{31.37}$$

Specializing this metric, one gets in the case of a *conformally recurrent* space the line element (31.20) with $H = f(u, x) + g(u, y)$, and for a *conformally symmetric* space again the metric (31.20), but now with $H = x^2 - y^2 + h(x^2 + y^2)$, $h = \text{const}$.

Among all these spaces of type N, the only vacuum solutions are the *pp* waves (§ 21.5.), which may be characterized by the property of being complex recurrent vacuum solutions (Ehlers and Kundt (1962)).

31.2.4. Space-times of type O

Starting from the definition (31.24) of a *recurrent* space, one easily obtains

$$R_{,a} = RK_a, \qquad 2S_{ab;[cd]} = S_{eb}R^e{}_{acd} + S_{ae}R^e{}_{bcd} = 0. \tag{31.38}$$

The Weyl tensor being zero, from this equation follows

$$S_{ac}S^c{}_b - \frac{1}{4}g_{ab}S_{cd}S^{cd} + \frac{1}{6}RS_{ab} = 0. \tag{31.39}$$

Inserting into (31.39) the canonical forms (§ 5.1.) of the traceless tensor S_{ab} and substituting the results back into (31.27), one sees that the following cases can occur.

Space-times with a non-vanishing recurrence vector K_a must be plane waves

$$\begin{gathered} ds^2 = dx^2 + dy^2 - 2\,du\,dv - \frac{1}{2}\varkappa_0\Phi^2(u)\,(x^2+y^2)\,du^2, \\ K_a = \Phi_{,a}/2\Phi, \qquad k_{a;b} = 0, \qquad R_{ab} = \varkappa_0\Phi^2 k_a k_b. \end{gathered} \tag{31.40}$$

Symmetric space-times ($K_a = 0$) are either spaces of constant curvature (§ 8.5.), or they possess a constant timelike or spacelike vector field orthogonal to a three-dimensional space of constant curvature,

$$ds^2 = \frac{dx^2 + dy^2 + \varepsilon\,dw^2}{\left[1 + \frac{R}{24}(x^2 + y^2 + \varepsilon w^2)\right]^2} - \varepsilon dz^2, \qquad \varepsilon = \pm 1, \tag{31.41}$$

or they are a product of two 2-spaces of constant curvature

$$ds^2 = \frac{2\,d\zeta\,d\bar\zeta}{[1 + \lambda\zeta\bar\zeta]^2} - \frac{2\,du\,dv}{[1 + \lambda uv]^2} \tag{31.42}$$

(which is a Bertotti-Robinson-like metric (10.8)), or they are plane waves (31.40) with $\Phi = \text{const}$, $K_a = 0$.

31.3. Killing tensors of order two

31.3.1. The basic definitions

A Killing tensor of order m is a symmetric tensor $K_{a_1 \ldots a_m}$, which satisfies

$$K_{(a_1 \ldots a_m;c)} = 0. \tag{31.43}$$

Killing tensors are generalizations of Killing vectors ($m = 1$). In this chapter we will confine ourselves to *Killing tensors of order two*, which in accordance with (31.43) obey

$$K_{ab} = K_{ba}, \qquad K_{(ab;c)} = 0. \tag{31.44}$$

Splitting the Killing tensor into its traceless part P_{ab} and the trace K,

$$K_{ab} = P_{ab} + g_{ab}K/4, \tag{31.45}$$

one easily gets

$$4P^a{}_{b;a} = -3K_{,b} \tag{31.46}$$

and

$$P_{(ab;c)} - \frac{1}{3} g_{(ab}P^d{}_{c);d} = 0. \tag{31.47}$$

A symmetric, traceless tensor P_{ab} is said to be a *conformal Killing tensor* if it satisfies the equation (31.47). If, in addition to (31.47), $P^a{}_{b;a}$ is a gradient, then K_{ab} constructed according to (31.46) and (31.45) is a Killing tensor.

Trivial examples of a Killing tensor are the metric tensor g_{ab} and all products

$$K_{ab} = \underset{1}{\xi}_{(a}\underset{2}{\xi}_{b)} \tag{31.48}$$

of Killing vectors $\underset{1}{\xi}_a$, $\underset{2}{\xi}_b$ (not necessarily different), and linear combinations of these with constant coefficients. Killing tensors *not* admitting this type of representation are referred to as non-trivial or non-redundant or irreducible.

Some authors use the term Killing or *Killing-Yano tensor* to describe skew-symmetric tensors $a_{bc\ldots de}$ satisfying

$$a_{bc\ldots de;f} + a_{bc\ldots df;e} = 0, \tag{31.49}$$

which is also a generalization of the Killing equation. At first glance this seems to be a quite different type of tensor, but it can easily be shown that symmetric Killing tensors can be constructed from Killing-Yano tensors. In the cases of Killing-Yano tensors $a_b = \xi_b$ and $a_{bcde} = \text{const}\ \varepsilon_{bcde}$, the corresponding Killing tensors $\xi_a\xi_b$ and $\varepsilon_{bcde}\varepsilon^{cde}{}_a = -6g_{ab}$ are trivial. For second rank tensors $\underset{A}{a}_{bc}$ the corresponding (symmetric) Killing tensors are

$$K_{nm} = \underset{A}{a}_{ni}\underset{B}{a}^i{}_m + \underset{B}{a}_{ni}\underset{A}{a}^i{}_m \tag{31.50}$$

and linear combinations of these. A third-rank Killing-Yano tensor a_{bcd} can be replaced by the vector $a_e = \varepsilon_e{}^{bcd}a_{bcd}$, which because of (31.49) satisfies

$$4a_{e;f} = g_{ef}a^c{}_{;c}. \tag{31.51}$$

If $\underset{A}{a}_b$ and $\underset{B}{a}_c$ are two such vectors, then

$$K_{bc} = \underset{A}{a}_b\underset{B}{a}_c + \underset{A}{a}_c\underset{B}{a}_b - 2\underset{A}{a}^d\underset{B}{a}_d g_{bc} \tag{31.52}$$

and linear combinations of such terms are symmetric Killing tensors.

31.3.2. Properties of space-times admitting Killing tensors

From the more physical point of view, the interest in Killing tensors originated in their connection with quadratic first integrals of geodesic motion and the separability of various partial differential equations, e.g. the Hamilton-Jacobi equation.

Let $\boldsymbol{t}$ be the tangent vector of an affinely parametrized geodesic,

$$\mathrm{D}t^a/\mathrm{D}\lambda = t^a{}_{;b}t^b = 0. \tag{31.53}$$

Then, because of (31.44),

$$\frac{\mathrm{d}}{\mathrm{d}\lambda} K_{ab}t^a t^b = K_{ab;c}t^a t^b t^c = 0 \tag{31.54}$$

holds, so $K_{ab}t^a t^b$ is a quadratic first integral of motion. For a conformal Killing tensor P_{ab}, satisfying the weaker condition (31.47), the analogous statement is true only for null geodesics.

Skew-symmetric Killing-Yano tensors a_{bc} induce the existence of a vector $a_{bc}t^b$, which because of (31.49) is parallelly propagated along the (arbitrary) geodesic, and to each vector $\boldsymbol{a}$ satisfying (31.51) corresponds a skew-symmetric tensor $a_c t_b - a_b t_c$ which is constant along the geodesic. Each of these cases gives rise to quadratic first integrals, arising, respectively, from the invariants of the vector and the tensor, in accordance with (31.50), (31.51) and (31.54).

For the connection between Killing tensors and separability of partial differential equations, we refer the reader to the literature (cf. Havas (1975), Dietz (1976), Collinson and Fugère (1977), Boyer et al. (1978), Benenti and Francaviglia (1979)). We mention only the result of Woodhouse (1975), that the separable coordinates for the Hamilton-Jacobi equation in a Lorentzian manifold are adapted either to a Killing vector or to an eigenvector of a (symmetric) Killing tensor of order two.

31.3.3. Theorems on Killing tensors in four-dimensional space-times

In analogy with the treatment of Killing vectors and groups of motions, the final goal of studying Killing tensors and their properties is (a) to find all Killing tensors of a given space-time and/or (b) to classify space-times with respect to the non-trivial Killing tensors they admit. Neither problem is yet solved, and we can only give some known theorems.

I. Theorems on symmetric second order Killing tensors K_{ab}

Theorem 31.1. *A four-dimensional space-time admits at most* 50 (*linearly independent*) *Killing tensors of order two. The maximum number of* 50 *is attained if and only if the space-time is of constant curvature; in that case, all* 50 *Killing tensors are reducible* (Hauser and Malhiot (1975)).

Theorem 31.2. *Every type D vacuum solution, the Weyl tensor being*

$$C^*{}_{abmn} = 2\Psi_2(V_{ab}U_{mn} + U_{ab}V_{mn} + W_{ab}W_{mn}), \tag{31.55}$$

admits a conformal Killing tensor

$$P_{ab} = -\frac{1}{2}\,\Phi\overline{\Phi}W_{ac}\overline{W}^c{}_b = (A^2 + B^2)\left[k_a l_b + l_a k_b + \frac{1}{2}\,g_{ab}\right], \quad \Phi = A + \mathrm{i}B = \mathrm{const}\,(\Psi_2)^{-1/3}. \tag{31.56}$$

This conformal Killing tensor is irreducible provided the space-time admits fewer than four Killing vectors (Walker and Penrose (1970)).

Theorem 31.3. *Every type D vacuum solution, with the exception of the C-metric and its generalization (Case III of Kinnersley's* (1969b) *classification), admits a Killing tensor*

$$K_{ab} = (A^2 + B^2)\,(l_a k_b + k_a l_b) + B^2 g_{ab}, \quad A + \mathrm{i}B = \mathrm{const}\,(\Psi_2)^{-1/3}, \tag{31.57}$$

the complex constant being adjusted in such a way that

$$DA = \Delta A = \delta B = 0 \tag{31.58}$$

(Walker and Penrose (1970), Hughston et al. (1972), Hughston and Sommers (1973)).

Theorem 31.4. *If a space-time admits a Killing tensor of Segré characteristic* [(1 1) (1, 1)],

$$K_{ab} = A^2(l_a k_b + k_a l_b) + B^2(m_a\overline{m}_b + \overline{m}_a m_b), \tag{31.59}$$

then its two null eigenvectors $\boldsymbol{k}$ *and* $\boldsymbol{l}$ *are both shearfree and geodesic (and, therefore, principal null vectors of the Weyl tensor if* $R_{11} = R_{14} = R_{44} = 0$, *cf. Theorem* 7.1). *Furthermore, the functions* A *and* B *have to satisfy*

$$DA = \Delta A = \delta B = 0, \qquad \delta A^2 = (\overline{\pi} - \tau)\,(A^2 + B^2), \quad DB^2 = -(\varrho + \overline{\varrho})\,(A^2 + B^2), \qquad \Delta B^2 = (\mu + \overline{\mu})\,(A^2 + B^2). \tag{31.60}$$

If the space-time is a vacuum solution, then (31.57) holds.

If a space-time admits a Killing tensor (31.59) with $(A_{,a}B_{,b} - A_{,b}B_{,a})\,(A^{,a}B^{,b} - A^{,b}B^{,a}) \neq 0$ and $R^{ab}A_{,a}B_{,b} = 0$, then it admits a two-parameter Abelian isometry group commuting with K_{ab} (Hauser and Malhiot (1976)).

Ref.: Petrov (1966), p. 319; Williams (1968); Sommers (1973); Hauser and Malhiot (1974, 1978); Cosgrove (1978); □ Dietz and Rüdiger (1979).

II. Theorems on Killing-Yano tensors

Theorem 31.5. *If a space-time admits a non-degenerate skew-symmetric Killing-Yano tensor, then this tensor can be written as*

$$a_{bc} = A(l_b k_c - k_b l_c) + \mathrm{i}B(m_b\overline{m}_c - \overline{m}_b m_c), \tag{31.61}$$

the Weyl tensor is of type D (or O), the degenerate, geodesic and shearfree null eigenvectors being $\boldsymbol{l}$ *and* $\boldsymbol{k}$, *the Ricci tensor satisfies*

$$R_{bc}a^c{}_d + R_{dc}a^c{}_b = 0, \tag{31.62}$$

and the real functions A and B have to obey

$$\begin{aligned} &DA = \Delta A = \delta B = 0, \\ &D(A + \mathrm{i}B) = -\varrho(A + \mathrm{i}B), \qquad \Delta(A + \mathrm{i}B) = \mu(A + \mathrm{i}B), \\ &\delta(A + \mathrm{i}B) = -\tau(A + \mathrm{i}B), \qquad \bar{\delta}(A + \mathrm{i}B) = \pi(A + \mathrm{i}B), \end{aligned} \tag{31.63}$$

(Collinson (1974, 1976b), Stephani (1978)).

Theorem 31.6. *All type D vacuum solutions which admit a symmetric Killing tensor* K_{bc} *(see Theorem* 31.3), *also admit a skew-symmetric Killing tensor* a_{bc}, *the two tensors being connected by* (31.50). (Collinson (1976b), Stephani (1978)).

Theorem 31.7. *If a space-time admits a degenerate skew-symmetric Killing tensor* a_{bc}, *then this tensor can be written as*

$$a_{bc} = k_b p_c - p_b k_c, \tag{31.64}$$

$\boldsymbol{k}$ *is a (null) Killing vector, the Weyl tensor is of type N (or O),* $\boldsymbol{k}$ *being the multiple eigenvector, and the Ricci tensor has to satisfy* (31.62) (Collinson (1974), Stephani (1978)).

Theorem 31.8. *If a space-time admits a third-rank skew-symmetric Killing tensor, i.e. a vector* $\boldsymbol{a}$ *satisfying* (31.51), *then* $\boldsymbol{a}$ *is a constant null vector or space-time is conformally flat* (Collinson (1974), Stephani (1978)).

31.4. Some remarks concerning space-times with other special properties

Besides the vector and tensor fields dealt with up to now, there are several others which may equally well be used to characterize space-times. In addition, more general concepts of classification of space-times can be used. We would like to mention a few of these and give some key references to more detailed information.

The (infinitesimal) *holonomy group* at a point p is the group of linear transformations generated by parallel transport of a vector along a closed curve (homotopic to zero) through p. It is, therefore, a subgroup of the Lorentz group at p. Its properties are connected with the Petrov classification and the possibility of constant tensor fields in the space-time and also with the uniqueness of the metric given $R^a{}_{bcd}$ (Goldberg and Kerr (1961), Beiglböck (1964), Ihrig (1975)).

In Riemannian spaces, various symmetries described by properties of infinitesimal transformations $x^i = x_0{}^i + \xi^i(x)\,\delta\tau$ exist; some of them have an obvious geometric interpretation. As a generalization of motions, i.e. (Killing) vectors $\boldsymbol{\xi}$ satisfying

$$\pounds_{\xi} g_{nm} = \xi_{n;m} + \xi_{m;n} = 0 \tag{31.65}$$

one can consider:

conformal motions: $\pounds_\xi g_{nm} = 2\lambda(x)\, g_{nm}$ (31.66)

homothetic motions: $\pounds_\xi g_{nm} = 2a g_{nm}, \qquad a = \text{const}$ (31.67)

projective collineations: $\pounds_\xi \begin{Bmatrix} i \\ jk \end{Bmatrix} = \delta^i_j \varphi_{,k} + \delta^i_k \varphi_{,j}$ (31.68)

affine collineations: $\pounds_\xi \begin{Bmatrix} i \\ jk \end{Bmatrix} = 0$ (31.69)

Ricci collineations: $\pounds_\xi R_{nm} = 0$ (31.70)

curvature collineations: $\pounds_\xi R_{abnm} = 0.$ (31.71)

Conformal motions (Petrov (1966), p. 272) preserve angles between two directions at a point, and map null geodesics into null geodesics. *Homothetic motions* (McIntosh (1976, 1979)), in addition, scale all distances by the same constant factor (thus leading to self-similar space-times), and preserve the null geodesic affine parameters. *Projective collineations* map geodesics into geodesics; *affine collineations* preserve, in addition, the affine parameters on geodesics (Katzin and Levine (1972)). Obviously, motions, affine collineations, and homothetic motions are automatically curvature collineations. However, there are Riemannian spaces V_n which admit curvature collineations not generated by a Killing vector (Katzin et al. (1969), Collinson (1970)). *Ricci collineations* are dealt with in Davis et al. (1976).

If for a space-time a *Cauchy problem* can properly be stated, then the Cauchy data set contains all information about the space-time, and space-time symmetries may be characterized in terms of these Cauchy data (Coll (1975b), Berger (1976), O'Murchadha and York (1976)).

Riemannian space-times may be characterized by the *scalar algebraic invariants* which can be constructed from the curvature tensor and its covariant derivatives up to the $(k-2)$-order. For a V_n, the number of independent invariants is

$$N(n, k) = \frac{n(n+1)\,(n+k)!}{2n!\,k!} - \frac{(n+k+1)!}{(n-1)!\,(k+1)!} + n, \tag{31.72}$$

$$N(n, 1) = 0, \qquad N(2, 2) = 1.$$

In a V_4, the number of invariants of the curvature tensor is consequently $N(4, 2) = 14$, cp. § 5.1., and for vacuum this number reduces to 4; there exist non-flat metrics, where all 14 invariants vanish, namely special plane waves (Strunz (1974), Kerr (1963b), Singh et al. (1969), Géhéniau and Debever (1956a)).

Chapter 32. Lokal isometric embedding of four-dimensional Riemannian manifolds

32.1. The why of embedding

It is a well known theorem of differential geometry (Eisenhart (1949), p. 143) that one can regard every (analytic) four-dimensional space-time V_4 (at least locally) as a subspace of a flat pseudo-euclidean space E_N of $N \leqq 10$ dimensions. If we choose Cartesian coordinates y^A in describing E_N,

$$\overset{(N)}{\mathrm{d}s}{}^2 = \sum_{A=1}^{N} e_A (\mathrm{d}y^A)^2 = \eta_{AB}\,\mathrm{d}y^A\,\mathrm{d}y^B, \qquad e_A = \pm 1, \tag{32.1}$$

then the subspace V_4 (coordinates x^a) will be given by the parametric representation

$$y^A = y^A(x^a) \tag{32.2}$$

and the metric of this subspace as induced by (32.1) is

$$\begin{aligned} \mathrm{d}s^2 &= g_{ab}\,\mathrm{d}x^a\,\mathrm{d}x^b = \eta_{AB} y^A{}_{,a} y^B{}_{,b}\,\mathrm{d}x^a\,\mathrm{d}x^b, \\ A, B, \ldots &= 1 \ldots N, \qquad a, b, \ldots = 1 \ldots 4. \end{aligned} \tag{32.3}$$

If these three equations (32.1)–(32.3) hold, they describe a local isometric embedding of V_4 into E_N. The *minimum* number of extra dimensions is called the *embedding class p* (or just class) of the V_4 in question, $0 \leqq p \leqq 6$.

Attempts have been made to give a physical meaning to the flat embedding space, or to use it as an auxiliary space for visualizing or deriving physical properties of the embedded space-time. In the context of this book, our point of view is a more pragmatic one. The invariance of the embedding class gives rise to a classification scheme of all solutions of Einstein's field equations according to their respective embedding class. From a mathematical point of view, this classification scheme is on an equal footing with the classifications with respect to groups of motions or to the Petrov types, and it will give a refinement of both these schemes. Moreover, there is some hope of obtaining exact solutions by the method of embedding, at least for some simple cases of low embedding class, solutions which are not readily available by other methods. We are *not* interested in the embedding itself, the functions $y^A(x^a)$ will not be determined or given here. A large number of explicit embeddings can be found in Rosen (1965) and Collinson (1968a). Other aspects of the embedding problem are discussed in Goenner (1980).

32.2. The basic formulae governing embedding

To get deeper insight into the geometrical properties of the embedding described by (32.1)–(32.3), we introduce an N-leg at every point of V_4 and consider the change of this N-leg along V_4, i.e. we consider the covariant derivative with respect to the coordinates x^n of V_4.

The N-leg in question consists of four vectors $y^A{}_{,a}$ (vectors in E_N, $a = 1 \dots 4$) tangent to V_4 and p unit vectors $n^{\alpha A}$ ($\alpha = 1 \dots p$) orthogonal to V_4 and to each other,

$$\eta_{AB} n^{\alpha A} n^{\beta B} = e^\alpha \delta^{\alpha\beta}, \qquad e^\alpha = \pm 1, \tag{32.4}$$

$$\eta_{AB} n^{\alpha A} y^B{}_{,a} = 0. \tag{32.5}$$

(In these and the following formulae, summation over Greek indices takes place only if explicitly indicated.)

The covariant derivatives (covariant with respect to coordinates x^n and metric g_{ab}) of the basic vectors $n^{\alpha A}$ and $y^B{}_{,a}$ are vectors and tensors (respectively) in V_4, but again vectors in the embedding space E_N and are, therefore, linear combinations of the basic vectors. Starting from the metric (32.3), we get

$$g_{ab;c} = 0 = \eta_{AB}(y^A{}_{,a;c} y^B{}_{,b} + y^A{}_{,a} y^B{}_{,b;c}), \tag{32.6}$$

and subtracting from this the expressions obtained by substituting $(c\,a\,b)$ and $(a\,b\,c)$ for $(a\,c\,b)$, we conclude that

$$\eta_{AB} y^A{}_{,c} y^B{}_{,a;b} = 0 \tag{32.7}$$

holds. Equation (32.7) tells us that $y^B{}_{,a;b}$ is a vector orthogonal to V_4, which can therefore be expressed as a linear combination of the normal vectors $n^{\alpha B}$,

$$y^B{}_{,a;b} = \sum_\alpha e^\alpha \Omega^\alpha{}_{ab} n^{\alpha B}, \qquad \Omega^\alpha{}_{ab} = \Omega^\alpha{}_{ba}. \tag{32.8}$$

The p symmetric tensors $\Omega^\alpha{}_{ab}$ (tensors in V_4) defined by this equation are generalizations to higher dimensions of the tensor of the second fundamental form used in the theory of hypersurfaces.

In a similar way we conclude from (32.4), (32.5) and (32.8) that

$$\begin{aligned} n^{\alpha A}{}_{,a} &= -\Omega^\alpha{}_{ab} g^{bc} y^A{}_{,c} + \sum_\beta e^\beta t^{\beta\alpha}{}_a n^{\beta A}, \\ t^{\beta\alpha}{}_b + t^{\alpha\beta}{}_b &= 0, \end{aligned} \tag{32.9}$$

holds. The $p(p-1)/2$ vectors $t^{\beta\alpha}{}_b$ (vectors in V_4) defined by (32.9) are sometimes called torsion vectors.

The tensors $\Omega^\alpha{}_{ab}$ and the vectors $t^{\beta\alpha}{}_b$ cannot be prescribed arbitrarily. They have to satisfy the conditions of integrability of the system (32.8)–(32.9), which turn out to be

$$R_{abcd} = \sum_\alpha e^\alpha (\Omega^\alpha{}_{ac} \Omega^\alpha{}_{bd} - \Omega^\alpha{}_{ad} \Omega^\alpha{}_{bc}), \tag{32.10}$$

(Gauss)

$$\Omega^\alpha{}_{ab;c} - \Omega^\alpha{}_{ac;b} = \sum_\beta e^\beta (t^{\beta\alpha}{}_c \Omega^\beta{}_{ab} - t^{\beta\alpha}{}_b \Omega^\beta{}_{ac}), \tag{32.11}$$

(Codazzi)

$$t^{\beta\alpha}{}_{a;b} - t^{\beta\alpha}{}_{b;a} = \sum_\nu e^\nu (t^{\nu\beta}{}_b t^{\nu\alpha}{}_a - t^{\nu\beta}{}_a t^{\nu\alpha}{}_b) + g^{cd}(\Omega^\beta{}_{cb}\Omega^\alpha{}_{da} - \Omega^\beta{}_{ca}\Omega^\alpha{}_{db}). \tag{32.12}$$

(Ricci)

The last three equations (32.10)–(32.12) are the most important equations of embedding theory. They are written entirely in terms of the V_4 in question, using only vector and tensor fields on V_4.

If a V_4 is of embedding class p, then it must admit p tensor fields $\Omega^\alpha{}_{ab}$ and $p(p-1)/2$ vector fields $t^{\alpha\beta}{}_b$ which satisfy (32.10)–(32.12), the constants $e^\alpha = \pm 1$ being suitably chosen; and if p is the minimum number enabling (32.10)–(32.12) to be satisfied, then V_4 is of embedding class p. For $p > 1$, the tensors $\Omega^\alpha{}_{ab}$ and the vectors $t^{\alpha\beta}{}_b$ are not uniquely defined by the embedding because of the possibility of performing a rotation (pseudo-rotation) at each point of V_4 of the basic unit vectors $n^{\alpha A}$ orthogonal to V_4. These degrees of freedom may be used to simplify the equations (32.10)–(32.12) in special cases, see § 32.5.

If one inserts the curvature tensor (32.10) into the Bianchi identities $R_{ab[cd;e]} = 0$, one gets identities involving the derivatives of the tensors $\Omega^\alpha{}_{ab}$. It turns out (Gupta and Goel (1975), Goenner (1977)) that as a consequence of these identities parts of the Codazzi equations (32.11) are automatically satisfied. In this sense, the Gauss equations (32.10) and the Codazzi equations (32.11) are not completely independent of each other. In some exceptional cases, only the Gauss equations need to be satisfied to ensure the embedding property of a given V_4.

32.3. Some theorems on local isometric embedding

Unfortunately, no practical way of solving the Gauss-Codazzi-Ricci equations (32.10)–(32.12) in the general case is known, either in the sense of determining all solutions of Einstein's equations of a given embedding class or in the sense of determining the embedding class of a given metric. Progress made so far concentrates on three points, namely (a) solutions of embedding class one and two, (b) explicit embedding of certain metrics or classes of metrics, and (c) the connections between embedding class and other properties of the metric, e.g. special vector and tensor fields, groups of motions. We will postpone (a) and (b) to later sections and deal here only with (c).

32.3.1. General theorems

For the sake of completeness, we start by repeating the theorem mentioned in § 32.1. If we denote by $V_n(s, t)$ a Riemannian space with s spacelike and t timelike directions, and by $E_N\,(S, T)$ the pseudo-euclidean embedding space, then the theorem is (Eisenhart (1949), Friedman (1965)):

Theorem 32.1. *Any analytic Riemannian manifold $V_n(s, t)$ can be (locally) isometrically embedded in E_N (S, T) with $s + t = n$, $S + T = N$, $n \leqq N \leqq n(n+1)/2$, $s \leqq S$, $t \leqq T$.*

Correspondingly, the embedding class p of a V_n is at most $n(n-1)/2$. For space-time $n = 4$, and $p \leqq 6$.

If two spaces $\hat{V}_n$ and V_n are conformally related, $\hat{g}_{ab} = e^{2U} g_{ab}$, then their respective embedding classes are related too. This can be seen by starting from the embedding (32.3), with $N = n + p$, of V_n into a flat $(p + n)$-dimensional space. The relations

$$\begin{aligned} z^A &= \mathrm{e}^U y^A, \qquad z^{N+1} = \mathrm{e}^U(\eta_{AB} y^A y^B - 1/4), \\ z^{N+2} &= \mathrm{e}^U(\eta_{AB} y^A y^B + 1/4) \end{aligned} \tag{32.13}$$

then give $\hat{V}_n$ as embedded in a flat $(n + p + 2)$-dimensional space. Thus we have established

Theorem 32.2. *If two space are conformally related, then their respective embedding classes differ by at most 2. In particular, the class of conformally flat spaces is at most 2.*

An example of a conformally flat space of embedding class 1 is the Robertson-Walker metric

$$\mathrm{d}s^2 = f^2(t)\,[\mathrm{d}\varphi^2 + \sin^2\varphi(\mathrm{d}\psi^2 + \sin^2\psi\,\mathrm{d}\alpha^2)] - \mathrm{d}t^2, \tag{32.14}$$

the embedding being given by

$$\begin{aligned} y^1 &= f(t)\cos\varphi, \qquad\qquad y^2 = f(t)\sin\varphi\cos\psi, \\ y^3 + \mathrm{i}y^4 &= f(t)\sin\varphi\sin\psi\,\mathrm{e}^{\mathrm{i}\alpha}, \qquad y^5 = \int \sqrt{f'^2(t) + 1}\,\mathrm{d}t, \\ \mathrm{d}s^2 &= (\mathrm{d}y^1)^2 + (\mathrm{d}y^2)^2 + (\mathrm{d}y^3)^2 + (\mathrm{d}y^4)^2 - (\mathrm{d}y^5)^2. \end{aligned} \tag{32.15}$$

32.3.2. Vector and tensor fields and embedding class

The existence of special vector and tensor fields in a V_4 may reduce its embedding class below the maximum value $p = 6$.

As an example we consider a V_4 which admits a non-null vector field $\boldsymbol{v}$ satisfying

$$\pounds_{\boldsymbol{v}} h_{ab} = \pounds_{\boldsymbol{v}}(g_{ab} - v_a v_b/v^2) = 0, \qquad h_{ab} v^a = 0. \tag{32.16}$$

In the frame of reference defined by $v^i = (0, 0, 0, v^4)$, the condition (32.16) implies $h_{4i} = 0$, $h_{ji,4} = 0$, i.e.

$$\mathrm{d}s^2 = h_{\alpha\beta}(x^\nu)\,\mathrm{d}x^\alpha\,\mathrm{d}x^\beta + \varepsilon v^{-2}(v_i\,\mathrm{d}x^i)^2, \quad \alpha, \beta = 1, 2, 3, \quad \varepsilon = \pm 1. \tag{32.17}$$

(In the case of a timelike unit vector field, $v^a/v = u^a$, the field would be called shear- and expansionfree; it describes a rigid congruence.)

If, in addition, v_a/v is a gradient, $v_i\,\mathrm{d}x^i/v = \mathrm{d}x^4$, then the metric admits a (covariantly) constant vector field, cf. (31.8). As the $h_{\alpha\beta}$-part of the metric is a V_3 and can be embedded in at most 6 dimensions, we get

Theorem 32.3. *If a V_4 admits a non-null constant vector field, then its embedding class is $p \leqq 3$.*

By inspection of the metric (31.9) admitting a constant null vector field, and using the representation $dx^3\,dx^4 = du^2 - dv^2$, one gets in a similar way

Theorem 32.4. *If a V_4 admits a constant null vector field, then its embedding class is $p \leqq 4$.*

As every V_2 can be embedded in a E_3, in the case of the metrics (31.6) and (31.7) which possess two constant vector fields we get

Theorem 32.5. *If a V_4 admits two constant non-null vector fields, then its embedding class is $p \leqq 1$.*

Theorem 32.6. *If a V_4 admits a constant null vector field and a constant non-null vector field orthogonal to each other, then its embedding class is $p \leqq 2$.*

If in (32.17) v_a/v is proportional to a gradient, the metric can be written as

$$ds^2 = h_{\alpha\beta}(x^\gamma)\,dx^\alpha\,dx^\beta + \varepsilon f^2(x^i)\,(dx^4)^2. \tag{32.18}$$

With respect to the embedding class, two different cases may occur. For $f_{,4} = 0$, the metric

$$ds^2 = h_{\alpha\beta}(x^\nu)\,dx^\alpha\,dx^\beta + \varepsilon f^2(x^\nu)\,(dx^4)^2, \qquad \alpha, \beta, \nu = 1, 2, 3, \tag{32.19}$$

is that of a space-time admitting a normal Killing vector field $\xi^i = (0, 0, 0, \xi^4)$. By introducing (Szekeres (1966a))

$$u + iv = f\,e^{ix^4}, \qquad \varepsilon(du^2 + dv^2) = \varepsilon[df^2 + f^2(dx^4)^2], \tag{32.20}$$

we see that the problem has been reduced to finding an embedding of the three-dimensional metric $h_{\alpha\beta}\,dx^\alpha\,dx^\beta - \varepsilon\,df^2$, which can be done in at most six dimensions:

Theorem 32.7. *If a V_4 admits a non-null normal Killing vector field, then its embedding class is $p \leqq 4$.*

In the general case $f_{,4} \neq 0$, the $f^2(dx^4)^2$-part of the metric (32.18) needs, according to Theorem 32.2, at most three dimensions for embedding, which add to the six dimensions of the $h_{\alpha\beta}\,dx^\alpha\,dx^\beta$ part. Hence we get

Theorem 32.8. *If a V_4 admits a normal non-null vector field $\boldsymbol{v}$ satisfying* (32.16), *then its embedding class is $p \leqq 5$.*

Besides metrics satisfying (32.16), we may consider metrics characterized by the existence of a vector field $v_a = (x^4)_{,a}$ with

$$\begin{aligned} v_{a;b} &= \frac{\Theta}{3}\,(g_{ab} - \varepsilon v_a v_b), \qquad v_a v^a = \varepsilon \neq 0, \\ ds^2 &= f^2(x^\alpha)\,h_{\alpha\beta}(x^\nu)\,dx^\alpha\,dx^\beta + \varepsilon(dx^4)^2, \qquad \alpha, \beta, \nu = 1, 2, 3. \end{aligned} \tag{32.21}$$

For the embedding of the three-dimensional space with metric $h_{\alpha\beta}$ we need at most three additional dimensions; application of Theorem 32.2 then leads to

Theorem 32.9. *If a V_4 admits a non-null vector field $\boldsymbol{v}$ satisfying* (32.21), *then its embedding class is $p \leqq 5$.*

32.3.3. Groups of motions and embedding class

If groups of motions of a V_4 induce the existence of subspaces of low embedding class, e.g. of subspaces of constant curvature, then the existence of the group will induce a low class for V_4 too. The details will depend on the order r of the groups G_r (A_r, if Abelian) and on the dimensions of their orbits (S = spacelike, T = timelike, N = null). Some of the results (Goenner (1973)) are given in Table 32.1. We see that, to a certain extent, high symmetry induces low embedding class. The converse is not true; metrics of class one without any symmetry are known.

Table 32.1 Upper limits for the embedding class p of various metrics admitting groups (T_{ab} not specified)

Group → Orbit ↓	G_{10}	G_7	G_6		G_3	A_3	A_2
			$G_6 \supset G_5$	$G_6 \not\supset G_5$			
V_4	$p \leqq 1$	$p \leqq 2$	$p \leqq 4$	$p \leqq 2$	—	—	—
S_3, T_3	—	—	—	$p \leqq 1$		$p \leqq 3$	—
N_3	—	—	$p \leqq 2$			$p \leqq 3$	—
S_2, T_2	—	—	—		flat: $p \leqq 3$ non-flat: $p \leqq 2$	—	$p \leqq 4$

To give an example of the method of reasoning, we consider an arbitrary spherically symmetric line element

$$\mathrm{d}s^2 = b^2(r, t)\,(\mathrm{d}\vartheta^2 + \sin^2\vartheta\,\mathrm{d}\varphi^2) + a^2(r, t)\,\mathrm{d}r^2 - c^2(r, t)\,\mathrm{d}t^2. \tag{32.22}$$

If we put

$$\begin{gathered} y^1 = b\cos\vartheta, \qquad y^2 = b\sin\vartheta\cos\varphi, \qquad y^3 = b\sin\vartheta\sin\varphi, \\ (\mathrm{d}y^1)^2 + (\mathrm{d}y^2)^2 + (\mathrm{d}y^3)^2 = b^2(\mathrm{d}\vartheta^2 + \sin^2\vartheta\,\mathrm{d}\varphi^2) + (\mathrm{d}b)^2, \end{gathered} \tag{32.23}$$

and take into account that the 2-space with metric $a^2(r, t)\,\mathrm{d}r^2 - c^2(r, t)\,\mathrm{d}t^2 - [\mathrm{d}b(r, t)]^2$ can be embedded into an E_3, we see that spherically symmetric space-times are of class $p \leqq 2$ ($p = 1$ is possible for special functions a, b, c (Ikeda et al. (1963), Karmarkar (1948)).

If one compares the theorems given in this section (and especially the methods of the proofs outlined or mentioned) with the general theory of embedding sketched in the preceding section, one may feel that the two sections are disconnected: no use has been made of the Gauss-Codazzi-Ricci equations. The reason for this incoherence is the fact, already stated above, that no systematic treatment of these equations has been carried out, with the exception of metrics of class one or two. The following sections will be devoted to these metrics.

32.4. Exact solutions of embedding class one

32.4.1. The Gauss and Codazzi equations and the possible types of Ω_{ab}

Application of the general theory outlined in § 32.2. to class one yields: a V_4 is of class one if and only if there is a symmetric tensor Ω_{ab} satisfying

$$R_{abcd} = e(\Omega_{ac}\Omega_{bd} - \Omega_{ad}\Omega_{bc}), \qquad e = \pm 1, \quad \text{(Gauss)} \tag{32.24}$$

and

$$\Omega_{ab;c} = \Omega_{ac;b}, \qquad \text{(Codazzi)} \tag{32.25}$$

$e = \pm 1$ being suitably chosen.

If $\Omega^{-1}{}_{ab}$ exists, then the Codazzi equations are a consequence of the Gauss equations and the Bianchi identities (cf. Goenner (1977)):

Theorem 32.10. *If a non-singular symmetric tensor Ω_{ab} satisfying* (32.24) *exists, then space-time is of embedding class $p = 1$.*

The Gauss equations (32.24) and the field equations yield

$$\varkappa_0 \left(T_{ab} - \frac{1}{2}\, g_{ab} T^c{}_c\right) = R_{ab} = e(\Omega_{ab}\Omega^c{}_c - \Omega_{ac}\Omega^c{}_b). \tag{32.26}$$

Due to the algebraic simplicity of this equation, all possible tensors Ω_{ab} which correspond to an energy-momentum tensor of the perfect fluid or Maxwell type can be determined. The calculations are straightforward, starting with a suitable tetrad representation of T_{ab} and Ω_{ab}. If Ω_{ab} is known, the Petrov type can easily be obtained from (32.24). Four different cases occur, namely (Stephani (1967a)):

(a) *Perfect fluid metrics, Petrov type O (conformally flat)*

$$\begin{gathered} T_{ab} = (\mu + p)\, u_a u_b + p g_{ab}, \qquad \Omega_{ab} = A u_a u_b + C g_{ab}, \\ \varkappa_0 \mu = 3C^2 > 0, \qquad \varkappa_0 p = C(2A - 3C), \qquad e = +1. \end{gathered} \tag{32.27}$$

(b) *Perfect fluid metrics, Petrov type D*

$$\begin{gathered} T_{ab} = (\mu + p)\, u_a u_b + p g_{ab}, \\ \Omega_{ab} = 2C u_a u_b + C g_{ab} + A v_a v_b, \qquad v_a v^a = 1, \qquad u_a v^a = 0, \\ \varkappa_0 \mu = e(3C + 2A)\, C, \qquad \varkappa_0 p = eC^2, \qquad AC \neq 0. \end{gathered} \tag{32.28}$$

(c) *Pure radiation fields, Petrov type N*

$$\begin{gathered} T_{ab} = \Phi^2 k_a k_b, \qquad k_a k^a = 0, \\ \Omega_{ab} = A k_a k_b + C z_a z_b, \qquad z_a z^a = 1, \qquad z^a k_a = 0, \qquad eAC = \varkappa_0 \Phi^2, \end{gathered} \tag{32.29}$$

and

$$\begin{gathered} T_{ab} = \Phi^2 k_a k_b, \qquad k^a k_a = 0, \\ \Omega_{ab} = B(k_a z_b + z_a k_b), \qquad z_a z^a = 1, \qquad z^a k_a = 0, \qquad -eB^2 = \varkappa_0 \Phi^2. \end{gathered} \tag{32.30}$$

The corresponding space-times will be determined in the following sections. In each of the four cases the functions A, B, C and the vector fields $\boldsymbol{u}$, $\boldsymbol{v}$, $\boldsymbol{k}$, $\boldsymbol{z}$ have to be chosen so that the Gauss and Codazzi equations (32.24) and (32.25) are satisfied. To get the metric, one uses the Codazzi equations (and parts of the Gauss equations) to find preferred vector fields, adjusts the coordinates to these vector fields and tries to solve the remaining Gauss equations. In the case of an electromagnetic null field, the Maxwell equations have to be satisfied too.

Concerning the remaining types of energy-momentum tensor, the question is answered by

Theorem 32.11. *There are no class one solutions of the Einstein-Maxwell equations with non-null electromagnetic field* (Collinson (1968b)), *and no class one vacuum solutions.*

32.4.2. Conformally flat perfect fluid solutions of class one

The problem we have to solve is the following: find all metrics with curvature tensor (32.24) and Ω_{ab} given by (32.27), i.e. all metrics with the curvature tensor

$$\begin{aligned} R_{abcd} = {} & C^2(g_{ac}g_{bd} - g_{ad}g_{bc}) \\ & + CA(g_{ac}u_bu_d + g_{bd}u_au_c - g_{ad}u_bu_c - g_{bc}u_au_d). \end{aligned} \tag{32.31}$$

(Because of Theorem 32.10, the Codazzi equations are a consequence of the Bianchi identities for $C \neq A$; $C = A$ is either (if $\Theta = 0$) the Einstein universe (32.41) or (if $\Theta \neq 0$) it is a special case of (32.46)). For $C = 0$, space-time is flat.

An almost trivial solution of (32.31) is

$$A = 0, \quad \varkappa_0 T_{ab} = -3C^2 g_{ab}, \qquad C^2 = \text{const}, \tag{32.32}$$

which corresponds to a space of constant curvature (de Sitter space). Allowing for negative μ ($C^2 < 0$), we have thus shown that the spaces of constant curvature are of embedding class one (or zero, if they are flat). From now on we assume $A \neq 0$.

Starting from (32.31), the Bianchi identities give us

$$u_{a;b} = -\dot{u}_a u_b + \Theta h_{ab}/3, \qquad h_{ab} \equiv g_{ab} + u_a u_b; \tag{32.33}$$

the velocity field is normal and shearfree. In the comoving frame of reference $u_i = (0, 0, 0, u_4)$ the metric takes the form

$$ds^2 = h_{\mu\nu}\,dx^\mu\,dx^\nu - (u_4)^2\,dt^2 \qquad \mu, \nu = 1, 2, 3. \tag{32.34}$$

Furthermore, the Bianchi identities yield

$$\begin{aligned} C = C(t), \qquad & \partial_t C = \Theta A u_4/3, \\ \Theta = \Theta(t), \qquad & A_{,a} = \dot{A}u_a - A\dot{u}_a. \end{aligned} \tag{32.35}$$

The calculations now depend on whether the velocity field is expansionfree ($\Theta = 0$) or not ($\Theta \neq 0$).

Expansionfree solutions. If Θ vanishes, (32.33) and (32.35) lead to

$$C = \text{const}, \qquad \partial_t h_{\mu\nu} = 0, \qquad A = u^4 f(t). \tag{32.36}$$

Because of (32.36), the spatial part of the Gauss equations (32.31) reads

$$\overset{3}{R}_{\mu\nu\sigma\tau} = \overset{4}{R}_{\mu\nu\sigma\tau} = C^2(h_{\mu\sigma}h_{\nu\tau} - h_{\mu\tau}h_{\sigma\nu}), \tag{32.37}$$

so the 3-space $h_{\mu\nu}$ is a space of constant curvature,

$$\mathrm{d}s^2 = \mathrm{d}r^2/(1 - C^2r^2) + r^2(\mathrm{d}\vartheta^2 + \sin^2\vartheta\,\mathrm{d}\varphi^2) - (u_4)^2\,\mathrm{d}t^2. \tag{32.38}$$

The remaining Gauss equations

$$C(A - C)\,(u_4)^2\,h_{\sigma\tau} = -u_4(u_{\sigma;4;\tau} - u_{\sigma;\tau;4}) = u_4(u_{4,\sigma\tau} - \Gamma^{\nu}_{\sigma\tau}u_{4,\nu}) \tag{32.39}$$

are a system of differential equations for the function u_4 which can be completely integrated. The result (Stephani (1967a), Kramer et al. (1972)) is the metric

$$\begin{aligned} \mathrm{d}s^2 &= \frac{\mathrm{d}r^2}{1 - C^2r^2} + r^2(\mathrm{d}\vartheta^2 + \sin^2\vartheta\,\mathrm{d}\varphi^2) - (u_4)^2\,\mathrm{d}t^2, \\ u_4 &= rf_1(t)\sin\vartheta\sin\varphi + rf_2(t)\sin\vartheta\cos\varphi + rf_3(t)\cos\vartheta \\ &\quad + f_4(t)\sqrt{1 - C^2r^2} - C^{-1}, \\ \varkappa_0\mu &= 3C^2 = \text{const}, \qquad \varkappa_0 p = -\varkappa_0\mu + 2Cu^4, \qquad A = u^4\ (\neq \text{const}). \end{aligned} \tag{32.40}$$

We see that all these expansionfree conformally flat solutions are generalizations of the interior Schwarzschild solution (14.14) ($f_1 = f_2 = f_3 = 0$, $f_4 = \text{const}$), containing four arbitrary functions of time. In the general case, the metric admits no Killing vector at all.

No dust solutions ($p = 0$) are included in the class (32.40): $p = 0$, $\mu \neq 0$ leads to $2A = 3C = \text{const}$, and $\dot{\boldsymbol{u}}$ vanishes because of (32.35), i.e. $\boldsymbol{u}$ is covariantly constant. This is compatible with the form (32.31) of the curvature tensor only if $A = C = 0$; i.e. space-time must be flat.

The case $A = C$ ist the Einstein universe

$$\mathrm{d}s^2 = \mathrm{d}r^2/(1 - C^2r^2) + r^2(\mathrm{d}\vartheta^2 + \sin^2\vartheta\,\mathrm{d}\varphi^2) - \mathrm{d}t^2. \tag{32.41}$$

Solutions with non-vanishing expansion. For $\Theta \neq 0$, we get from (32.33)–(32.35) the system

$$3\,\partial_t h_{\mu\nu} = -2\Theta u_4 h_{\mu\nu}, \qquad A_{,\mu} = Au^4u_{4,\mu}, \qquad 3\,\partial_t C = \Theta A u_4, \tag{32.42}$$

which is integrated by

$$\begin{aligned} h_{\mu\nu} &= \frac{\bar{h}_{\mu\nu}(x^\sigma)}{V^2(x^\sigma, t)}, \qquad \frac{\partial_t V}{V} = \frac{\Theta}{3}u_4, \qquad A = u^4 f(t), \\ \Theta &= \Theta(t), \qquad C = C(t), \qquad \partial_t C = -\Theta f/3. \end{aligned} \tag{32.43}$$

The spatial part $h_{\mu\nu}$ of the metric tensor is time-dependent; hence we have to calculate the curvature tensor of the hypersurfaces $t = \text{const}$ by means of

$$\overset{3}{R}_{\mu\nu\sigma\tau} = \overset{4}{R}_{\mu\nu\sigma\tau} - \frac{1}{4}\left[(\partial_t h_{\mu\sigma})(\partial_t h_{\nu\tau}) - (\partial_t h_{\mu\tau})(\partial_t h_{\sigma\nu})\right](u^4)^2. \tag{32.44}$$

Using (32.42) and (32.31), we get from this representation

$$\overset{3}{R}_{\mu\nu\sigma\tau} = (C^2 - \Theta^2/9)\,(h_{\mu\sigma}h_{\nu\tau} - h_{\mu\tau}h_{\nu\sigma}); \tag{32.45}$$

the hypersurfaces $t = \text{const}$ are spaces of constant curvature, but with time-dependent metric. Transforming the metric $h_{\mu\nu}$ for $t = 0$ into the canonical form $h_{\mu\nu} = V^{-2}\delta_{\mu\nu}$, we find that, due to (32.42), this form is preserved in time, and the complete metric takes the form (Stephani (1967a))

$$\begin{aligned} \mathrm{d}s^2 &= V^{-2}(\mathrm{d}x^2 + \mathrm{d}y^2 + \mathrm{d}z^2) - \left(\frac{3V_{,4}}{\Theta(t)\,V}\right)^2 \mathrm{d}t^2, \\ V &= V_0(t) + \frac{C^2(t) - \Theta^2(t)/9}{4V_0(t)}\,\{[x - x_0(t)]^2 + [y - y_0(t)]^2 + [z - z_0(t)]^2\}, \\ \varkappa_0\mu &= 3C^2(t), \qquad \varkappa_0 p = -\varkappa_0\mu + 2CC_{,4}V/V_{,4} = C(2A - 3C). \end{aligned} \tag{32.46}$$

The remaining Gauss equations being satisfied, the metrics (32.46) are all of class one, with arbitrary functions $x_0(t)$, $y_0(t)$, $z_0(t)$, $C(t)$, $\Theta(t)$ and $V_0(t)$. This class is a generalization of the Robertson-Walker cosmological models (§ 12.2.).

Dust solutions of this class are obtained by putting $2A = 3C$. This implies $u_4 = -1$ and $3\,\partial_t V = -V\Theta$, which exactly characterizes the Friedmann dust models.

32.4.3. Type D perfect fluid solutions of class one

The type D solutions of class one are characterized by (32.28), i.e. by

$$\begin{aligned} \Omega_{ab} &= 2Cu_a u_b + Cg_{ab} + Av_a v_b, \qquad AC \neq 0, \\ \varkappa_0 T_{ab} &= e2C(A + 2C)\,u_a u_b + eC^2 g_{ab}, \end{aligned} \tag{32.47}$$

$$R_{abcd} = e(\Omega_{ac}\Omega_{bd} - \Omega_{ad}\Omega_{bc}). \tag{32.48}$$

For $A + 2C = 0$, the Codazzi equations (32.25) give $u_{a;b} = v_a p_b$, $v_{a;b} = u_a p_b$. The Ricci identities $u_{a;bc} - u_{a;cb} = u^d R_{dabc}$, together with the Gauss equations (32.48), then imply $C = 0$; i.e. space-time is flat. We can therefore confine the discussion to the case $2C + A \neq 0$.

As with the conformally flat solutions, two different cases occur, depending now on whether the acceleration $\dot{u}_a = u_{a;b}u^b$ vanishes or not. The two cases will be treated separately.

Solutions with vanishing acceleration. After some lengthy but straightforward calculation, the Codazzi equations and parts of the Gauss equations (via the Ricci identity for u_a) give the expressions

$$\begin{aligned} u_{a;b} &= a_1 v_a v_b + a_2(g_{ab} - v_a v_b + u_a u_b), \\ v_{a;b} &= a_1 u_a v_b + p_a v_b, \qquad p_a u^a = p_a v^a = 0, \\ A_{,a} &= [(2C + A)\,a_1 - 2Ca_2]\,u_a + a_3 v_a + Ap_a, \\ C_{,a} &= 2Ca_2 u_a, \end{aligned} \tag{32.49}$$

for the derivatives of u_a, v_a, C and A. In adapted coordinates $u_i = (0, 0, 0, -1)$, $v_i = (0, 0, V, 0)$, the metric then reads

$$\mathrm{d}s^2 = F^2(t)\,[\mathrm{d}x^2 + H^2(x, y)\,\mathrm{d}y^2] + V^2(x, y, z, t)\,\mathrm{d}z^2 - \mathrm{d}t^2. \tag{32.50}$$

The system of embedding equations, i.e. (32.49) and the Gauss equations, can be completely integrated. It turns out that the $x - y$-space is a space of constant curvature.

If this space is flat, then only $e = 1$ is possible and we get the metric (Stephani (1968))

$$\begin{aligned}
&\mathrm{d}s^2 = t(\mathrm{d}r^2 + r^2\,\mathrm{d}\varphi^2) + V^2\,\mathrm{d}z^2 - \mathrm{d}t^2,\\
&V(r, \varphi, z, t) = t\sqrt{t}\,G_1(z) + \sqrt{t}\,[G_2(z)\,r\cos\varphi + G_3(z)\,r\sin\varphi\\
&\qquad + \frac{3}{4}\,G_1(z)\,r^2 + G_4(z)] + G_5(z),\\
&\varkappa_0 p = \frac{1}{4t^2}, \quad \varkappa_0\mu = 3\varkappa_0 p - \frac{4t\sqrt{t}\,G_1(z) + G_5(z)}{2t^2 V(r, \varphi, z, t)}, \quad G_1{}^2 + G_5{}^2 \neq 0.
\end{aligned} \tag{32.51}$$

This metric contains five arbitrary functions $G_i(z)$. The subclass $G_1 = G_2 = G_3 = 0$ is a metric of plane symmetry.

If the $x - y$-space is non-flat, one finally gets (Barnes (1974), Stephani (1968))

$$\begin{aligned}
&\mathrm{d}s^2 = F^2(t)\left[\frac{\mathrm{d}r^2}{1 + \varepsilon r^2} + r^2\,\mathrm{d}\varphi^2\right] + V^2(r, \varphi, z, t)\,\mathrm{d}z^2 - \mathrm{d}t^2,\\
&V = G_1(z)\int F^{-1}\,\mathrm{d}t + G_2(z) + F\Big[G_3(z)\,r\cos\varphi + G_4(z)\,r\sin\varphi\\
&\qquad + G_5(z)\sqrt{(1 + \varepsilon r^2)}\Big],\\
&\varkappa_0\mu = 3\varkappa_0 p + \frac{2eb}{VF^4}\left[G_1(z)\left\{\varepsilon tF - eb^2\int F^{-1}\,\mathrm{d}t\right\} - eb^2 G_2(z)\right],\\
&\varkappa_0 p = eb^2F^{-4}, \qquad F^2 = \varepsilon(t^2 - eb^2), \qquad \varepsilon = \pm 1,\\
&b = \text{const}, \qquad G_1{}^2 + G_2{}^2 \neq 0.
\end{aligned} \tag{32.52}$$

The metrics (32.51) and (32.52) cover all type D class one perfect fluid solutions without acceleration.

Solutions with acceleration. Evaluation of the Codazzi equations (32.25) for the tensor Ω_{ab} given in (32.47) and partial use of the Ricci identities yields

$$\begin{aligned}
u_{a;b} &= a_1 v_a u_b + a_2 v_a v_b + a_3(g_{ab} - v_a v_b + u_a u_b), \qquad a_1 \neq 0,\\
v_{a;b} &= a_1 u_a u_b + a_2 u_a v_b + (2C + A)\,A^{-1}a_1(g_{ab} - v_a v_b + u_a u_b),\\
C_{,b} &= 2Ca_3 u_b + (2C + A)\,a_1 v_b,\\
A_{,b} &= [(2C + A)\,a_2 - 2Ca_3]\,u_b + a_4 v_b.
\end{aligned} \tag{32.53}$$

The preferred vector field $\boldsymbol{v}$ is parallel to the acceleration $\dot{\boldsymbol{u}}$ and both fields are normal; the space orthogonal to them turns out (via the Gauss equations) to be of constant curvature. So we are led to introduce a coordinate system

$$ds^2 = F^2(z, t)\left[\frac{d\varrho^2}{1 - \varepsilon\varrho^2} + \varrho^2 \, d\varphi^2\right] + V^2(z, t) \, dz^2 - U^2(z, t) \, dt^2, \qquad (32.54)$$

$$u_i = (0, 0, 0, -U), \qquad v_i = (0, 0, V, 0).$$

These metrics obviously admit a G_3 on V_2, cf. Chapter 13. The most general class one metric of this form is not known; the Gauss and Codazzi equations have been solved only for a shearfree velocity field, i.e. for $a_2 = a_3$. In that case, it turns out that a_2 and a_3 must vanish; the metric is necessarily static. The embedding equations then give (Kohler and Chao (1965), Stephani (1967a))

$$ds^2 = \varepsilon \frac{(K + 2kr^2)}{K + kr^2} \, dr^2 + r^2 \left(\frac{d\varrho^2}{1 - \varepsilon\varrho^2} + \varrho^2 \, d\varphi^2\right) - (K + kr^2) \, dt^2,$$

$$\varkappa_0 p = \frac{\varepsilon ek}{K + 2kr^2}, \qquad \varkappa_0 \mu = \varepsilon ek \frac{3K + 2kr^2}{(K + 2kr^2)^2}, \qquad (32.55)$$

$$K, k = \text{const}, \qquad e = \pm 1, \qquad \varepsilon = \pm 1.$$

32.4.4. Pure radiation field solutions of class one

As mentioned above, a null field solution, $T_{ab} = \Phi^2 k_a k_b$, can be of class one only if either

$$\Omega_{ab} = A k_a k_b + C z_a z_b, \qquad eAC = \varkappa_0 \Phi^2 > 0, \qquad (32.56)$$

or

$$\Omega_{ab} = B(k_a z_b + z_a k_b), \qquad -eB^2 = \varkappa_0 \Phi^2 > 0, \qquad (32.57)$$

holds ($k_a z^a = 0 = k_a k^a$, $z_a z^a = 1$). In both cases the Gauss equations give

$$R_{abcd} = \varkappa_0 \Phi^2 (k_a k_c z_b z_d + z_a z_c k_b k_d - k_a k_d z_b z_c - z_a z_d k_b k_c), \qquad (32.58)$$

and because of

$$C_{abcd} k^d = 0 = R_{ab} k^b \qquad (32.59)$$

these solutions are of Petrov type N.

The null vector field $\boldsymbol{k}$, being the eigenvector field of a type N metric satisfying (32.58), is geodesic and shearfree. Moreover, in both the cases (32.56) and (32.57) the Codazzi and Gauss equations imply that $\boldsymbol{k}$ is also normal and non-diverging. Hence all null field solutions of embedding class one necessarily belong to Kundt's class (cf. Chapter 27.).

The case $\Omega_{ab} = A k_a k_b + C z_a z_b$. Here the Codazzi and Gauss equations show that (in a suitable gauge) the null vector $\boldsymbol{k}$ is constant. As shown in § 21.5., pure radiation fields of type N possessing a constant null vector can be transformed to

$$ds^2 = dx^2 + dy^2 - 2 \, du \, dv - 2H(x, y, u) \, du^2,$$

$$k_i = (0, 0, 0, -1), \qquad \varkappa_0 \Phi^2 = H_{,11} + H_{,22}, \qquad z_i = (z_1, z_2, 0, z_4). \qquad (32.60)$$

Table 32.2 Class one solutions

T_{ab}	Petrov type	Ω_{ab}	Kinematical classification		Metrics	Type of metric	General solution known
Perfect fluid:			—		(8.35)	de Sitter	Yes
			$\omega_{ab} = 0$	$\Theta = 0$	(32.40)	generalized interior Schwarzschild	Yes
	O	$Au_a u_b + Cg_{ab}$	$\sigma_{ab} = 0$	$\Theta \neq 0$	(32.46)	generalized Friedmann	Yes
$pg_{ab} + (\mu + p)u_a u_b$	D	$2Cu_a u_b + Cg_{ab} + Av_a v_b$	$\omega_{ab} = 0$	$\dot{u}_a = 0$	(32.51), (35.52)	inhomogeneous cosmological models	Yes
				$\dot{u}_a \neq 0$ $\sigma_{ab} = 0$	(32.55)	static, spherical symmetry	Yes
				$\dot{u}_a \neq 0$ $\sigma_{ab} \neq 0$	?	?	No
Pure radiation:	N	$Ak_a k_b + Cz_a \dot{z}_b$	$k_{a;b} = 0$		(32.61)–(32.62)	Einstein-Maxwell plane wave	No
$\Phi^2 k_a k_b$		$B(k_a z_b + z_a k_b)$	$k_{a;b} = 0$		(32.63)–(32.64)	Einstein-Maxwell plane wave	Yes
			$\Theta = \sigma = \omega = \varkappa = 0$		(32.65)	plane wave, no Maxwell field	Yes

The general solution of the Gauss-Codazzi equations is not known. The metric

$$\begin{aligned} ds^2 &= dr^2 + r^2\, d\varphi^2 - 2\, du\, dv - 2[r\alpha(u) + \beta(u)]\, du^2, \\ k_i &= (0, 0, 0, -1), \qquad z_i = (0, r, 0, 0), \\ eA &= \alpha(u), \qquad C = r^{-1}, \qquad \varkappa_0\Phi^2 = \alpha r^{-1}, \end{aligned} \tag{32.61}$$

is a special solution, the electromagnetic field tensor being

$$\begin{aligned} F_{ab} &= k_a p_b - k_b p_a, \qquad \sqrt{r}\, p_i = \sqrt{\alpha}\, (\cos\psi, -r\sin\psi, 0, 0), \\ \psi &= \varphi/2 + \delta(u). \end{aligned} \tag{32.62}$$

The case $\Omega_{ab} = B(k_a z_b + z_a k_b)$. Two different types belong to this class, depending on whether the null vector $\boldsymbol{k}$ is constant or not.

If $\boldsymbol{k}$ is constant, then the Gauss-Codazzi equations admit only the metric (Collinson (1968b))

$$\begin{aligned} ds^2 &= dx^2 + dy^2 - 2\, du\, dv - [\alpha(u)\, x + \beta(u)\, y]^2\, du^2, \\ e &= -1, \qquad \varkappa_0\Phi^2 = \alpha^2 + \beta^2. \end{aligned} \tag{32.63}$$

The corresponding electromagnetic null field is

$$F_{ab} = k_a p_b - p_a k_b, \qquad p_i = (\alpha\cos\varphi, \beta\sin\varphi, 0, 0), \qquad \varphi = \varphi(u). \tag{32.64}$$

If $\boldsymbol{k}$ is not constant, then the metric is necessarily of the form (Collinson (1968b))

$$\begin{aligned} ds^2 &= dx^2 + dy^2 + \frac{4v}{x}\, du\, dx - 2\, du\, dv \\ &\quad - 2\left[\alpha(u) + \beta(u)\, xy + \delta(u)\, x - \frac{1}{2}\frac{v^2}{x^2}\right] du^2, \\ \varkappa_0\Phi^2 &= 2\alpha/x^2. \end{aligned} \tag{32.65}$$

No Einstein-Maxwell field with metric (32.65) exists, because the condition of integrability $\Delta \ln \Phi^2 = 0$ cannot be satisfied.

32.5. Exact solutions of embedding class two

32.5.1. The Gauss-Codazzi-Ricci equations and theorems on class two metrics

According to the general theory outlined in § 35.2., a V_4 is of embedding class two if and only if there exist two symmetric tensors $\Omega_{ab}(= \Omega^1{}_{ab})$, $\Lambda_{ab}(= \Omega^2{}_{ab})$, and a vector t_a satisfying

$$R_{abcd} = e_1(\Omega_{ac}\Omega_{bd} - \Omega_{ad}\Omega_{bc}) + e_2(\Lambda_{ac}\Lambda_{bd} - \Lambda_{ad}\Lambda_{bc}), \tag{32.66}$$

$$\Omega_{ab;c} - \Omega_{ac;b} = e_2(t_c\Lambda_{ab} - t_b\Lambda_{ac}),$$
$$\Lambda_{ab;c} - \Lambda_{ac;b} = -e_1(t_c\Omega_{ab} - t_b\Omega_{ac}), \tag{32.67}$$

$$t_{a;b} - t_{b;a} = \Omega_{ac}\Lambda^c{}_b - \Lambda_{ac}\Omega^c{}_b. \tag{32.68}$$

The tensors Ω_{ab}, Λ_{ab} and t_a are not uniquely determined. If a set $(\Omega_{ab}, \Lambda_{ab}, t_a, e_1, e_2)$ satisfies the embedding equations (32.66)–(32.68), then so does $(\bar{\Omega}_{ab}, \bar{\Lambda}_{ab}, \bar{t}_a, \bar{e}_1, \bar{e}_2)$ given by

$$\bar{\Omega}_{ab} = A\Omega_{ab} + B\Lambda_{ab}, \qquad \bar{\Lambda}_{ab} = C\Omega_{ab} + D\Lambda_{ab},$$
$$\bar{t}_a = (AD - BC)\, t_a + e_1 CA_{,a} + e_2 DB_{,a}, \tag{32.69a}$$

if the functions A, B, C, D satisfy

$$e_1 = \bar{e}_1 A^2 + \bar{e}_2 C^2, \qquad e_2 = \bar{e}_1 B^2 + \bar{e}_2 D^2, \qquad e_1 AC + e_2 BD = 0,$$
$$\bar{e}_1 = e_1 A^2 + e_2 B^2, \qquad \bar{e}_2 = e_1 C^2 + e_2 D^2, \qquad \bar{e}_1 AB + \bar{e}_2 CD = 0. \tag{32.69b}$$

This transformation corresponds to a rotation (pseudo-rotation) in the two-dimensional space orthogonal to the V_4 in the embedding space E_6; it can be used to simplify Ω_{ab} and Λ_{ab}, the curvature tensor being given.

Some simple purely algebraic conditions can be derived from the embedding equations. Using the property

$$\varepsilon^{abcd}\Omega_{an}\Omega_{bm}\Omega_{cp}\Omega_{dq} = \varepsilon_{nmpq}\Omega/g \tag{32.70}$$

of the ε-tensor (Ω being the determinant of Ω_{ab}), one obtains (Yakupov (1968a, 1968b))

$$\varepsilon^{abcd}\varepsilon^{nmrs}\varepsilon^{pqik}R_{abnm}R_{cdpq}R_{rsik} = 0. \tag{32.71}$$

Combining the Gauss and Ricci equations, one gets (Matsumoto (1950))

$$-\frac{1}{2}\, e_1 e_2 \varepsilon^{idmn} R^{ab}{}_{cd} R_{mnab} = \varepsilon^{idmn}(t_{c;d} - t_{d;c})\,(t_{m;n} - t_{n;m}). \tag{32.72}$$

Because of the skew symmetry of the tensors involved, the only information contained in (32.72) is

$$\varepsilon^{cdmn} R_{abcd} R^{ab}{}_{mn} = -e_1 e_2 8 \varepsilon^{cdmn} t_{c;d} t_{m;n}. \tag{32.73}$$

As the curvature tensor and the Weyl tensor always satisfy

$$\varepsilon^{edmn} R_{abcd} R^{ab}{}_{mn} = \varepsilon^{edmn} C_{abcd} C^{ab}{}_{mn}, \tag{32.74}$$

the curvature tensor may be replaced in (32.73) by the Weyl tensor (Goenner (1973)).

Theorem 32.12. *If a V_4 is of class two, it necessarily satisfies* (32.71) *and* (32.73).

A thorough investigation of class two space-times has been carried out only for vacuum solutions and conformally flat solutions. We will summarize the results in the following sections.

32.5.2. Vacuum solutions of class two

In addition to the algebraic conditions (32.71) and (32.73) imposed on the curvature tensor R_{abcd} and on $t_{c;d}$, another simple condition can be derived in the vacuum case from the Gauss-Codazzi-Ricci equations (Yakupov (1968a, 1968b)). Calculating the derivatives $\Omega_{a[b;c;i]}$ and $\Lambda_{a[b;c;i]}$ either by means of the Codazzi and Ricci equations or by means of the Ricci identity, one gets

$$\begin{aligned} 2e_2\Lambda_{a[b}t_{c;d]} &= R_{ea[bc}\Omega^e{}_{d]}\,, \\ -2e_1\Omega_{a[b}t_{c;d]} &= R_{ea[bc}\Lambda^e{}_{d]}\,, \end{aligned} \tag{32.75}$$

and after summation over a, c and insertion of $R_{nm} = 0$,

$$\begin{aligned} \Omega^a_a\,(t_{c;b} - t_{b;c}) + \Omega^a_b(t_{a;c} - t_{c;a}) - \Omega^a_c(t_{a;b} - t_{b;a}) &= 0, \\ \Lambda^a_a(t_{c;b} - t_{b;c}) + \Lambda^a_b(t_{a;c} - t_{c;a}) - \Lambda^a_c(t_{a;b} - t_{b;a}) &= 0. \end{aligned} \tag{32.76}$$

Multiplication of these equations with Ω^c_d and Λ^c_d, respectively, and antisymmetrization leads to

$$\begin{aligned} -(\Omega^a_a)^2(t_{b;d} - t_{d;b}) + (t_{a;c} - t_{c;a})\,(\Omega^a_b\Omega^c_d - \Omega^a_d\Omega^c_b) \\ -\Omega^a_c\Omega^c_d(t_{a;b} - t_{b;a}) + \Omega^a_c\Omega^c_b(t_{a;d} - t_{d;a}) = 0 \end{aligned} \tag{32.77}$$

and an equivalent equation for Λ_{ab}. Making use of the Gauss equations (32.66) and of

$$R_{ab} = e_1(\Omega^c_c\Omega_{ab} - \Omega^c_a\Omega_{cb}) + e_2(\Lambda^c_c\Lambda_{ab} - \Lambda_{ac}\Lambda^c{}_b) = 0, \tag{32.78}$$

one finally gets from (32.77)

Theorem 32.13. *A vacuum space-time of class two necessarily satisfies*

$$R^{ac}{}_{bd}t_{a;c} = 0. \tag{32.79}$$

Starting with the normal form (Table 4.2) of the curvature tensor for each Petrov type, the purely algebraic relations (32.71), (32.73) and (32.79) can be used to get further information about the algebraic structure of the curvature tensor and the tensor $t_{a;b}$, which will simplify the evaluation of the Gauss-Codazzi-Ricci equations.

Following this line of investigation, Yakupov (1973) arrived at the following results (published without proof): (a) there are no class two vacuum solutions of Petrov type *III*; (b) in all class two vacuum solutions $\boldsymbol{t}$ is a gradient, i.e. Ω_{ab} and Λ_{ab} commute, cf. (32.68).

If one evaluates the Gauss equations (32.66) in the Newman-Penrose formalism and partially uses the Codazzi equations (32.67), one gets (Collinson (1966))

Theorem 32.14. *If an algebraically special vacuum solution is of class two, then its multiple null eigenvectors are normal (and geodesic and shearfree); if it is non-degenerate (type II or III), then the null congruence has zero divergence too.*

The methods and results stated above should provide the necessary tools to find all class two vacuum solutions. No *complete* list of these metrics has yet been published.

The following vacuum solutions are known to be of class two (Rosen (1965), Collinson (1968a)): (i) Petrov type D: the Kasner solutions (11.50) and (11.51); the degenerate static solutions (classes A and B of Table 16.2), including the Schwarzschild solution (all admitting a G_4 on V_3). (ii) Petrov type N: plane-fronted waves with parallel rays (§ 21.5.).

32.5.3. Conformally flat solutions

By definition, conformally flat metrics can always be transformed to

$$ds^2 = e^{2U(x,y,z,t)}\,[dx^2 + dy^2 + dz^2 - dt^2]. \tag{32.80}$$

They can be characterized by the property that their Weyl tensor C_{abcd} vanishes, which in view of (3.50) is equivalent to

$$R_{abcd} = \frac{1}{2}\,(g_{ac}R_{bd} + R_{ac}g_{bd} - g_{ad}R_{bc} - R_{ad}g_{bc}) - \frac{R}{6}\,(g_{ac}g_{bd} - g_{ad}g_{bc}). \tag{32.81}$$

One easily sees from (32.81) that non-flat conformally flat vacuum solutions cannot exist. Even if a space-time is known to satisfy (32.81), the explicit transformation of its metric into the form (32.80) can be rather difficult to find.

As stated in Theorem 32.2, all conformally flat metrics are at most of embedding class two. In this section, we will use the technique of embedding to determine all conformally flat gravitational fields created by a perfect fluid or an electromagnetic field (Stephani (1967b)). The conformally flat Einstein-Maxwell fields have also been found by other techniques (Cahen and Leroy (1966), McLenaghan et al. (1975)).

Conformally flat solutions with a perfect fluid.

In the case of a perfect fluid

$$T_{ab} = (\mu + p)\,u_a u_b + p g_{ab}, \tag{32.82}$$

the condition (32.81) of zero Weyl tensor reads

$$\begin{aligned} 6R_{abcd} &= 3\varkappa_0(\mu + p)\,(g_{ac}u_b u_d + u_a u_c g_{bd} - g_{ad}u_b u_c - u_a u_d g_{bc}) \\ &\quad + 2\varkappa_0\mu(g_{ac}g_{bd} - g_{ad}g_{bc}). \end{aligned} \tag{32.83}$$

Comparing (32.82) with (32.31) together with (32.27), we see that the curvature tensor (32.83) can be written as

$$R_{abcd} = \Omega_{ac}\Omega_{bd} - \Omega_{ad}\Omega_{bc}, \tag{32.84}$$

with

$$\Omega_{ab} = A u_a u_b + C g_{ab}, \qquad \varkappa_0\mu = 3C^2, \qquad \varkappa_0 p = 2CA - 3C^2. \tag{32.85}$$

Equation (32.84), which is exactly the Gauss equation for a class one space-time, indicates that the conformally flat perfect fluid solutions may be of class one. To prove this conjecture, we have to show that the Codazzi equations $\Omega_{ab;c} = \Omega_{ac;b}$ are satisfied too.

If $A = C$, i.e. $\mu = -3p$, then the curvature tensor is simply

$$R_{abcd} = \varkappa_0\mu(h_{ac}h_{bd} - h_{ad}h_{bc})/3, \qquad h_{ab} = g_{ab} + u_a u_c, \tag{32.86}$$

and the Bianchi identities yield

$$3u_{a;b} = \Theta h_{ab}, \qquad 3\mu_{,a} = 2\mu\Theta u_a. \tag{32.87}$$

Because of these equations, Ω_{ab} satisfies the Codazzi equations and this implies that space-time is of embedding class one. In detail, for $\Theta = 0$ the corresponding solution is the Einstein universe (32.41), and for $\Theta \neq 0$ the solution is a special case of the generalized Friedmann universes (32.46).

If $A \neq C$ and $C \neq 0$, the tensor Ω_{ab} is non-singular, and because of Theorem 32.10 the Codazzi equations are satisfied and space-time is of class one. We summarize the results in

Theorem 32.15. *All conformally flat perfect fluid solutions* ($\mu \neq 0$) *are of embedding class one, and are therefore all contained either in the generalized interior Schwarzschild type metrics* (32.40) *or in the generalized Friedmann type metrics* (32.46) (Stephani (1967b)).

All conformally flat solutions with $\mu = 0$ (Barnes (1974)) are of embedding class two, because if μ was zero in (32.27), this would imply $C = 0$ and space-time would be flat.

Conformally flat solutions with electromagnetic non-null fields.

In the case of a non-null field (5.13), i.e.

$$T_{ab} = \Phi^2 \overset{(1)}{g}_{ab} - \Phi^2 \overset{(2)}{g}_{ab}, \tag{32.88}$$

$$g_{ab} = \overset{(1)}{g}_{ab} + \overset{(2)}{g}_{ab} = (x_a x_b + y_a y_b) - (z_a z_b - u_a u_b), \tag{32.89}$$

the curvature tensor has the form

$$R_{abcd} = \varkappa_0 \Phi^2 \left[\left(\overset{(1)}{g}_{ac}\overset{(1)}{g}_{bd} - \overset{(1)}{g}_{ad}\overset{(1)}{g}_{bc}\right) - \left(\overset{(2)}{g}_{ac}\overset{(2)}{g}_{bd} - \overset{(2)}{g}_{ad}\overset{(2)}{g}_{bc}\right)\right]. \tag{32.90}$$

If we start with a tetrad representation of Ω_{ab} and Λ_{ab}, use the identities

$$\begin{aligned}
&\varepsilon^{abcd}[(R_{abnm}\Omega_{cp}\Omega_{dq} - \Omega_{cq}\Omega_{dp}) + (\Omega_{an}\Omega_{bm} - \Omega_{am}\Omega_{bn}) R_{cdpq}] = 0,\\
&\varepsilon^{abcd}[(R_{abnm}\Lambda_{cp}\Lambda_{dq} - \Lambda_{cq}\Lambda_{dp}) + (\Lambda_{an}\Lambda_{bm} - \Lambda_{am}\Lambda_{bn}) R_{cdpq}] = 0,\\
&n, m, p, q \text{ not all different},
\end{aligned} \tag{32.91}$$

which follow from the Gauss equations (32.66), together with (32.70) and (32.90), and perform some suitable gauge transformations (32.69), a lengthy but straightforward calculation yields

$$\begin{aligned}
\Lambda_{ab} &= G u_a u_b + H(u_a z_b + z_a u_b) + K z_a z_b,\\
\Omega_{ab} &= E x_a x_b + F y_a y_b, \qquad e_1 EF = e_2(GK - H^2) = \varkappa_0 \Phi^2.
\end{aligned} \tag{32.92}$$

The Codazzi and Ricci equations (32.67)—(32.68) then tell us that t_a vanishes and

$$\overset{(1)}{g}_{ab;c} = 0 = \overset{(2)}{g}_{ab;c}, \qquad \Phi_{,c} = 0, \tag{32.93}$$

holds, i.e. the curvature tensor is constant:

$$R_{abcd;e} = 0. \tag{32.94}$$

The solution in question is a symmetric, decomposable, conformally flat space-time. As shown in § 31.2., the only metric with these properties is the Bertotti-Robinson metric (31.42), i.e.

$$ds^2 = dx^2 + \cos^2\left(\sqrt{\varkappa_0}\,\Phi x\right) dy^2 + \cos^2\left(\sqrt{\varkappa_0}\,\Phi t\right) dz^2 - dt^2. \tag{32.95}$$

The corresponding Maxwell field (determined only up to a duality rotation) is constant too, and can be written as

$$F_{ab} = \Phi\sqrt{2}\,(u_a z_b - z_a u_b), \qquad u_i = -\delta_i^4, \qquad z_i = \delta_i^3 \cos\left(\sqrt{\varkappa_0}\,\Phi t\right). \tag{32.96}$$

Theorem 32.16. *The only conformally flat solution with non-null electromagnetic field is the metric* (32.95)—(32.96), *which is the product of two-dimensional spaces of constant curvature. Both the curvature tensor and the electromagnetic field tensor are constant* (Cahen and Leroy (1966), Stephani (1967b)).

Conformally flat solutions with pure radiation or electromagnetic null fields. In the case of

$$T_{ab} = \Phi^2 k_a k_b \tag{32.97}$$

the curvature tensor of a conformally flat metric is given by

$$2R_{abcd} = \varkappa_0 \Phi^2 (k_a k_c g_{bd} + g_{ac} k_b k_d - k_a k_d g_{bc} - g_{ad} k_b k_c). \tag{32.98}$$

One can now use a tetrad representation of Ω_{ab} and Λ_{ab} with respect to the tetrad x_a, y_a, l_a, k_a, and evaluate the Gauss equations along the lines indicated above. The result is that only the following structures are admissible:

$$\begin{aligned} \Omega_{ab} &= C k_a k_b + D(x_a x_b + y_a y_b), \qquad 2e_1 CD = \varkappa_0 \Phi^2, \\ \Lambda_{ab} &= F x_a x_b + G y_a y_b, \qquad e_1 D^2 + e_2 FG = 0, \end{aligned} \tag{32.99}$$

and

$$\begin{aligned} \Omega_{ab} &= C k_a k_b + D(k_a y_b + y_a k_b) + E y_a y_b, \qquad -2e_2 F^2 = \varkappa_0 \Phi^2, \\ \Lambda_{ab} &= F(k_a x_b + x_a k_b), \qquad 2e_1(EC - D^2) = \varkappa_0 \Phi^2. \end{aligned} \tag{32.100}$$

In both cases the Codazzi equations yield

$$k_{a;b} = 0, \qquad \Phi_{,b} = \Phi' k_b; \tag{32.101}$$

the null eigenvector is constant (the same result can be obtained more easily if one evaluates the Bianchi identities for the curvature tensor (32.98) and uses $k_{a;[b;c]} = 0$).

Equation (32.101) is the key to finding the explicit solution; because of (32.101) and (32.97), the metric is a plane wave with parallel rays, cf. § 21.5., which is conformally flat only if

$$H_{,xx} = H_{,yy} = \varkappa_0 \Phi^2/2, \qquad H_{,xy} = 0. \tag{32.102}$$

Thus we arrive at the metric

$$ds^2 = dx^2 + dy^2 - 2\,du\,dv - \varkappa_0 \Phi^2(u)\,(x^2 + y^2)\,du^2/2. \tag{32.103}$$

The corresponding Maxwell field always exists and is given by

$$F_{ab} = \Phi(u)\,(k_a p_b - p_a k_b)$$
$$p_a = (\cos\varphi, \sin\varphi, 0, 0), \qquad \varphi = \varphi(u). \tag{32.104}$$

Theorem 32.17. *The only conformally flat solutions with a pure radiation or a null electromagnetic field are the special plane waves* (32.103)—(32.104) *with a constant null eigenvector* $\boldsymbol{k}$ (McLenaghan et al. (1975)).

These waves are the only conformally flat recurrent space-times with non-vanishing recurrence vector (§ 31.2.)

Table 32.3 Metrics known to be of class two. (+: Some of them have class one)

Type of metric	Metrics	Reference (for embedding)
Vacuum type D with G_4 on V_3	(24.20)	Rosen (1965) Collinson (1968a)
All conformally flat Einstein-Maxwell fields	(32.95)—(32.96) (32.103)—(32.104)	§ 32.5.3.
All pp waves (vacuum or not)+	$dx^2 + dy^2 - 2\,du\,dv - 2H\,du^2$	Collinson (1968a)
Melvin universe (Geon)	(20.10)	Collinson (1968a)
All metrics with a G_3 on a non-flat S_2 or T_2 +	Chapters 13., 14.	§ 32.2.3.

32.6. Exact solutions of embedding class $p > 2$

So far no systematic research has been done to find solutions of embedding class greater than two. The reason for this failure is probably the lack of any usable method for a systematic treatment of the embedding equations. In particular, the trick of first solving the Gauss equations by purely algebraical methods (thus getting the tensors $\Omega^{\alpha}{}_{ac}$ in terms of the energy-momentum tensor or in terms of the curvature tensor), and then solving the rest of the embedding equations, does not work here because the Gauss equations are too complicated.

We can only give some metrics where the embedding class is known either from general theorems as given in § 32.3. or from the explicit embedding. These metrics are:

$\underline{p = 3}$: (a) Static axisymmetric vacuum solutions (Weyl's class, cf. § 18.1. (Szekeres (1966a), Collinson (1968a)). Some of them have class $p = 2$. (b) The Petrov type *III* vacuum metric (27.40) with a G_2 (Collinson (1968a)). (c) The Gödel cosmos (10.25) (Collinson (1968a)).

$\underline{p = 4}$: All Robinson-Trautman solutions (Chapter 24.) (Collinson (1968a)).

32.7. Remarks on global embedding

Nearly all the work summarized in this chapter deals with local embedding only, i.e. the embedding of an open and simply connected neighbourhood of a point of the given V_4. In contrast, the global embedding of a V_4 can be considered. The number of extra dimensions needed for the embedding can be considerably higher than that for local embedding.

Global embedding of a V_4 may give a deeper insight into the geometrical properties of space-time. In fact, the maximal analytic extension of the Schwarzschild solution was found by the method of embedding (Fronsdal (1959)). No systematic analysis of global embedding of exact solutions has yet been done. Only an upper limit on the embedding class is known: a compact (non-compact) space-time V_4 is at most of embedding class $p = 46$ ($p = 87$).

Ref.: Friedman (1965), Penrose (1965), Clarke (1970), Greene (1970).

PART V. TABLES

Chapter 33. The interconnections between the main classification schemes

33.1. Introduction

As already pointed out in Chapter 1., the solutions of Einstein's field equations could be (and have been) classified according to (at least) four main classification schemes, namely with respect to symmetry groups, Petrov types, energy-momentum tensors, and special vector and tensor fields. Whereas the first two schemes have been used in extenso in this book, the others played only a secondary role, and the connections between Petrov types and groups of motions were also treated only occasionally.

This last chapter is devoted to the interconnections of the first three of the classification schemes mentioned above. It consists mainly of tables. § 33.2. gives the (far from complete) classification of the algebraically special solutions in terms of symmetry groups. § 33.3. contains four tables, wherein the solutions (and their status of existence or/and knowledge) are tabulated by all three combinations of energy-momentum tensors, Petrov types, and groups of motions.

For perfect fluid solutions, the connection between the kinematical properties of the four-velocity (see § 6.1.) and groups of motions was discussed e.g. by Ehlers (1961) and Wainwright (1979).

The reader who is interested in an introductory survey of the solutions as classified by some invariant properties should consult (besides the subject index and the table of contents)

§ 9.1. and		
Table 9.1, p. 116	for	metrics with groups of motions,
Table 10.1, p. 125	for	homogeneous solutions,
§§ 11.1. and 11.5.	for	hypersurface-homogeneous solutions (admitting a group G_3 on orbits V_3),
§§ 22.1.—22.3.	for	algebraically special solutions,
Table 25.1, p. 268	for	twisting algebraically special vacuum solutions,
Table 28.1, p. 312	for	Kerr-Schild-solutions,
Table 32.2, p. 366	for	solutions of embedding class 1,
Table 32.3, p. 373	for	solutions of embedding class 2.

In the tables of this chapter the following symbols are used:

∄: does not exist

A: all solutions are known

S: some (special) solutions are known.

Th., Ch. are abbreviations for "Theorem" and "Chapter" respectively.

33.2. The connection between Petrov types and groups of motions

In Parts II. and III. of this book, two methods for the invariant classification of gravitational fields, namely groups of motions and Petrov types, were treated quite independently. We have seen that many solutions admitting an isometry group are algebraically special and vice versa. Occasionally, if it was known to us, we mentioned the Petrov type of a solution classified according to the underlying group structure or referred to the group of motions of a solution of a certain Petrov type. In this section and in Table 33.3 we want to collect the known results on the connection between these two invariant classifications.

If one knows the group of motions and asks for the possible Petrov types, the following facts impose some restrictions.

(i) The existence of an *isotropy group* H_s (see § 9.2.) implies that the Weyl tensor is degenerate, i.e. the Petrov type is N, D or O. In particular, a group G_3 on non-null orbits V_2 implies type D or O (Theorem 13.1).

(ii) The *static* solutions are of type I, D or O (cp. § 16.6.1.).

(iii) The *stationary axisymmetric* vacuum solutions cannot be of type III (see § 17.5.). For all admissible Petrov types, the subclasses admitting a group G_r, $r \geqq 3$, were determined by Collinson and Dodd (1971).

Some *hypersurface-homogeneous* (cp. Chapter 11.) algebraically special Einstein spaces ($R_{ab} = \Lambda g_{ab}$) were determined by Siklos (1978). Apart from special plane waves (§ 21.5.) and the Λ-term solutions of (10.8), (11.42), (11.46), and (11.47), Siklos obtained the Petrov type III Robinson-Trautman solution (24.15), i.e.

$$\mathrm{d}s^2 = r^2 x^{-3}(\mathrm{d}x^2 + \mathrm{d}y^2) - 2\,\mathrm{d}u\,\mathrm{d}r + \frac{3}{2}\,x\,\mathrm{d}u^2, \tag{33.1}$$

where the Killing vectors ∂_y, ∂_u and $2(x\,\partial_x + y\,\partial_y) + r\,\partial_r - u\,\partial_u$ generate a group G_3VI_h, $h = -1/9$, its $\Lambda \neq 0$ generalization (11.45), and the homogeneous non-diverging solutions (10.33) and (10.34). These two latter solutions belong to Kundt's class (Chapter 27.) and are, in the metric form (27.7), given by $\left(\lambda = \sqrt{|\Lambda|},\ \Lambda < 0\right)$

$$\mathrm{d}s^2 = 2(\lambda x)^{-2}\,(\mathrm{d}x^2 + \mathrm{d}y^2) - 2\,\mathrm{d}u(\mathrm{d}v + 2v\,\mathrm{d}x/x + x\,\mathrm{d}u) \tag{33.2}$$

and

$$\mathrm{d}s^2 = 2(\lambda x)^{-2}\,(\mathrm{d}x^2 + \mathrm{d}y^2) - 2\,\mathrm{d}u(\mathrm{d}v + 2v\,\mathrm{d}x/x + (4x/3\lambda)\,\mathrm{d}y + 2x^4\,\mathrm{d}u). \tag{33.3}$$

They were found by Kaigorodov (1962) to be the type N and III Einstein spaces of maximum mobility (maximum order of G_r): the groups of motions are G_5 for (33.2) and G_4 for (33.3); in both cases there is a subgroup G_3VI_h, $h = -1/9$.

We now start with a specified Petrov type and ask for the groups of motions. For the various Petrov types of the algebraically special, diverging ($\varrho \neq 0$) *vacuum* fields, the dimension of the maximal group G_r and the corresponding solutions are given in Table 33.1.

The result that twisting type N vacuum solutions admit at most a G_1 is due to Collinson (1969). The other restrictions on the dimension of the maximal group were obtained by Kerr and Debney (1970), and confirmed by Held (1976a, b) using the modified version of the Newman-Penrose formalism (see § 7.3.).

The type D vacuum solutions admit either a G_2 or a G_4. The same holds true for the type D Einstein-Maxwell fields (19.6).

Table 33.1 The algebraically special, diverging vacuum solutions of maximum mobility

Petrov type	Twist	Maximal group	Solutions
N	$\omega \neq 0$	G_1	S (25.71)
	$\omega = 0$	G_2	S Table 33.2
III	$\omega \neq 0$	G_2	S (33.4), $m + \mathrm{i}M = 0$
	$\omega = 0$	G_3	A (33.1)
II	$\omega \neq 0$	G_2	S (25.60), (33.4)–(33.6)
	$\omega = 0$	G_2	S Table 33.2
D	$\omega \neq 0$	G_2	A § 25.5.
	$\omega = 0$	G_4	A (24.20)

Kerr and Debney (1970) determined all algebraically special, diverging, vacuum solutions admitting a group G_r, $r > 2$, and some solutions admitting a G_2. These latter solutions are (i) the Demiański solution (25.60), (ii) the solution

$$\begin{aligned} & m + \mathrm{i}M = m_0 + \mathrm{i}M_0 = \text{const}, \qquad P^2 = \left(\sqrt{2}\,x\right)^3 = (\zeta + \bar\zeta)^3, \\ & L = \mathrm{i}\,\frac{M_0}{6\sqrt{2}}\,x^{-3} + \mathrm{i}x^{-3/2}\left(C_1 x^{\sqrt{13}/2} + C_2 x^{-\sqrt{13}/2}\right), \\ & \qquad\qquad C_1,\ C_2 \text{ real constants}, \end{aligned} \tag{33.4}$$

which is a special member of (25.47) (the transformation (25.27), $f = \zeta$, $F_{,u} = 1$, relates (33.4) and (25.47)), and (iii) the solution

$$\begin{aligned} & P = 1, \qquad -m + \mathrm{i}M = 2A(1 - 3/\alpha)\,\zeta^{-3/\alpha}, \\ & L = A\bar\zeta^2\zeta^{1-3/\alpha} + B\bar\zeta\zeta^{-1/\alpha}, \qquad \operatorname{Re}\alpha = 1, \end{aligned} \tag{33.5}$$

where A and B are complex constants. (33.4) and (33.5) respectively admit groups G_2I and G_2II and belong to the classes (25.43)–(25.47) and (25.58)–(25.59). A vacuum solution of the class (25.43)–(25.47), with a G_2II, was found by □ Lun (1978): it reads

$$
\begin{aligned}
m + \mathrm{i}M &= (m_0 + \mathrm{i}M_0)\,\bar{\zeta}^{3/2}, \qquad P^2 = (\zeta + \bar{\zeta})^3, \\
L &= \frac{m + \mathrm{i}M}{3P^2} + x^{-3/2}\left[A\left(s + \sqrt{1+s^2}\right)^{\sqrt{13}/2}\left(s - \frac{\sqrt{13}}{2}\sqrt{1+s^2}\right)\right. \\
&\quad \left. + B\left(s + \sqrt{1+s^2}\right)^{-\sqrt{13}/2}\left(s + \frac{\sqrt{13}}{2}\sqrt{1+s^2}\right)\right], \\
s &\equiv y/x
\end{aligned}
\tag{33.6}
$$

(m_0, M_0, A, B real constants).

The classes (25.43)–(25.47) and (25.58)–(25.59) cover all algebraically special diverging vacuum solutions which admit a G_1 generated by an asymptotically timelike Killing vector field $\xi = \partial_u$ (Held (1976), Zenk and Das (1978)). The algebraically special vacuum solutions ($\varrho \neq 0$) with an orthogonally transitive G_2I (cp. § 8.6.) are of Petrov type D (Weir and Kerr (1977)).

The groups of motions of the Robinson-Trautman vacuum solutions (Chapter 24.) were systematically analysed by Collinson and French (1967) using a null tetrad formulation of the Killing equations. The results are given in Table 33.2.

Table 33.3 Petrov types versus groups G_r of motions (on non-null orbits V_d)

	V_4				V_3	
	G_7	G_6	G_5	G_4	G_6	G_4
I	$\nexists$ § 9.2.	$\nexists$ § 9.2.	$\nexists$ § 9.2.	(10.14), (10.26) –(10.31)	$\nexists$ § 9.2.	$\nexists$ § 9.2.
D	$\nexists$ § 9.2.	(10.8)	(10.25)		$\nexists$ § 9.2.	§§ 11.4., 12.3., 13.4., 14.1. (11.42), (13.47), (20.10) (24.42), (29.26)
II	$\nexists$ § 9.2.	$\nexists$ § 9.2.	$\nexists$ § 9.2.		$\nexists$ § 9.2.	$\nexists$ § 9.2.
N	$\nexists$ § 9.2.	(10.12)	(10.33)		$\nexists$ § 9.2.	
III	$\nexists$ § 9.2.	$\nexists$ § 9.2.	$\nexists$ § 9.2.	(10.34)	$\nexists$ § 9.2.	$\nexists$ § 9.2.
O	(10.23) (10.35)	(10.7) $A = 0$, (10.16)			§ 12.2.	(14.14)

Table 33.2 Robinson-Trautman vacuum solutions admitting two or more Killing vectors. The gauge $m = \text{const}$ is used. C.F. refers to equations in Collinson and French (1967). Type *III* solutions with G_2 exist in all three cases, but only (24.16) is explicitly known

	Structure of P	*II*	*D*	*III*	*N*
G_4	$P(\zeta, \bar{\zeta}) = 1 + \frac{K}{2} \zeta\bar{\zeta}$	$\nexists$	A (24.20)	$\nexists$	$\nexists$
G_3	$P(\zeta, \bar{\zeta}) = (\zeta + \bar{\zeta})^{3/2}$, $m = 0$	$\nexists$	$\nexists$	A (33.1)	$\nexists$
	$P(\zeta, \bar{\zeta})$	S (24.24)	$\nexists$		$\nexists$
G_2	$P(\zeta + \bar{\zeta}, u)$	S (24.25)	A (24.23)	S (24.16)	A C.F. (6.17)
	$P(\zeta + u, \bar{\zeta} + u)$ $P_{,\zeta} \neq P_{,\bar{\zeta}}$	$\nexists$	$\nexists$		A C.F. (6.20)–(6.22)

Table 33.3 (cont.)

V_3	V_2		V_1
G_3	G_3	G_2	G_1
§§ 11.3.3., 11.4. §§ 12.4., 20.2. (11.50)–(11.59)	$\nexists$ § 9.2.	§§ 15.1.–15.3., 18.6., §§ 19.2.1., 20.4., 20.5., 30.5. (18.2), (18.17), (18.21) (18.29), (19.2), (19.4) (19.38), (20.25), (20.40)	§ 15.4. (16.49), (16.50), (16.59), (16.70), (16.73), (22.17), $b = 0$, (22.18)
(16.67)	§§ 13.5., 14.2. (13.20), (13.48) (24.41), (32.51)	§ 25.5. (16.66), (16.68), (19.6) (19.19), (19.40), (24.43) (27.57), (30.55), (20.44)	
(11.46), (29.19)	$\nexists$ § 9.2.	§ 25.2.6. (18.23), (20.25), (24.24) (24.25), (25.60), (29.17) (33.4)–(33.6)	§§ 25.2.3., 25.2.5. (24.54), (26.44), (26.45) (28.92), (29.20), (29.36) (30.73)
(11.47), Table 21.1	$\nexists$ Th. 13.1	Table 21.1	(25.71)
(11.45), (11.47), (33.1)	$\nexists$ § 9.2.	(24.16), (27.40)	
			(29.36), $q = 1/2$

Table 33.4 Energy-momentum tensors versus groups G_r of motions (on null-non orbits V_d)

	V_4				V_3	
	G_7	G_6	G_5	G_4	G_6	G_4
Vacuum	$\nexists$ Th. 10.1	A (10.12), $a = 0$	$\nexists$ Th. 10.1	A (10.14)	$\nexists$ § 11.1.	A (24.20) §§ 11.3.1., 13.4.
Einstein-Maxwell non-null fields	$\nexists$ Th. 10.3	A (10.16)	$\nexists$ Th. 10.3	$\nexists$ Th. 10.3	$\nexists$ § 11.1.	A (24.42) §§ 11.3.1., 13.4.
null fields	$\nexists$ Th. 10.1	A (10.12)	$\nexists$ Th. 10.1	$\nexists$ Th. 10.1	$\nexists$ § 11.1.	
Pure radiation	A (10.35)	A (10.12)			$\nexists$ § 11.1.	
Λ-term	$\nexists$ § 10.5	A (10.8)	A (10.33)	A (10.34)	$\nexists$ § 11.1.	A § 11.3.1.
Perfect fluid	A (10.23)	$\nexists$ Th. 10.4	A (10.25)	A (10.26) –(10.31)	S § 12.2.	S § 12.3. § 14.1. (13.47)

To the authors' knowledge, the groups of motions of the algebraically special diverging non-empty spaces have not been investigated and the symmetries of the (non-diverging) solutions of Kundt's class (Chapter 27.) are also unknown except for the vacuum *pp* waves (cp. Table 21.1).

Table 33.3 gives the solutions listed in this book for which both the Petrov type and the symmetry group are known. Special cases of some solutions may admit a higher-dimensional group or/and their Petrov type may be more special. A variety of solutions are not contained in Table 33.3 because the group or the Petrov type is not known, but these solutions are partly covered by Tables 33.4–33.6.

33.3. Tables

The tables refer to the equation numbers of the solutions given in this book and/or the relevant sections. As a rule, reference to a section (§) includes all (or most of the) solutions given therein.

The symbols $\nexists$ (= non-existence), A (= All), S (= Some) are used as explained at the end of § 33.1.

Table 33.4 (cont.)

V_3	V_2		V_1
G_3	G_3	G_2	G_1
S § 11.3.2. (20.7), (33.1) Table 21.1	$\nexists$ Th. 13.5	S §§ 18.6., 25.2.6., 25.5. §§ 15.1.—15.3., 30.5.; Table 21.1 (18.2), (18.17), (18.21), (18.23), (18.29), (20.24), (20.25), (22.20), (24.16), (24.23)—(24.25), (25.60), (27.40), (33.4), (33.5)	S §§ 15.4., 25.2.3., § 25.2.5. (16.49), (16.50), (16.59), (22.17), $b = 0$, (22.18) (25.71)
S § 11.3.3. (20.9), (20.11), (20.12)	$\nexists$ Th. 13.5	S §§ 19.1.2, 26.5., (30.5); Th. 20.1 (19.2), (19.4), (19.6), (19.19), (24.43), (27.57)	S (16.73), (24.54)
	A (24.41)		
	A Table 13.1	S (20.40)	S (26.44), (26.45) (28.92)
S § 11.3.2.	$\nexists$ Th. 13.5		S (30.73)
S §§ 11.4., 12.4. (16.67), (20.13) (20.16), (20.18) (29.19)	S § 13.5., § 14.2., (13.48) (32.51)	S Th. 15.1; §§ 19.2., 20.5. (16.66), (16.68), (29.17)	S (16.70), (29.20) (29.36)

Table 33.5 Algebraically special vacuum, Einstein-Maxwell and pure radiation fields (non-aligned or with $\varkappa\bar{\varkappa} + \sigma\bar{\sigma} \neq 0$). No conformally flat solutions of these types exist

		II	*D*	*III*	*N*
Vacuum	$\varrho \neq \bar{\varrho}$	S Ch. 25 (25.42) (25.43)–(25.47) (25.50)–(25.55) (25.58)–(25.59)	A § 25.5., § 19.1. (25.73)–(25.74) (25.62), (25.65). (25.68) (25.75)	S § 25.4. (25.43)–(25.47) with $m + \mathrm{i}M = 0$	S § 25.3. (25.71)
	$\varrho = \bar{\varrho} \neq 0$	S Ch. 24, Th. 24.1 (24.24), (24.25)	A § 24.1. (24.20), (24.23)	S Th. 24.2 (24.15)	A § 24.1.
	$\varrho = 0$	S § 27.5., Th. 27.1	A § 27.5. (27.41)	A § 27.5. (27.34), (27.38)	A § 27.5., § 21.5. (21.38), (21.42) ($F = 0$) (27.38) with $W_0 = 0$
Aligned $\varkappa\bar{\varkappa} + \sigma\bar{\sigma} \neq 0$	E – M non-null	? § 22.1.	? § 22.1.	$\nexists$ § 22.1.	$\nexists$ § 22.1.
	E – M null	$\nexists$ Th. 7.4	$\nexists$ Th. 7.4	$\nexists$ Th. 7.4	$\nexists$ Th. 7.4
	pure radiation	$\nexists$ § 7.5.	$\nexists$ § 7.5.	$\nexists$ § 7.5.	S § 22.1. (22.10), (22.12) (22.13)–(22.14)
Non-aligned	E – M non-null	?	?	S § 22.1.	A § 22.1. (22.5)
	E – M null	$\nexists$ Th. 7.4	$\nexists$ Th. 7.4	$\nexists$ Th. 7.4	$\nexists$ Th. 7.4
	pure radiation	?	?	?	$\nexists$ § 22.1.

Table 33.6 Algebraically special (non-vacuum) Einstein-Maxwell and pure radiation fields, aligned and with $\varkappa\bar{\varkappa} + \sigma\bar{\sigma} = 0$

		II	*D*	*III*	*N*	*O*
E − M non-null	$\varrho \neq \bar{\varrho}$	S Th. 26.2 §§ 26.2.–26.4. (26.26)–(26.28)	S §§ 19.1., 26.5. (19.6), (26.29)	∄ § 26.5.	∄ Th. 26.3	∄ Th. 26.3, 32.16
	$\varrho = \bar{\varrho} \neq 0$	S Th. 24.3, 24.5 (24.53)–(24.57)	A § 24.2. (24.42)–(24.44)	∄ § 24.2.	∄ Th. 26.3	∄ Th. 26.3, 32.16
	$\varrho = 0$	S § 27.7.	S § 27.7. (27.55)–(27.57)	S § 27.7. (27.54)	A § 27.7. (27.46)	A Th. 32.16 (32.95)–(32.96)
E − M null	$\varrho \neq \bar{\varrho}$			§ 26.5.	∄ Th. 26.3	∄ Th. 26.3, 32.17
	$\varrho = \bar{\varrho} \neq 0$	S Th. 24.3 (24.51)	A § 24.2. (24.41)	S § 24.2. (24.40)	∄ Th. 26.3	∄ Th. 26.3, 32.17
	$\varrho = 0$	§ 27.6.			S §§ 27.6., 28.3. (28.73)–(28.74)	A Th. 32.17 (32.103)–(32.104)
pure radiation	$\varrho \neq \bar{\varrho}$	S § 26.1. (26.38)–(26.42) (26.43)–(26.45), (26.51)			∄ § 26.6. Th. 26.3	∄ Th. 32.17
	$\varrho = \bar{\varrho} \neq 0$	S Th. 24.6	A § 24.3. (24.62)–(24.64)	S § 24.3. (24.61)	∄ § 24.3.	∄ Th. 32.17
	$\varrho = 0$	S §§ 27.6., 28.4. (28.92)		A § 27.6.	A § 27.6.	A Th. 32.17 (32.103)–(32.104)

Bibliography

Abdussattar. *See* Singh and Abdussattar (1973, 1974)

Adler, R. J. *See* Schiffer et al. (1973)

Adler, R. J., and Sheffield, C. (1973). *Classification of space-times in general relativity*, J. Math. Phys. **14**, 465. *See* § 4.3.

Alekseev, G. A., and Khlebnikov, V. I. (1978). *Newmann-Penrose formalism and its application in general relativity* (in Russian), Fiz. Elem. Chast. i Atom. Yadra. **9** No. 5, 790. *See* § 7.2.

Armenti, A. *See* Gautreau et al. (1972)

Assad, M. J. D. *See* Damião Soares and Assad (1978).

Avakyan, R. M., and Horský, J. (1975). *The gravitational field of the homogeneous plane disk* (in Russian), Akad. Nauk Armjan. SSR, Astrofiz. **11**, 689. *See* § 13.6.

Baldwin, O. R., and Jeffery, G. B. (1926). *The relativity theory of plane waves*, Proc. Roy. Soc. Lond. **A 111**, 95. *See* § 21.5.

Banerjee, A. (1970a). *Null electromagnetic fields in general relativity admitting timelike or null Killing vectors*, J. Math. Phys. **11**, 51. *See* §§ 16.6., 21.4.

Banerjee, A. (1970b). *Null electromagnetic fields in general relativity*, J. Phys. **A 3**, 501. *See* § 21.4.

Banerjee, A. *See also* Chakravarty et al. (1976), De and Banerjee (1972)

Banerji, S. *See* Das and Banerji (1978)

Bardeen, J. M. (1970). *A variational principle for rotating stars in general relativity*, Astrophys. J. **162**, 71. *See* § 17.3.

Barnes, A. (1972). *Static perfect fluids in general relativity*, J. Phys. **A 5**, 374. *See* §§ 11.4., 16.6.

Barnes, A. (1973a). *On shearfree normal flows of a perfect fluid*, GRG **4**, 105. *See* § 29.3.

Barnes, A. (1973b). *On Birkhoff's theorem in general relativity*, Commun. Math. Phys. **33**, 75. *See* §§ 8.5., 13.4., 21.1.

Barnes, A. (1974). *Space-times of embedding class one in general relativity*, GRG **5**, 147. *See* §§ 5.1., 32.4., 32.5.

Barnes, A. (1977). *A class of algebraically general non-null Einstein-Maxwell fields. II*, J. Phys. **A 10**, 755. *See* § 11.3.

Barnes, A. (1978). *A class of homogeneous Einstein-Maxwell fields*, J. Phys. **A 11**, 1303. *See* §§ 11.2., 11.3.

Barnes, A. (1979). *On space-times admitting a three-parameter isometry group with two-dimensional null orbits*, J. Phys. **A 12**, 1493. *See* §§ 11.1., 21.1.

Barrow, J. D. (1978). *Quiescent cosmology*, Nature **272**, 211. *See* §§ 12.3., 12.4.

Bartrum, P. C. (1967). *Null electromagnetic field in the form of spherical radiation*, J. Math. Phys. **8**, 1464. *See* § 24.2.

Basey, T. C. (1975). *A note on the generation of twisting 'D' metrics from non-twisting 'D' metrics*, Z. Naturforsch. **30a**, 1200. *See* § 30.6.

Bashkov, V. I. (1976). *Classes of exact solutions of Einstein's equations* (in Russian), Itogi Nauki i Tekhniki, Ser. Algebra, Topologiya, Geometriya, **14**, VINITI, Moscow. *See* § 1.4.

Batakis, N., and Cohen, J. M. (1972). *Closed anisotropic cosmological models*, Ann. Phys. (USA) **73**, 578. *See* § 12.3.

Beck, G. (1925). *Zur Theorie binärer Gravitationsfelder*, Z. Phys. **33**, 713. *See* § 20.3.
Behr, C. G. *See* Estabrook et al. (1968)
Beiglböck, W. (1964). *Zur Theorie der infinitesimalen Holonomiegruppe in der Allgemeinen Relativitätstheorie*, Z. Phys. **179**, 148. *See* § 31.4.
Bel, L. (1958). *Étude algébrique d'un certain type de tenseurs de courbure. Le cas 3 de Petrov*, C. R. Acad. Sci. (Paris) **247**, 2096. *See* § 4.3.
Bel, L. (1959). *Quelques remarques sur la classification de Petrov*, C.R. Acad. Sci. (Paris) **248**, 2561. *See* § 4.3.
Bell, P., and Szekeres, P. (1972). *Some properties of higher spin rest-mass zero fields in general relativity*, Int. J. Theor. Phys. **6**, 111. *See* § 7.5.
Bell, P., and Szekeres, P. (1974). *Interacting electromagnetic shock waves in general relativity*, GRG **5**, 275. *See* § 15.2.
Benenti, S., and Francaviglia, M. (1979). *Remarks on certain separability structures and their applications to general relativity*, GRG **10**, 79. *See* § 31.3.
Berger, B. K. (1976). *Homothetic and conformal motions in spacelike slices of solutions of Einstein's equations*, J. Math. Phys. **17**, 1268. *See* § 31.4.
Bergmann, O. *See* Knight and Bergmann (1974)
Bergmann, P. G., Cahen, M., and Komar, A. B. (1965). *Spherically symmetric gravitational fields*, J. Math. Phys. **6**, 1. *See* § 13.4.
Bertotti, B. (1959). *Uniform electromagnetic field in the theory of general relativity*, Phys. Rev. **116**, 1331. *See* §§ 10.3., 19.1.
Bespal'ko, R. M. *See* Gutman and Bespal'ko (1967)
Białas, A. (1963). *Electromagnetic waves in general relativity as the source of information*, Acta Phys. Polon. **24**, 465. *See* § 28.3.
Bianchi, L. (1897). *Sugli spazii a tre dimensioni che ammettono un gruppo continuo di movimenti*, Soc. Ital. Sci. Mem. di Mat. **11**, 267. *See* § 8.2.
Birkhoff, G. D. (1923). *Relativity and Modern Physics*, Harvard Univ. Press, Cambridge, p. 255. *See* § 13.4.
Bitsadze, A. V. (1977). *On some classes of exact solutions to the gravitational field equations* (in Russian), Dokl. Akad. Nauk SSSR **233**, 517. *See* § 30.3.
Bonanos, S. (1976). *Hamilton-Jacobi separable, axisymmetric, perfect-fluid solutions of Einstein's equations*, Commun. Math. Phys. **49**, 53. *See* § 19.2.
Bondi, H. (1947). *Spherically symmetrical models in general relativity*, Mon. Not. Roy. Astr. Soc. **107**, 410. *See* § 13.5.
Bondi, H., Pirani, F. A. E., and Robinson, I. (1959). *Gravitational waves in general relativity. III. Exact plane waves*, Proc. Roy. Soc. Lond. **A 251**, 519. *See* § 21.5.
Bonnor, W. B. (1954). *Static magnetic fields in general relativity*, Proc. Phys. Soc. Lond. **A 67**, 225. *See* §§ 19.1., 20.2., 30.3.
Bonnor, W. B. (1961). *Exact solutions of the Einstein-Maxwell equations*, Z. Phys. **161**, 439. *See* §§ 17.5., 30.4.
Bonnor, W. B. (1966). *An exact solution of the Einstein-Maxwell equations referring to a magnetic dipole*, Z. Phys. **190**, 444. *See* §§ 19.1., 30.5.
Bonnor, W. B. (1969a). *A new interpretation of the NUT metric in general relativity*, Proc. Camb. Phil. Soc. **66**, 145. *See* § 18.3.
Bonnor, W. B. (1969b). *The gravitational field of light*, Commun. Math. Phys. **13**, 163. *See* § 21.5.
Bonnor, W. B. (1970). *On a Robinson-Trautman solution of Einstein's equations*, Phys. Letters **A 31**, 269. *See* § 24.1.
Bonnor, W. B. (1976). *Non-radiative solutions of Einstein's equations for dust*, Commun. Math. Phys. **51**, 191. *See* § 29.3.
Bonnor, W. B. (1977). *A rotating dust cloud in general relativity*, J. Phys. **A 10**, 1673. *See* § 19.2.
Bonnor, W. B., and Swaminarayan, N. S. (1964). *An exact solution for uniformly accelerated particles in general relativity*, Z. Phys. **177**, 240. *See* § 18.2.
Bonnor, W. B., and Swaminarayan, N. S. (1965). *An exact stationary solution of Einstein's equations*, Z. Phys. **186**, 222. *See* § 18.3.

Bonnor, W. B., and Tomimura, N. (1976). *Evolution of Szekeres's cosmological models*, Mon. Not. Roy. Astr. Soc. **175**, 85. *See* § 29.3.
Bonnor, W. B., and Vaidya, P. C. (1972). *Exact solutions of the Einstein-Maxwell equations for an accelerated charge*, in: O'Raifeartaigh, L. (Ed.), General Relativity (Papers in Honour of J. L. Synge), Clarendon Press, Oxford, p. 119. *See* §§ 28.1., 28.4.
Bonnor, W. B., Sulaiman, A. H., and Tomimura, N. (1977). *Szekeres's space-times have no Killing vectors*, GRG **8**, 549. *See* § 29.3.
Bose, S. K. (1975). *Studies in the Kerr-Newman metric*, J. Math. Phys. **16**, 772. *See* § 19.1.
Boyer, C. B., Kalnins, E. G., and Miller, W. (1978). *Separable coordinates for four-dimensional Riemannian spaces*, Commun. Math. Phys. **59**, 285. *See* § 31.3.
Boyer, R. H., and Lindquist, R. W. (1967). *Maximal analytic extension of the Kerr metric*, J. Math. Phys. **8**, 265. *See* § 18.5.
Brans, C. H. (1974). *Complex structures and representations of the Einstein equations*, J. Math. Phys. **15**, 1559. *See* § 4.2.
Brans, C. H. (1975). *Some restrictions on algebraically general vacuum metrics*, J. Math. Phys. **16**, 1008. *See* § 4.2.
Bray, M. (1971). *Sur certaines solutions du problème intérieur en relativité générale*, C. R. Acad. Sci. (Paris) **A 272**, 841. *See* § 20.5.
Brdička, M. (1951). *On gravitational waves*, Proc. Roy. Irish Acad. **A 54**, 137. *See* § 21.5.
Breuer, R. A. (1975). *Gravitational perturbation theory and synchrotron radiation*, Lect. Notes Phys. **44**, 66, Springer-Verlag, Berlin, Heidelberg, New York. *See* § 7.3.
Brickell, F., and Clark, R. S. (1970). *Differentiable manifolds: an introduction*, Van Nostrand Reinhold, London. *See* §§ 2.1., 8.1.
Brill, D. R. (1964). *Electromagnetic fields in a homogeneous, nonisotropic universe*, Phys. Rev. **B 133**, 845. *See* §§ 11.3., 19.1.
Brinkmann, M. W. (1923). *On Riemann spaces conformal to Euclidean spaces*, Proc. Natl. Acad. Sci. U.S. **9**, 1. *See* § 21.5.
Brown, E. *See* Zund and Brown (1971)
Buchdahl, H. A. (1954). *Reciprocal static solutions of the equations* $G_{\mu\nu} = 0$, Quart. J. Math. Oxford **5**, 116. *See* § 30.3.
Buchdahl, H. A. (1956). *Reciprocal static solutions of the equations of the gravitational field*, Austral. J. Phys. **9**, 13. *See* § 30.5.
Buchdahl, H. A. (1959). *General relativistic fluid spheres*, Phys. Rev. **116**, 1027. *See* § 14.1.
Buchdahl, H. A. (1964). *A relativistic fluid sphere resembling the Emden polytrope of index 5*, Astrophys. J. **140**, 1512. *See* § 14.1.
Buchdahl, H. A. (1967). *General-relativistic fluid spheres. III. A static gaseous model*, Astrophys. J. **147**, 310. *See* § 14.1.
Buchdahl, H. A., and Land, W. J. (1968). *The relativistic incompressible sphere*, J. Austr. Math. Soc. **8**, 6. *See* § 14.1.

Cahen, M. (1964). *On a class of homogeneous spaces in general relativity*, Bull. Acad. Roy. Belgique Cl. Sci. **50**, 972. *See* § 10.5.
Cahen, M. *See also* Bergmann et al. (1965), Debever and Cahen (1960, 1961)
Cahen, M., and Debever, R. (1965). *Sur le théorème de Birkhoff*, C.R. Acad. Sci. (Paris) **260**, 815. *See* § 13.4.
Cahen, M., and Defrise, L. (1968). *Lorentzian 4-dimensional manifolds with "local isotropy"*, Commun. Math. Phys. **11**, 56. *See* §§ 9.2., 10.3., 11.1., 11.3.
Cahen, M., and Leroy, J. (1965). *Ondes gravitationnelles en présence d'un champ électromagnétique*, Bull. Acad. Roy. Belgique Cl. Sci. **51**, 996. *See* §§ 22.1., 27.7.
Cahen, M., and Leroy, J. (1966). *Exact solutions of Einstein-Maxwell equations*, J. Math. Mech. **16**, 501. *See* §§ 22.1., 27.7., 32.5.
Cahen, M., and McLenaghan, R. G. (1968). *Métriques des espaces lorentziens symétriques à quatre dimensions*, C.R. Acad. Sci. (Paris) **A 266**, 1125. *See* § 31.2.
Cahen, M., and Sengier, J. (1967). *Espaces de classe D admettant un champ électromagnétique*, Bull. Acad. Roy. Belgique Cl. Sci. **53**, 801. *See* § 24.2.

Cahen, M., and Spelkens, J. (1967). *Espaces de type III solutions des équations de Maxwell-Einstein,* Bull. Acad. Roy. Belgique Cl. Sci. **53**, 817. *See* §§ 22.1, 27.5.

Cahen, M., Debever, R., and Defrise, L. (1967). *A complex vectorial formalism in general relativity,* J. Math. Mech. **16**, 761. *See* § 3.5.

Campbell, S. J., and Wainwright, J. (1977). *Algebraic computing and the Newman-Penrose formalism in general relativity,* GRG **8**, 987. *See* § 7.1.

Caplan, T. A. *See* Davies and Caplan (1971)

Caporali, A. (1978). *Non-existence for stationary, axially symmetric, asymptotically flat solutions of the Einstein equations for dust,* Phys. Letters **A 66**, 5. *See* § 19.2.

Carlson, G. T., and Safko, J. L. (1978). *An investigation of some of the kinematical aspects of plane symmetric space-times,* J. Math. Phys. **19**, 1617. *See* § 13.4.

Carmeli, M. (1977). *Group Theory and General Relativity,* McGraw-Hill, New York, p. 244. *See* § 22.3.

Carter, B. (1968a). *Hamilton-Jacobi and Schrödinger separable solutions of Einstein's equations,* Commun. Math. Phys. **10**, 280. *See* § 19.1.

Carter, B. (1968b). *A new family of Einstein spaces,* Phys. Letters **A 26**, 399. *See* §§ 13.4., 18.5., 19.1., 25.5., 27.5.

Carter, B. (1970). *The commutation property of a stationary, axisymmetric system,* Commun. Math. Phys. **17**, 233. *See* § 17.1.

Carter, B. (1972). *Black hole equilibrium states,* in: DeWitt, B., and D. (Eds.), Black Holes (Les Houches Lectures 1972), Gordon and Breach, New York. *See* §§ 16.4., 17.2., 18.5., 19.1.

Case, L. A. (1968). *Changes of the Petrov type of a space-time,* Phys. Rev. **176**, 1554. *See* § 4.2.

Chakravarty, N., Dutta Choudhoury, S. B., and Banerjee, A. (1976). *Nonstatic spherically symmetric solutions for a perfect fluid in general relativity,* Austral. J. Phys. **29**, 113. *See* § 14.2.

Chandrasekhar, S. (1978). *The Kerr metric and stationary axisymmetric gravitational fields,* Proc. Roy. Soc. Lond. **A 358**, 405. *See* §§ 17.5., 18.5.

Chao, K. L. *See* Kohler and Chao (1965)

Chazy, J. (1924). *Sur la champ de gravitation de deux masses fixes dans la théorie de la relativité,* Bull. Soc. Math. France **52**, 17. *See* § 18.1.

Chernov, A. S. *See* Kompaneets and Chernov (1964)

Chinnapared, K. *See* Newman et al. (1965)

Chitre, D. M. *See* Kinnersley and Chitre (1977—1978, 1978)

Chitre, D. M., Güven, R., and Nutku, Y. (1975). *Static cylindrically symmetric solutions of the Einstein-Maxwell equations,* J. Math. Phys. **16**, 475. *See* § 20.2.

Churchill, R. V. (1932). *Canonical forms for symmetric linear vector functions in pseudo-euclidean space,* Trans. Amer. Math. Soc. **34**, 784. *See* § 5.1.

Clark, R. S. *See* Brickell and Clark (1970)

Clarke, C. J. (1970). *On the isometric global embedding of pseudo-Riemannian manifolds,* Proc. Roy. Soc. Lond. **A 314**, 417. *See* § 32.7.

Cohen, J. M. *See* Batakis and Cohen (1972)

Cohn, P. M. (1957). *Lie Groups,* Cambridge Univ. Press. *See* § 8.1.

Coll, B. (1975a). *Sur l'invariance du champ électromagnétique dans un espace-temps d'Einstein-Maxwell admettant un groupe d'isométries.* C.R. Acad. Sci. (Paris) **A 280**, 1773. *See* § 9.1.

Coll, B. (1975b). *Sur la détermination, par des données de Cauchy des champs de Killing admis par un espace-temps d'Einstein-Maxwell,* C.R. Acad. Sci. (Paris) **A 281**, 1109. *See* § 31.4.

Collins, C. B. (1971). *More qualitative cosmology,* Commun. Math. Phys. **23**, 137. *See* §§ 11.2., 11.3., 12.3., 12.4.

Collins, C. B. (1972). *Qualitative magnetic cosmology,* Commun. Math. Phys. **27**, 37. *See* §§ 11.3., 12.4.

Collins, C. B. (1974). *Tilting at cosmological singularities,* Commun. Math. Phys. **39**, 131. *See* §§ 11.4., 12.3.

Collins, C. B. (1977). *Global structure of the 'Kantowski-Sachs' cosmological models,* J. Math. Phys. 18, 2116. *See* §§ 8.2., 11.1., 12.3., 13.1.

Collins, C. B., Glass, E. N., and Wilkinson, D. (1980). *Exact homogeneous cosmologies,* GRG (to appear). *See* § 12.3.

Collinson, C. D. (1964). *Symmetry properties of Harrison space-times.* Proc. Camb. Phil. Soc. **60**, 259. *See* § 15.4.

Collinson, C. D. (1966). *Empty space-times of embedding class two,* J. Math. Phys. **7**, 608. *See* § 32.5.

Collinson, C. D. (1967). *Empty space-times algebraically special on a given world line or hypersurface,* J. Math. Phys. **8**, 1547. *See* § 7.5.

Collinson, C. D. (1968a). *Embedding of the plane-fronted waves and other space-times,* J. Math. Phys. **9**, 403. *See* §§ 32.1., 32.5., 32.6.

Collinson, C. D. (1968b). *Einstein-Maxwell fields of embedding class one,* Commun. Math. Phys. **8**, 1. *See* § 32.4.

Collinson, C. D. (1969). *Symmetries of type N empty space-times possessing twisting geodesic rays,* J. Phys. **A 2**, 621. *See* § 33.2.

Collinson, C. D. (1970). *Curvature collineations in empty space-times,* J. Math. Phys. **11**, 818. *See* § 31.4.

Collinson, C. D. (1974). *The existence of Killing tensors in empty space-times,* Tensor **28**, 173. *See* § 31.3.

Collinson, C. D. (1976a). *The uniqueness of the Schwarzschild interior metric,* GRG **7**, 419. *See* § 19.2.

Collinson, C. D. (1976b). *On the relationship between Killing tensors and Killing-Yano tensors,* Int. J. Theor. Phys. **15**, 311. *See* § 31.3.

Collinson, C. D., and Dodd, R. K. (1969). *Petrov classification of stationary axisymmetric empty space-time,* N. Cim. **B 62**, 229. *See* § 17.5.

Collinson, C. D., and Dodd, R. K. (1971). *Symmetries of stationary axisymmetric empty space-times,* N. Cim. **B 3**, 281. *See* § 33.2.

Collinson, C. D., and French, D. C. (1967). *Null tetrad approach to motions in empty space-time,* J. Math. Phys. **8**, 701. *See* §§ 22.3, 24.1., 33.2.

Collinson, C. D., and Fugère, J. (1977). *Empty space-times with separable Hamilton-Jacobi equation,* J. Phys. **A 10**, 745. *See* § 31.3.

Collinson, C. D., and Shaw, R. (1972). *The Rainich conditions for neutrino fields,* Int. J. Theor. Phys. **6**, 347. *See* § 5.1.

Cook, M. W. (1975). *On a class of exact spherically symmetric solutions to the Einstein gravitational field equations,* Austral. J. Phys. **28**, 413. *See* § 14.2.

Cosgrove, C. M. (1977). *New family of exact stationary axisymmetric gravitational fields generalising the Tomimatsu-Sato solutions.* J. Phys. **A 10**, 1481. *See* § 18.6.

Cosgrove, C. M. (1978). *A new formulation of the field equations for the stationary axisymmetric vacuum gravitational field, I. General theory, II. Separable solutions,* J. Phys. **A 11**, 2389, 2405. *See* §§ 18.6., 31.3.

Couch, E. *See* Newman et al. (1965)

Cox, D., and Flaherty, E. J. (1976). *A conventional proof of Kerr's theorem,* Commun. Math. Phys. **47**, 75. *See* § 28.1.

Curzon, H. E. J. (1924). *Cylindrical solutions of Einstein's gravitation equations.* Proc. London Math. Soc. **23**, 477. *See* § 18.1.

Dale, P. (1978). *Axisymmetric gravitational fields: a nonlinear differential equation that admits a series of exact eigenfunction solutions,* Proc. Roy. Soc. Lond. **A 362**, 463. *See* § 18.6.

Damião Soares, I., and Assad, M. J. D. (1978). *Anisotropic Bianchi VIII/IX cosmological models with matter and electromagnetic fields,* Phys. Letters **A 66**, 359. *See* § 12.3.

Darmois, G. (1927). *Les équations de la gravitation einsteinienne,* Mémorial des sciences mathématique, Fasc. XXV, Gauthier-Villars, Paris. *See* § 18.1.

Das, A. (1973). *Static gravitational fields. II. Ricci rotation coefficients,* J. Math. Phys. **14**, 1099. *See* § 16.6.

Das, A. *See also* Zenk and Das (1978)

Das, K. C., and Banerji, S. (1978). *Axially symmetric stationary solutions of Einstein-Maxwell equations,* GRG **9**, 845. *See* § 30.5.

Datt, B. (1938). *Über eine Klasse von Lösungen der Gravitationsgleichungen der Relativität,* Z. Phys. **108**, 314. *See* § 13.5.

Datta, B. K. (1965). *Homogeneous nonstatic electromagnetic fields in general relativity*, N. Cim. **36**, 109. *See* § 11.3.

Datta, B. K. (1967). *Nonstatic electromagnetic fields in general relativity*, N. Cim. **A 47**, 568. *See* § 13.4.

Dautcourt, G. (1964). *Gravitationsfelder mit isotropem Killingvektor*, in: Infeld, L. (Ed.), Relativistic Theories of Gravitation, Pergamon Press, Gauthier-Villars, PWN, p. 300. *See* § 21.4.

Davies, H., and Caplan, T. A. (1971). *The space-time metric inside a rotating cylinder*, Proc. Camb. Phil. Soc. **69**, 325. *See* § 20.2.

Davis, T. M. (1976). *A simple application of the Newman-Penrose spin coefficient formalism. I and II*, Int. J. Theor. Phys. **15**, 315 (I), **15**, 319 (II). *See* § 7.1.

Davis, W. R. *See* Katzin et al. (1969)

Davis, W. R., Green, L. H., and Norris, L. K. (1976). *Relativistic matter fields admitting Ricci collineations and related conservation laws*, N. Cim. **B 34**, 256. *See* § 31.4.

De, U. K., and Banerjee, A. (1972). *Two classes of solutions of null electromagnetic radiation*, Prog. Theor. Phys. **47**, 1204. *See* § 21.4.

Debever, R. (1959). *Tenseur de super-énergie, tenseur de Riemann: cas singuliers*. C.R. Acad. Sci. (Paris) **249**, 1744. *See* § 4.3.

Debever, R. (1960). *Espaces-temps du type III de Petrov*, C.R. Acad. Sci. (Paris) **251**, 1352. *See* § 27.5.

Debever, R. (1964). *Le rayonnement gravitationnel: Le tenseur de Riemann en relativité générale*, Cahiers Phys. **18**, 303. *See* § 4.1.

Debever, R. (1965). *Local tetrad in general relativity*. In: Atti del convegno sulla relatività generale: problemi dell'energia e ondi gravitationali, Barbèra, Firenze. *See* § 10.2.

Debever, R. (1966). *Représentation vectorielle de la courbure en relativité générale: transformations conformes*, in: Perspectives in Geometry and Relativity, Ind. Univ. Press, p. 96. *See* §§ 3.4., 3.7.

Debever, R. (1969). *Sur les espaces de Brandon Carter*, Bull. Acad. Roy. Belgique Cl. Sci. **55**, 8. *See* § 19.1.

Debever, R. (1971). *On type D expanding solutions of Einstein-Maxwell equations*, Bull. Soc. Math. Belg. **23**, 360. *See* §§ 19.1., 24.2., 25.5.

Debever, R. (1974). *Sur une classe d'espaces lorentziens*, Bull. Acad. Roy. Belgique Cl. Sci. **60**, 998. *See* § 28.1.

Debever, R. (1976). *Structures pré-maxwelliennes involutives en relativité générale*, Bull. Acad. Roy. Belgique Cl. Sci. **62**, 662. *See* §§ 19.1., 26.5.

Debever, R. *See also* Cahen and Debever (1965), Cahen et al. (1967), Géhéniau and Debever (1956a, b)

Debever, R., and Cahen, M. (1960). *Champs électromagnétiques constants en relativité générale*, C.R. Acad. Sci. (Paris) **251**, 1160. *See* § 31.1.

Debever, R., and Cahen, M. (1961). *Sur les espaces-temps, qui admettent un champ de vecteurs isotropes parallèles*, Bull. Acad. Roy. Belgique Cl. Sci. **47**, 491. *See* §§ 6.1., 6.2.

Debney, G. (1971). *On vacuum space-times admitting a null Killing bivector*. J. Math. Phys. **12**, 2372. *See* §§ 21.4., 27.5.

Debney, G. (1972a). *Null Killing vectors in general relativity*, N. Cim. Lett. **5**, 954. *See* § 21.4.

Debney, G. (1972b). *Symmetry in Einstein-Maxwell space-time*, J. Math. Phys. **13**, 1469. *See* § 21.4.

Debney, G. (1973). *Expansion-free Kerr-Schild fields*, N. Cim. Lett. **8**, 337. *See* §§ 27.6., 28.2.

Debney, G. (1974). *Expansion-free electromagnetic solutions of the Kerr-Schild class*, J. Math. Phys. **15**, 992. *See* §§ 21.4., 27.6., 28.3.

Debney, G. *See also* Kerr and Debney (1970)

Debney, G., Kerr, R. P., and Schild, A. (1969). *Solutions of the Einstein and Einstein-Maxwell equations*, J. Math. Phys. **10**, 1842. *See* §§ 23.1., 25.1., 28.2., 28.3.

Defrise, L. (1969). *Groupes d'isotropie et groupes de stabilité conforme dans les espaces lorentziens*, Thése, Université Libre de Bruxelles. *See* §§ 8.1., 8.4., 9.2., 10.1., 11.1.

Defrise, L. *See also* Cahen and Defrise (1968), Cahen et al. (1967)

Demiański, M. (1972). *New Kerr-like space-time*, Phys. Letters **A 42**, 157. *See* §§ 25.2., 30.6.

Demiański, M. (1973). *Some new solutions of the Einstein equations of astrophysical interest*, Acta Astron. (Poland) **23**, 197. *See* §§ 18.5., 30.6.

Demiański, M. (1976). *Method of generating stationary Einstein-Maxwell fields*, Acta Phys. Polon. **B 7**, 567. *See* § 30.3.

Demiański, M. *See also* Plebański and Demiański (1976)

Demiański, M., and Grishchuk, L. P. (1972). *Homogeneous rotating universe with flat space*, Commun. Math. Phys. **25**, 233. *See* § 12.4.

Demiański, M. and Newman, E. T. (1966). *A combined Kerr-NUT solution of the Einstein field equations*, Bull. Acad. Polon. Sci. Ser. Math. Astron. Phys. **14**, 653. *See* §§ 18.5. 19.1., 30.5.

de Sitter, W. *See* Einstein and de Sitter (1932)

Dietz, W. (1976). *Separable coordinate systems for the Hamilton-Jacobi, Klein-Gordon and wave equations in curved spaces*, J. Phys. **A 9**, 519. *See* § 31.3.

d'Inverno, R. A., and Russell-Clark, R. A. (1971). *Classification of the Harrison metrics*, J. Math. Phys. **12**, 1258. *See* §§ 4.4., 15.4.

Dodd, R. K. *See* Collinson and Dodd (1969, 1971)

Dolan, P. (1968). *A singularity-free solution of the Maxwell-Einstein-equation*, Commun. Math. Phys. **9**, 161. *See* § 10.3.

Doroshkevich, A. G. (1965). *Model of a universe with a uniform magnetic field* (in Russian), Astrofiz. **1**, 255. *See* §§ 12.1., 12.3.

Dowker, J. S. *See* Roche and Dowker (1968)

Dozmorov, I. M. (1971). *Solutions of the Einstein equations related by null vectors. II, III* (in Russian), Izv. VUZ. Fiz. no. 11, 68 (II), no. 11, 76 (III). *See* § 28.5.

Dozmorov, I. M. (1973). *The algebraic classification of the matter tensor* (in Russian), Izv. VUZ. Fiz. no. 12, 101. *See* § 5.1.

Dunn, K. A., and Tupper, B. O. J. (1976). *A class of Bianchi type VI cosmological models with electromagnetic field*, Astrophys. J. **204**, 322. *See* §§ 12.1., 12.4.

Dutta Choudhoury, S. B. *See* Chakravarty et al. (1976)

Eddington, A. S. (1924). *A comparison of Whitehead's and Einstein's formulae*, Nature **113**, 192. *See* § 13.4.

Edwards, D. (1972). *Exact expressions for the properties of the zero-pressure Friedmann models*, Mon. Not. Roy. Astr. Soc. **159**, 51. *See* § 12.2.

Egorov, I. P. (1955). *Riemannian spaces V_4 of nonconstant curvature and maximum mobility* (in Russian), Dokl. Akad. Nauk. SSSR **103**, 9. *See* § 30.3.

Ehlers, J. (1957). *Konstruktionen und Charakterisierungen von Lösungen der Einsteinschen Gravitationsfeldgleichungen*, Dissertation, Hamburg. *See* § 30.3.

Ehlers, J. (1961). *Beiträge zur relativistischen Mechanik kontinuierlicher Medien*, Akad. Wiss. Lit. Mainz, Abhandl. Math.-Nat. Kl., Nr. 11. *See* §§ 6.1., 33.1.

Ehlers, J. (1962). *Transformations of static exterior solutions of Einstein's gravitational field equations into different solutions by means of conformal mappings*, Colloques Internationaux C.N.R.S. No. 91 (Les théories relativistes de la gravitation), 275. *See* §§ 19.2., 30.5.

Ehlers, J. *See also* Jordan et al. (1960, 1961)

Ehlers, J., and Kundt, W. (1962). *Exact solutions of the gravitational field equations*, in: Witten, L. (Ed.), Gravitation: an introduction to current research, Wiley, New York, London. *See* §§ 1.4., 5.1., 6.1., 16.6., 21.5., 31.1., 31.2.

Ehlers, J., Rosenblum, A., Goldberg, J. N., and Havas, P. (1976). *Comments on gravitational radiation damping and energy loss in binary systems*, Astrophys. J. **208**, L. 77. *See* § 1.1.

Einstein, A., and Rosen, N. J. (1937). *On gravitational waves*, J. Franklin Inst. **223**, 43. *See* § 20.3.

Einstein, A., and de Sitter, W. (1932). *On the relation between the expansion and mean velocity of the universe*, Proc. Natl. Acad. Sci. U.S. **18**, 213. *See* § 12.2.

Eisenhart, L. P. (1933). *Continuous Groups of Transformations*, Princeton Univ. Press. *See* §§ 8.1., 8.4., 8.5., 8.6.

Eisenhart, L. P. (1949). *Riemannian Geometry*, Princeton Univ. Press. *See* §§ 3.1., 3.7., 8.4., 31.1., 32.1., 32.3.

Ellis, G. F. R. (1967). *Dynamics of pressure-free matter in general relativity*, J. Math. Phys. 8, 1171. *See* §§ 9.2., 11.1., 11.4., 13.5.

Ellis, G. F. R. *See also* Hawking and Ellis (1973), King and Ellis (1973), Stewart and Ellis (1968)

Ellis, G. R. F. and King, A. R. (1974). *Was the big bang a whimper*? Commun. Math. Phys. **38**, 119. *See* § 12.1.

Ellis, G. F. R., and MacCallum, M. A. H., (1969). *A class of homogeneous cosmological models*, Commun. Math. Phys. **12**, 108. *See* §§ 8.2., 11.2., 11.3., 12.4.

Eltgroth, P. G. *See* Vajk and Eltgroth (1970)

Ernst, F. J. (1968a). *New formulation of the axially symmetric gravitational field problem*, Phys. Rev. **167**, 1175. *See* §§ 17.5., 18.5.

Ernst, F. J. (1968b). *New formulation of the axially symmetric gravitational field problem II*, Phys. Rev. **168**, 1415. *See* § 16.4.

Ernst, F. J. (1973). *Charged version of Tomimatsu-Sato spinning mass field*, Phys. Rev. **D 7**, 2520. *See* § 30.5.

Ernst, F. J. (1976a). *Removal of the nodal singularity of the C-metric*, J. Math. Phys. **17**, 515. *See* § 19.1.

Ernst, F. J. (1976b). *New representation of the Tomimatsu-Sato solution*, J. Math. Phys. **17**, 1091. *See* § 18.5.

Ernst, F. J. (1977). *A new family of solutions of the Einstein field equations*, J. Math. Phys. **18**, 233. *See* § 18.6.

Ernst, F. J. (1978). *Coping with different languages in the null tetrad formulation of general relativity*, J. Math. Phys. **19**, 489. *See* § 7.1.

Ernst, F. J. *See also* Hauser and Ernst (1978a, b)

Ernst, F. J., and Hauser, I. (1978). *Field equations and integrability conditions for special type N twisting gravitational fields*, J. Math. Phys. **19**, 1816. *See* § 25.3.

Ernst, J. F., and Wild, W. J. (1976). *Kerr black holes in a magnetic universe*, J. Math. Phys. **17**, 182. See § 30.5.

Esposito, F. P., and Glass, E. N. (1976). *The Weyl tensor and stationary electrovac space-times*, J. Math. Phys. **17**, 282. *See* § 16.6.

Esposito, F. P., and Witten, L. (1973). *Five-parameter exterior solution of the Einstein-Maxwell field equations*, Phys. Rev. **D 8**, 3302. *See* § 30.5.

Estabrook, F. B. *See* Wahlquist and Estabrook (1966)

Estabrook, F. B., Wahlquist, H. D., and Behr, C. G. (1968). *Dyadic analysis of spatially homogeneous world models*, J. Math. Phys. **9**, 497. *See* § 8.2.

Evans, A. B. (1977). *Static fluid cylinders in general relativity*, J. Phys. **A 10**, 1303. *See* § 20.2.

Evans, A. B. (1978). *Correlation of Newtonian and relativistic cosmology*, Mon. Not. Roy. Astr. Soc. **183**, 727. *See* §§ 11.3., 12.4.

Exton, E. *See* Newman et al. (1965)

Farnsworth, D. L. (1967). *Some new general relativistic dust metrics possessing isometries*, J. Math. Phys. **8**, 2315. *See* § 12.3.

Farnsworth, D. L., and Kerr, R. P. (1966). *Homogeneous dust-filled cosmological solutions*, J. Math. Phys. **7**, 1625. *See* §§ 8.2., 10.4.

Finkelstein, D. (1958). *Past-future asymmetry of the gravitational field of a point particle*, Phys. Rev. **110**, 965. *See* § 13.4.

Finkelstein, R. J. (1975). *The general relativistic fields of a charged rotating source*, J. Math. Phys. **16**, 1271. *See* § 30.6.

Flaherty, E. J. *See* Cox and Flaherty (1976)

Flanders, H. (1963). *Differential Forms with Applications to the Physical Sciences*, Acad. Press, New York. *See* §§ 2.1., 2.7.

Foster, J., and Newman, E. T. (1967). *Note on the Robinson-Trautman solutions*, J. Math. Phys. **8**, 189. *See* § 24.1.

Foyster, J. M. *See* McIntosh and Foyster (1972)

Foyster, J. M. and McIntosh, C. B. G. (1972). *A class of solutions of Einstein's equations which admit a 3-parameter group of isometries*, Commun. Math. Phys. **27**, 241. *See* § 13.4.

Francaviglia, M. *See* Benenti and Francaviglia (1979)

Frehland, E. (1972). *Exact gravitational field of the infinitely long rotating hollow cylinder*, Commun. Math. Phys. **26**, 307. *See* § 20.2.

French, D. C. *See* Collinson and French (1967)

Friedman, A. (1965). *Isometric embedding of Riemannian manifolds into euclidean spaces*, Rev. Mod. Phys. **37**, 201. *See* §§ 32.3., 32.7.

Friedmann, A. (1922). *Über die Krümmung des Raumes*, Z. Phys. **10**, 377. *See* § 12.2.

Friedmann, A. (1924). *Über die Möglichkeit einer Welt mit konstanter negativer Krümmung des Raumes*, Z. Phys. **21**, 326. *See* § 12.2.

Frolov, V. P. (1974). *Kerr and Newman-Unti-Tamburino type solutions of Einstein's equations with cosmological term* (in Russian), Teor. Mat. Fiz. **21**, 213. *See* § 18.5.

Frolov, V. P., and Khlebnikov, V. I. (1975). *Gravitational field of radiating systems I. Twisting free type D metrics*, Preprint no. 27, Lebedev Phys. Inst., Akad. Nauk, Moscow. *See* § 24.3.

Frolov, V. P. (1977). *Newman-Penrose formalism in general relativity* (in Russian), Problems in General Relativity and Theory of Group Representations, Trudy Lebedev Inst., Akad. Nauk SSR, **96**, 72–180. *See* §§ 7.2., 24.1.

Fronsdal, C. (1959). *Completion and embedding of the Schwarzschild solution*, Phys. Rev. **116**, 778. *See* § 32.7.

Fugère, J. *See* Collinson and Fugère (1977)

Gautreau, R., and Hoffman, R. B. (1970). *Class of exact solutions of the Einstein-Maxwell equations*, Phys. Rev. **D 2**, 271. *See* § 30.5.

Gautreau, R., and Hoffman, R. B. (1972). *Generating potential for the NUT metric in general relativity*, Phys. Letters **A 39**, 75. *See* § 18.3.

Gautreau, R., and Hoffman, R. B. (1973). *The structure of the sources of Weyl-type electrovac fields in general relativity*, N. Cim. **B 16**, 162. *See* § 16.6.

Gautreau, R., Hoffman, R. B., and Armenti, A. (1972). *Static multiparticle systems in general relativity*, N. Cim. **B 7**, 71. *See* § 19.1.

Géhéniau, J. (1956). *Les invariants de courbure des espaces riemanniens de la relativité*, Bull. Acad. Roy. Belgique Cl. Sci. **42**, 253. *See* § 5.1.

Géhéniau, J. (1957). *Une classification des espaces einsteiniens* C.R. Acad. Sci. (Paris) **244**, 723. *See* § 4.3.

Géhéniau, J., and Debever, R. (1956a). *Les quatorze invariants de courbure de l'espace riemannien à quatre dimensions*, Helv. Phys. Acta Suppl. **4**, 101. *See* § 31.4.

Géhéniau, J., and Debever, R. (1956b). *Les invariants de courbure de l'espace de Riemann à quatre dimensions*, Bull. Acad. Roy. Belgique Cl. Sci. **42**, 114. *See* § 5.1.

Geroch, R. (1969). *Limits of spacetimes*, Commun. Math. Phys. **13**, 180. *See* § 30.6.

Geroch, R. (1971). *A method for generating solutions of Einstein's equations*, J. Math. Phys. **12**, 918. *See* §§ 16.1., 30.3.

Geroch, R. (1972), *A method for generating new solutions of Einstein's equation. II*, J. Math. Phys. **13**, 394. *See* §§ 17.3., 30.4.

Geroch, R., Held, A., and Penrose, R. (1973). *A space-time calculus based on pairs of null directions*. J. Math. Phys. **14**, 874. *See* § 7.3.

Gibbons, G. W., and Russell-Clark, R. A. (1973). *Note on the Sato-Tomimatsu solution of Einstein's equations*, Phys. Rev. Lett. **30**, 398. *See* § 18.5.

Glass, E. N. (1975). *The Weyl tensor and shear-free perfect fluids*, J. Math. Phys. **16**, 2361. *See* §§ 6.2., 16.6.

Glass, E. N. *See also* Collins et al. (1980), Esposito and Glass (1976)

Glass, E. N., and Goldman, S. P. (1978). *Relativistic spherical stars reformulated*, J. Math. Phys. **19**, 856. *See* § 14.1.

Gödel, K. (1949). *An example of a new type of cosmological solution of Einstein's field equation of gravitation*. Rev. Mod. Phys. **21**, 447. *See* § 10.4.

Goel, P. *See* Gupta and Goel (1975)

Goenner, H. (1970). *Einstein tensor and generalizations of Birkhoff's theorem*. Commun. Math. Phys. **16**, 34. *See* § 13.4.

Goenner, H. (1973). *Local isometric embedding of Riemannian manifolds and Einstein's theory of gravitation*, Habilitationsschrift, Göttingen. *See* §§ 32.3., 32.5.

Goenner, H. (1977). *On the interdependence of the Gauss-Codazzi-Ricci equations of local isometric embedding*, GRG **8**, 139. *See* §§ 32.2., 32.4.

Goenner, H. (1980). *Local isometric embedding of Riemannian manifolds and Einstein's theory of gravitation*, in Held, A. (Ed.), General Relativity and Gravitation one hundred years after the birth of Albert Einstein, Vol. I, Plenum Press, New York. *See* § 32.1.

Goenner, H., and Stachel, J. (1970). *Einstein tensor and 3-parameter groups of isometries with 2-dimensional orbits*, J. Math. Phys. **11**, 3358. *See* §§ 13.1., 13.2., 13.4.

Goldberg, J. N. *See* Ehlers et al. (1976), Kerr and Goldberg (1961)

Goldberg, J. N., and Kerr, R. P. (1961). *Some applications of the infinitesimal-holonomy group to the Petrov classification of Einstein spaces*, J. Math. Phys. **2**, 327. *See* § 31.4.

Goldberg, J. N., and Sachs, R. K. (1962). *A theorem on Petrov types*, Acta Phys. Polon., Suppl. **22**, 13. *See* § 7.5.

Goldman, S. P. *See* Glass and Goldman (1978)

Goodinson, P. A. (1969). *The electromagnetic tensor in Riemannian space-time*, Anales de Fisica **65**, 351. *See* § 5.2.

Goodinson, P. A., and Newing, R. A. (1968). *Einstein-Maxwell fields with non-zero charge-current distribution*, J. Inst. Math. Appl. **4**, 270. *See* § 5.4.

Gowdy, R. H. (1971). *Gravitational waves in closed universes*, Phys. Rev. Lett. **27**, 826. *See* § 15.3.

Gowdy, R. H. (1975). *Closed gravitational-wave universes: analytic solutions with two-parameter symmetry*, J. Math. Phys. **16**, 224. *See* § 15.3.

Green, L. H. *See* Davis et al. (1976)

Greenberg, P. J. (1970). *The general theory of space-like congruences with an application to vorticity in relativistic hydrodynamics*, J. Math. Anal. Applic. (USA) **30**, 128. *See* § 6.1.

Greenberg, P. J. (1972a). *The algebra of the Riemann curvature tensor in general relativity: preliminaries*, Stud. Appl. Math. **51**, 277. *See* § 3.4.

Greenberg, P. J. (1972b). *The algebra of the Riemann curvature tensor in general relativity: The relation of the invariants of the Einstein curvature tensor to the invariants describing matter*, Stud. Appl. Math. **51**, 369. *See* § 3.4.

Greene, R. E. (1970). *Isometric embedding of Riemannian and pseudo-Riemannian manifolds*, Memoirs Am. Math. Soc. no. 97. *See* § 32.7.

Griffiths, J. B. (1975). *On the superposition of gravitational waves*, Proc. Camb. Phil. Soc. **77**, 559. *See* § 15.2.

Griffiths, J. B. (1976a). *Interacting electromagnetic waves in general relativity*, J. Phys. **A 9** 1273. *See* §§ 11.3., 15.2.

Griffiths, J. B. (1976b). *The collision of plane waves in general relativity*, Ann. Phys. (USA), **102**, 388. *See* § 15.2.

Grishchuk, L. P. *See* Demiański and Grishchuk (1972)

Gupta, P. S. (1959). *Pulsating fluid sphere in general relativity*, Ann. Physik 2, 421. *See* § 14.2.

Gupta, Y. K., and Goel, P. (1975). *Class II analogue of T. Y. Thomas's theorem and different types of embeddings of static spherically symmetric space-times*, GRG **6**, 499. *See* § 32.2.

Gürses, M. (1977). *Some solutions of stationary, axially-symmetric gravitational field equations*, J. Math. Phys. **18**, 2356. *See* § 19.1.

Gürses, M., and Gürsey, F. (1975). *Lorentz covariant treatment of the Kerr-Schild geometry*, J. Math. Phys. **16**, 2385. *See* § 28.1.

Gürsey, F. *See* Gürses and Gürsey (1975)

Güven, R. *See* Chitre et al. (1975)

Gutman, I. I., and Bespal'ko, R. M. (1967). *Some exact spherically symmetric solutions of Einstein's equations* (in Russian), Sbornik Sovrem. Probl. Grav., Tbilissi, p. 201. *See* § 14.2.

Hacyan, S., and Plebański, J. F. (1975). *Some type III solutions of the Einstein-Maxwell equations*, Int. J. Theor. Phys. **14**, 319. *See* § 27.7.

Hacyan, S. *See also* Plebański and Hacyan (1979)

Hall, G. S. (1973). *On the Petrov classification of gravitational fields*, J. Phys. **A 6**, 619. *See* § 4.3.

Hall, G. S. (1974). *On null Maxwell fields*, N. Cim. Lett. **9**, 667. *See* § 27.2.

Hall, G. S. (1976a). *Ricci recurrent space times*, Phys. Letters **A 56**, 17. *See* §§ 5.1., 31.2.

Hall, G. S. (1976b). *The classification of the Ricci tensor in general relativity theory*, J. Phys. **A 9**, 541. *See* § 5.1.

Hall, G. S. (1978). *Riemannian curvature and the Petrov classification*, Z. Naturforsch. **33a**, 559. *See* § 27.2.

Hamoui, A. (1969). *Deux nouvelles classes de solutions non-statiques à symétrie sphérique des équations d'Einstein-Maxwell*, Ann. Inst. H. Poincaré **A 10**, 195. *See* § 13.4.

Harris, R. A., and Zund, J. D. (1975). *Some Bel-Petrov type III solutions of the Einstein equations*, Tensor **29**, 277. *See* § 27.5.

Harris, R. A., and Zund, J. D. (1978). *Multi-variable Kasner type solutions of the source-free Einstein equations*, Tensor **32**, 39. *See* § 15.4.

Harrison, B. K. (1959). *Exact three-variable solutions of the field equations of general relativity*, Phys. Rev. **116**, 1285. *See* § 15.4.

Harrison, B. K. (1965). *Electromagnetic solutions of the field equations of general relativity*, Phys. Rev. **B 138**, 488. *See* §§ 20.4., 30.3.

Harrison, B. K. (1968). *New solutions of the Einstein-Maxwell equations from old*, J. Math. Phys. **9**, 1744. *See* §§ 16.4., 30.3.

Harrison, B. K. (1978). *Bäcklund transformation for the Ernst equation of general relativity*, Phys. Rev. Lett. **41**, 119. *See* § 30.4.

Harrison, E. R. (1967). *Classification of uniform cosmological models*, Mon. Not. Roy. Astr. Soc. **137**, 69. *See* § 12.2.

Hartle, J. B., and Hawking, S. W. (1972). *Solutions of the Einstein-Maxwell equations with many black holes*, Commun. Math. Phys. **26**, 87. *See* § 19.1.

Harvey, A., and Tsoubelis, D. (1977). *Exact Bianchi IV cosmological model*, Phys. Rev. **D 15**, 2734. *See* § 10.2.

Hauser, I. (1974). *Type N gravitational field with twist*, Phys. Rev. Lett. **33**, 1112. *See* § 25.3.

Hauser, I. (1978). *Type N gravitational field with twist. II*, J. Math. Phys. **19**, 661. *See* § 25.3.

Hauser, I. *See also* Ernst and Hauser (1978)

Hauser, I., and Ernst, F. J. (1978). *On the generation of new solutions of the Einstein-Maxwell field equations from electrovac spacetimes with isometries*, J. Math. Phys. **19**, 1316. *See* § 30.3.

Hauser, I., and Ernst, F. J. (1979). *SU (2,1) generation of electrovacs from Minkowski space*, J. Math. Phys. **20**, 1041. *See* § 30.5.

Hauser, I., and Malhiot, R. J. (1974). *Spherically symmetric static space-times which admit stationary Killing tensors of rank two*, J. Math. Phys. **15**, 816. *See* § 31.3.

Hauser, I., and Malhiot, R. J. (1975). *Structural equations for Killing tensors of order two. I, II*, J. Math. Phys. **16**, 150, 1625. *See* § 31.3.

Hauser, I., and Malhiot, R. J. (1976). *On space-time Killing tensors with a Segré characteristic* [(11) (11)], J. Math. Phys. **17**, 1306. *See* § 31.3.

Hauser, I., and Malhiot, R. J. (1978). *Forms of all space-time metrics which admit* [(11)(11)] *Killing tensors with nonconstant eigenvalues*, J. Math. Phys. **19**, 187. *See* § 31.3.

Havas, P. (1975). *Separation of variables in the Hamilton-Jacobi, Schrödinger, and related equations. I, II*, J. Math. Phys. **16**, 1461, 2476. *See* § 31.3.

Havas, P. *See also* Ehlers et al. (1976)

Hawking, S. W. *See* Hartle and Hawking (1972)

Hawking, S. W., and Ellis, G. F. R. (1973). *The large scale structure of space-time*, Cambridge Univ. Press. *See* §§ 2.1., 5.3., 10.4.

Heintzmann, H. (1969). *New exact static solutions of Einstein's field equations*, Z. Phys. **228**, 489. *See* § 14.1.

Held, A. (1974a). *A type-(3,1) solution to the vacuum Einstein equations*, N. Cim. Lett. **11**, 545. *See* §§ 25.2., 25.4.

Held, A. (1974b). *A formalism for the investigation of algebraically special metrics. I*, Commun. Math. Phys. **37**, 311. *See* § 7.3.

Held, A. (1975). *A formalism for the investigation of algebraically special metrics. II*, Commun. Math. Phys. **44**, 211. *See* § 7.3.

Held, A. (1976a). *Killing vectors in empty space algebraically special metrics. I*, GRG **7**, 177. *See* §§ 7.3., 26.4., 33.2.

Held, A. (1976b). *Killing vectors in empty space algebraically special metrics. II*, J. Math. Phys. **17**, 39. *See* 33.2.

Held, A. *See also* Geroch et al. (1973)

Helgason, S. (1962). *Differential Geometry and Symmetric Spaces*, Acad. Press, New York. *See* § 2.1.

Herlt, E. (1972). *Über eine Klasse innerer stationärer axial-symmetrischer Lösungen der Einsteinschen Feldgleichungen mit idealem fluidem Medium*, Wiss. Z. Univ. Jena, Math.-Nat. R. **21**, 19. *See* § 19.2.

Herlt, E. (1978). *Static and stationary axially symmetric gravitational fields of bounded sources. I. Solutions obtainable from the van Stockum metric*, GRG **9**, 711. *See* §§ 19.1., 30.3., 30.5.

Herlt, E. (1979). *Static and stationary axially symmetric gravitational fields of bounded sources. II. Solutions obtainable from Weyl's class*, GRG **11**, 337. *See* § 19.1.

Hiromoto, R. E., and Ozsváth, I. (1978). *On homogeneous solutions of Einstein's field equations.* GRG **9**, 299. *See* §§ 10.1., 10.2.

Hoenselaers, C. (1976). *On generation of solutions of Einstein's equations*, J. Math. Phys. **17**, 1264. *See* § 30.2.

Hoenselaers, C. (1978a). *A new solution of Ernst's equation*, J. Phys. **A 11**, L 75. *See* § 18.6.

Hoenselaers, C. (1978b). *On the effect of motions on energy-momentum tensor*, Prog. Theor. Phys. **59**, 1518. *See* § 9.1.

Hoenselaers, C. (1978c). *Algebraically special one Killing vector solutions of Einstein's equations*, Prog. Theor. Phys. **60**, 747. *See* § 15.4.

Hoenselaers, C., and Vishveshwara, C. V. (1979). *Interiors with relativistic dust-flow*, J. Phys. **A 12**, 209. *See* § 19.2.

Hoffman, R. B. (1969a). *Stationary axially symmetric generalizations of the Weyl solution in general relativity*, Phys. Rev. **182**, 1361. *See* § 18.4.

Hoffman, R. B. (1969b). *Stationary 'noncanonical' solutions of the Einstein vacuum field equations*, J. Math. Phys. **10**, 953. *See* § 21.5.

Hoffman, R. B. *See also* Gautreau and Hoffman (1970, 1972, 1973), Gautreau et al. (1972)

Hori, S. (1978) *On the exact solution of Tomimatsu-Sato family for an arbitrary integral value of the deformation parameter*, Prog. Theor. Phys. **59**, 1870. *See* § 18.5.

Horský, J. (1975). *The gravitational field of a homogeneous plate with a non-zero cosmological constant*, Czech. J. Phys. **B 25**, 1081. *See* § 13.6.

Horský, J. *See also* Novotny and Horský (1974), Avakyan and Horský (1975)

Horský, J., and Novotný, J. (1972). *The generalized Taub solution*, Bull. Astron. Inst. Czech. **23**, 266. *See* § 12.4.

Horský, J., Lorenc, P., and Novotný, J. (1977). *A non-static source of the Taub solution of Einstein's gravitational equations*, Phys. Letters **A 63**, 79. *See* § 13.5.

Hughston, L. P. (1971). *Generalized Vaidya metrics*, Int. J. Theor. Phys. **4**, 267. *See* § 26.6.

Hughston, L. P., and Sommers, P. (1973). *Spacetimes with Killing tensors*, Commun. Math. Phys. **32**, 147. *See* § 31.3.

Hughston, L. P., Penrose, R., Sommers, P., and Walker, M. (1972). *On a quadratic first integral for the charged particle orbits in the charged Kerr solution*, Commun. Math. Phys. **27**, 303. *See* § 31.3.

Ihrig, E. (1975). *The uniqueness of g_{ij} in terms of $R^l{}_{ijk}$*, Int. J. Theor. Phys. **14**, 23. *See* § 31.4.

Ikeda, M., Kitamura, S., and Matsumoto, M. (1963). *On the embedding of spherically symmetric space-times*, J. Math. Kyoto Univ. **3**, 71. *See* § 32.3.

Israel, W. (1958). *Discontinuities in spherically symmetric gravitational fields and shells of radiation*, Proc. Roy. Soc. Lond. **A 248**, 404. *See* § 13.1.

Israel, W. (1970). *Differential forms in general relativity*, Commun. Dublin Inst. Adv. Stud. **A 19**. *See* §§ 2.1., 3.4.

Israel, W., and Khan, K. A. (1964). *Collinear particles and Bondi dipoles in general relativity,* N. Cim. **33**, 331. *See* § 18.1.

Israel, W., and Spanos, J. T. J. (1973). *Equilibrium of charged spinning masses in general relativity,* N. Cim. Lett. **7**, 245. *See* § 16.7.

Israel, W., and Wilson, G. A. (1972). *A class of stationary electromagnetic vacuum fields,* J. Math. Phys. **13**, 865. *See* § 16.7.

Ince, W. C. W. *See* Wainwright et al. (1979)

Jacobs, K. C. (1968). *Spatially homogeneous and Euclidean cosmological models with shear,* Astrophys. J. **153**, 661. *See* §§ 12.3., 12.4.

Jacobs, K. C. (1969). *Cosmologies of Bianchi type I with a uniform magnetic field,* Astrophys. J. **155**, 379. *See* §§ 11.3., 12.1., 12.4.

Jeffery, G. B. *See* Baldwin and Jeffery (1926)

Jordan, P., and Kundt, W. (1961). *Geometrodynamik im Nullfall,* Akad. Wiss. Lit. Mainz. Abhandl. Math.-Nat. Kl. 1961 no. 3. *See* § 5.4.

Jordan, P., Ehlers, J., and Kundt, W. (1960). *Strenge Lösungen der Feldgleichungen der Allgemeinen Relativitätstheorie,* Akad. Wiss. Lit. Mainz. Abhandl., Math.-Nat. Kl. **1960** no. 2. *See* §§ 3.4., 3.6., 3.7., 16.1., 21.5.

Jordan, P., Ehlers, J., and Sachs, R. K. (1961). *Beiträge zur Theorie der reinen Gravitationsstrahlung,* Akad. Wiss. Lit. Mainz, Abhandl. Math.-Nat. Kl. **1960** no. 1. *See* §§ 4.3., 6.1.

Joseph, V. (1966). *A spatially homogeneous gravitational field,* Proc. Camb. Phil. Soc. **62**, 87. *See* § 11.3.

Kaigorodov, V. R. (1962). *Einstein spaces of maximum mobility* (in Russian), Dokl. Akad. Nauk SSSR **146**, 793. *See* §§ 10.5., 33.2.

Kaigorodov, V. R. (1967). *Exact type III solutions of the field equations* $R_{ij} = 0$ (in Russian), Grav. i Teor. Otnos., Univ. Kazan **3**, 155. *See* § 27.5.

Kaigorodov, V. R. (1972). *Petrov classification and recurrent spaces* (in Russian), Gravitatsiya, Nauk dumka, Kiev, 52. *See* § 31.2.

Kalnins, E. G. *See* Boyer et al. (1978)

Kammerer, J. B. (1966). *Sur les directions principales du tenseur de courbure,* C.R. Acad. Sci. (Paris) **263**, 533. *See* § 7.4.

Kantowski, R. (1966). *Some relativistic cosmological models,* Ph. D. Thesis, Univ. of Texas. *See* §§ 11.1., 12.3.

Kantowski, R., and Sachs, R. K. (1966). *Some spatially homogeneous anisotropic relativistic cosmological models,* J. Math. Phys. **7**, 443. *See* §§ 11.1., 12.3.

Karmarkar, K. R. (1948). *Gravitational metrics of spherical symmetry and class one,* Proc. Indian Acad. Sci. **A 27**, 56. *See* § 32.3.

Karmarkar, K. R. *See also* Narlikar and Karmarkar (1946)

Kasner, E. (1921). *Geometrical theorems on Einstein's cosmological equations,* Amer. J. Math. **43**, 217. *See* § 11.3.

Katzin, G. H., and Levine, J. (1972). *Applications of Lie derivatives to symmetries, geodesic mappings, and first integrals in Riemannian spaces,* Colloquium Math. **26**, 21. *See* § 31.4.

Katzin, G. H., Levine, J., and Davis, W. R. (1969). *Curvature collineations: A fundamental symmetry property of the space-times of general relativity defined by the vanishing Lie derivative of the Riemann curvature tensor,* J. Math. Phys. **10**, 617. *See* § 31.4.

Kelley, E. F. *See* Kinnersley and Kelley (1974)

Kellner, A. (1975). *1-dimensionale Gravitationsfelder,* Dissertation Göttingen. *See* §§ 11.1., 11.3.

Kerr, R. P. (1963a). *Gravitational field of a spinning mass as an example of algebraically special metrics,* Phys. Rev. Lett. **11**, 237. *See* §§ 18.5., 25.1., 25.5.

Kerr, R. P. (1963b). *Scalar invariants and groups of motions in a four dimensional Einstein space,* J. Math. Mech. **12**, 33. *See* §§ 8.4., 31.4.

Kerr, R. P. *See also* Debney et al. (1969), Farnsworth and Kerr (1966), Goldberg and Kerr (1961), Weir and Kerr (1977)

Kerr, R. P., and Debney, G. (1970). *Einstein spaces with symmetry groups,* J. Math. Phys. **11**, 2807. *See* §§ 11.3., 25.2., 33.2.

Kerr, R. P., and Goldberg, J. N. (1961). *Einstein spaces with four-parameter holonomy groups*, J. Math. Phys. **2**, 332. *See* § 27.5.

Kerr, R. P., and Schild, A. (1965). *Some algebraically degenerate solutions of Einstein's gravitational field equations*, Proc. Symp. Appl. Math. **17**, 199. *See* § 28.2.

Kerr, R. P., and Schild, A. (1967). *A new class of vacuum solutions of the Einstein field equations*, in: Atti del convegno sulla relatività generale, Firenze, p. 222. *See* § 28.2.

Khan, K. A. *See* Israel and Khan (1964)

Khan, K. A., and Penrose, R. (1971). *Scattering of two impulsive gravitational plane waves*, Nature **229**, 185. *See* § 15.2.

Kharbediya, L. I. (1976). *Some exact solutions of the Friedmann equations with the cosmological term* (in Russian), Astron. Zh. (USSR) **53**, 1145. *See* § 12.2.

Khlebnikov, V. I. *See* Alekseev and Khlebnikov (1978), Frolov and Khlebnikov (1975)

King, A. R. (1974). *New types of singularity in general relativity: the general cylindrically symmetric stationary dust solution*, Commun. Math. Phys. **38**, 157. *See* § 20.2.

King, A. R. *See also* Ellis and King (1974)

King, A. R., and Ellis, G. F. R. (1973). *Tilted homogeneous cosmological models*, Commun. Math. Phys. **31**, 209. *See* § 12.1.

Kinnersley, W. (1969a). *Field of an arbitrarily accelerating point mass*, Phys. Rev. **186**, 1335. *See* §§ 28.1., 28.4.

Kinnersley, W. (1969b). *Type D vacuum metrics*, J. Math. Phys. **10**, 1195. *See* §§ 7.1., 11.3., 19.1., 24.1., 25.5., 27.5., 31.3.

Kinnersley, W. (1973). *Generation of stationary Einstein-Maxwell fields*, J. Math. Phys. **14**, 651. *See* § 30.3.

Kinnersley, W. (1975). *Recent progress in exact solutions*, in: Shaviv, G., and Rosen, J. (Ed.), General Relativity and Gravitation (Proceedings of GR7, Tel-Aviv 1974), Wiley, New York, London. *See* §§ 1.1., 1.4., 11.3., 24.1., 30.3.

Kinnersley, W. (1977). *Symmetries of the stationary Einstein-Maxwell field equations. I*, J. Math. Phys. **18**, 1529. *See* § 30.4.

Kinnersley, W. *See also* Walker and Kinnersley (1972)

Kinnersley, W., and Chitre, D. M. (1977—78). *Symmetries of the stationary Einstein-Maxwell field equations. II—IV*, J. Math. Phys. **18**, 1538; **19**, 1926, 2037. *See* § 30.4.

Kinnersley, W., and Chitre, D. M. (1978). *Group transformation that generates the Kerr and Tomimatsu-Sato metrics*, Phys. Rev. Lett. **40**, 1608. *See* §§ 30.4., 30.5.

Kinnersley, W., and Kelley, E. F. (1974). *Limits of the Tomimatsu-Sato gravitational field*, J. Math. Phys. **15**, 2121. *See* § 18.6.

Kinnersley, W., and Walker, M. (1970). *Uniformly accelerating charged mass in general relativity*, Phys. Rev. **D 2**, 1359. *See* §§ 19.1., 28.1.

Kitamura, S. *See* Ikeda et al. (1963), Takeno and Kitamura (1968, 1969)

Klein, O. (1947). *On a case of radiation equilibrium in general relativity theory and its bearing on the early stage of stellar evolution*, Ark. Mat. Astr. Fys. **A 34**, 1. *See* § 14.1.

Klein, O. (1953). *On a class of spherically symmetric solutions of Einstein's gravitational equations*, Ark. Fys. **7**, 487. *See* § 14.1.

Klekowska, J., and Osinovsky, M. E. (1973). *Group-theoretical analysis of some type N solutions of the Einstein field equations in vacuo*, Preprint ITP-73-II7E, Kiev. *See* § 10.2.

Knight, W. R., and Bergmann, O. (1974). *Interacting matter and radiation in homogeneous isotropic world models*, Int. J. Theor. Phys. **9**, 47. *See* § 12.2.

Kobayashi, S., and Nomizu, K. (1969). *Foundations of Differential Geometry* (2 vols.) Interscience Tracts in Pure and Applied Mathematics, **15**, Interscience, New York. *See* § 2.1.

Kobiske, R. A., and Parker, L. (1974). *Solution of the Einstein-Maxwell equations for two unequal spinning sources in equilibrium*, Phys. Rev. **D 10**, 2321. *See* § 19.1.

Kohler, M., and Chao, K. L. (1965). *Zentralsymmetrische statische Schwerefelder mit Räumen der Klasse I*, Z. Naturforsch. **20a**, 1537. *See* § 32.4.

Köhler, E., and Walker, M. (1975). *A remark on the generalized Goldberg-Sachs theorem*, GRG **6**, 507. *See* § 22.1.

Komar, A. B. *See* Bergmann et al. (1965)

Kompaneets, A. S. (1958). *Strong gravitational waves in vacuum* (in Russian), Zh. Eksper. Teor. Fiz. **34**, 953. *See* § 20.3.

Kompaneets, A. S. (1959). *Propagation of a strong electromagnetic-gravitational wave in vacuum* (in Russian), Zh. Eksper. Teor. Fiz. **37**, 1722. *See* § 20.4.

Kompaneets, A. S., and Chernov, A. S. (1964). *Solution of the gravitation equations for a homogeneous anisotropic model* (in Russian), Zh. Eksper. Teor. Fiz. **47**, 1939. *See* § 12.3.

Kóta, J., and Perjés, Z. (1972). *All stationary vacuum metrics with shearing geodesic eigenrays*, J. Math. Phys. **13**, 1695. *See* § 16.5.

Kottler, F. (1918). *Über die physikalischen Grundlagen der Einsteinschen Gravitationstheorie*, Annalen Physik **56**, 410. *See* § 13.4.

Kowalczyński, J. K. (1978). *Charged tachyon in general relativity: can it be detected?*, Phys. Letters **A 65**, 269. *See* § 24.2.

Kowalczyński, J. K., and Plebański, J. F. (1977). *Metric and Ricci tensors for a certain class of space-times of D type*, Int. J. Theor. Phys. **16**, 371. *See* § 27.7.

Kozarzewski, B. (1965). *Asymptotic properties of the electromagnetic and gravitational fields*, Acta Phys. Polon. **27**, 775. *See* § 22.1.

Kramer, D. (1972). *Gravitational field of a rotating radiating source*, Third Sov. Grav. Conf. Erevan, Tezisy, 321. *See* § 26.6.

Kramer, D. (1977). *Einstein-Maxwell fields with null Killing vector*, Acta Phys. Hung. **43**, 125. *See* § 30.5.

Kramer, D. *See also* Neugebauer and Kramer (1969)

Kramer, D., and Neugebauer, G. (1968a). *Algebraisch spezielle Einstein-Räume mit einer Bewegungsgruppe*, Commun. Math. Phys. **7**, 173. *See* § 27.5.

Kramer, D., and Neugebauer, G. (1968b). *Zu axialsymmetrischen stationären Lösungen der Einsteinschen Feldgleichungen für das Vakuum*, Commun. Math. Phys. **10**, 132. *See* §§ 17.5., 30.4.

Kramer, D., and Neugebauer, G. (1969). *Eine exakte stationäre Lösung der Einstein-Maxwell-Gleichungen*, Annalen Physik **24**, 59. *See* § 30.5.

Kramer, D., and Neugebauer, G. (1971). *Innere Reissner-Weyl-Lösung*, Annalen Physik **27**, 129. *See* § 30.5.

Kramer, D., Neugebauer, G., and Stephani, H. (1972). *Konstruktion und Charakterisierung von Gravitationsfeldern*, Fortschr. Physik **20**, 1. *See* §§ 30.3., 30.5., 32.4.

Krasiński, A. (1974). *Solutions for the Einstein field equations for a rotating perfect fluid. I. Presentation of the flow-stationary and vortex-homogeneous solutions*, Acta Phys. Polon. **B 5**, 411. *See* §§ 5.5., 20.2.

Krasiński, A. (1975). *Solutions of the Einstein field equations for a rotating perfect fluid. II. Properties of the flow-stationary and vortex-homogeneous solutions*, Acta Phys. Polon. **B 6**, 223. *See* § 20.2.

Krasiński, A. (1978). *All flow-stationary cylindrically symmetric solutions of the Einstein field equations for a rotating isentropic perfect fluid*, Rep. Math. Phys. **14**, 225. *See* § 20.2.

Krishna Rao, J. (1963). *Type-N gravitational waves in non-empty space-time*, Curr. Sci. **32**, 350. *See* § 20.4.

Krishna Rao, J. (1964). *Cylindrical waves in general relativity*, Proc. Nat. Inst. Sci. India **A 30**, 439. *See* § 20.4.

Krishna Rao, J. (1970). *Cylindrically symmetric null fields in general relativity*, Indian J. Pure & Appl. Math. **1**, 367. *See* § 20.4.

Kruskal, M. D. (1960). *Maximal extension of Schwarzschild metric*, Phys. Rev. **119**, 1743. *See* § 13.4.

Kuchowicz, B. (1966). *Some new solutions of the gravitational field equations*, Dept. Radiochemistry, Univ. of Warsaw, Preprint. *See* § 14.1.

Kuchowicz, B. (1967). *Exact formulae for a general relativistic fluid sphere*. Phys. Letters **A 25**, 419. *See* § 14.1.

Kuchowicz, B. (1968). *General relativistic fluid spheres. I. New solutions for spherically symmetric matter distributions*, Acta Phys. Polon. **33**, 541. *See* § 14.1.

Kuchowicz, B. (1970). *General relativistic fluid spheres. III. A simultaneous solving of two equations,* Acta Phys. Polon. **B 1**, 437. *See* § 14.1.

Kuchowicz, B. (1972). *General relativistic fluid spheres. V. Non-charged static spheres of perfect fluid in isotropic coordinates,* Acta Phys. Polon. **B 3**, 209. *See* § 14.1.

Kundt, W. (1961). *The plane-fronted gravitational waves,* Z. Phys. **163**, 77. *See* §§ 21.4., 27.1., 27.2., 27.5.

Kundt, W. (1962). *Exact solutions of the field equations: twist-free pure radiation fields,* Proc. Roy. Soc. Lond. **A 270**, 328. *See* § 27.1.

Kundt, W. *See also* Ehlers and Kundt (1962), Jordan and Kundt (1961), Jordan et al. (1960)

Kundt, W., and Thompson, A. (1962). *Le tenseur de Weyl et une congruence associée de géodésiques isotropes sans distorsion,* C.R. Acad. Sci. (Paris) **254**, 4257. *See* § 7.5.

Kundt, W., and Trümper, M. (1962). *Beiträge zur Theorie der Gravitations-Strahlungsfelder,* Akad. Wiss. Lit. Mainz, Abhandl. Math.-Nat. Kl. 1962 no. 12. *See* §§ 7.5., 21.4., 27.1., 29.4.

Kundt, W., and Trümper, M. (1966). *Orthogonal decomposition of axi-symmetric stationary spacetimes,* Z. Phys. **192**, 419. *See* § 17.2.

Kustaanheimo, P. (1947). *Some remarks concerning the connexion between two spherically symmetric relativistic metrics,* Comment. Phys. Math., Helsingf. **13**, 8. *See* § 14.2.

Kustaanheimo, P., and Qvist, B. (1948). *A note on some general solutions of the Einstein field equations in a spherically symmetric world,* Comment. Phys. Math., Helsingf. **13**, 12. *See* §§ 14.1., 14.2.

Lal, K. B., and Prasad, H. (1969). *Cylindrical wave solutions of field equations of general relativity containing electromagnetic fields,* Tensor **20**, 45. *See* § 20.4.

Lanczos, C. (1962). *The splitting of the Riemann tensor,* Rev. Mod. Phys. **34**, 379. *See* §§ 3.5., 3.6.

Land, W. J. *See* Buchdahl and Land (1968)

Lauten III, W. T., and Ray, J. R. (1975). *Space-times with groups of motions on null hypersurfaces,* N. Cim. Lett. **14**, 63. *See* § 21.2.

Lauten III, W. T., and Ray, J. R. (1977). *Investigations of space-times with four-parameter groups of motions acting on null hypersurfaces,* J. Math. Phys. **18**, 885. *See* § 21.2.

Leibovitz, C. (1971). *Time-dependent solutions of Einstein's equations,* Phys. Rev. **D 4**, 2949. *See* § 14.2.

Lemaître, G. (1927). *Un univers homogène de masse constante et de rayon croissant, rendant compte de la vitesse radiale de nebuleuses extra-galactiques,* Ann. Soc. Sci. Bruxelles **A 47**, 49. *See* § 12.2.

Lemaître, G. (1933). *L'univers en expansion,* Ann. Soc. Sci. Bruxelles I **A 53**, 51. *See* §§ 13.4., 13.5.

Leroy, J. (1970). *Un espace d'Einstein de type N à rayons non intégrables,* C.R. Acad. Sci. (Paris) **A 270**, 1078. *See* § 11.3.

Leroy, J. (1976). *Champs électromagnétiques à rayons intégrables, divergents et sans distorsion,* Bull. Acad. Roy. Belgique Cl. Sci. **62**, 259. *See* § 24.2.

Leroy, J. (1978). *Sur une classe d'espaces-temps solutions des equations d'Einstein-Maxwell,* Bull. Acad. Roy. Belgique Cl. Sci. **64**, 130. *See* § 26.5.

Leroy, J. *See also* Cahen and Leroy (1965, 1966), McLenaghan and Leroy (1972)

Letelier, P. S. (1975). *Self-gravitating fluids with cylindrical symmetry,* J. Math. Phys. **16**, 1488. *See* § 20.5.

Letelier, P. S., and Tabensky, R. R. (1974). *The general solution to Einstein-Maxwell equations with plane symmetry,* J. Math. Phys. **15**, 594. *See* § 13.4.

Letelier, P. S., and Tabensky, R. R. (1975a). *Cylindrical self-gravitating fluids with pressure equal to energy density,* N. Cim. **B 28**, 407. *See* § 20.5.

Letelier, P. S., and Tabensky, R. R. (1975b). *Singularities for fluids with $p = \omega$ equations of state,* J. Math. Phys. **16**, 8. *See* § 20.5.

Levi-Civita, T. (1917–1919). *ds^2 einsteiniani in campi newtoniani,* Rend. Acc. Lincei **26**, 307 (1917), **27**, 3, 183, 220, 240, 283, 343 (1918), **28**, 3, 101 (1919). *See* §§ 16.6., 20.2.

Levine, J. (1936). *Groups of motions in conformally flat spaces. I,* Bull. Amer. Math. Soc. **42**, 418. *See* § 22.2.

Levine, J. (1939). *Groups of motions in conformally flat spaces. II*, Bull. Amer. Math. Soc. **45**, 766. *See* § 22.2.

Levine, J. *See also* Katzin and Levine (1972), Katzin et al. (1969)

Levy, H. (1968). *Classification of stationary axisymmetric gravitational fields*, N. Cim. **B 56**, 253. *See* § 18.4.

Lewis, T. (1932). *Some special solutions of the equations of axially symmetric gravitational fields*, Proc. Roy. Soc. Lond. **A 136**, 176. *See* §§ 17.3., 18.4.

Lichnerowicz, A. (1955). *Théories relativistes de la gravitation et de l'électromagnétisme*, Masson, Paris. *See* § 16.1.

Lind, R. W. (1974). *Shear-free, twisting Einstein-Maxwell metrics in the Newman-Penrose formalism*, GRG **5**, 25. *See* §§ 7.1., 23.1., 26.3.

Lind, R. W. (1975a). *Gravitational and electromagnetic radiation in Kerr-Maxwell spaces*, J. Math. Phys. **16**, 34. *See* §§ 26.4., 26.5.

Lind, R. W. (1975b). *Stationary Kerr-Maxwell spaces*, J. Math. Phys. **16**, 39. *See* § 26.4.

Lindquist, R. W. *See* Boyer and Lindquist (1967)

Lorenc, P. *See* Horský et al. (1977)

Lovelock, D. (1967). *A spherically symmetric solution of the Maxwell-Einstein equations*, Commun. Math. Phys. **5**, 257. *See* § 10.3.

Ludwig, G. (1969). *Classification of electromagnetic and gravitational fields*, Amer. J. Phys. **37**, 1225. *See* § 4.2.

Ludwig, G. (1970). *Geometrodynamics of electromagnetic fields in the Newman-Penrose formalism*, Commun. Math. Phys. **17**, 98. *See* § 5.4.

Ludwig, G., and Scanlan, G. (1971). *Classification of the Ricci tensor*, Commun. Math. Phys. **20**, 291. *See* § 5.1.

Lukács, B. (1973). *All vacuum metrics with space-like symmetry and shearing geodesic timelike eigenrays*, Report KFKI-1973-38, Centr. Res. Inst. Phys. Acad. Sci., Budapest. *See* § 16.5.

Lukács, B. (1974). *All stationary, rigidly rotating incoherent fluid metrics with geodesic and/or shearfree eigenrays*, Report KFKI-1974-87, Central Res. Inst. Phys. Acad. Sci., Budapest. *See* §§ 16.5., 19.2.

Lukács, B., and Perjés, Z. (1973). *Electrovac fields with geodesic eigenrays*, GRG **4**, 161. *See* § 16.5.

Lukash, V. N. (1974). *Gravitational waves that conserve the homogeneity of space* (in Russian), Zh. Eksper. Teor. Fiz. **67**, 1594. *See* § 11.3.

Maartens, R., and Nel, S. D. (1978). *Decomposable differential operators in a cosmological context*, Commun. Math. Phys. **59**, 273. *See* §§ 12.1., 12.3., 12.4.

MacCallum, M. A. H. (1971). *A class of homogeneous cosmological models. I. Asymptotic behaviour*, Commun. Math. Phys. **20**, 57. *See* § 11.3.

MacCallum, M. A. H. (1972). *On 'diagonal' Bianchi cosmologies*, Phys. Letters **A 40**, 385. *See* § 11.2.

MacCallum, M. A. H. (1973). *Cosmological models from a geometric point of view*, in: Cargèse Lectures in Physics, Vol. 6, Gordon and Breach, New York. *See* § 11.2.

MacCallum, M. A. H. (1979a). *Anisotropic and inhomogeneous relativistic cosmologies*, in: Hawking, S. W., and Israel, W. (Ed.), General Relativity: an Einstein centenary survey, Cambridge Univ. Press. *See* § 11.2.

MacCallum, M. A. H. (1979b). *The mathematics of anisotropic cosmologies*, in: Demiański, M. (Ed.), Proceedings of the First International Cracow School of Cosmology, Springer-Verlag, Berlin, Heidelberg, New York. *See* § 11.2.

MacCallum, M. A. H. (1979c). *On the classification of the real four-dimensional Lie algebras*, Preprint. *See* § 8.2.

MacCallum, M. A. H. (1980). *Locally isotropic spacetimes with non-null homogeneous hypersurfaces*, in: Tipler, F. J. (Ed.), Essays in General Relativity: a Festschrift for Abraham Taub, Academic Press. *See* § 11.1.

MacCallum, M. A. H. *See also* Ellis and MacCallum (1969)

MacCallum, M. A. H., and Taub, A. H. (1972). *Variational principles and spatially homogeneous universes including rotation*, Commun. Math. Phys. **25**, 173. *See* § 11.2.

MacCallum, M. A. H., Stewart, J. M., and Schmidt, B. G. (1970). *Anisotropic stresses in homogeneous cosmologies*, Commun. Math. Phys. **17**, 343. *See* § 11.2.

Maison, D. (1978). *Are the stationary, axially symmetric Einstein equations completely integrable?*, Phys. Rev. Lett. **41**, 521. *See* § 30.4.

Maitra, S. G. (1966). *Stationary dust-filled cosmological solution with $\Lambda = 0$ and without closed timelike lines*, J. Math. Phys. **7**, 1025. *See* §§ 19.2., 20.2.

Majumdar, S. D. (1947). *A class of exact solutions of Einstein's field equations*, Phys. Rev. **72**, 390. *See* §§ 16.7., 19.1.

Malhiot, R. J. *See* Hauser and Malhiot (1974, 1975, 1976, 1978)

Marder, L. (1958a). *Gravitational waves in general relativity. I. Cylindrical waves*, Proc. Roy. Soc. Lond. **A 244**, 524. *See* § 20.3.

Marder, L. (1958b). *Gravitational waves in general relativity. II. The reflexion of cylindrical waves*, Proc. Roy. Soc. Lond. **A 246**, 133. *See* § 20.3.

Marder, L. (1969). *Gravitational waves in general relativity. XI. Cylindrical-spherical waves*, Proc. Roy. Lond. **A 313**, 83. *See* § 20.3.

Marek, J. J. J. (1968). *Some solutions of Einstein's equations in general relativity*, Proc. Camb. Phil. Soc. **64**, 167. *See* § 18.6.

Mark, J. *See* Schiffer et al. (1973)

Mariot, L. (1954). *Le champ électromagnétique singulier*, C.R. Acad. Sci. (Paris) **238**, 2055. *See* § 7.5.

Marshman, B. J. *See* Wainwright et al. (1979)

Matsumoto, M. (1950). *Riemann spaces of class two and their algebraic characterization*, J. Math. Soc. (Japan) **2**, 67. *See* § 32.5.

Matsumoto, M. *See also* Ikeda et al. (1963)

May, T. L. (1975). *Uniform model universes containing interacting blackbody radiation and matter with internal energy*, Astrophys. J. **199**, 322. *See* § 12.2.

McIntosh, C. B. G. (1968). *Relativistic cosmological models with both radiation and matter*, Mon. Not. Roy. Astr. Soc. **140**, 461. *See* § 12.2.

McIntosh, C. B. G. (1972). *Cosmological models with two fluids. I. Robertson-Walker metric*, Austral. J. Phys. **25**, 75. *See* § 12.2.

McIntosh, C. B. G. (1976). *Homothetic motions in general relativity*, GRG **7**, 199. *See* § 31.4.

McIntosh, C. B. G. (1978). *Einstein-Maxwell space-times with symmetries and with non-null electromagnetic fields*, GRG **9**, 277. *See* § 30.3.

McIntosh, C. B. G. (1979) *Symmetry mappings in Einstein-Maxwell space times*, GRG **10**, 61. *See* § 31.4.

McIntosh, C. B. G. *See also* Foyster and McIntosh (1972)

McIntosh, C. B. G., and Foyster, J. M. (1972). *Cosmological models with two fluids. II. Conformal and conformally flat metrics*, Austral. J. Phys. **25**, 83. *See* § 12.2.

McLenaghan, R. G., and Leroy, J. (1972). *Complex recurrent space-times*, Proc. Roy. Soc. Lond. **A 327**, 229. *See* § 31.2.

McLenaghan, R. G., and Tariq, N. (1975). *A new solution of the Einstein-Maxwell equations*, J. Math. Phys. **16**, 11. *See* § 10.3.

McLenaghan, R. G., Tariq, N., and Tupper, B. O. J. (1975). *Conformally flat solutions of the Einstein-Maxwell equations for null electromagnetic fields*, J. Math. Phys. **16**, 829. *See* § 32.5.

McLenaghan, R. G. *See* Cahen and McLenaghan (1968)

McVittie, G. C. (1929). *On Einstein's unified field theory*, Proc. Roy. Soc. Lond. **124**, 366. *See* § 13.4.

McVittie, G. C. (1933). *The mass-particle in an expanding universe*, Mon. Not. Roy. Astr. Soc. **93**, 325. *See* § 14.2.

McVittie, G. C. (1956). *General Relativity and Cosmology*, Chapman and Hall, London. *See* § 12.1.

McVittie, G. C. (1967). *Gravitational motions of collapse or of expansion in general relativity*. Ann. Inst. H. Poincaré **A 6**, 1. *See* § 14.2.

McVittie, G. C., and Wiltshire, R. J. (1975). *Fluid spheres and R- and T-regions in general relativity*, Int. J. Theor. Phys. **14**, 145. *See* §§ 13.1., 14.2.

McVittie, G. C., and Wiltshire, R. J. (1977). *Relativistic fluid spheres and noncomoving coordinates. I*, Int. J. Theor. Phys. **16**, 121. *See* § 14.2.

Mehra, A. L. (1966). *Radially symmetric distribution of matter*, J. Austr. Math. Soc. **6**, 153. *See* § 14.1.

Melnick, J., and Tabensky, R. R. (1975). *Exact solutions to Einstein field equations*, J. Math. Phys. **16**, 958. *See* § 16.6.

Melvin, M. A. (1964). *Pure magnetic and electric geons*, Phys. Letters **8**, 65. *See* § 20.2.

Melvin, M. A. (1975). *Homogeneous axial cosmologies with electromagnetic fields and dust.* Ann. N.Y. Acad. Sci. **262**, 253. *See* § 12.1.

Michalski, H., and Wainwright, J. (1975). *Killing vector fields and the Einstein-Maxwell field equations in general relativity*, GRG **6**, 289. *See* § 17.2.

Miller, W. *See* Boyer et al. (1978)

Misner, C. W. (1968). *The isotropy of the universe*, Astrophys. J. **151**, 431. *See* § 11.2.

Misner, C. W., and Sharp, D. H. (1964). *Relativistic equations for adiabatic, spherically symmetric gravitational collapse*, Phys. Rev. **B 136**, 571. *See* § 14.2.

Misner, C. W., and Taub, A. H. (1968). *A singularity-free empty universe* (in Russian), Zh. Eksper. Teor. Fiz. **55**, 233. *See* § 11.3.

Misner, C. W., and Wheeler, J. A. (1957). *Classical physics as geometry: Gravitation, electromagnetism, unquantized charge, and mass as properties of curved empty space*, Ann. Phys. (USA) **2**, 525. *See* § 5.4.

Misner, C. W., Thorne, K. S., and Wheeler, J. A. (1973). *Gravitation*, Freeman, San Francisco. *See* §§ 2.1., 18.5., 19.1.

Misra, M. (1962a). *Electrovac universes*, Proc. Nat. Inst. Sci. India **A 28**, 105. *See* § 19.1.

Misra, M. (1962b). *A non-singular electromagnetic field*, Proc. Camb. Phil. Soc. **58**, 711. *See* § 20.4.

Misra, M. (1966). *Some exact solutions of the field equations of general relativity*, J. Math. Phys. **7**, 155. *See* § 20.4.

Misra, M., and Radhakrishna, L. (1962). *Some electromagnetic fields of cylindrical symmetry*, Proc. Nat. Inst. Sci. India **A 28**, 632. *See* § 20.4.

Misra, R. M. (1969). *The gravitational field and the type of matter*, Proc. Nat. Inst. Sci. India **A 35**, 590. *See* § 5.1.

Misra, R. M., and Pandey, D. P. (1973). *Class of electromagnetic fields*, J. Phys. **A 6**, 924. *See* § 19.1.

Misra, R. M., and Singh, R. A. (1967). *Three-dimensional formulation of gravitational null fields. II*, J. Math. Phys. **8**, 1065. *See* § 3.5.

Misra, R. M., and Srivastava, D. C. (1973). *Dynamics of fluid spheres of uniform density*, Phys. Rev. **D 8**, 1653. *See* § 14.2.

Misra, R. M., and Srivastava, D. C. (1974). *Charged dust spheres in general relativity*, Phys. Rev. **D 9**, 844. *See* § 13.5.

Montgomery, W., O'Raifeartaigh, L., and Winternitz, P. (1969). *Two-variable expansions of relativistic amplitudes and the subgroups of the SU(2, 1)group*, Nucl. Phys. **B 11**, 39. *See* § 30.3.

Mukherji, B. C. (1938). *Two cases of exact gravitational fields with axial symmetry*, Bull. Calc. Math. Soc. **30**, 95. *See* § 20.2.

Murphy, G. L. (1973). *Big-bang model without singularities*, Phys. Rev. **D 8**, 4231. *See* § 12.1.

Narain, U. (1970). *Some properties of gravitational fields of magnetic type*, Phys. Rev. **D 2**, 278 *See* § 4.2.

Nariai, H. (1950). *On some static solutions of Einstein's gravitational field equations in a spherically symmetric case*, Sci. Rep. Tôhoku Univ., I, **34**, 160. *See* § 14.1.

Nariai, H. (1951). *On a new cosmological solution of Einstein's field equations of gravitation*, Sci. Rep. Tôhoku Univ., I, **35**, 62. *See* § 13.4.

Nariai, H. (1967). *A simple model for gravitational collapse with pressure gradient*, Prog. Theor. Phys. **38**, 92. *See* § 14.2.

Narlikar, V. V. (1947). *Some new results regarding spherically symmetrical fields in relativity*, Curr. Sci. **16**, 113. *See* § 14.2.

Narlikar, V. V., and Karmarkar, K. R. (1946). *On a curious solution of relativistic field equations,* Curr. Sci. **15,** 69. *See* § 11.3.

Nel, S. D. *See* Maartens and Nel (1978)

Neugebauer, G. (1969). *Untersuchungen zu Einstein-Maxwell-Feldern mit eindimensionaler Bewegungsgruppe,* Habilitationsschrift, Jena. *See* § 16.7.

Neugebauer, G. (1979). *Bäcklund transformations of axially symmetric stationary gravitational fields,* J. Phys. **A 12,** L 67. *See* § 30.4.

Neugebauer, G., private communication. *See* § 30.3.

Neugebauer, G. *See also* Kramer and Neugebauer (1968a, b, 1969, 1971), Kramer et al. (1972)

Neugebauer, G., and Kramer, D. (1969). *Eine Methode zur Konstruktion stationärer Einstein-Maxwell-Felder,* Annalen Physik **24,** 62. *See* §§ 16.4., 30.2., 30.3.

Neugebauer, G., and Sust, M. (1975). *Reduktion von Krümmungs- und Ricci-Tensoren n-dimensionaler Riemannscher Räume bezüglich eines beliebigen, nicht isotropen Vektorfeldes,* Wiss. Z. Univ. Jena, Math.-Nat. R. **24,** 547. *See* § 6.2.

Newing, R. A. *See* Goodinson and Newing (1968)

Newman, E. T. (1974). *Lienard-Wiechert fields and general relativity,* J. Math. Phys. **15,** 44. *See* § 24.1.

Newman, E. T. *See also* Demiański and Newman (1966), Foster and Newman (1967)

Newman, E. T., and Penrose, R. (1962). *An approach to gravitational radiation by a method of spin coefficients,* J. Math. Phys. **3,** 566. *See* §§ 7.1., 7.5.

Newman, E. T., and Tamburino, L. A. (1962). *Empty space metrics containing hypersurface orthogonal geodesic rays,* J. Math. Phys. **3,** 902. *See* § 22.3.

Newman, E. T., and Unti, T. W. J. (1963). *A class of null flat-space coordinate systems,* J. Math. Phys. **4,** 1467. *See* §§ 24.3., 28.1.

Newman, E., Tamburino, L., and Unti, T. (1963). *Empty-space generalization of the Schwarzschild metric,* J. Math. Phys. **4,** 915. *See* §§ 11.3., 25.5.

Newman, E. T., Couch, E., Chinnapared, K., Exton, A., Prakash, A., and Torrence, R. (1965). *Metric of a rotating, charged mass,* J. Math. Phys. **6,** 918. *See* §§ 19.1., 28.3., 30.6.

Nightingale, J. D. (1973). *Independent investigations concerning bulk viscosity in relativistic homogeneous isotropic cosmologies,* Astrophys. J. **185,** 105. *See* § 12.1.

Nomizu, K. *See* Kobayashi and Nomizu (1969)

Nordström, G. (1918). *On the energy of the gravitational field in Einstein's theory,* Proc. Kon. Ned. Akad. Wet. **20,** 1238. *See* §§ 13.4., 19.1.

Norris, L. K. *See* Davis et al. (1976)

Novikov, I. D. (1963). *On the evolution of a semiclosed universe* (in Russian), Zh. Astrof. **40,** 772. *See* § 13.4.

Novoseller, D. E. (1975). *Type-N gravitational field with twist — analytical addendum to Hauser's solution,* Phys. Rev. **D 11,** 2330. *See* § 25.3.

Novotný, J. *See* Horský and Novotný (1972), Horský et al. (1977)

Novotný, J., and Horský, J. (1974). *On the plane gravitational condensor with the positive gravitational constant,* Czech. J. Phys. **B 24,** 718. *See* 13.4.

Nutku, Y. *See* Chitre et al. (1975)

Öktem, F. (1976). *On parallel null 1-planes in space-time,* N. Cim. **B 34,** 169. *See* § 6.1.

Oleson, M. (1971). *A class of type [4] perfect fluid space-times,* J. Math. Phys. **12,** 666. *See* § 29.4.

O'Murchadha, N., and York, J. W. (1976). *Gravitational potentials: A constructive approach to general relativity,* GRG **7,** 257. *See* § 31.4.

Önengüt, G., and Serdaroğlu, M. (1975). *Two-parameter static and five-parameter stationary solutions of the Einstein-Maxwell equations,* N. Cim. **B 27,** 213. *See* § 30.5.

O'Raifeartaigh, L. *See* Montgomery et al. (1969)

Osinovsky, M. E. *See* Klekowska and Osinovsky (1973)

Ozsváth, I. (1965a). *All homogeneous solutions of Einstein's field equations with incoherent matter and electromagnetic radiation,* J. Math. Phys. **6,** 1265. *See* § 10.5.

Ozsváth, I. (1965b). *New homogeneous solutions of Einstein's field equations with incoherent matter,* Akad. Wiss. Lit. Mainz, Math.-Nat. Kl. 1965 no. 1. *See* §§ 10.4., 10.5.

Ozsváth, I. (1965c). *Homogeneous solutions of the Einstein-Maxwell equations*, J. Math. Phys. **6**, 1255. *See* § 10.3.

Ozsváth, I. (1965d). *New homogeneous solutions of Einstein's field equations with incoherent matter obtained by a spinor technique*. J. Math. Phys. **6**, 590. *See* § 10.4.

Ozsváth, I. (1966). *Two rotating universes with dust and electromagnetic field*, in: Hoffmann, B. (Ed.), Perspectives in Geometry and Relativity, Indiana University Press p. 245. *See* §§ 10.5., 11.1.

Ozsváth, I. (1970). *Dust-filled universes of class II and class III*, J. Math. Phys. **11**, 2871. *See* § 10.4.

Ozsváth, I. (1977). *Spatially homogeneous Lichnerowicz universes*, GRG **8**, 737. *See* § 12.1.

Ozsváth, I. *See also* Hiromoto and Ozsváth (1978)

Ozsváth, I., and Schücking, E. (1962). *An anti-Mach metric*, in : Recent Developments in General Relativity, Pergamon Press — PWN, p. 339. *See* § 10.2.

Ozsváth, I., and Schücking, E. (1969). *The finite rotating universe*, Ann. Phys. (USA) **55**, 166. *See* § 10.4.

Pandey, D. B. *See* Misra and Pandey (1973)

Pandya, I. M., and Vaidya, P. C. (1961). *Wave solutions in general relativity. I*, Proc. Nat. Inst. Sci. India **A 27**, 620. *See* § 27.5.

Papapetrou, A. (1947). *A static solution of the equations of the gravitational field for an arbitrary charge distribution*, Proc. Roy. Irish Acad. **A 51**, 191. *See* §§ 16.7., 19.1.

Papapetrou, A. (1953). *Eine rotationssymmetrische Lösung in der Allgemeinen Relativitätstheorie*, Annalen Physik **12**, 309. *See* § 18.3.

Papapetrou, A. (1963). *Quelques remarques sur les champs gravitationnels stationnaires*, C.R. Acad. Sci. (Paris) **257**, 2797. *See* § 16.4.

Papapetrou, A. (1966). *Champs gravitationnels stationnaires à symétrie axiale*, Ann. Inst. H. Poincaré **A 4**, 83. *See* §§ 17.2., 17.3., 20.3.

Papapetrou, A. (1971a). *Quelques remarques sur le formalisme de Newman-Penrose*, C.R. Acad. Sci. (Paris) **A 272**, 1537. *See* § 7.2.

Papapetrou, A. (1971b). *Les relations identiques entre les équations du formalisme de Newman-Penrose*, C.R. Acad. Sci. (Paris) **A 272**, 1613. *See* § 7.2.

Parker, L. *See* Kobiske and Parker (1974)

Parker, L., Ruffini, R., and Wilkins, D. (1973). *Metric of two spinning charged sources in equilibrium*, Phys. Rev. **D 7**, 2874. *See* § 19.1.

Patel, L. K. *See* Vaidya and Patel (1973)

Patel, L. K., and Vaidya, P. C. (1969). *On plane-symmetric cosmological models*, Progress of Math. **3**, 158. *See* § 12.4.

Patera, J., and Winternitz, P. (1977). *Subalgebras of real three and four-dimensional Lie algebras*, J. Math. Phys. **18**, 1449. *See* § 8.2.

Patnaik, S. (1970). *Einstein-Maxwell fields with plane symmetry*, Proc. Camb. Phil. Soc. **67**, 127. *See* § 13.4.

Peebles, P. J. (1971). *Physical Cosmology*, Princeton Univ. Press. *See* § 12.1.

Penrose, R. (1960). *A spinor approach to general relativity*, Ann. Phys. (USA) **10**, 171. *See* §§ 3.6., 4.3.

Penrose, R. (1965). *A remarkable property of plane waves in general relativity*, Rev. Mod. Phys. **37**, 215. *See* § 32.7.

Penrose, R. (1968). *Structure of space-time*, in: DeWitt, C. M., and Wheeler, J. A. (Eds.) Battelle Rencontre, 1967 Lectures in Math. and Phys., Chapter VII, Benjamin, New York, Amsterdam. *See* §§ 3.6., 4.3.

Penrose, R. (1972). *Spinor classification of energy tensors*, Gravitatsiya, Nauk dumka, Kiev, p. 203. *See* § 5.1.

Penrose, R. *See also* Geroch et al. (1973), Hughston et al. (1972), Khan and Penrose (1971), Newman and Penrose (1962), Walker and Penrose (1970)

Peres, A. (1960a). *Invariants of general relativity. II. Classification of spaces*, N. Cim. **18**, 36. *See* § 5.1.

Peres, A. (1960b). *Null electromagnetic fields in general relativity theory,* Phys. Rev. **118**, 1105. *See* § 21.5.

Perjés, Z. (1968). *A method for constructing certain axially-symmetric Einstein-Maxwell fields,* N. Cim. **B 55**, 600. *See* § 19.1.

Perjés, Z. (1970). *Spinor treatment of stationary space-times,* J. Math. Phys. **11**, 3383. *See* § 16.5.

Perjés, Z. (1971). *Solutions of the coupled Einstein-Maxwell equations representing the fields of spinning sources,* Phys. Rev. Lett. **27**, 1668. *See* § 16.7.

Perjés, Z. (1976). *Introduction to spinors and Petrov types in general relativity,* Acta Phys. Acad. Sci. Hung. **41**, 173. *See* § 4.3.

Perjés, Z. *See also* Kóta and Perjés (1972), Lukács and Perjés (1973)

Petrov, A. Z. (1962). *Gravitational field geometry as the geometry of automorphisms,* in: Recent Developments in General Relativity, Pergamon Press — PWN, p. 379. *See* §§ 10.2., 27.5.

Petrov, A. Z. (1963a). *Types of field and energy-momentum tensor* (in Russian), Uchonie Zapiski Kazan. Gos. Univ. **123**, 3. *See* § 5.1.

Petrov, A. Z. (1963b). *On Birkhoff's Theorem* (in Russian), Uchonie Zapiski Kazan. Gos. Univ. **123**, 61. *See* § 13.4.

Petrov, A. Z. (1963c). *Central-symmetric gravitational fields* (in Russian), Zh. Eksper. Teor. Fiz. **44**, 1525. *See* § 13.4.

Petrov, A. Z. (1966). *New methods in General Relativity* (in Russian), Nauka, Moscow. [English edition of Petrov's book: Einstein spaces, Pergamon Press (1969)]. *See* §§ 4.2., 8.1., 8.2., 8.4., 8.5., 9.1., 10.5., 11.1., 15.1., 20.3., 21.1., 31.1., 31.3., 31.4.

Pirani, F. A. E. (1957). *Invariant formulation of gravitational radiation theory,* Phys. Rev. **105**, 1089. *See* § 4.3.

Pirani, F. A. E. (1965). *Introduction to gravitational radiation theory,* in: Brandeis (1964) Lectures on General Relativity, vol. 1, Prentice-Hall, Englewood Cliffs., New Jersey, p. 249. *See* §§ 3.6., 7.2.

Pirani, F. A. E. *See also* Bondi et al. (1959)

Plebański, J. (1964). *The algebraic structure of the tensor of matter,* Acta Phys. Polon. **26**, 963. *See* § 5.1.

Plebański, J. (1967). *On conformally equivalent Riemannian spaces,* Centro de investigation y de estudios, Mexico. *See* § 8.5.

Plebański, J. (1975). *A class of solutions of Einstein-Maxwell equations,* Ann. Phys. (USA) **90**, 196. *See* § 19.1.

Plebański, J. (private communication). *Solutions of Einstein equations* $G_{\mu\nu} = -\varrho K_\mu K_\nu$ *determined by the condition that* K_μ *is the quadruple Debever-Penrose vector. See* § 22.1.

Plebański, J. *See also* Hacyan and Plebański (1975), Kowalczyński and Plebański (1977)

Plebański, J., and Demiański, M. (1976). *Rotating, charged and uniformly accelerating mass in general relativity,* Ann. Phys. (USA) **98**, 98. *See* §§ 19.1., 25.5.

Plebański, J., and Hacyan, S. (1979). *Some exceptional electro-vac type D metrics with cosmological constant,* J. Math. Phys. **20**, 1004. *See* § 22.1.

Plebański, J., and Stachel, J. (1968). *Einstein tensor and spherical symmetry,* J. Math. Phys. **9**, 269. *See* §§ 5.1., 13.4.

Plebański, J., and Schild, A. (1976). *Complex relativity and double KS metrics,* N. Cim. **B 35**, 35. *See* § 28.5.

Polishchuk, R. F. (1970). *Petrov classification of gravitational fields and chronometrical invariants* (in Russian), Vestn. Mosk. Univ. Fiz. Astr. **3**, 350. *See* § 4.2.

Prakash, A. *See* Newman et al. (1965)

Prasad, H. *See* Lal and Prasad (1969)

Qvist, B. *See* Kustaanheimo and Qvist (1948)

Radhakrishna, L. (1963). *Some exact non-static cylindrically symmetric electrovac universes,* Proc. Nat. Inst. Sci. India **A 29**, 588. *See* § 20.4.

Radhakrishna, L. *See also* Misra and Radhakrishna (1962), Singh et al. (1965)

Rainich, G. Y. (1925). *Electrodynamics in general relativity*. Trans. Amer. Math. Soc. **27**, 106. *See* § 5.4.

Rastall, P. (1960). *A solution of the Einstein field equations*, Canad. J. Phys. **38**, 1661. *See* § 15.4.

Ray, J. R. *See* Lauten III and Ray (1975, 1977)

Ray, J. R., and Thompson, E. L. (1975). *Space-time symmetries and the complexion of the electromagnetic field*, J. Math. Phys. **16**, 345. *See* § 9.1.

Ray, J. R., and Wei, M. S. (1977). *A solution-generating theorem with applications in general relativity*, N. Cim. **B 42**, 151. *See* § 30.5.

Raychaudhuri, A. (1955). *Relativistic cosmology, I.*, Phys. Rev. **98**, 1123. *See* § 6.2.

Raychaudhuri, A. (1958). *An anisotropic cosmological solution in general relativity*, Proc. Phys. Soc. Lond. **72**, 263. *See* § 12.4.

Raychaudhuri, A. (1975). *Spherically symmetric charged dust distributions in general relativity*, Ann. Inst. H. Poincaré **22**, 229. *See* § 13.5.

Reina, C., and Treves, A. (1975). *Gyromagnetic ratio of Einstein-Maxwell fields*, Phys. Rev. **D 11**, 3031. *See* § 30.5.

Reissner, H. (1916). *Über die Eigengravitation des elektrischen Feldes nach der Einsteinschen Theorie*, Annalen Physik **50**, 106. *See* §§ 13.4., 19.1.

Reuss, J.-D. (1968). *Un champ gravitationnel cylindrique et non stationnaire*, C.R. Acad. Sci. (Paris) **A 266**, 794. *See* § 20.3.

Robertson, H. P. (1935). *Kinematics and world-structure*, Astrophys. J. **82**, 284. *See* § 10.1.

Robertson, H. P. (1936). *Kinematics and world-structure*, Astrophys. J. **83**, 137. *See* § 10.1.

Robinson, B. B. (1961). *Relativistic universes with shear*, Proc. Natl. Acad. Sci. U.S. **47**, 1852. *See* § 12.4.

Robinson, I (1959). *A solution of the Einstein-Maxwell equations*, Bull. Acad. Polon. Sci., Ser. Math. Astr. Phys. **7**, 351. *See* §§ 10.3., 19.1.

Robinson, I. (1961). *Null electromagnetic fields*, J. Math. Phys. **2**, 290. *See* § 7.5.

Robinson, I. (1975). *On vacuum metrics of type (3, 1)*, GRG **6**, 423. *See* §§ 24.1., 25.2., 25.4.

Robinson, I. *See also* Bondi et al. (1959)

Robinson, I., and Robinson, J. R. (1969). *Vacuum metrics without symmetry*, Int. J. Theor. Phys. **2**, 231. *See* §§ 25.1., 25.2., 25.4.

Robinson, I., and Schild, A. (1963). *Generalization of a theorem by Goldberg and Sachs*, J. Math. Phys. **4**, 484. *See* § 7.5.

Robinson, I., and Trautman, A. (1962). *Some spherical gravitational waves in general relativity*, Proc. Roy. Soc. Lond. **A 265**, 463. *See* §§ 23.1., 23.2., 24.1., 24.2.

Robinson, I., Robinson, J. R., and Zund, J. D. (1969a). *Degenerate gravitational fields with twisting rays*, J. Math. Mech. **18**, 881. *See* §§ 23.1., 25.1.

Robinson, I., Schild, A., and Strauss, H. (1969b). *A generalized Reissner-Nordström solution*, Int. J. Theor. Phys. **2**, 243. *See* §§ 26.2., 26.3., 26.4.

Robinson, J. R. *See* Robinson and Robinson (1969), Robinson et al. (1969a)

Roche, J. A., and Dowker, J. S. (1968). *The algebraic structure of the vacuum Riemann tensor*, J. Phys. **A 1**, 527. *See* § 4.2.

Roos, W. (1976). *On the existence of interior solutions in general relativity*, GRG **7**, 431. *See* § 19.2.

Roos, W. (1977). *The matter-vacuum matching problem in general relativity: general methods and special cases*, GRG **8**, 753. *See* § 19.2.

Rosen, G. (1962). *Symmetries of the Einstein-Maxwell equations*, J. Math. Phys. **3**, 313. *See* § 11.3.

Rosen, J. (1965). *Embedding of various relativistic Riemannian spaces in pseudo-euclidean spaces*, Rev. Mod. Phys. **37**, 204. *See* §§ 32.1., 32.5.

Rosen, N. J. *See* Einstein and Rosen (1937)

Rosenblum, A. *See* Ehlers et al. (1976)

Roy, S. R. *See* Singh and Roy (1972)

Roy, S. R., and Singh, P. N. (1977). *A plane symmetric universe filled with disordered radiation*, J. Phys. **A 10**, 49. *See* § 13.6.

Roy, S. R., and Tripathi, V. N. (1972). *Cylindrically symmetric electromagnetic field of the second class*, Indian J. Pure & Appl. Math. **3**, 926. *See* § 20.4.

Ruban, V. A. (1978). *The dynamics of the spatially homogeneous (axisymmetric) cosmological models. II*, Preprint no. 412, Leningrad. *See* § 12.3., 12.4.
Ruffini, R. *See* Parker et al. (1973)
Russell-Clark, R. A. *See* d'Inverno and Russell-Clark (1971), Gibbons and Russell-Clark (1973)
Ryan, M. P., and Shepley, L. C. (1975). *Homogeneous Relativistic Cosmologies*, Princeton Univ. Press. *See* §§ 10.4., 11.2., 11.3.

Sachs, R. K. (1962). *Gravitational waves in general relativity. VIII. Waves in asymptotically flat space-time*, Proc. Roy. Soc. **A 270**, 103. *See* § 25.1.
Sachs, R. K. *See also* Goldberg and Sachs (1962), Jordan et al. (1961), Kantowski and Sachs (1966)
Sackfield, A. (1971). *Physical interpretation of NUT metric*, Proc. Camb. Phil. Soc. **70**, 89. *See* § 18.3.
Safko, J. L. (1977). *A "superposition" of static, cylindrically symmetric solutions of the Einstein-Maxwell field*, Phys. Rev. **D 16**, 1678. *See* § 20.2.
Safko, J. L. *See also* Carlson and Safko (1978)
Sapar, A. (1970). *Basic dependences for cosmological models with matter and radiation* (in Russian), Astron. Zh. (USSR) **47**, 503. *See* § 12.2.
Sato, H. *See* Tomimatsu and Sato (1972, 1973)
Saunders, P. T. (1967). *Non-isotropic model universes*, Ph. D. Thesis, London. *See* § 11.3., 12.4.
Sbytov, Y. G. (1976). *Interaction of plane gravitational waves* (in Russian), Zh. Eksper. Teor. Fiz. **71**, 2001. *See* § 15.2.
Scanlan, G. *See* Ludwig and Scanlan (1971)
Schiffer, M. M., Adler, R. J., Mark, J., and Sheffield, C. (1973). *Kerr geometry as complexified Schwarzschild geometry*, J. Math. Phys. **14**, 52. *See* § 30.6.
Schild, A. *See* Debney et al. (1969), Kerr and Schild (1965, 1967), Plebański and Schild (1976), Robinson and Schild (1963), Robinson et al. (1969b)
Schimming, R. (1974). *Riemannsche Räume mit ebenfrontiger und mit ebener Symmetrie*, Math. Nachr. **59**, 129. *See* § 21.5.
Schmidt, B. G. (1967). *Isometry groups with surface-orthogonal trajectories*, Z. Naturforsch. **22a**, 1351. *See* § 8.6.
Schmidt, B. G. (1968). *Riemannsche Räume mit mehrfach transitiver Isometriegruppe*, Ph. D. Thesis, Hamburg. *See* §§ 8.6., 9.2., 10.1.
Schmidt, B. G. (1971). *Homogeneous Riemannian spaces and Lie algebras of Killing fields*, GRG **2**, 105. *See* 8.6.
Schmidt, B. G. *See also* MacCallum et al. (1970)
Schmutzer, E. (1968). *Relativistische Physik* (Klass. Theorie), Teubner, Leipzig. *See* § 3.6.
Schouten, J. A. (1954). *Ricci-Calculus*, Springer-Verlag, Berlin, Göttingen, Heidelberg. *See* §§ 2.8., 3.1.
Schrank, G. *See* Thompson and Schrank (1969)
Schücking, E. *See* Ozsváth and Schücking (1962, 1969)
Schwarzschild, K. (1916a). *Über das Gravitationsfeld eines Massenpunktes nach der Einsteinschen Theorie*, Sitz. Preuss. Akad. Wiss., 189. *See* § 13.4.
Schwarzschild, K. (1916b). *Über das Gravitationsfeld einer Kugel aus inkompressibler Flüssigkeit nach der Einsteinschen Theorie*, Sitz. Preuss. Akad. Wiss., 424. *See* § 14.1.
Sciama, D. W. (1961). *Recurrent radiation in general relativity*, Proc. Camb. Phil. Soc. **57**, 436. *See* § 31.2.
Send, W. (1972). *Lösungen und asymptotisches Verhalten einer Klasse von kosmologischen Modellen*, Diplomarbeit, Hamburg. *See* § 12.4.
Sengier, J. *See* Cahen and Sengier (1967)
Serdaroğlu, M. *See* Önengüt and Serdaroğlu (1975)
Sharan, R. *See* Singh et al. (1965), Singh et al. (1969)
Sharp, D. H. *See* Misner and Sharp (1964)
Shaw, R. *See* Collinson and Shaw (1972)

Sheffield, C. *See* Adler and Sheffield (1973), Schiffer et al. (1973)

Shepley, L. C. *See* Ryan and Shepley (1975)

Shikin, I. S. (1966). *Homogeneous anisotropic cosmological model with magnetic field* (in Russian), Dokl. Akad. Nauk SSSR **171**, 73. *See* § 12.1., 12.3.

Shikin, I. S. (1972). *Gravitational fields with groups of motion on two-dimensional transitivity hypersurfaces in a model with matter and a magnetic field*, Commun. Math. Phys. **26**, 24. *See* § 11.1.

Shikin, I. S. (1975). *Anisotropic cosmological model of Bianchi type-V in general (axially symmetric) case with moving matter* (in Russian), Zh. Eksp. Teor. Fiz. **68**, 1583. *See* §§ 11.4., 12.3.

Siklos, S. T. C. (1976a). *Two completely singularity-free NUT space-times*, Phys. Letters **A 59**, 173. *See* § 11.3.

Siklos, S. T. C. (1976b). *Singularities, invariants and cosmology*, Ph. D. Thesis, Cambridge, *See* §§ 5.1., 8.2., 9.2., 10.2., 11.2.

Siklos, S. T. C. (1978). *Algebraically special homogeneous space-times*, Preprint Univ. of Oxford, *See* §§ 10.2., 10.5., 11.2., 11.3., 33.2.

Singatullin, R. S. (1973). *Exact wave solutions of Einstein-Maxwell equations defined by Einstein-Rosen solutions* (in Russian), Grav. i Teor. Otnos., Univ. Kazan **9**, 67. *See* § 20.4.

Singh, D. N. *See* Singh and Singh (1968), Singh et al. (1969)

Singh, K. P., and Abdussattar (1973). *Plane-symmetric cosmological model. II*, J. Phys. **A 6**, 1090. *See* § 12.4.

Singh, K. P., and Abdussattar (1974). *A plane-symmetric universe filled with perfect fluid*, Curr. Sci. **43**, 372. *See* § 12.4.

Singh, K. P., and Roy, S. R. (1972). *Classification of electromagnetic fields*, Indian J. Pure & Appl. Math. **3**, 532. *See* § 5.2.

Singh, K. P., and Singh, D. N. (1968). *A plane symmetric cosmological model*, Mon. Not. Roy. Astr. Soc. **140**, 453. *See* § 12.4.

Singh, K. P., Radhakrishna, L., and Sharan, R. (1965). *Electromagnetic fields and cylindrical symmetry*, Ann. Phys. (USA) **32**, 46. *See* § 20.4.

Singh, K. P., Sharan, R., and Singh, D. N. (1969). *Scalar differential invariants in general relativity*, Proc. Nat. Inst. Sci. India **A 35**, 94. *See* § 31.4.

Singh, P. N. *See* Roy and Singh (1977)

Singh, R. A. *See* Misra and Singh (1967)

Sistero, R. F. (1972). *Relativistic non-zero pressure cosmology*, Astrophys. & Space Sci. **17**, 150. *See* § 12.2.

Skripkin, V. A. (1960). *Point explosion in an ideal incompressible fluid in the general theory of relativity* (in Russian), Dokl. Akad. Nauk SSSR **135**, 1072. *See* § 14.2.

Sneddon, G. E. (1975). *Change of Petrov type under generation of solutions of Einstein's equations*, J. Math. Phys. **16**, 740. *See* § 30.5.

Sneddon, G. E. (1976). *Hamiltonian cosmology: a further investigation*, J. Phys. **A 9**, 229. *See* § 11.2.

Som, M. M. *See* Teixeira et al. (1977a, b)

Sommers, P. (1973). *On Killing tensors and constants of motion*, J. Math. Phys. **14**, 787. *See* § 31.3.

Sommers, P. (1976). *Properties of shear-free congruences of null geodesics*, Proc. Roy. Soc. Lond. **A 349**, 309. *See* § 25.6.

Sommers, P. (1977). *Type N vacuum space-times as special functions in* C^2, GRG **8**, 855. *See* § 25.3.

Sommers, P. *See also* Hughston and Sommers (1973), Hughston et al. (1972)

Sommers, P., and Walker, M. (1976). *A note on Hauser's type N gravitational field with twist*, J. Phys. **A 9**, 357. *See* § 25.3.

Spanos, J. T. J. *See* Israel and Spanos (1973)

Spelkens, J. *See* Cahen and Spelkens (1967)

Srivastava, D. C. *See* Misra and Srivastava (1973, 1974)

Stachel, J. J. (1966). *Cylindrical gravitational news*, J. Math. Phys. **7**, 1321. *See* § 20.3.

Stachel, J. *See also* Plebański and Stachel (1968), Goenner and Stachel (1970)

Stephani, H. (1967a). *Über Lösungen der Einsteinschen Feldgleichungen, die sich in einen fünfdimensionalen flachen Raum einbetten lassen,* Commun. Math. Phys. **4**, 137. *See* § 32.4.

Stephani, H. (1967b). *Konform flache Gravitationsfelder,* Commun. Math. Phys. **5**, 337. *See* § 32.5.

Stephani, H. (1968). *Einige Lösungen der Einsteinschen Feldgleichungen mit idealer Flüssigkeit, die sich in einen fünfdimensionalen flachen Raum einbetten lassen,* Commun. Math. Phys. **9**, 53. *See* § 32.4.

Stephani, H. (1978). *A note on Killing tensors,* GRG **9**, 789. *See* § 31.3.

Stephani, H. (1979). *A method to generate algebraically special pure radiation field solutions from the vacuum,* J. Phys. **A 12**, 1045. *See* § 26.6.

Stephani, H. *See also* Kramer et al. (1972)

Sternberg, S. (1964). *Lectures on Differential Geometry,* Prentice-Hall, Englewood Cliffs, N.J. *See* §§ 2.1., 2.7.

Stewart, J. M. *See* MacCallum et al. (1970)

Stewart, J. M., and Ellis, G. F. R. (1968). *Solutions of Einstein's equations for a fluid which exhibit local rotational symmetry,* J. Math. Phys. **9**, 1072. *See* §§ 9.2., 11.1., 11.2., 11.4., 29.1., 29.2., 29.3.

Stewart, J. M., and Walker, M. (1973). *Black holes: the outside story,* Springer Tracts in Modern Physics **69**, Springer Verlag, Berlin, Heidelberg, New York. *See* § 18.5.

Stoker, J. J. (1969). *Differential Geometry,* Pure and Applied Math. A Series of Texts and Monographs, Wiley, NewYork—London—Sidney—Toronto. *See* § 2.1.

Strauss, H. *See* Robinson et al. (1969b)

Strunz, H. C. (1974). *On the problem of the determination of general relativistic spaces by their scalar differential invariants,* Ph. D. Thesis, Univ. of Washington. *See* § 31.4.

Suhonen, E. (1968). *General relativistic fluid sphere at mechanical and thermal equilibrium,* Kgl. Danske Vidensk. Sels., Mat.-Fys. Medd. **36**, 1. *See* § 14.1.

Sulaiman, A. H. *See* Bonnor et al (1977)

Sust, M. *See* Neugebauer and Sust (1975)

Swaminarayan, N. S. *See* Bonnor and Swaminarayan (1964, 1965)

Synge, J. L. (1960), *Relativity: The General Theory,* North-Holland, Amsterdam. *See* § 16.5.

Synge, J. L. (1964). *The Petrov classification of gravitational fields,* Commun. Dublin Inst. Adv. Stud. **A**, no. 15. *See* §§ 4.1., 4.2.

Szafron, D. A. (1977). *Inhomogeneous cosmologies: New exact solutions and their evolution,* J. Math. Phys. **18**, 1673. *See* § 29.3.

Szafron, D. A., and Wainwright, J. (1977). *A class of inhomogeneous perfect fluid cosmologies,* J. Math. Phys. **18**, 1668. *See* § 29.3.

Szekeres, G. (1960). *On the singularities of a Riemannian manifold.* Publ. Mat. Debrecen **7**, 285. *See* § 13.4.

Szekeres, P. (1963). *Spaces conformal to a class of spaces in general relativity,* Proc. Roy. Soc. Lond. **A 274**, 206. *See* § 3.7.

Szekeres, P. (1966a). *Embedding properties of general relativistic manifolds,* N. Cim. **43**, 1062. *See* § 32.3., 32.6.

Szekeres, P. (1966b). *On the propagation of gravitational fields in matter,* J. Math. Phys. **7**, 751. *See* §§ 3.5., 7.5., 22.1., 27.7., 29.4.

Szekeres, P. (1968). *Multipole particles in equilibrium in general relativity,* Phys. Rev. **176**, 1446. *See* § 18.1.

Szekeres, P. (1970). *Colliding gravitational waves,* Nature **228**, 1183. *See* § 15.2.

Szekeres, P. (1972). *Colliding plane gravitational waves,* J. Math. Phys. **13**, 286. *See* § 15.2.

Szekeres, P. (1975). *A class of inhomogeneous cosmological models,* Commun. Math. Phys. **41**, 55. *See* § 29.3.

Szekeres, P. *See also* Bell and Szekeres (1972, 1974)

Tabensky, R. R. *See* Letelier and Tabensky (1974, 1975a, b), Melnick and Tabensky (1975)

Tabensky, R. R., and Taub, A. H. (1973). *Plane symmetric self-gravitating fluids with pressure equal to energy density,* Commun. Math. Phys. **29**, 61. *See* § 13.6., 15.1.

Takeno, H. (1961). *The mathematical theory of plane gravitational waves in general relativity*, Sci. Rep. Res. Inst. Theor. Phys. Hiroshima Univ. no. **1**. *See* § 21.5.

Takeno, H. (1966). *The theory of spherically symmetric space-times*, Sci. Rep. Res. Inst. Theor. Phys. Hiroshima Univ. no. **5**. *See* § 13.1.

Takeno, H., and Kitamura, S. (1968). *On the space-time admitting a parallel null vector field*, Tensor **19**, 207. *See* § 6.2.

Takeno, H., and Kitamura, S. (1969). *On the Einstein tensors of spherically symmetric space-times*, Progress of Math. **3**, 7. *See* § 13.1.

Talbot, C. J. (1969). *Newman-Penrose approach to twisting degenerate metrics*, Commun. Math. Phys. **13**, 45. *See* §§ 7.1., 23.1., 30.6.

Tamburino, L. *See* Newman and Tamburino (1962), Newman et al. (1963)

Tanabe, Y. (1974). *A comment on physical interpretation of the Tomimatsu-Sato metrics*, Prog. Theor. Phys., **52**, 727. *See* § 18.5.

Tanabe, Y. (1977). *SU(2,1) symmetry of the Einstein-Maxwell fields*, Prog. Theor. Phys. **57**, 840. *See* § 30.3.

Tanabe, Y. (1978). *Exact solutions of the stationary axisymmetric Einstein-Maxwell equations*, Prog. Theor. Phys. **60**, 142. *See* § 30.3.

Tariq, N. *See* McLenaghan and Tariq (1975), McLenaghan et al. (1975)

Tariq, N., and Tupper, B. O. J. (1975). *A class of algebraically general solutions of the Einstein-Maxwell equations for non-null electromagnetic fields*, GRG **6**, 345. *See* § 11.2., 11.3.

Taub, A. H. (1951). *Empty space-times admitting a three parameter group of motions*, Ann. Math. **53**, 472. *See* §§ 11.2., 11.3., 13.4.

Taub, A. H. (1956). *Isentropic hydrodynamics in plane symmetric space-times*, Phys. Rev. **103**, 454. *See* § 13.6.

Taub, A. H. (1968). *Restricted motions of gravitating spheres*. Ann. Inst. H. Poincaré **A 9**, 153. *See* § 14.2.

Taub, A. H. (1972). *Plane-symmetric similarity solutions for self-gravitating fluids*, in: O'Raifeartaigh, L. (Ed.), General Relativity (Papers in Honour of J. L. Synge), Clarendon Press, Oxford, p. 133. *See* § 13.6.

Taub, A. H. (1976). *High frequency gravitational radiation in Kerr-Schild space-times*, Commun. Math. Phys. **47**, 185. *See* § 24.3.

Taub, A. H. *See also* MacCallum and Taub (1972), Misner and Taub (1968), Tabensky and Taub (1973)

Tauber, G. E. (1957). *The gravitational fields of electric and magnetic dipoles*, Canad. J. Phys. **35**, 477. *See* § 19.1.

Tauber, G. E. (1967). *Expanding universe in conformally flat coordinates*, J. Math. Phys. **8**, 118. *See* § 12.2.

Teixeira, A. F. da F., Wolk, I., and Som, M. M. (1977a). *Exact relativistic cylindrical solution of disordered radiation*, N. Cim. **B 41**, 387. *See* § 20.2.

Teixeira, A. F. da F., Wolk, I., Som, M. M. (1977b). *Exact relativistic solution of disordered radiation with planar symmetry*, J. Phys. **A 10**, 1679. *See* § 13.6.

Thompson, A. (1966). *A class of related space-times*, Tensor **17**, 92. *See* § 28.5.

Thompson, A. *See also* Kundt and Thompson (1962)

Thompson, E. L. *See* Ray and Thompson (1975)

Thompson, I. H., and Whitrow, G. J. (1967). *Time-dependent internal solutions for spherically symmetrical bodies in general relativity*, Mon. Not. Roy. Astr. Soc. **136**, 207. *See* § 14.2.

Thompson, J., and Schrank, G. (1969). *Algebraic classification of four-dimensional Riemann spaces*, J. Math. Phys. **10**, 766. *See* § 5.1.

Thorne, K. S. (1967). *Primordial element formation, primordial magnetic fields and the isotropy of the universe*, Astrophys. J. **148**, 51. *See* §§ 12.1., 12.3.

Thorne, K. S. *See also* Misner et al. (1973)

Thorpe, J. A. (1969). *Curvature and the Petrov canonical forms*, J. Math. Phys. **10**, 1. *See* § 4.2.

Tolman, R. C. (1934a). *Effect of inhomogeneity in cosmological models*, Proc. Nat. Acad. Sci. U.S. **20**, 69. *See* § 13.5.

Tolman, R. C. (1934b). *Relativity, Thermodynamics, and Cosmology,* Oxford Univ. Press. *See* § 12.2.
Tolman, R. C. (1939). *Static solutions of Einstein's field equations for spheres of fluid,* Phys. Rev. **55**, 364. *See* § 14.1.
Tomimatsu, A., and Sato, H. (1972). *New exact solution for the gravitational field of a spinning mass,* Phys. Rev. Lett. **29**, 1344. *See* § 18.5.
Tomimatsu, A., and Sato, H. (1973). *New series of exact solutions for gravitational fields of spinning masses,* Prog. Theor. Phys. **50**, 95. *See* § 18.5.
Tomimura, N. (1977). *Particular exact inhomogeneous solution with matter and pressure,* N. Cim. **B 42**, 1. *See* § 29.3.
Tomimura, N. *See also* Bonnor and Tomimura (1976), Bonnor et al. (1977)
Tooper, R. F. (1964). *General relativistic polytropic fluid spheres,* Astrophys. J. **140**, 434. *See* § 14.1.
Torrence, R. *See* Newman et al. (1965), Unti and Torrence (1966)
Trautman, A. (1962). *On the propagation of information by waves,* in: Recent Developments in General Relativity, Pergamon Press — PWN, p. 459. *See* §§ 28.1., 28.2.
Trautman, A. *See also* Robinson and Trautman (1962)
Treves, A. *See* Reina and Treves (1975)
Trim, D. W., and Wainwright, D. J. (1974). *Nonradiative algebraically special space-times,* J. Math. Phys. **15**, 535. *See* §§ 25.1., 25.2., 26.2., 26.3., 26.4.
Tripathi, V. N. *See* Roy and Tripathi (1972)
Trollope, R. *See* Wyman and Trollope (1965)
Trümper, M. (1962). *Zur Bewegung von Probekörpern in Einsteinschen Gravitations-Vakuumfeldern,* Z. Phys. **168**, 55. *See* § 16.6.
Trümper, M. (1965). *On a special class of type-I gravitational fields,* J. Math. Phys. **6**, 584. *See* § 6.2.
Trümper, M. (1967), *Einsteinsche Feldgleichungen für das axialsymmetrische, stationäre Gravitationsfeld im Innern einer starr rotierenden idealen Flüssigkeit,* Z. Naturforsch. **22a**, 1347. *See* § 19.2.
Trümper, M. *See* also Kundt and Trümper (1962, 1966)
Tsoubelis, D. *See* Harvey and Tsoubelis (1977)
Tupper, B. O. J. (1976). *A class of algebraically general solutions of the Einstein-Maxwell equations for non-null electromagnetic fields. II.* GRG **7**, 479. *See* § 10.3.
Tupper, B. O. J. *See also* Dunn and Tupper (1976), McLenaghan et al. (1975), Tariq and Tupper (1975)

Unti, T. *See* Newman and Unti (1963), Newman et al. (1963)
Unti, T. W. J., and Torrence, R. J. (1966). *Theorem on gravitational fields with geodesic rays,* J. Math. Phys. **7**, 535. *See* § 22.3.
Urbantke, H. (1972). *Note on Kerr-Schild type vacuum gravitational fields,* Acta Phys. Austr. **35**, 396. *See* § 28.2.
Urbantke, H. (1975). *Der metrische Ansatz von Trautman, Kerr und Schild und einige seiner Anwendungen,* Acta Phys. Austr. **41**, 1. *See* §§ 22.1., 28.4.

Vaidya, P. C. (1951). *The gravitational field of a radiating star,* Proc. Indian Acad. Sci. **A 33**, 264. *See* § 13.4.
Vaidya, P. C. (1968). *Nonstatic analogues of Schwarzschild's interior solution in general relativity,* Phys. Rev. (2) **174**, 1615. *See* § 14.2.
Vaidya, P. C. (1973). *Some algebraically special solutions of Einstein's equations. II.,* Tensor **27**, 276. *See* § 28.4.
Vaidya, P. C. (1974). *A generalized Kerr-Schild solution of Einstein's equations,* Proc. Camb. Phil. Soc. **75**, 383. *See* § 28.4.
Vaidya, P. C. (1977). *The Kerr metric in cosmological background,* Pramana **8**, 512. *See* § 19.2.
Vaidya, P. C. *See also* Bonnor and Vaidya (1972), Pandya and Vaidya (1961), Patel and Vaidya (1969)

Vaidya, P. C., and Patel, L. K. (1973). *Radiating Kerr metric*, Phys. Rev. **D 7**, 3590. *See* § 28.4.
Vajk, J. P. (1969). *Exact Robertson-Walker cosmological solutions containing relativistic fluids*, J. Math. Phys. **10**, 1145. *See* § 12.2.
Vajk, J. P., and Eltgroth, P. G. (1970). *Spatially homogeneous anisotropic cosmological models containing relativistic fluid and magnetic field*, J. Math. Phys. **11**, 2212. *See* § 12.3.
van Stockum, W. J. (1937). *The gravitational field of a distribution of particles rotating about an axis of symmetry*, Proc. Roy. Soc. Edinburgh **A 57**, 135. *See* §§ 18.4., 19.2.
Vickers, P. A. (1973). *Charged dust spheres in general relativity*, Ann. Inst. H. Poincaré **A 18**, 137. *See* § 13.5.
Vishveshwara, C. V. *See* Hoenselaers and Vishveshwara (1978)
Vishveshwara, C. V., and Winicour, J. (1977). *Relativistically rotating dust cylinders*, J. Math. Phys. **18**, 1280. *See* §§ 19.2., 20.2.
Volkoff, G. M. (1939). *On the equilibrium of massive spheres*, Phys. Rev. **55**, 413. *See* § 14.1.
Voorhees, B. H. (1970). *Static axially symmetric gravitational fields*, Phys. Rev. **D 2**, 2119. *See* § 18.1.
Voorhees, B. H. (1972). *Electric Weyl spacetimes and the stationary problem*, Phys. Letters **A 39**, 114. *See* § 30.5.

Wagh, R. V. (1955). *On some spherically symmetrical models in relativity*, J. Univ. Bombay **24**, 5. *See* § 14.2.
Wahlquist, H. D. (1968). *Interior solution for a finite rotating body of perfect fluid*, Phys. Rev. **172**, 1291. *See* § 19.2.
Wahlquist, H. D. *See also* Estabrook et al. (1968)
Wahlquist, H. D., and Estabrook, F. B. (1966). *Rigid motions in Einstein spaces*, J. Math. Phys. **7**, 894. *See* § 6.2.
Wainwright, J. (1970). *A class of algebraically special perfect fluid space-times*, Commun. Math. Phys. **17**, 42. *See* §§ 21.4., 29.2.
Wainwright, J. (1974). *Algebraically special fluid space-times with hypersurface-orthogonal shearfree rays*, Int. J. Theor. Phys. **10**, 39. *See* § 29.1.
Wainwright, J. (1977a). *Classification of the type D perfect fluid solutions of the Einstein equations*, GRG **8**, 797. *See* § 29.3.
Wainwright, J. (1977b). *Characterization of the Szekeres inhomogeneous cosmologies as algebraically special space-times*, J. Math. Phys. **18**, 672. *See* § 29.3.
Wainwright, J. (1979). *A classification scheme for non-rotating inhomogeneous cosmologies*, J. Phys. **A 12**, 2015. *See* § 33.1.
Wainwright, J. *See also* Campbell and Wainwright (1977), Michalski and Wainwright (1975), Szafron and Wainwright (1977), Trim and Wainwright (1974)
Wainwright, J., Ince, W. C. W., and Marshman, B. J. (1979). *Spatially homogeneous and inhomogeneous cosmologies with equation of state* $p = \mu$, GRG **10**, 259. *See* §§ 5.5., 12.1., 12.3., 12.4., 15.1.
Walker, A. G. (1936). *On Milne's theory of world-structure*, Proc. London Math. Soc. **42**, 90. *See* § 10.1.
Walker, M. *See* Hughston et al. (1972), Kinnersley and Walker (1970), Köhler and Walker (1975), Sommers and Walker (1976)
Walker, M., and Kinnersley, W. (1972). *Some remarks on a radiating solution of the Einstein-Maxwell equations*, Lect. Notes Phys. **14**, 48 (Springer-Verlag, Berlin, Heidelberg, New York). *See* § 19.1.
Walker, M., and Penrose, R. (1970). *On quadratic first integrals of the geodesic equations for type* {22} *spacetimes*, Commun. Math. Phys. **18**, 265. *See* § 31.3.
Wang, M. Y. (1974). *Class of solutions of axial-symmetric Einstein-Maxwell equations*, Phys. Rev. **D 9**, 1835. *See* § 30.5.
Ward, J. P. (1976). *Equilibrium — its connection to global integrability conditions for stationary Einstein-Maxwell fields*, Int. J. Theor. Phys. **15**, 293. *See* § 16.7.
Warner, F. W. (1971). *Foundations of Differentiable Manifolds and Lie Groups*, Scott, Foresman, Glenview, Illinois. *See* § 8.1.

Wei, M. S. *See* Ray and Wei (1977)

Weinberg, S. (1972). *Gravitation and Cosmology: Principles and Applications of the General Theory of Relativity*, Wiley, New York. *See* §§ 8.5., 12.1.

Weir, G. J., and Kerr, R. P. (1977). *Diverging type-D metrics*, Proc. Roy. Soc. Lond. **A 355**, 31. *See* §§ 19.1., 25.1., 25.2., 25.5., 33.2.

Wesson, P. S. (1978). *An exact solution to Einstein's equations with a stiff equation of state*, J. Math. Phys. **19**, 2283. *See* § 14.2.

Weyl, H. (1917). *Zur Gravitationstheorie*, Annalen Physik **54**, 117. *See* §§ 18.1., 19.1.

Wheeler, J. A. *See* Misner and Wheeler (1957), Misner et al. (1973)

Whittaker, J. M. (1968). *An interior solution in general relativity*, Proc. Roy. Soc. Lond. **A 306**, 1. *See* § 14.1.

Whitrow, G. J. *See* Thompson and Whitrow (1967)

Wild, W. J. *See* Ernst and Wild (1976)

Wilkins, D. *See* Parker et al. (1973)

Wilkinson, D. *See* Collins et al. (1980)

Williams, G. (1968). *Relativistic space-times having corresponding geodesics*, J. Phys. **A 1**, 457. *See* § 31.3.

Wilson, G. A. *See* Israel and Wilson (1972)

Wilson, S. J. (1968). *A Gödel-type solution of the Einstein-Maxwell equation*, Canad. J. Phys. **46**, 2361. *See* § 20.2.

Wiltshire, R. J. *See* McVittie and Wiltshire (1975, 1977)

Winicour, J. (1975). *All stationary axisymmetric rotating dust metrics*, J. Math. Phys. **16**, 1806. *See* § 19.2.

Winicour, J. *See* also Vishveshwara and Winicour (1977)

Winternitz, P. *See* Montgomery et al. (1969), Patera and Winternitz (1977)

Witten, L. (1959). *Invariants of general relativity and the classification of spaces*, Phys. Rev. **113**, 357. *See* §§ 3.6., 5.1.

Witten, L. (1962). *A geometric theory of the electromagnetic and gravitational fields*, in: Witten, L. (Ed.) Gravitation: an introduction to current research, Wiley, New York, London, p. 382. *See* § 20.2.

Witten, L. *See also* Esposito and Witten (1973)

Wolk, I. *See* Teixeira et al. (1977a, b)

Woodhouse, N. M. J. (1975). *Killing tensors and the separation of the Hamilton-Jacobi equation*, Commun. Math. Phys. **44**, 9. *See* § 31.3.

Woolley, M. L. (1973a). *On the eigenvectors of the energy-momentum tensor in the Einstein-Maxwell theory*, Proc. Camb. Phil. Soc. **74**, 281. *See* § 5.2.

Woolley, M. L. (1973b). *On the role of the Killing tensor in the Einstein-Maxwell theory*, Commun. Math. Phys. **33**, 135. *See* § 30.3.

Wyman, M. (1946). *Equations of state for radially symmetric distributions of matter*, Phys. Rev. **70**, 396. *See* § 14.2.

Wyman, M. (1948). *Radially symmetric distributions of matter*, Phys. Rev. **75**, 1930. *See* § 14.1.

Wyman, M. (1978). *Nonstatic spherically symmetric isotropic solutions for a perfect fluid in general relativity*, Austral. J. Phys. **31**, 111. *See* § 14.2.

Wyman, M., and Trollope, R. (1965). *Null fields in Einstein-Maxwell field theory*, J. Math. Phys. **6**, 1965. *See* §§ 27.5., 27.6.

Yakupov, M. Sh. (1968a). *Algebraic characterization of second order Einstein spaces* (in Russian), Grav. i Teor. Otnos., Univ. Kazan **4/5**, 78. *See* § 32.5.

Yakupov, M. Sh. (1968b). *On Einstein spaces of embedding class two* (in Russian), Dokl. Akad. Nauk SSSR **180**, 1096. *See* § 32.5.

Yakupov, M. Sh. (1973). *Einstein spaces of class two* (in Russian), Grav. i Teor. Otnos., Univ. Kazan **9**, 109. *See* § 32.5.

Yamazaki, M. (1977a). *On the Kerr and the Tomimatsu-Sato spinning mass solutions*, Prog. Theor. Phys. **57**, 1951. *See* § 18.5.

Yamazaki, M. (1977b). *On the Kerr-Tomimatsu-Sato family of spinning mass solutions*, J. Math. Phys. **18**, 2502. *See* § 18.5.

Yano, K. (1955). *The Theory of Lie Derivatives and Its Application*, North-Holland, Amsterdam. *See* § 2.8.

York, J. W. *See* O'Murchadha and York (1976)

Zaikov, R. G. (1971). *Periodic model of the evolving universe*, C.R. Acad. Bulg. Sci. **24**, 1301. *See* § 12.2.

Zakharov, V. D. (1965). *A physical characteristic of Einsteinian spaces of degenerate type II in the classification of Petrov* (in Russian), Dokl. Akad. Nauk SSSR, **161**, 563. *See* § 4.2.

Zakharov, V. D. (1970). *Algebraical and group theoretical methods in general relativity: Invariant Petrov type characterisation of the type of Einstein spaces* (in Russian), Probl. Teor. Grav. i. Elem. Chastits **3**, 128. *See* § 4.2.

Zakharov, V. D. (1972). *Gravitational waves in Einstein's theory of gravitation* (in Russian), Nauka, Moscow. *See* §§ 4.2., 21.5.

Zenk, L. G., and Das, A. (1978). *An algebraically special subclass of vacuum metrics admitting a Killing motion*, J. Math. Phys. **19**, 535. *See* § 33.2.

Zipoy, D. M. (1966). *Topology of some spheroidal metrics*, J. Math. Phys. **7**, 1137. *See* § 18.1.

Zund, J. D., and Brown, E. (1971). *The theory of bivectors*, Tensor **22**, 179. *See* § 3.4.

References added in proof

☐ Allnutt, J. A. (1980). *An approach to perfect-fluid space-times by a method of spin-coefficients*, GRG (to appear). *See* § 20.5.

☐ Bampi, F., and Cianci, R. (1979). *Generalized axisymmetric space-times*, Comm. Math. Phys. **70**, 69. *See* § 21.3.

☐ Bayin, S. S. (1978). *Solutions of Einstein's field equations for static fluid spheres*, Phys. Rev. **D 18**, 2745. *See* § 14.1.

☐ Belinski, V. A., and Zakharov, V. E. (1978). *Integration of the Einstein equations by the inverse scattering method and calculation of exact soliton solutions* (in Russian), Zh. Eksp. Teor. Fiz. **75**, 1953. *See* § 30.4.

☐ Belinski, V. A., and Zakharov, V. E. (1979). *Stationary gravitational solutions with axial symmetry* (in Russian), Zh. Eksp. Teor. Fiz. **77**, 3. *See* § 30.4.

☐ Bonnor, W. B. (1979a). *A source for Petrov's homogeneous vacuum space-time*, Phys. Letters **A 75**, 25. *See* § 10.2.

☐ Bonnor, W. B. (1979b). *A three-parameter solution of the static Einstein-Maxwell equations*, J. Phys. **A 12**, 853. *See* § 19.1.

☐ Bronnikov, K. A. (1979). *Static fluid cylinders and plane layers in general relativity*, J. Phys. **A 12**, 201. *See* § 20.2.

☐ Charach, Ch. (1979). *Electromagnetic Gowdy universe*, Phys. Rev. **D 19**, 3516. *See* § 15.3.

☐ Choquet-Bruhat, Y., DeWitt-Morette, C., and Dillard-Bleick, M. (1978). *Analysis, Manifolds and Physics*, North-Holland Publishing Company, Amsterdam, New York, Oxford. *See* § 2.1.

☐ Cianci, R. *See* ☐ Bampi and Cianci (1979)

☐ Cohen, J. M. *See* ☐ Flaclas and Cohen (1978, 1979)

☐ Collins, C. B., and Szafron, D. A. (1979). *A new approach to inhomogeneous cosmologies*, J. Math. Phys. **20**, 2347. *See* § 29.3.

☐ Cosgrove, C. M. (1979a). *Stationary axisymmetric gravitational fields: an asymptotic flatness preserving transformation*, Lecture given at the Einstein Centenary Summer School on Gravitational Radiation and Collapsed Objects, Perth, Australia. *See* § 30.4.

☐ Cosgrove, C. M. (1979b). *Continuous groups and Bäcklund transformations generating asymptotically flat solutions*, Lecture presented at The Second Marcel Grossmann Meeting on the Recent Developments of General Relativity, Trieste, Italy. *See* §§ 30.4., 30.5.

☐ Cosgrove, C. M. (1980). *Relationships between the group-theoretic and soliton-theoretic techniques for generating stationary axisymmetric gravitational solutions*, Preprint, Moltana State University. *See* § 30.4.

☐ Cox, D., and Kinnersley, W. (1979). *Yet another formulation of the Einstein equations for stationary axisymmetry*, J. Math. Phys. **20**, 1225. *See* §§ 17.5., 20.3.

☐ Crade, R. F., and Hall, G. S. (1979). *Energy-momentum tensors in locally isotropic space-times*, Phys. Letters **A 75**, 17. *See* § 5.1.

☐ DeWitt-Morette, C. *See* ☐ Choquet-Bruhat et al. (1978).

☐ Dietz, W., and Rüdiger, R. (1979). *Shearfree congruences of null geodesics and Killing tensors*, Preprint, University of Würzburg. *See* § 31.3.

☐ Dillard-Bleick, M. *See* ☐ Choquet-Bruhat et al. (1978)

☐ Ernst, F. J. *See* ☐ Hauser and Ernst (1979a, b)

☐ Flaclas, C., and Cohen, J. M. (1978). *Locally rotationally symmetric cosmological model containing a nonrotationally symmetric electromagnetic field*, Phys. Rev. **D 18**, 4373. *See* § 9.1.

☐ Flaclas, C., and Cohen, J. M. (1979). *Tilted Bianchi type-III cosmologies*, Phys. Rev. **D 19**, 1051. *See* § 12.3.

☐ Glass, E. N. (1979). *Shear-free gravitational collapse*, J. Math. Phys. **20**. 1508. *See* § 14.2.

☐ Halil, M. (1979). *Colliding plane gravitational waves*, J. Math. Phys. **20**, 120. *See* § 15.2.

☐ Hall, G. S. *See* ☐ Crade and Hall (1979)

☐ Hauser, I., and Ernst, F. J. (1979a). *Integral equation method for effecting Kinnersley-Chitre transformations*, Phys. Rev. **D 20**, 362, 1783. *See* § 30.4.

☐ Hauser, I., and Ernst, F. J. (1979b). *A homogeneous Hilbert problem for the Kinnersley-Chitre transformations*, Preprint, Institute of Technology, Chicago. *See* § 30.4.

☐ Herlt, E. (1980). *Kerr-Schild pure radiation fields with axial symmetry* GRG (to appear). *See* § 28.4.

☐ Herlt, E., and Hermann, H. Ch. (1980). *A note on the equivalence of two axially symmetric stationary perfect fluid equations*, Exp. Technik Physik **28**, 97. *See* § 19.2.

☐ Hermann, H. Ch. *See* ☐ Herlt and Hermann (1980)

☐ Hoenselaers, C., Kinnersley, W., and Xanthopoulos, B. C. (1979). *Generation of asymptotically flat, stationary space-times with any number of parameters*, Phys. Rev. Lett. **42**, 481. *See* §§ 30.4., 30.5.

☐ Hoenselaers, C., Kinnersley, W., and Xanthopoulos, B. C. (1980). *Symmetries of the stationary Einstein-Maxwell equations. VI. Transformations which generate asymptotically flat space-times with arbitrary multipole moments*, J. Math. Phys. (to appear). *See* §§ 30.4., 30.5.

☐ Kinnersley, W. *See* ☐ Cox and Kinnersley (1979), ☐ Hoenselaers et al. (1979, 1980)

☐ Koppel, A. (1967). *On the exact semigeodetic two-variable solutions of Einstein's empty space equations* (in Russian), Trudy Inst. Fiz. Astron., Akad. Nauk Eston. SSR, **33**, 18. *See* § 15.4.

☐ Kundu, P. (1978). *Projection tensor formalism for stationary axisymmetric gravitational fields*, Phys. Rev. **D 18**, 4471. *See* § 17.3.

☐ Kramer, D. (1978). *Homogeneous Einstein-Maxwell fields*, Acta Phys. Acad. Sci. Hung. **44**, 353. *See* § 10.3.

☐ Kramer, D. (1980). *Space-times with a group of motions on null hypersurfaces*, J. Phys. **A 13**, L 43. *See* § 21.2.

□ Kramer, D., and Neugebauer, G. (1980). *The superposition of two Kerr solutions,* Phys. Letters **A 75**, 259. *See* § 30.5.

□ Lun, A. W. C. (1978). *A class of twisting type II and type III solutions admitting two Killing vectors,* Phys. Letters **A 69**, 79. *See* § 33.2.

□ MacCallum, M. A. H. *See* □ Siklos and MacCallum (1980)
□ Maison, D. (1979). *On the complete integrability of the stationary, axially symmetric Einstein equations,* J. Math. Phys. **20**, 871. *See* § 30.4.
□ Marshman, B. J. *See* □ Wainwright and Marshman (1979)
□ McIntosh, C. B. G. (1978). *Self-similar cosmologies with equation of state $p = \mu$,* Phys. Letters **A 69**, 1. *See* § 15.1.

□ Neugebauer, G. (1980a). *A general integral of the axially symmetric stationary Einstein equations,* J. Phys. **A 13**, L 19. *See* § 30.4.
□ Neugebauer, G. (1980b). *Recursive calculation of axially symmetric stationary Einstein fields,* J. Phys. **A** (to appear). *See* §§ 30.4., 30.5.
□ Neugebauer, G. *See also* □ Kramer and Neugebauer (1980)

□ Oleson, M. K. (1972). *Algebraically special fluid space-times with shearing rays,* Ph. D. Thesis, University of Waterloo. *See* §§ 29.1., 29.4.

□ Rüdiger, R. *See* □ Dietz and Rüdiger (1979)

□ Siklos, S. T. C., and MacCallum, M. A. H. (1980). *The algebraically special hypersurface-homogeneous Einstein spaces, Preprint. See* § 11.3.
□ Stephani, H. (1980). *Some new algebraically special radiation field solutions connected with Hauser's type N vacuum solution,* Abstracts of GR9, Jena. *See* §§ 25.4., 26.6.
□ Szafron, D. A. *See* □ Collins and Szafron (1979)

□ Tanabe, Y. (1979). *Generation of the Lewis solutions from the Weyl solutions,* J. Math. Phys. **20,** 1486. *See* § 30.5.

□ Wainwright, J., and Marshman, B. J. (1979). *Some exact cosmological models with gravitational waves,* Phys. Letters **A 72**, 275. *See* § 15.1.

□ Xanthopoulos, B. C. *See* □ Hoenselaers et al. (1979, 1980)

□ Zakharov, V. E. *See* □ Belinski and Zakharov (1978, 1979)

Index